Fuhrhop/Penzlin **Organic Synthesis**

Distribution:

VCH Verlagsgesellschaft, P.O. Box 1260/1280, D-6940 Weinheim (Federal Republic of Germany)
USA and Canada: VCH Publishers, 303 N.W. 12th Avenue, Deerfield Beach, FL 33442-1705 (USA)

ISBN 3-527-25879-5 (VCH Verlagsgesellschaft)
ISBN 0-89573-246-7 (VCH Publishers)

Jürgen Fuhrhop/Gustav Penzlin

Organic Synthesis

Concepts, Methods, Starting Materials

VCH

Prof. Dr. Jürgen Fuhrhop
Freie Universität Berlin
Institut für Organische Chemie
Takustraße 3
D-1000 Berlin 33

Dr. Gustav Penzlin
Beilstein-Institut für Literatur
Varrentrappstraße 40–42
D-6000 Frankfurt/M. 90

First edition 1983
First reprint of the first edition 1984
Second reprint (with corrections) of the first edition 1986

Editor: Dr. Gerd Giesler
Production manager: Dipl.-Ing. (FH) Hans Jörg Maier

This book contains 21 tables

Deutsche Bibliothek Cataloguing in Publication Data

Fuhrhop, Jürgen:

Organic synthesis: concepts, methods, starting materials / Jürgen Fuhrhop; Gustav Penzlin. — Weinheim; Deerfield Beach, Fl.: VCH, 1986
    ISBN 3-527-25879-5
NE: Penzlin, Gustav:

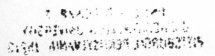

Dedicated to the City of West Berlin

Preface

This book was written for the advanced chemistry student and for the research chemist. Its purpose is to convey knowledge about concepts, methods, starting materials, and target molecules that play important roles in modern organic syntheses. Moreover, it shows the application of this knowledge to retrosynthetic analysis and gives an introduction to the design of synthetic plans.

Important concepts are summarized at the beginning of each chapter. They include

- the synthon approach,
- systematic evaluation of the arrangement of functionality,
- strategies for achieving regio- and stereoselectivity in carbon-carbon bond formation and functional group conversions,
- strategies for enforcing thermodynamically unfavourable reactions, and
- retrosynthetic analysis.

The synthetic methods described were selected under the aspects of applicability, simplicity, and didactic value. Applicability implies that the method has been used in complex syntheses. Simplicity means that the method is not too time consuming. And didactic value is judged by some principles of selectivity control which we wanted to demonstrate.

Finally, a list of commercially available starting materials tells the synthetic chemist what industry can do for him, and a selection of complex synthetic procedures shows what he may aim for.

We should like to acknowledge the help and advice we have received from many students and colleagues, in particular Prof. Kevin Smith and Prof. A. Gossauer, who read the whole of the manuscript and made helpful suggestions. The "Lektorat Verlag Chemie" was giving us vigorous support throughout the time of writing and correcting. Above all we must thank our typist, Mrs. Jutta Fuhrhop, who has transferred difficult to read manuscripts to several intermediate and a final typescript with indulgence and great skill.

Berlin, in January 1983   Jürgen Fuhrhop
                                                          Gustav Penzlin

Contents

1 Synthons in the Synthesis of Carbon Chains and Carbocycles

Introduction and Summary

- Most synthetic reactions, which produce carbon-carbon bonds, are polar: a negatively polarized ("electronegative") carbon atom (electron **d**onor, **d**) of one reagent is combined with a positively polarized ("electropositive") carbon atom (electron **a**cceptor, **a**) of another reagent. A new covalent carbon-carbon bond is formed.

- The formal carbanions and carbocations used as units in synthesis are called donor synthons and acceptor synthons. They are derived from reagents with functional groups.

Example:

CH_3^- methyl donor synthon (from the reagent CH_3Li)
CH_3^+ methyl acceptor synthon (from the reagent CH_3I)
The combination (= synthetic reaction) of both synthons would yield ethane.

- The change of a donor reagent into an acceptor reagent and vice versa is called "umpolung" (= dipole inversion).

Example:
$$\begin{array}{cc} \text{(a)} & \text{(d)} \\ CH_3I + 2\,Li & \rightarrow CH_3Li + LiI \end{array}$$

- Synthons are numbered according to the relative positions of a functional group (FG) and the reactive carbon atom.

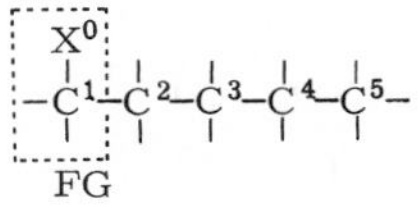

X = hetero atom
FG = functional group

If the carbon atom C-1 of the functional group itself is reacting, one has a d^1- or a^1-synthon. If the carbon atom C-2 next to the functional group (the α-carbon atom) is the reaction centre, we call it a d^2- or a^2-synthon. If the β-carbon atom C-3 is the reactive one, we assign d^3 or a^3 to the corresponding synthon, etc. Alkyl synthons without functional groups are called alkylating synthons. The electronegative hetero atom of the functional group may also form covalent bonds with acceptor synthons. In such cases we speak of d^0-synthons.

Examples

synthon type	example	reagent	functional group
d^0	$CH_3S^{\ominus}$	CH_3SH	$-\overset{\mid}{\underset{\mid}{C}}-S-$
d^1	$^{\ominus}C\equiv N$	K^+CN^-	$-C\equiv N$
d^2	$H_2\overset{\ominus}{C}-CHO$	CH_3CHO	$-CHO$
d^3	$\overset{\ominus}{C}\equiv C-\overset{\mid}{\underset{\mid}{C}}-NH_2$	$Li^{\oplus}\ ^{\ominus}C\equiv C-\overset{\mid}{\underset{\mid}{C}}-NH_2$	$-\overset{\mid}{\underset{\mid}{C}}-NH_2$
alkyl d	$^{\ominus}CH_3$	$LiCH_3$	$-$
a^0	$^{\oplus}P(CH_3)_2$	$(CH_3)_2P-Cl$	$-P(CH_3)_2$
a^1	$H_3C-\overset{OH}{\underset{\oplus}{C}}-CH_3$	$H_3C-\overset{O}{\overset{\|}{C}}-CH_3$	$>C=O$
a^2	$H_2\overset{\oplus}{C}-\overset{O}{\overset{\|}{C}}-CH_3$	$Br-CH_2-\overset{O}{\overset{\|}{C}}-CH_3$	$>C=O$
a^3	$H_2\overset{\oplus}{C}-\underset{H}{C}=\overset{O^{\ominus}}{C}-OR$	$H_2C=\underset{H}{C}-\overset{O}{\overset{\|}{C}}-OR$	$-C\overset{O}{\underset{OR}{\diagup}}$
alkyl a	$^{\oplus}CH_3$	$(CH_3)_3S^{\oplus}\ Br^{\ominus}$	$-$

● *The combination of two reagents corresponding to one d-synthon and one a-synthon under appropriate conditions yields an additional carbon-carbon bond (exception: d^0-synthons). The following obvious rules apply to the arrangement of functionality in the product ("target molecule"):*

reacting synthons:	products:
alkyl a + alkyl d	*non-functional*
alkyl a + d^1, alkyl d + a^1	*monofunctional*
$a^1 + d^1$	*1,2-difunctional*
$a^1 + d^2$, $a^2 + d^1$	*1,3-difunctional*
$a^1 + d^3$, $a^2 + d^2$, $a^3 + d^1$	*1,4-difunctional*
and so on.	

Examples for arrangements of functionality

non-functional products   ⬡ ▢ ⋋ $n\text{-}C_{100}H_{202}$

monofunctional
products

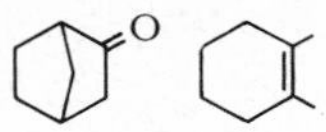

difunctional
products

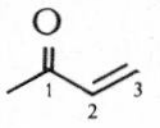

1,3-difunctional[*]

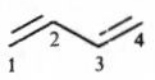

1,4-difunctional[*]

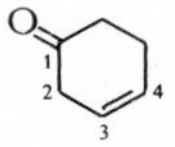

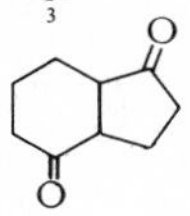

1,4-difunctional[*]

1,4- or 1,5-
difunctional

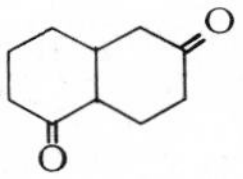

1,5-difunctional

● *If the target molecule is polyfunctional, the synthons must contain more than one functional group.*

● *If an open-chain organic molecule contains an electron acceptor* **and** *an electron donor site, two carbon atoms may be combined intramolecularly. This corresponds to the synthesis of a monocyclic compound.*

Example

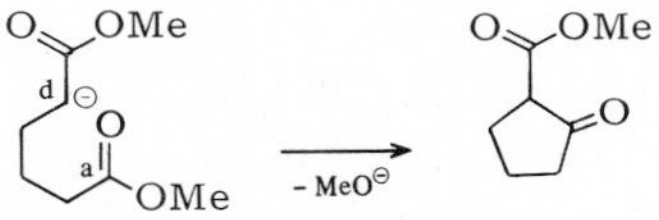

● *Intramolecular reactions of electron donor and acceptor sites in cyclic starting materials produce spirocyclic, fused, or bridged polycyclic compounds.*

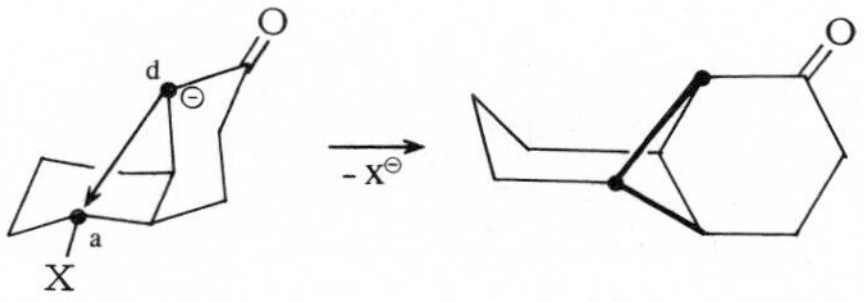

● *Chapter 1 first describes some common synthons and corresponding reagents. Emphasis is on regioselective carbanion formation. In the second part some typical synthetic*

---

* The ends of C—C multiple bonds are defined as the centers of functionality.

procedures in the following order of "arrangements of functionality in the target molecule" are given:

- *non-functional (= saturated), open-chain hydrocarbons;*
- *monofunctional (= unsaturated), open-chain hydrocarbons;*
- *hydrocarbons by special coupling reactions;*
- *monofunctional, open-chain C,H,O-compounds (in the order of rising oxidation number);*
- *difunctional open-chain C,H,O-compounds (in the order 1,2-; 1,3-; 1,4-; and 1,5-difunctional);*
- *cyclic hydrocarbons and C,H,O-compounds (in the order 3-, 4-, 5-, and 6-membered rings);*
- *1,6-difunctional C,H,O-compounds by cleavage reactions;*
- *bridged carbocycles.*

General problems with synthetic organic reactions are discussed together with some practical solutions for specific examples. These problems include

- *regio- and stereoselectivity by exploitation of the substrates' stereochemistry (p. 20, 22, 26) and differentiated nucleophilicity (p. 24f., 41, 53ff.)*
- *regioselectivity by reversible addition of activating groups (p. 40, 83)*
- *selectivity in competitive cyclization, polymerization, and dialkylation reactions (p. 23, 25)*
- *overcome of sterical hindrance (p. 33ff., 79) and of unfavorable entropy effects (p. 23, 40f., and 50f.)*

"Synthons" are defined as units which can be joined to (organic) molecules by known or conceivable synthetic operations (E.J. Corey, 1967A). A synthon may be as simple as a methyl anion or as complex as a steroid enolate anion. Since most synthetic reactions are polar, a synthon usually contains a nucleophilic electron donor centre (d) or an electrophilic electron acceptor centre (a). Synthons for the synthesis of carbon compounds are produced from organic reagents which contain functional groups such as $C-Br$, $C=C$, $C=O$, $C=N$ etc. The reactive site of the synthon may now either be on carbon atom C^1, which is part of the functional group (d^1 or a^1), or on remote carbon atoms C^n (d^n or a^n; $n \geq 2$). Important donor and acceptor synthons will be described in the order of increasing distance between the functional group and the polar, reactive carbon atom. Hetero atoms of functional groups are, with rare exceptions, electron donors (d^0).

1.1 Electron Donors (Nucleophiles)

1.1.1 Alkylating and d^1-Synthons

Carbanions are negatively charged organic species with an even number of electrons and the charge mainly concentrated on a carbon atom. In alkyl, alkenyl, and alkynyl anions all of the nonbonding electrons are localized on carbon atoms and these anions are particularly reactive. The ease of carbanion formation increases in the same order as the s-character of the CH bond: $C-CH < C=CH < \overset{\cdot}{C}\equiv CH$. Saturated alkyl anions are less stable when the carbon atom is highly substituted: tertiary $<$ secondary $<$ primary.

Alkyl, aryl, and alkenyl carbanions are usually produced from the corresponding halides by metal-halogen exchange*). Halides at sp^3-hybridized carbon atoms are more reactive than at sp^2-hybridized carbon atoms (for an example see p. 20). An ionic carbon-metal bond (e.g. C—Na, C—Li, or C—Mg) may subsequently be replaced by a more covalent carbon-metal bond (e.g. C—Cu) by metal-metal interchange. The metal ion may be strongly associated with the carbanion centre and modify its behaviour. However, carbanion chemistry and the chemistry of organometallic compounds, especially of lithium, magnesium, sodium, and potassium (M. Schlosser, 1973) are closely related. Their chemical reactions are similar. Synthetic chemistry, however, takes advantage of two important distinctions: lithium forms more covalent structures than other alkali metals (p. 11f., 22) and carbanions with a "soft" copper(I) counterion tend to be highly nucleophilic (p. 20f. and p. 35f.).

A saturated alkyl group does not exhibit functionality. It is not a d^1-synthon, because the functional groups, e.g. halide or metal ions, are lost in the course of the reaction. It functions as an alkyl synthon. Alkenyl anions (R. West, 1961) on the other hand, constitute $d^{1,2}$-synthons, because the C=C group remains in the product and may be subject to further synthetic operations.

$$\text{\textasciitilde\textasciitilde}Br + 2\ Li \longrightarrow \text{\textasciitilde\textasciitilde}Li + LiBr$$

$$\text{\textasciitilde}Cl + 2\ Li \xrightarrow[\text{(THF); 0-10 °C}]{\text{Li/Na 49 : 1}} \text{\textasciitilde}Li + LiCl$$

$$\text{\textasciitilde\textasciitilde}I + Mg \longrightarrow \text{\textasciitilde\textasciitilde}MgI \xrightarrow[-\ MgICl]{+\ ZnCl_2} \text{\textasciitilde\textasciitilde}ZnCl$$

$$CH_3Li + CuI \xrightarrow[-\ LiI]{} [CH_3Cu] \xrightarrow{+\ CH_3Li} Li^{\oplus}[Cu(CH_3)_2]^{\ominus}$$

$$2\ \text{\textasciitilde}Li + CuI \xrightarrow[-\ LiI]{} Li^{\oplus}\left[\text{\textasciitilde}Cu\text{\textasciitilde}\right]^{\ominus}$$

Alkynyl anions are more stable ($pK_a \approx 22$) than the more saturated alkyl or alkenyl anions ($pK_a \approx 40$-45). They may be obtained directly from terminal acetylenes by treatment with strong base, e.g. sodium amide (pK_a of $NH_3 \approx 35$). Frequently magnesium acetylides are made in proton-metal exchange reactions with more reactive Grignard reagents. Copper and mercury acetylides are formed directly from the corresponding metal acetates and acetylenes under neutral conditions (G.E. Coates, 1977; R.P. Houghton, 1979).

$$R-C\equiv CH + NH_2^{\ominus} \xrightarrow[\text{(liq. NH}_3)]{\text{NaNH}_2} R-C\equiv C^{\ominus} + NH_3$$

$$R-C\equiv CH + EtMgBr \xrightarrow[\text{(THF)}]{} R-C\equiv C-MgBr + EtH\uparrow$$

$$2\ R-C\equiv CH + M(OAc)_2 \longrightarrow M(-C\equiv C-R)_2 + 2\ AcOH$$
$$M = Hg,\ Cu$$

---

* For several standard procedures no references are given. They can be easily located in M. Fieser and L.F. Fieser, Reagents for Organic Synthesis, Vol 1, Wiley, N.Y., 1967 or standard textbooks.

There exist a number of d^1-synthons, which are stabilized by the delocalization of the electron pair into orbitals of hetero atoms, although the nucleophilic centre remains at the carbon atom. From nitroalkanes anions may be formed in aqueous solutions (e.g. CH_3NO_2: pK_a = 10.2). Nitromethane and -ethane anions are particularly useful in synthesis. The cyanide anion is also a classical d^1-synthon (HCN: pK_a = 9.1).

$$CH_3NO_2 + OH^\ominus \underset{(H_2O)}{\rightleftarrows} \left[\begin{array}{c} H_2\overset{\ominus}{C}-N\diagup^{\diagup O}_{\diagdown O} \\ \updownarrow \\ H_2C=N\diagup^{\diagup O^\ominus}_{\diagdown O} \end{array} \right] + H_2O$$

$$H-C\equiv N + OH^\ominus \underset{(H_2O)}{\rightleftarrows} \left[\begin{array}{c} ^\ominus C\equiv N \\ \updownarrow \\ C=N^\ominus \end{array} \right] + H_2O$$

More recent developments are based on the finding, that *the d-orbitals of silicon, sulfur, phosphorus and certain transition metals may also stabilize a negative charge on a carbon atom.* This is probably caused by a partial transfer of electron density from the carbanion into empty low-energy d-orbitals of the hetero atom (''backbonding'') or by the formation of ''ylides'', in which a positively charged ''onium centre'' is adjacent to the carbanion and stabilization occurs by ''ylene'' formation.

$$R_3Si-CH_2Cl \xrightarrow[- LiCl]{+ 2 Li} \left[R_3Si^{\cdots}\overset{\ominus}{C}H_2 \right] Li^\oplus$$

$$R-\left\langle \begin{array}{c} S \\ S \end{array} \right\rangle \xrightarrow[- BuH]{+ BuLi} \left[R-\overset{\ominus}{\left\langle \begin{array}{c} S \\ S \end{array} \right\rangle} \right] Li^\oplus \quad \textbf{1,3-Dithian-2-ide}$$

$$H_3C\overset{\overset{O}{\|}}{\underset{}{S}}CH_3 \xrightarrow[- H_2]{+ NaH} \left[H_2\overset{\ominus}{C}\overset{\overset{O}{\|}}{\underset{}{S}}CH_3 \right] Na^\oplus \quad \text{''Dimsyl'' anion}$$

$$Ph_3\overset{\oplus}{P}-CH_2-OMe \xrightarrow[- BuH]{+ BuLi} \left[Ph_3\overset{\oplus}{P}-\overset{\ominus}{C}\diagdown^H_{OMe} \longleftrightarrow Ph_3P=C\diagdown^H_{OMe} \right]$$

$$\qquad\qquad\qquad\qquad\qquad\qquad \text{''Ylide''} \qquad\qquad \text{''Ylene''}$$

Silylated carbanions are available from the corresponding silylated alkyl chlorides by Grignard techniques. The repulsive interaction between the electropositive silicon and metal atoms leads to particularly loose carbon-metal bonds, and metal-metal interchange is possible under a variety of conditions. If the silylated carbon atom carries a phenyl group, a sulfur atom, or a second silicon atom, deprotonation is also possible. It occurs with strong bases or with butyllithium. Allylic silanes are also deprotonated by butyllithium and form anions, in which the negative charge is delocalized over three carbon atoms. Vinyl silanes, however, tend to undergo nucleophilic additions to metal organyls. The negative charge of the resulting saturated anion is generally localized on the carbon atom adjacent to silicon (I. Fleming, 1979).

$$Me_3Si-CH_2Cl \xrightarrow[\text{(Et}_2\text{O)}]{+ Mg} Me_3Si-CH_2-MgCl$$

$$\downarrow HgCl_2$$

$$2\ Me_3Si-CH_2^{\ominus}\ K^{\oplus} \xleftarrow[- Hg]{+ 2\ K} (Me_3Si-CH_2-)_2Hg$$

$$Et_3Si \overset{Br}{\diagdown\!\!\diagdown} + Mg \longrightarrow Et_3Si \overset{MgBr}{\diagdown\!\!\diagdown}$$

$$Me_3Si-CH_2-SMe \xrightarrow[- BuH]{+ BuLi} \left[Me_3Si\overset{\ominus}{\cdots}\underset{H}{C}\cdots SMe \right] Li^{\oplus}$$

$$Me_3Si-CH_2-Ph \xrightarrow[- BuH]{+ BuLi} \left[Me_3Si\overset{\ominus}{\cdots}\underset{H}{C}\cdots Ph \right] Li^{\oplus}$$

$$Ph_3Si \diagup\!\!\diagdown\!\!\diagup \rightleftharpoons Ph_3Si \diagup\!\!\diagdown\!\!\diagdown \xrightarrow[- BuH]{+ BuLi} \left[Ph_3Si \diagup\!\!\diagdown\!\!\overset{\ominus}{\diagdown} \right] Li^{\oplus}$$

$$Me_3Si \diagdown\!\!\diagdown + Li \overset{}{\diagdown\!\!\diagup} \xrightarrow[\text{(TMEDA/C}_5\text{H}_{12})]{} Me_3Si \diagdown\!\!\overset{Li}{\diagdown}\!\!\diagup$$

Alkyl groups adjacent to a phosphorus atom in quaternary phosphonium salts lose a proton when treated with base. An ylide is formed, which is stabilized by (d-p) π-bonding or "ylene" formation. The strength of base required depends on the substituents on the carbon atom, which is to be deprotonated. The more these substituents are able to stabilize an adjacent negative charge, the more stable and the less reactive will the ylide be. Since the phosphorus atom is removed at the end of most synthetic operations, phosphorus ylides may, depending on the substituents, behave as non-functional (= alkyl) synthons or as d^1, d^2, ...d^n-synthons. Carbanions derived from phosphonic acids are also frequently used in synthesis (B.J. Walker, 1972; J.I.G. Cadogan, 1979).

alkylidene synthon

$$Ph_3\overset{\oplus}{P}-CH_2-CH_3 \xrightarrow[- H^{\oplus}]{\text{base}} \left[\begin{array}{l} Ph_3\overset{\oplus}{P}-\overset{\ominus}{\underset{H}{C}}-CH_3 \quad \text{Ylide} \\ \quad\quad \updownarrow \\ Ph_3P=\underset{H}{C}-CH_3 \quad \text{Ylene} \end{array} \right]$$

d^1 synthon

$$Ph_3\overset{\oplus}{P}\overset{\ominus}{\cdots}C\overset{H}{\underset{OMe}{\diagup}}$$

d^2 synthon

$$(EtO)_2\overset{O}{\underset{}{P}}\overset{\ominus}{\underset{H}{C}}\overset{O}{\underset{}{C}}\diagdown_{OR}$$

d^n synthon

$$Ph_3\overset{\oplus}{P}\overset{\ominus}{\cdots}\underset{H}{C}-(CH_2)_{n-2}\ COOR$$

Sulfur ylides contain a carbanion, which is stabilized by an adjacent positively-charged sulfur. Ylides derived from alkylsulfonium salts are usually generated and utilized at low temperatures. Oxosulfonium ylides are, however, stable near room temperature. The most common method of ylide formation is deprotonation of a sulfonium salt. What has been said about the applicability of phosphonium ylide synthons is also true for their sulfur analogues: in principle they may serve to introduce all kinds of substituents. So far they have mostly served as nonfunctional and d^2-synthons (B.M. Trost, 1975, A).

$$\left[H_3C - \overset{\oplus}{\underset{CH_3}{\overset{S}{S}}} - CH_3 \right] I^{\ominus} \xrightarrow[\substack{(DMSO/THF) \\ 5\,°C}]{H_3C-\overset{O}{\overset{\|}{S}}-CH_2^{\ominus} Na^{\oplus}} H_3C - \overset{\oplus}{\underset{CH_3}{\overset{S}{S}}} \overset{\ominus}{CH_2}$$

$$\left[H_3C - \overset{O}{\underset{CH_3}{\overset{\|\oplus}{S}}} - CH_3 \right] I^{\ominus} \xrightarrow[\substack{(H_2O)}]{NaOH} H_3C - \overset{O^{\ominus}}{\underset{CH_3}{\overset{|\oplus}{S}}} \overset{\ominus}{CH_2}$$

d^2 synthon

$$H_3C - \overset{CH_3}{\underset{\oplus}{\overset{|}{S}}} \overset{\ominus}{\cdots} \overset{O}{\underset{H}{\overset{\|}{C}}} - \overset{O}{\underset{}{C}} \diagdown OEt$$

The carbanions derived from thioacetals, however, are typical d^1-synthons. Most frequently used are 1,3-dithianes and C^α-silylated thioethers (see p. 33; D. Seebach, 1969, 1973; B.-T. Gröbel, 1974, 1977). In these derivatives the proton is removed by butyllithium in THF.

$$R-CHO + \overset{HS-}{\underset{HS-}{\Big\rangle}} \xrightarrow[-H_2O]{} R - \overset{S-}{\underset{S-}{\diagup}} \xrightarrow[\substack{(THF) \\ -30\,°C}]{BuLi} R - \overset{S}{\underset{S}{\diagup}} \ominus$$

In sulfoxides and sulfones an adjacent CH group is also deprotonated by strong bases. If one considers the sulfinyl (—SO—) or sulfonyl (—SO$_2$—) groups to be functional groups, then these carbanions are d^1-synthons. It will be shown later (p. 48 and 60ff.), that these anions may either serve as nonfunctional, d^1-, d^2- or d^3-synthons.

$$H_3C - \overset{O}{\underset{}{\overset{\uparrow}{S}}} - CH_3 \xrightarrow[-78\to25\,°C]{NaH(THF)} H_3C - \overset{O}{\underset{}{\overset{\|}{S}}} \overset{\ominus}{CH_2} \qquad \text{Dimsyl anion}$$

The large sulfur atom is a preferred reaction site in synthetic intermediates to introduce chirality into a carbon compound. Thermal equilibrations of chiral sulfoxides are slow, and carbanions with lithium or sodium as counterions on a chiral carbon atom adjacent to a sulfoxide group maintain their chirality. The benzylic proton of chiral sulfoxides is removed stereoselectively by strong bases. The largest groups prefer the *anti* conformation, e.g. phenyl and oxygen in the first example, phenyl and *tert*-butyl in the second. Deprotonation occurs at the methylene group on the least hindered site adjacent to the unshared electron pair of the sulfur atom (R.R. Fraser, 1972; F. Montanari, 1975).

In cyclic sulfoxides the diastereomeric product ratio is even higher, and the chirality of the sulfur atom has been efficiently transferred to the carbon atom in synthesis.

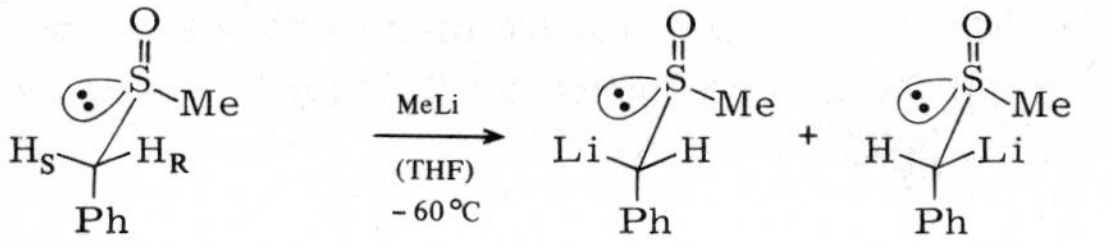

| (S)−(+) benzyl methyl sulfoxide | | (S,S) 94% | (R,S) 6% |

| (R)−(+) benzyl *t*-butyl sulfoxide | | (R,R) ~100% | (S,R) ~0% |

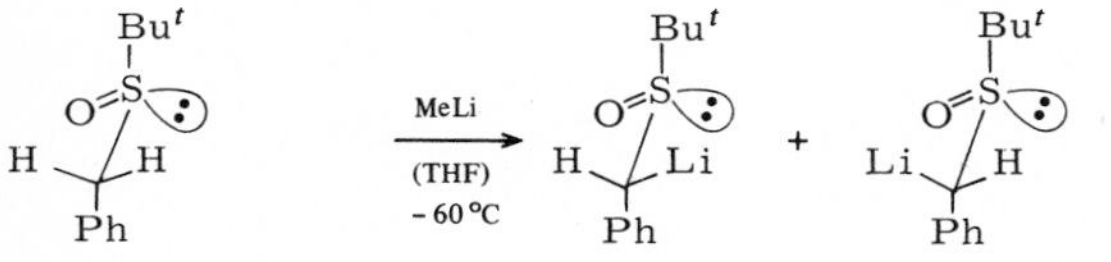

A final remark concerns the triorganylboranes, BR_3. These are electron-deficient Lewis acids and function therefore as acceptor synthons (a^0). If the boron atom is combined with an electron donor, it becomes negatively charged, and its carbon substituents now behave like carbanions. Rearrangement brings about neutralization of the boron atom and is energetically favored. A new C—C bond is formed in high yield, if an electron-accepting centre is present at an adjacent site (H.C. Brown, 1975; see pp. 21, 36f., and 44f.).

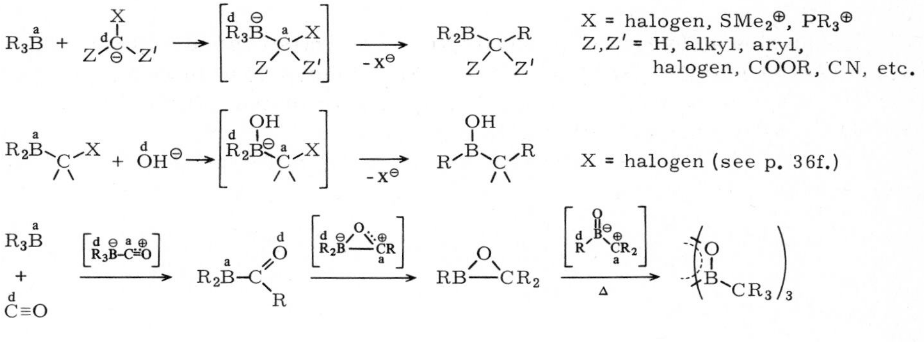

X = halogen, $SMe_2^{\oplus}$, $PR_3^{\oplus}$
Z,Z' = H, alkyl, aryl, halogen, COOR, CN, etc.

X = halogen (see p. 36f.)

1.1.2 d^2-Synthons

C—H bonds are polarized by attached unsaturated carbon substituents. Such groups "acti-vate" the neighbouring CH_3, CH_2, or CH groups in the following order: $CR=NR_2^+ >$ COR > CN > COOR > CR=NR > Ph > CR=CR_2. Two activating substituents reinforce each other.

1,3-Dioxo compounds are deprotonated at C-2 and C-4 by two equivalents of strong bases (e.g. LDA or BuLi). *Carbon atom C-4 of those dianions is much more nucleophilic than the less basic center C-2* (Hauser's rule; C.R. Hauser, 1958; K.G. Hampton, 1965). The for-mation of some typical d^2-synthons and their pK_a values are given below.

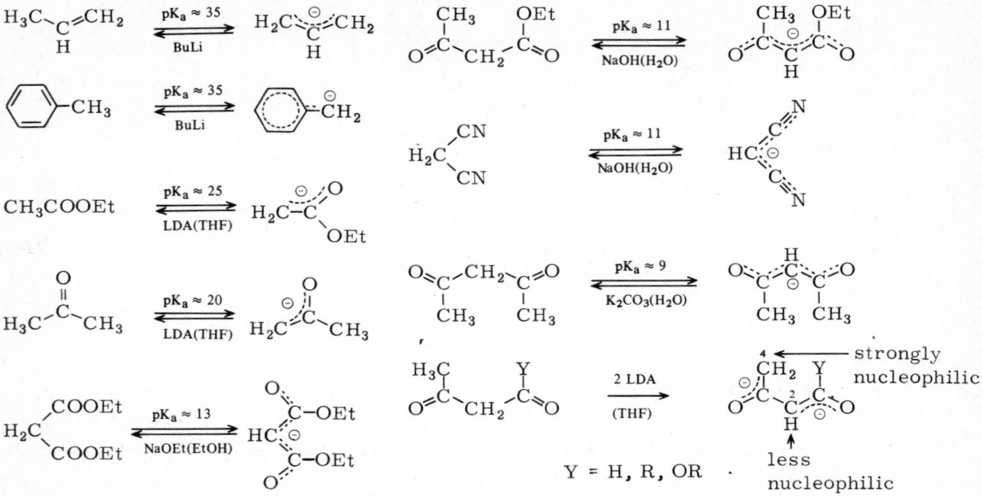

The formation of the above anions ("enolate" type) depend on equilibria between the carbon compounds, the base, and the solvent. To ensure a substantial concentration of the anionic synthons in solution the pK_a of both the conjugated acid of the base and of the solvent must be higher than the pK_a-value of the carbon compound. Alkali hydroxides in water ($pK_a \approx 16$), alkoxides in the corresponding alcohols ($pK_a \approx 20$), sodium amide in liquid ammonia ($pK_a \approx 35$), dimsyl sodium in dimethyl sulfoxide ($pK_a \approx 35$), sodium hydride, lithium amides, or lithium alkyls in ether or hydrocarbon solvents ($pK_a > 40$) are common combinations used in synthesis. Sometimes the bases (e.g. methoxides, amides, lithium alkyls) react as nucleophiles, in other words they do not abstract a proton, but their anion undergoes addition and substitution reactions with the carbon compound. If such is the case, sterically hindered bases are employed. A few examples are given below (H.O. House, 1972; I. Kuwajima, 1976).

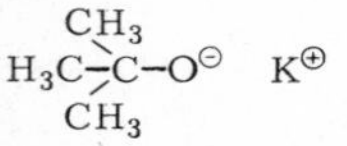

$H_3C-\underset{\underset{CH_3}{|}}{\overset{\overset{CH_3}{|}}{C}}-O^\ominus$ $K^\oplus$ KOBu^t potassium *t*-butoxide

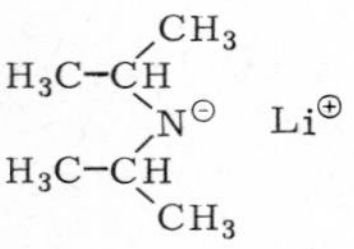

LDA   lithium diisopropylamide

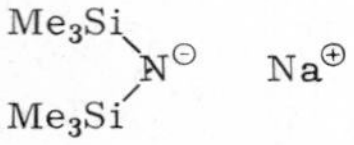

        sodium bis[trimethylsilyl]amide

$H_2N\diagdown\diagup\diagdown NH^\ominus$ $K^\oplus$ KAPA   potassium (3-aminopropyl)amide

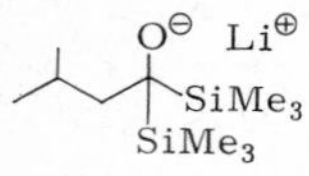

        lithium 1,1-bis (trimethylsilyl)-
        -3-methylbutoxide

weak bases

NMe$_2$

1-(dimethylamino)naphthalene,
"Hünig base"

$(CH_2-CH)_n$

poly [4-(diisopropylaminomethyl)styrene],
"polymeric Hünig base"

α-Carbanions of aldehydes, carboxylic esters and 1,3-dicarbonyl compounds are generally well defined, because only one carbon atom is activated. In unsymmetrically substituted ketones, however, both neighbouring C—H bonds are reactive and the problem of regioselective enolate formation arises. In favorable cases the regioselective deprotonation of only one carbon atom has been achieved.

The ketone is added to a large excess of a strong base at low temperature, usually LDA in THF at -78 °C. The more acidic and less sterically hindered proton is removed in a kinetically controlled reaction. The equilibrium with a thermodynamically more stable enolate (generally the one which is more stabilized by substituents) is only reached very slowly (H.O. House, 1977), and the "kinetic" enolates may be trapped and isolated as silyl enol ethers (J.K. Rasmussen, 1977; H.O. House, 1969). If, on the other hand, a weak acid is added to the solution, e.g. an excess of the non-ionized ketone or a non-nucleophilic alcohol such as *tert*-butanol, then the tautomeric enolate is preferentially formed (stabilized mostly by hyperconjugation effects). The rate of approach to equilibrium is particularly slow with lithium as the counterion and much faster with potassium or sodium.

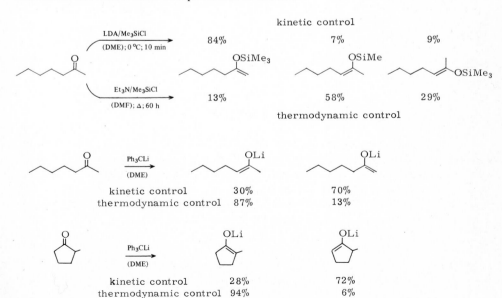

kinetic control	13%	87%
thermodynamic control	53%	47%

It should also be noted here, that under the conditions of "kinetic control" no non-ionized carbonyl compound is present. Therefore nucleophilic addition of the enolate to the carbonyl group is prevented and aldol type side reactions are suppressed. A carbonyl function may be converted to an imino group by condensation with a primary amine. The resulting N-alkylimine ("Schiff base") can be metallated with LDA in ether solvents at 0 °C. At -70 °C such lithiated Schiff bases are stable against self-addition. These carbanions are useful in selective aldol additions (see p. 53f., G. Wittig, 1968). Metallated dimethylhydrazones have also been used extensively. They allow stereo- and regioselective anion formation since the dimethylamino group orients away from large substituents and tends to locate the metal ion on the less substituted carbon atom (E.J. Corey, 1978A).

Ketones, in which one alkyl group R is sterically demanding, only give the *trans*-enolate* on deprotonation with LDA at -72 °C (W.A. Kleschick, 1977, see p. 57f.). Ketones also enolize regioselectively towards the less substituted carbon, and stereoselectively to the *trans*-enolate, if the enolates are formed by a bulky base and trapped with dialkyl boron triflates, $R_2BOSO_2CF_3$, at low temperatures (D.A. Evans, 1979). Both types of *trans*-enolates can be applied in stereoselective aldol reactions (see p. 57f.).

---

* *trans*-Enolate (*trans* relates to the carbon chain) is here equivalent to Z-enolate (Z is determined by application of the sequence rules)

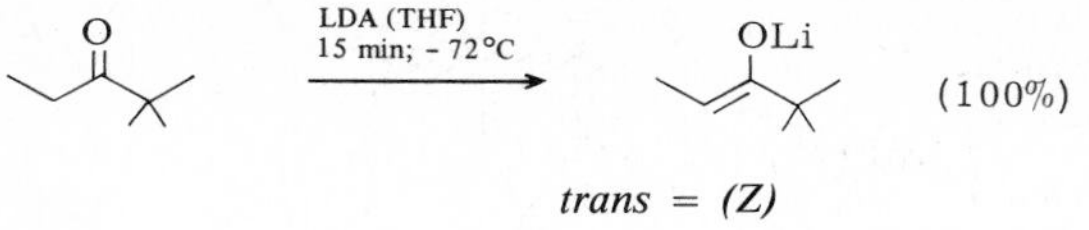

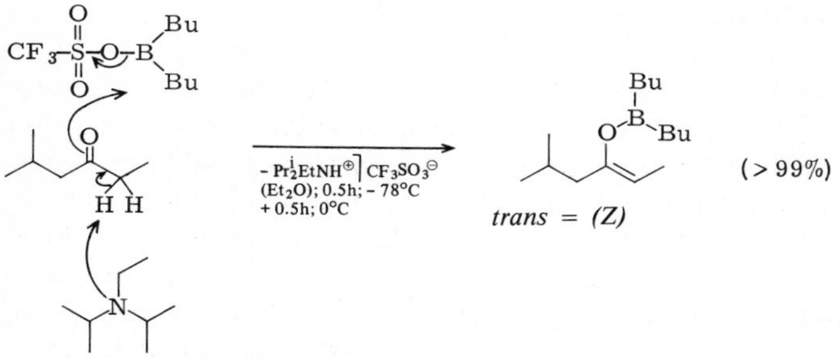

$$trans = (Z)$$

$$trans = (Z)$$

In readily available (see p. 22) cyclic imidoesters (e.g. 2-oxazolines) the α-carbon atom is metallated by LDA or butyllithium. The heterocycle may be regarded as a masked formyl or carboxyl group (see p. 22), and the alkyl substituent represents the carbon chain. The lithium ion is mainly localized on the nitrogen. Suitable chiral oxazolines form chiral chelates with the lithium ion, which are stable at -78 °C (A.I. Meyers, 1976; see p. 22).

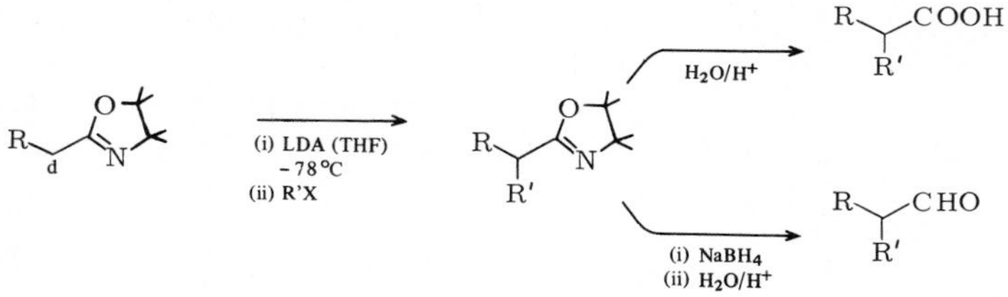

The condensation of aldehydes or ketones with secondary amines leads to "enamines" via N-hemiacetals and immonium hydroxides, when the water is removed. In these conjugated systems electron density and nucleophilicity are largely transferred from the nitrogen to the α-carbon atom, and thus enamines are useful electroneutral d^2-reagents (G.A. Cook, 1969; S.F. Dyke, 1973). A bulky heterocyclic substituent supports regio- and stereoselective reactions.

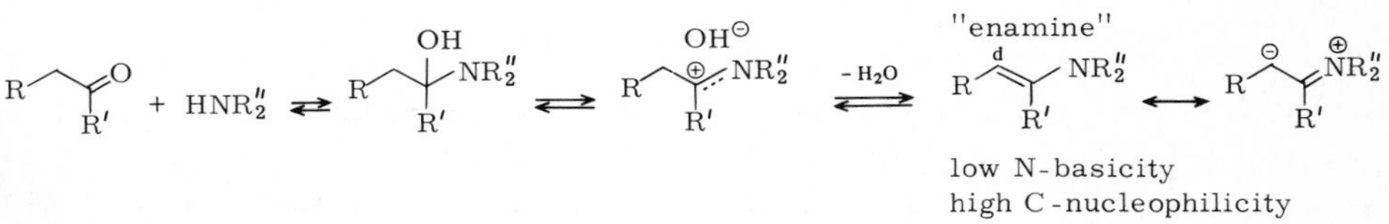

low N-basicity
high C-nucleophilicity

The N-basicity of the commonly used amines (pyrrolidine > piperidine > morpholine) drops by 2-3 orders of magnitude as a consequence of electron pair delocalization in the corresponding enamines. This effect is most pronounced in morpholino enamines (see table below). Furthermore there is a tendency of the five-membered ring to form an energetically favorable exocyclic double bond. This causes a much higher reactivity of pyrrolidino enamines as compared to the piperidino analogues towards electrophiles (G.A. Cook, 1969).

pKb values (N protonation)			Kb ratio
morpholine	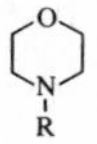 5.1	8.5	2500 : 1
piperidine	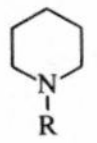 3.5	5.7	160 : 1
pyrrolidine	 3.2	5.2	100 : 1

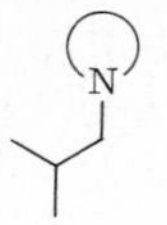

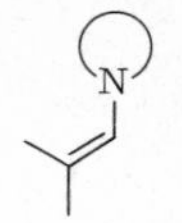

Benzylic anions, $ArCH_2^-$, are of little importance in the construction of carbon skeletons, and allylic anions, $R_2C=CR-CR_2^-$, are discussed in the d^3-synthons section below.

1.1.3 d^3-Synthons

A CH-group, which bears vinyl *and* sulfide substituents, is acidic enough to be metallated by strong bases. Other d^3-synthons may contain two activating functional groups in 1,4-position ("homoenolate"-equivalents). Only one of the α-carbons is deprotonated under appropiate conditions (e.g. succinic diesters). Another possibility is an acidic carbon and a non-acidic functional group in 1,3-positions (e.g. propargyl derivatives). Silyl ethers of α,β-unsaturated alcohols can also be converted to anions, which react as d^3-synthons (W. Oppolzer, 1976).

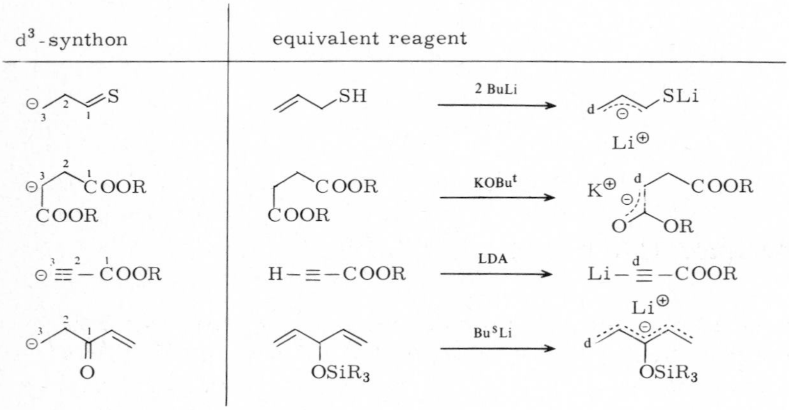

Substituted cyclopropanes readily available from addition of carbenoids to $C=C$ double bonds (see p. 68ff.) have become interesting d^3-reagents, since electron donating and withdrawing groups strongly facilitate ring opening by acids or bases. In the first example shown below ring opening occurs in a homoallylic rearrangement after the addition of a cyclopropyl anion to a carbonyl compound has taken place (E.J. Corey, 1975B), in the second the d^3-synthon is generated directly by Lewis acid promoted ring opening (E. Nakamura, 1977).

d^3-synthon	equivalent reagent, reaction

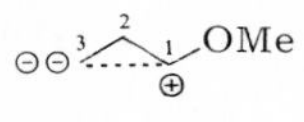

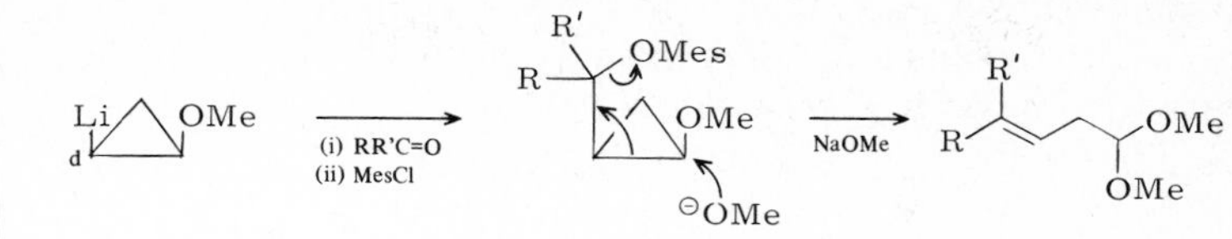

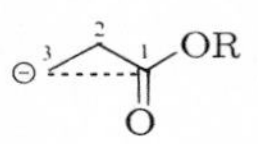

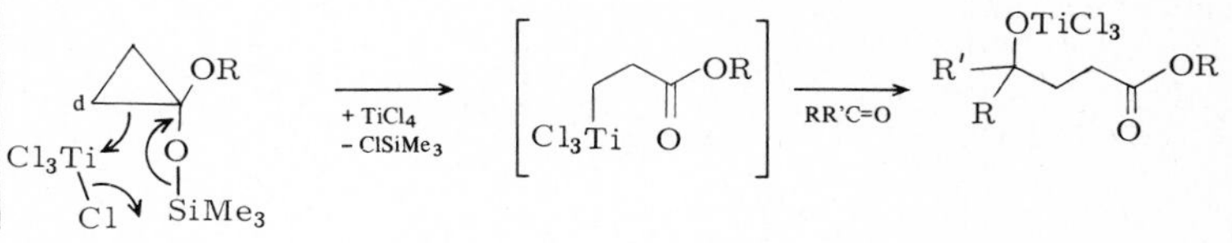

In the synthesis of complex organic molecules d^3-synthons have not become as important as their d^1 and d^2 counterparts.

1.1.4 d^n-Synthons

d^n-Synthons with $n > 3$ are individual species, in which the reactive and non-reactive functional groups are mostly non-interacting. Esoteric examples for d^5-synthons are the dianions
of unsaturated carbonyl compounds (M. Pohmakotr, 1977).

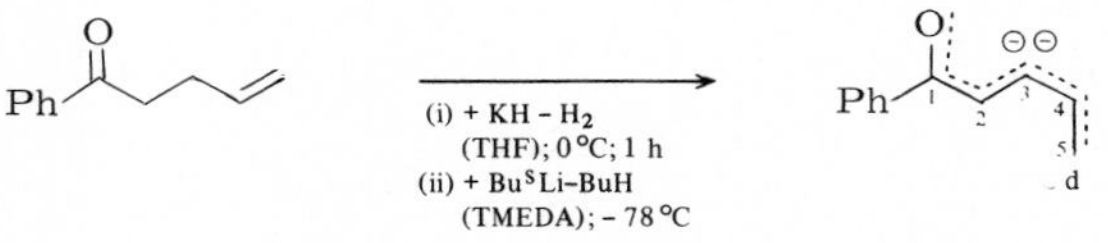

1.2 Electron Acceptors (Electrophiles, a-Synthons)

Alkyl halides and sulfonates are the most frequently used alkylating acceptor synthons. The
carbonyl group is used as the classical a^1-synthon. O-Silylated hemithioacetals (T.H. Chan,
1976) and formic acid orthoesters are examples for less common a^1-synthons. In most synthetic reactions carbon atoms with a partial positive charge (= positively polarized carbon) are
involved. More reactive, ''free'' carbocations as occurring in Friedel-Crafts type alkylations
and acylations are of comparably limited synthetic value, because they tend to react non-
selectively.

A a^2-Reagents are obtained by the halogenation or oxidation of carbonyl compounds or
olefins.

The electrophilicity of C=C double bonds conjugated with electron withdrawing
groupings leads to a^3-synthons. This is an important example of the ''vinylogous principle'':

Whenever functional groups are connected with a C=C double bond, their reactivity is often relayed through that double bond. Analogous rules can be applied to the corresponding cyclopropane derivatives.

A few examples of reactivity transfer through more than three carbon atoms are also known (a[6]-synthon: N. Miyaura, 1975, 1976).

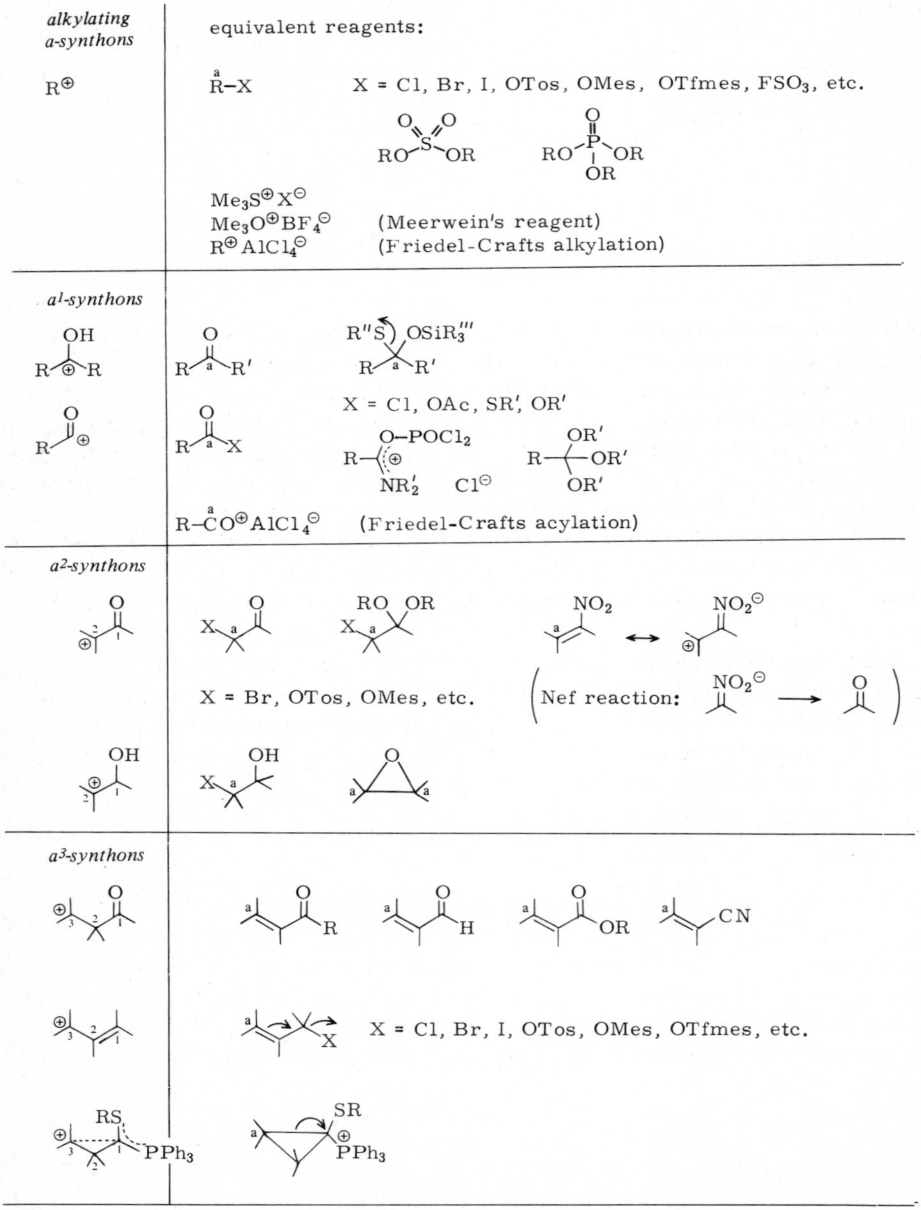

alkylating a-synthons	equivalent reagents:
$R^{\oplus}$	$\overset{a}{R}$–X X = Cl, Br, I, OTos, OMes, OTfmes, FSO$_3$, etc.
	RO–SO$_2$–OR RO–PO(OR)–OR
	Me$_3$S$^{\oplus}$X$^{\ominus}$
	Me$_3$O$^{\oplus}$BF$_4^{\ominus}$ (Meerwein's reagent)
	R$^{\oplus}$AlCl$_4^{\ominus}$ (Friedel-Crafts alkylation)

a[1]-synthons

X = Cl, OAc, SR', OR'

R–C(O–POCl$_2$)(⊕)NR$_2'$ Cl$^{\ominus}$ R–C(OR')(OR')OR'

R–$\overset{a}{C}$O$^{\oplus}$AlCl$_4^{\ominus}$ (Friedel-Crafts acylation)

a[2]-synthons

X = Br, OTos, OMes, etc.   (Nef reaction: →)

a[3]-synthons

X = Cl, Br, I, OTos, OMes, OTfmes, etc.

a⁴-synthons			$X = OH_2^{\oplus}$, Cl, Br, I, OMes, etc.
a⁵-synthons			} rarely used
a⁶-synthons			

1.3 Umpolung

Reagents with carbonyl type groupings exhibit a^1 or (if α,β-unsaturated) a^3 properties. In the presence of acidic or basic catalysts they may react as enol type electron donors (d^2 or d^4 reagents). This reactivity pattern is considered as "normal". It allows, for example, syntheses of 1,3- and 1,5-difunctional systems via aldol type ($a^1 + d^2$) or Michael type ($a^3 + d^2$) additions.

If hetero atoms are introduced or exchanged, the "normal" *reactivity of a carbon atom may be inverted* (e.g. $a^1 \rightarrow d^1$, $d^2 \rightarrow a^2$), *or a given reactivity may shift from one carbon atom to another* (e.g. $a^3 \rightarrow a^4$). It is also possible to change reactivity by adding carbon fragments (e.g. CN^-, $RC\equiv C^-$, carbenoids) to functional groups. All these processes leading to changes of the synthon type have been called "umpolung" (German: dipole inversion; G. Wittig, 1951). In rare cases the polarity of hetero atoms may also be inverted, e.g. $R-S^{(d)}-H \rightarrow R-S^{(a)}-S^{(a)}-R$.

In retro-synthetic analyses (see chapter 3) it is often useful to consider an "umpolung" of a given reagent, especially if the target molecule contains 1,2- or 1,4-difunctional systems. The schemes given below summarize some typical umpolung reactions and some specific synthons with their equivalent reagents. The review of D. Seebach (1979) is recommended for a more general survey of the method.

umpolung type	chemical reactions	examples on pages
	exchange of hetero atoms	
$a^1 \rightarrow d^1$	$\cancel{C}$Br + Ph₃P → − HBr → $d\bar{C}=\overset{\oplus}{P}Ph_3$	31f.
	$-\overset{a}{C}-X$ + Mg or + 2 M / − MX → $-\overset{d}{C}-MgX$, $-\overset{d}{C}-M$	7, 20
	$-\overset{a}{C}\overset{O}{\diagup}_H$ + HS SH / − H₂O / − H⊕ → $-\overset{d}{C}\overset{S}{\underset{S}{\diagup}}$	8, 21
	$\overset{O}{\underset{a}{\overset{\parallel}{C}}}X$ + Fe(CO)₄⊖⊖ / − X⊖ → $\overset{O}{\underset{d}{\overset{\parallel}{C}}}Fe(CO)_4$	43f.

	introduction of hetero atoms (oxidation)	
$d^{1,2} \longrightarrow a^{1,2}$	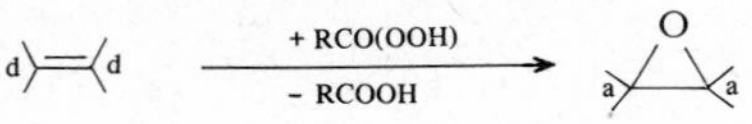 $\xrightarrow[- RCOOH]{+ RCO(OOH)}$	107, 113f.
$d^2 \longrightarrow a^2$	$\xrightarrow[- HBr]{+ Br_2}$	15f., 274
$a^3 \longrightarrow d^3$	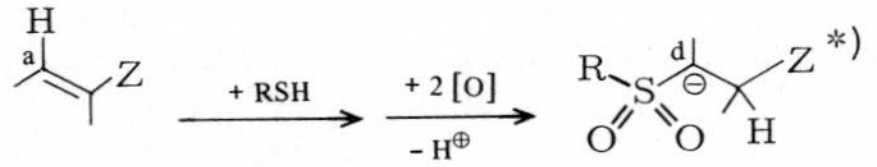 $\xrightarrow[- H^{\oplus}]{+ RSH \quad + 2\,[O]}$ $^{*)}$ $^*Z = CHO, COR, COOR, CN$	

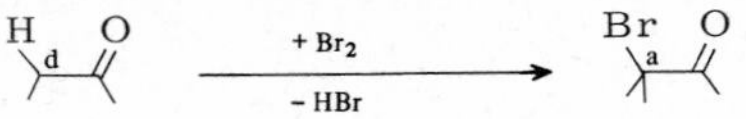

	addition of carbon fragments	
$a^1 \longrightarrow d^1$	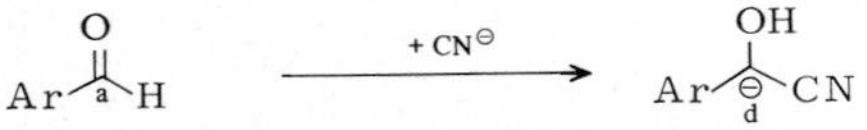 $\xrightarrow{+ CN^{\ominus}}$	47
$a^1 \longrightarrow d^3$	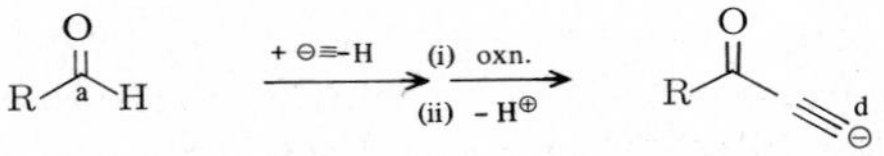 $\xrightarrow[\text{(ii)} - H^{\oplus}]{+ \ominus{\equiv}-H \quad \text{(i) oxn.}}$	49
$a^1 \longrightarrow a^3$	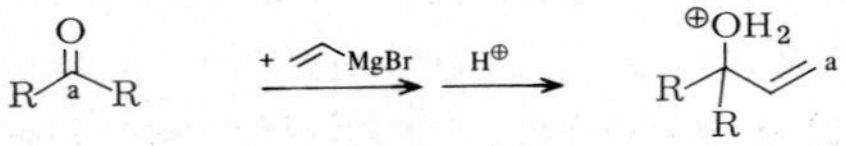 $\xrightarrow{+ \text{MgBr} \quad H^{\oplus}}$	58, 65
$a^3 \longrightarrow a^4$	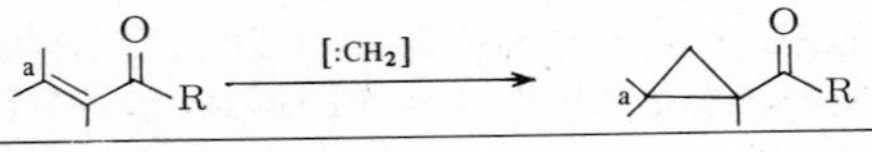 $\xrightarrow{[:CH_2]}$	69

Some specific acceptor and donor synthons

synthon   **equivalent reagents**

formyl

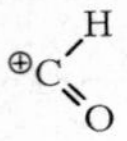

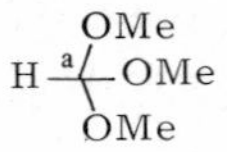

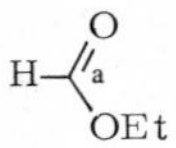

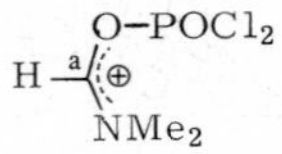

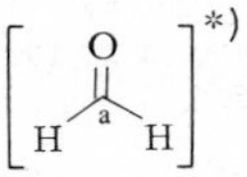

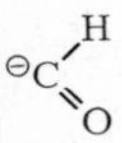

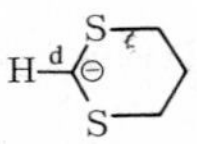

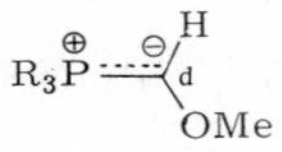

formylmethyl

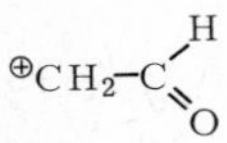

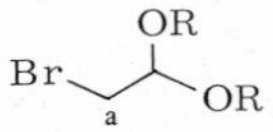

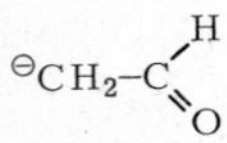

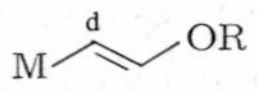

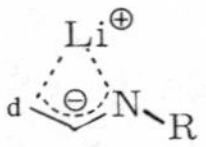

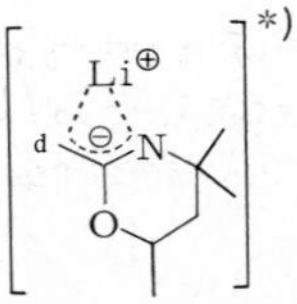

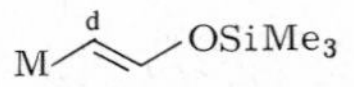

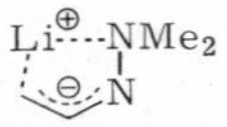

carboxymethyl

$^{\oplus}CH_2-COOH$

$^{\ominus}CH_2-COOH$

2-formylethyl

$^{\oplus}CH_2-CH_2-CHO$

$^{\ominus}CH_2-CH_2-CHO$

*[wrong oxidation state]

1.4 Introduction of Non-functional Alkyl Groups

The introduction of additional alkyl groups mostly involves the formation of a bond between a carbanion and a carbon attached to a suitable leaving group. S_N2-reactions prevail, although radical mechanisms are also possible, especially if organometallic compounds are involved. Since many carbanions and radicals are easily oxidized by oxygen, working under inert gas is advised, until it has been shown for each specific reaction that air has no harmful effect on yields.

The following aspects are important for alkylation reactions in the synthesis of complex molecules and will be exemplified:

- regio- and stereoselective alkylation of C—H, C—Hal, and C=C reaction centres in the presence of carbonyl groups
- selective introduction of only one (monoalkylation) or of two different (dialkylation) alkyl groups
- selective reaction of dibromides to yield either cyclic or open chain products.

The methods employed to solve such problems of regio- and stereoselectivity involve:

- variations of relative concentrations of educts and of reagents
- choice of different metal organyls
- thermodynamic or kinetic control by choice of reaction conditions
- utilization of specific sterical effects in substrates (usually cyclic)
- introduction of appropriate carbanion-stabilizing groups which are removed after alkylation.

The simplest case is the substitution of a halogen at a saturated carbon atom by an alkyl group. Organocopper reagents exhibit strong carbanionic capacity, and do attack ester groups only slowly (D.E. Bergbreiter, 1975). Ketones, however, should be protected. The relative reactivity of substrates toward lithium diorganocuprates is as follows: acid chlorides > aldehydes > tosylates > epoxides > bromides > ketones > esters > nitriles (G.H. Posner, 1975).

Some copper reagents bearing large alkyl groups also show pronounced stereoselectivity, when they react with dihalides. The halogen on the sterically more hindered side of a molecule is substituted much more slowly. An example is the conversion of a *gem*-dichloride into a chiral dialkyl compound as part of a synthetic sequence for sirenin (K. Kitatani, 1976). Substitution of a vinylic halide was used in the synthesis of juvenile hormones (E.J. Corey, 1968B). Michael type alkylations of activated C=C bonds are also possible with copper organyl anions. These reactions are regioselective, because the copper tends to form π-complexes with C=C bonds rather than σ-complexes with carbonyl groups (W.C. Still, 1976). They occur also stereoselective at the least hindered side of cyclic systems (G.H. Posner, 1972) or chiral crotonic esters (M. Kawana, 1966).

Most enones are reduced to anion radicals by organo cuprates. It is likely, that this reaction is connected with the alkylation. Both the formation of anion radicals and of conjugate adducts are not observed, when the redox potential of the enone becomes too negative (H.O. House, 1976).

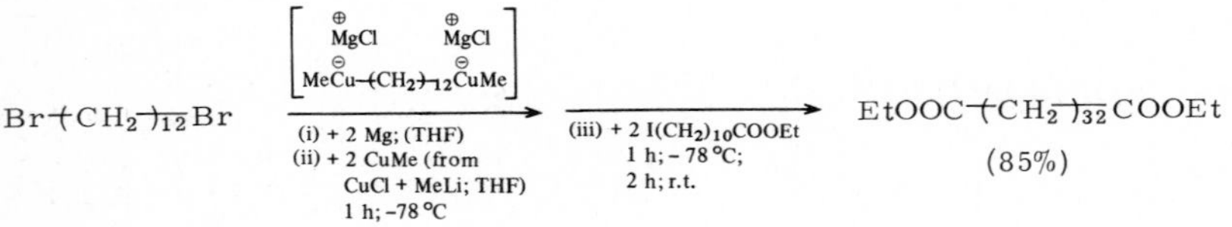

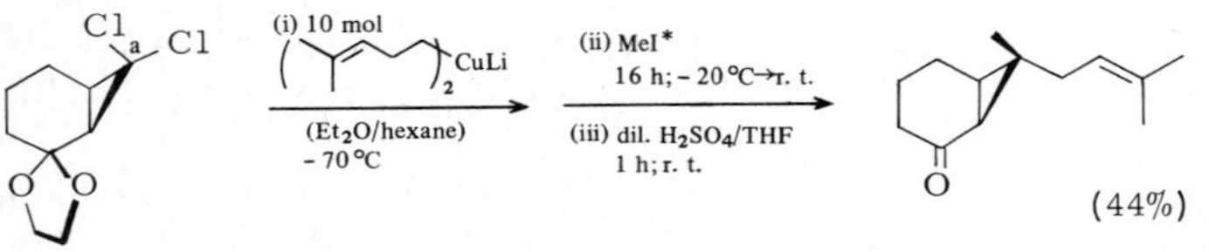

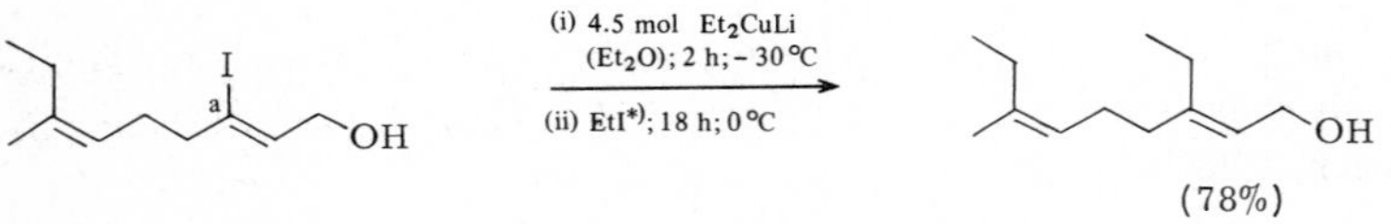

* metal-halogen exchange at ≥ 0 °C:

$$R_2Cu^\ominus + R'X \;\rightleftarrows\; RR'Cu^\ominus + RX \;\rightleftarrows\; \cdots$$

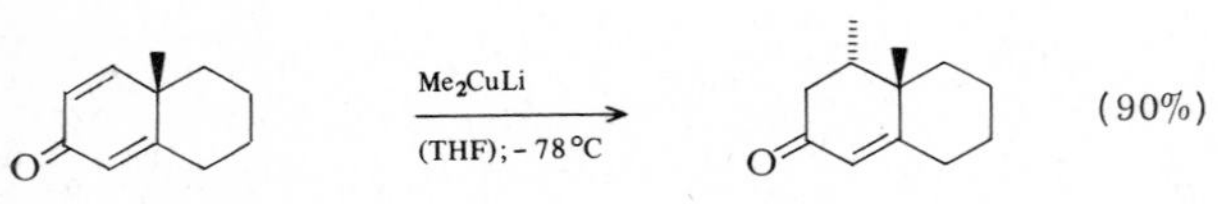

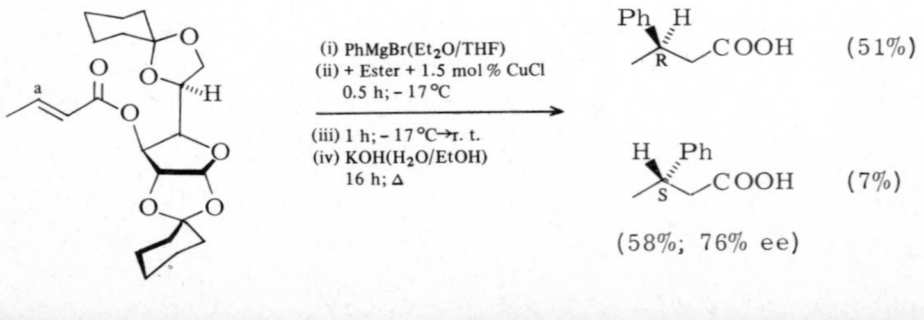

Substituted epoxides are attacked by organocopper reagents at the least hindered carbon atom and form alcohols (C.R. Johnson, 1973A). With α,β-unsaturated epoxides *trans*-allylic alcohols are produced selectively by 1,4-addition (W. Carruthers, 1973; G.H. Posner, 1972).

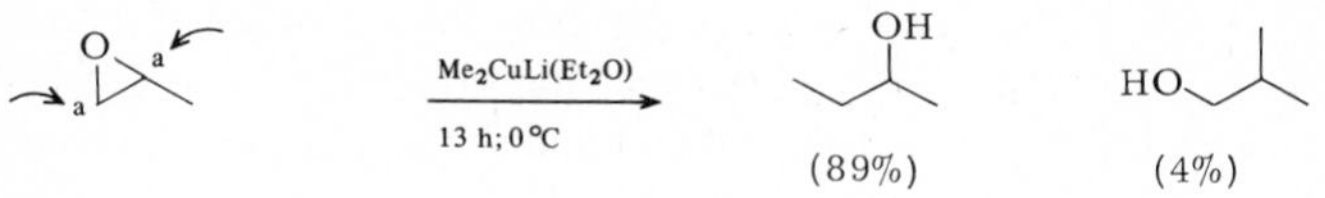

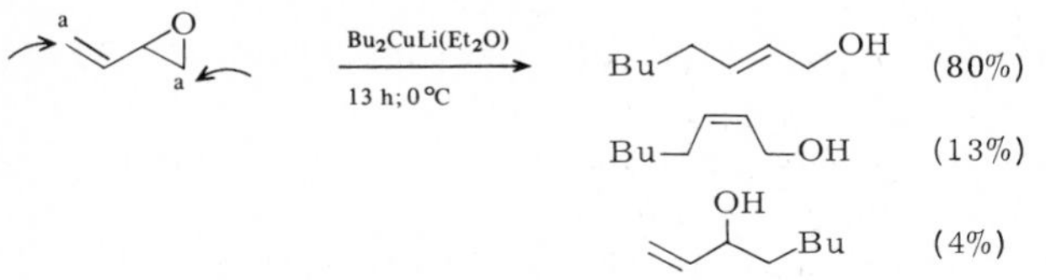

If boranes (K. Utimoto, 1973; H.C. Brown, 1975, 1980; A. Pelter, 1979) are used as donor synthons for the alkylation of α,β-unsaturated carbonyl compounds, no enolate anion is formed, and the β-position of the C=C bond is the only reaction site.

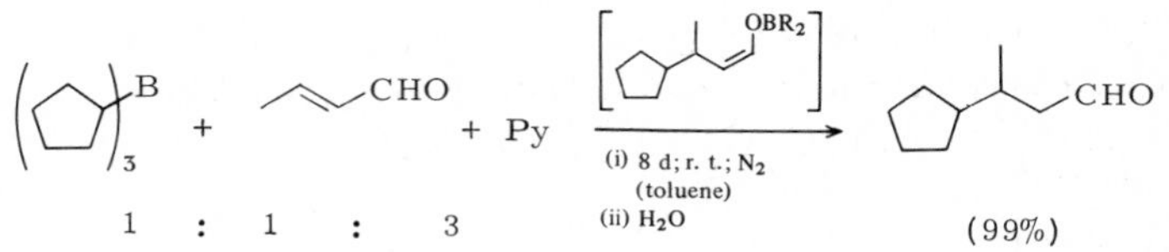

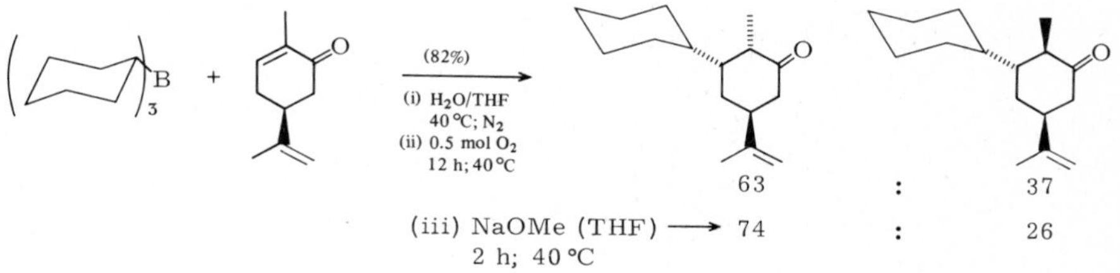

Lithium 1,3-dithian-2-ides (p. 6, 8) may be alkylated with alkyl bromides or iodides. Steric hindrance is usually of little importance and the resulting ketone can be easily liberated by hydrolysis (D. Seebach, 1969).

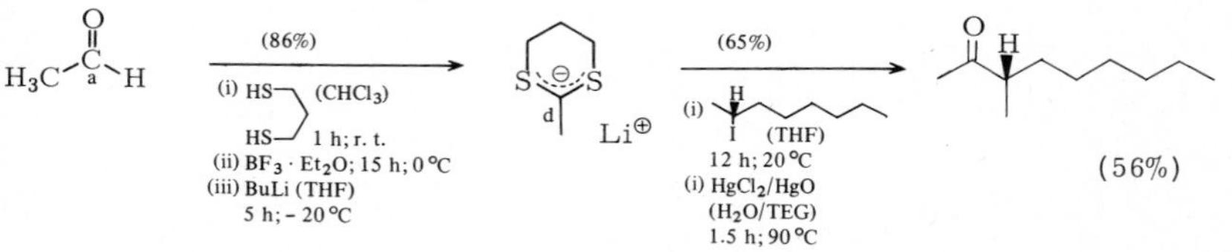

Esters are alkylated in the presence of strong bases in aprotic solvents. A common combination is LDA in tetrahydrofuran at low temperatures. Equimolar amounts of base are sufficient and only the mono-carbanion is formed. After addition of one mole of alkyl halide the products form rapidly, and no dialkylation, which is a problem in the presence of excess base, is possible. Addition of one more mole of LDA and of another alkyl halide leads to asymmetric dialkylation of one α-carbon atom in high yield (R.J. Cregge, 1973).

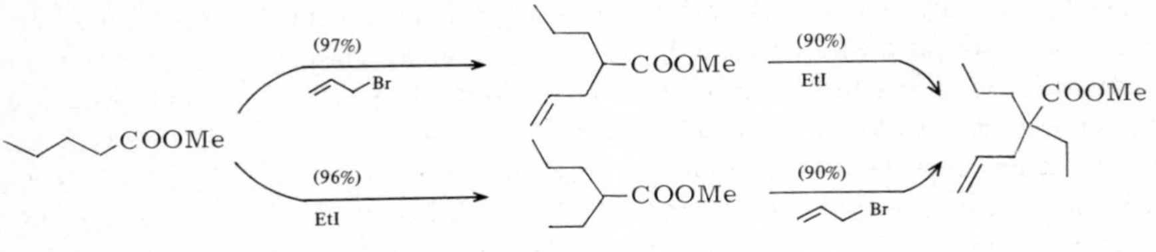

one-flask procedure: step 1 & 3 : + LDA/THF; 0.5 h; – 78 C
             step 2 & 4 : + RX/HMPTA; – 78 C

α-Alkylations of carboxylic acid derivatives lead to asymmetric carbon centres. The flexibility of a carbon chain, however, renders the formation of a rigid chiral carbanion, which would be necessary to induce asymmetric reactions, rather difficult. A chiral cyclic intermediate, in which the metal counterion would be rigidly fixed above or below the plane of the ring, would be more promising. Such anion has been obtained from 2-alkyl-Δ^2-oxazolines by Meyers (see p. 13; A.I. Meyers, 1976). The chiral Δ^2-oxazolines are prepared from nitriles and a chiral 2-aminoalcohol. The α-carbanion is alkylated, e.g. with butyl iodide at -78 °C. The oxazoline is then hydrolyzed and a chiral α-alkylcarboxylic acid is formed in about 80% optical purity. If the C=N bond is reduced with sodium borohydride before hydrolysis, α-alkylated aldehydes are formed instead.

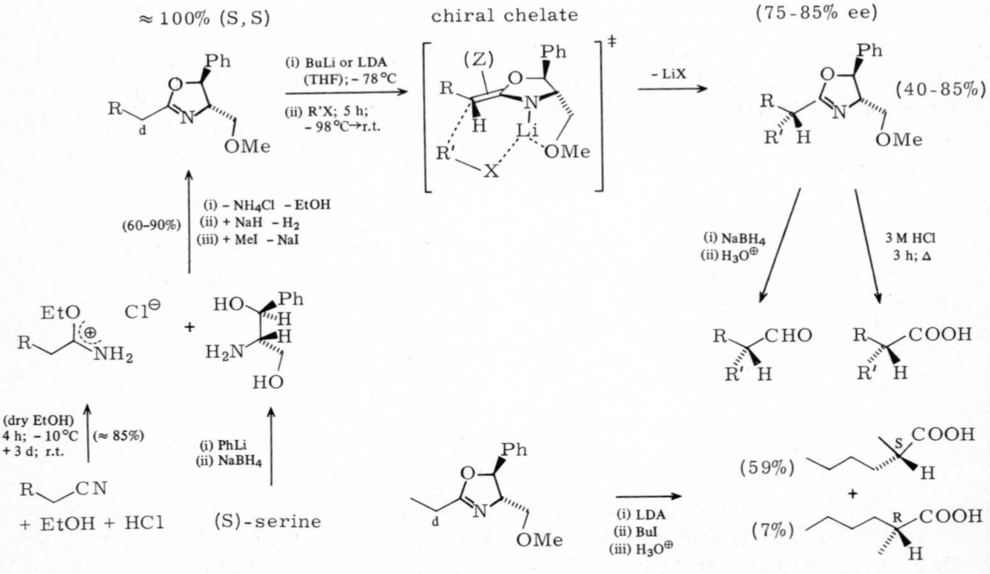

Less basic malonic ester anions may be employed for the twofold alkylation of dibromides. Cyclic 1,1-dicarboxylic esters are formed, if the reaction is executed in an appropriate manner. In the synthesis of cyclobutane diester **A** the undesired open-chain tetraester **B** was always a side product (J.A. Cason, 1949), the malonic ester and its monoalkylation product were always only partially ionized. Alkylation was therefore slow and intermolecular reactions of mono-alkyl intermediates with excess malonic ester prevailed. If the malonic ester was dissolved in ethanol containing sodium ethoxide, and 1,3-dibromopropane as well as more sodium ethoxide were added slowly to the solution, 63% of **A** and only 7% of **B** were isolated. The latter operations kept the malonic ester and its monoalkylated product in the ionic form, and the dibromide concentration low, so that the intramolecular reaction was favored against intermolecular reactions. The continuous addition of base during the reaction kept the ethoxide concentration low, which helped to prevent decomposition of the bromide by this nucleophile.

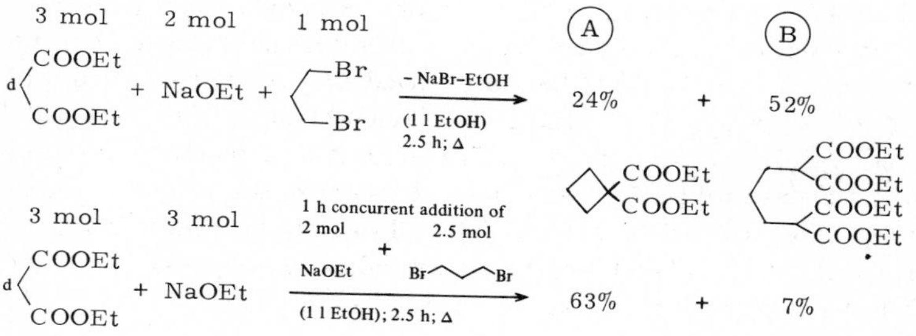

In a similar reaction the bis-tosylate shown below was to be converted into a bis-malonate derivative. In this cyclic derivative both groups are hold closely together and intramolecular reactions proved to be so much faster than intermolecular substitution under all experimental conditions, that only the undesired hydrindane carbocycle could be obtained (B.C. Ayres, 1958).

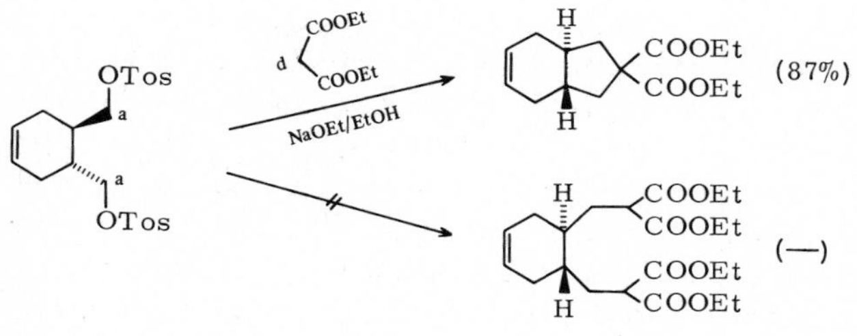

The direct α-alkylation of monoketones normally employs reaction of an alkyl halide or sulfonate with the enolate anion produced using a strong base. This method can be satisfactorily used with symmetrical ketones, which are to be dialkylated with a dihalide, and with intramolecular cyclization reactions, where the formation of five- and six-membered rings is often favored over the formation of three-, four-, seven-, and eight-membered rings (M. Mousseron, 1957; W.S. Johnson, 1963). Ketones such as 1-tetralone which can enolize in only one direction are excellent substrates (P.J. Hattersley, 1969). Regioselective alkylation of dianions according to Hauser's rule (see p. 9) is usually also a satisfactory procedure (F.W. Sum, 1979).

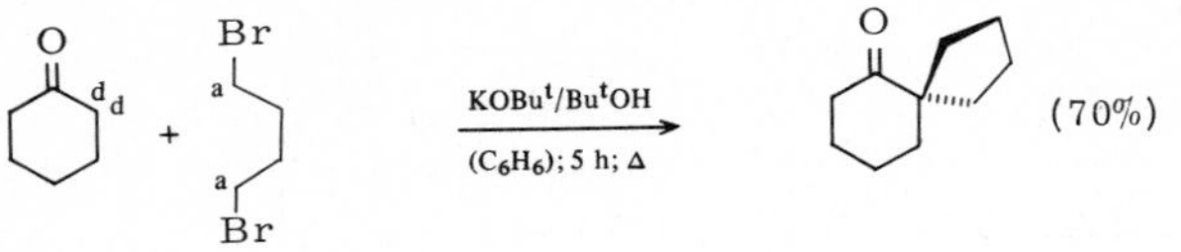

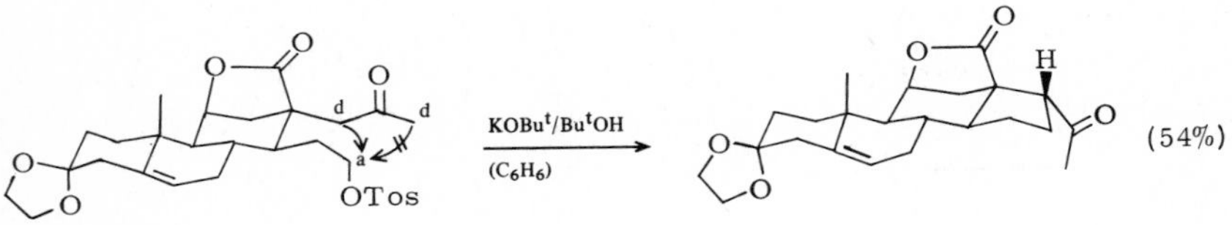

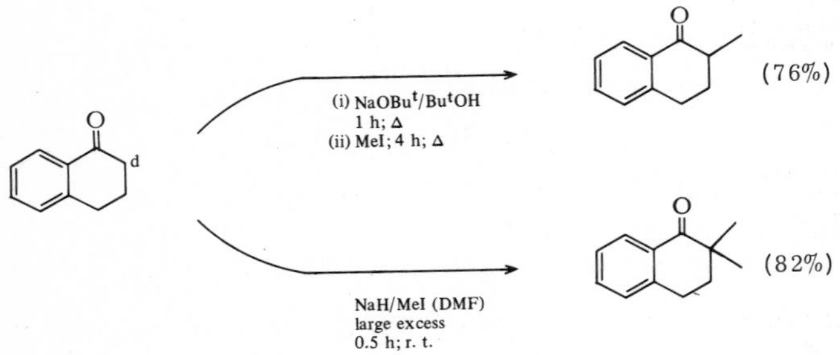

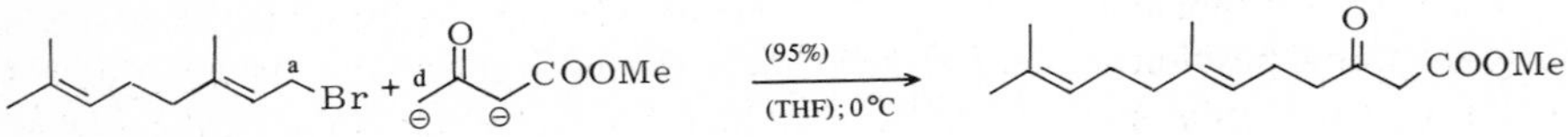

Normally, however, there are important problems with the regioselective monoalkylations of ketones and their α,β-unsaturated vinylogues. Di- and polyalkylations frequently occur and aldol addition of the enolate to the starting ketone cannot be avoided even under optimized conditions (see p. 11ff. and 53ff.). Most individual solutions of this problem have been found only after careful and tedious experimentation, although some general guidelines can be given. If the kinetic or thermodynamic control of enolate formation (see p. 11) is efficient enough to produce only one tautomer in good yield, the corresponding O-trimethylsilyl enols or enol acetates can be mono-alkylated in high yield. Methyllithium in 1,2-dimethoxyethane, a large excess of the alkyl halide, and short reaction times are applied.

An interesting case are the α,β-unsaturated ketones, which form carbanions, in which the negative charge is delocalized in a 5-centre-6-electron system. Alkylation, however, only occurs at the central, most nucleophilic position. This regioselectivity has been utilized by Woodward (R.B. Woodward, 1957; B.F. Mundy, 1972) in the synthesis of 4-dialkylated steroids. This reaction has been carried out at high temperature in a protic solvent. Therefore it yields the product, which is formed from the most stable anion (thermodynamic control).

In conjugated enones a proton adjacent to the carbonyl group, however, is removed much faster than a γ-proton. If the same alkylation, therefore, is carried out in an aprotic solvent, which does not catalyze tautomerizations, and if the temperature is kept low, the steroid is mono- or dimethylated at C-2 in comparable yield (L. Nedelec, 1974).

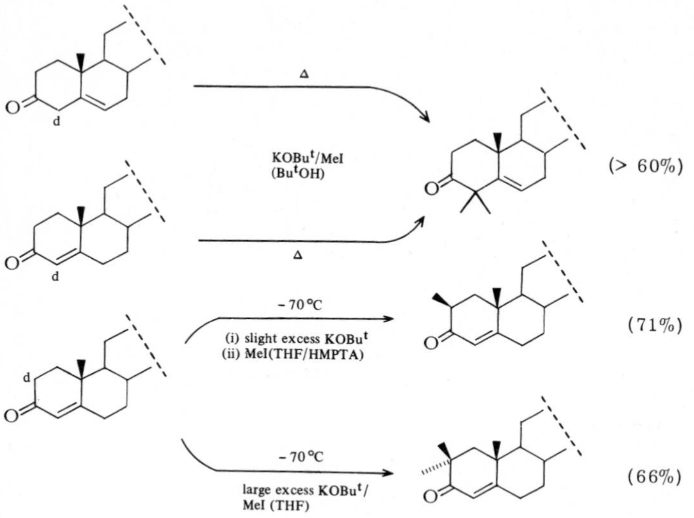

Enamines of monoketones, especially of cyclopentanone and cyclohexanone derivatives, have been attractive derivatives for monoalkylations (G.A. Cook, 1969; H.O. House, 1963; S.F. Dyke, 1973). The pyrrolidine enamine of cyclohexanone, for instance, yields with methyl iodide the α-monomethyl derivative exclusively. The large pyrrolidine substituent is fixed in an equatorial position and pushes the incoming methyl group into an axial position. A further alkylation in α- or α'-positions must now occur either in an equatorial position, which is largely blocked by the pyrrolidine substituent, or in the α'-axial position, where a large 1,3-diaxial interaction with the methyl group in the α-position builds up a high barrier (G. Stork, 1954). Under mild conditions only monomethylation occurs. Similar interactions occur in five-membered rings. Nowadays metallated imines and hydrazones are preferably used, since yields and selectivities tend to be higher with these anionic species. With anions of chiral hydrazones asymmetric induction was achieved in excellent yields (D. Enders, 1979).

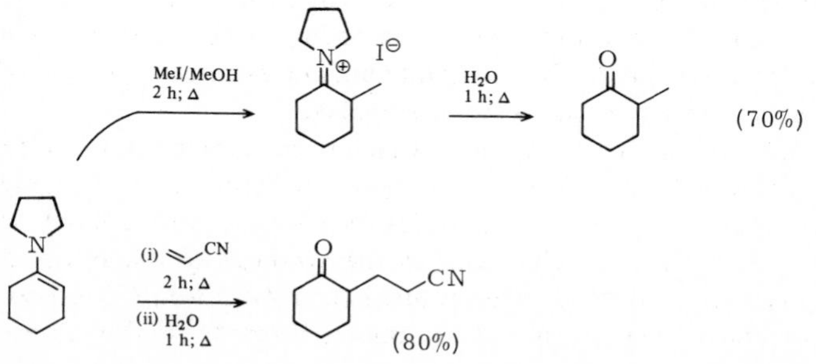

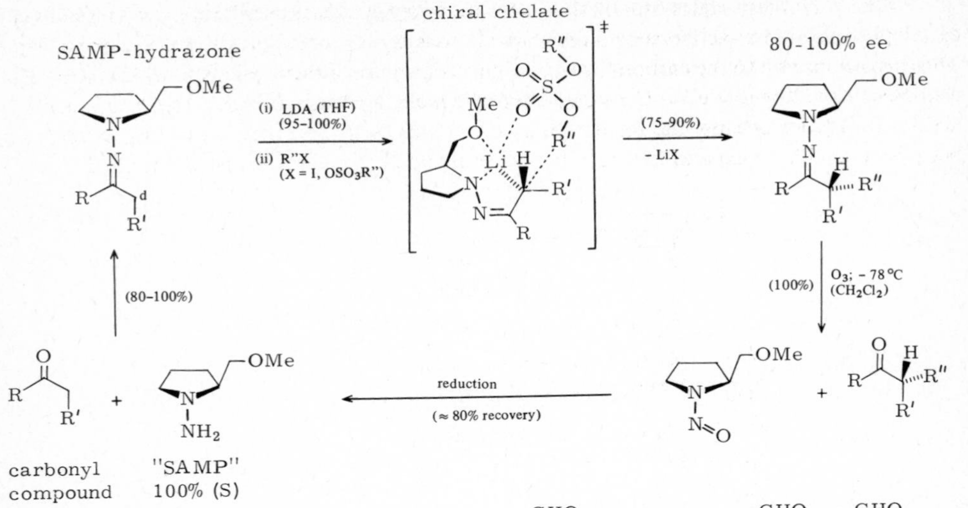

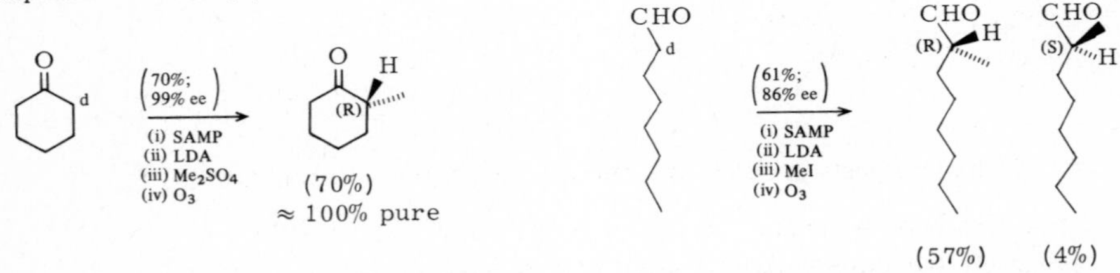

Anions of allylic thioethers may also be alkylated with alkyl bromides in high yield. The thioether groups can subsequently be removed by hydrogenolysis (F.W. Sum, 1979).

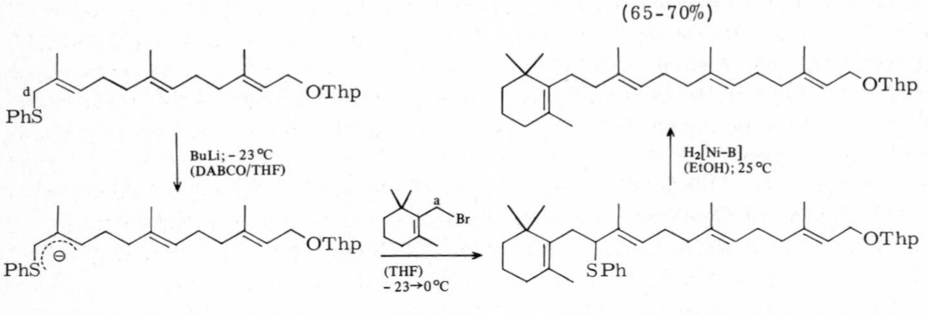

1.5 Formation of Alkenes and Alkynes

We shall only be concerned here with those reactions, which have been used in constructions of the carbon skeletons of complex compounds with concomitant regioselective incorporation of the double bond. 1,2-Eliminations are discussed on p. 123ff.

Olefin synthesis starts usually from carbonyl compounds and carbanions with relatively electropositive, redox-active substituents mostly containing phosphorus, sulfur, or silicon. The carbanions add to the carbonyl group and the oxy anion attacks the oxidizable atom Y intramolecularly. The oxide Y—O⁻ is then eliminated and a new C=C bond is formed. Such reactions take place because the formation of a Y=O bond is thermodynamically favored and because Y is able to expand its coordination sphere and to rise its oxidation number.

═Y = ═SiR$_3$	Peterson olefination
═Y = ═$\overset{\oplus}{P}$R$_3$, ═P$\overset{O}{\underset{OR}{-}}$OR	Wittig, Horner, Wadsworth-Emmons olefinations
═Y = ═SR, ═S$\overset{O}{\underset{R}{}}$	not useful, low tendency to YO⊖ elimination
═Y = ═$\overset{\oplus}{S}$R$_2$, ═$\overset{\oplus}{S}$$\overset{O}{\underset{R}{-}}$R	oxirane formation:
═Y = ═Cl	

Important synthetic aspects in such "olefinations" of carbonyl compounds are

● stereoselectivity in the formation of *cis*- or *trans*-olefins
● cyclization reactions in the absence of strong acids or bases

By far the most important reactions based on this scheme are the Wittig (G. Wittig, 1980; A. Maercker, 1965; J.I.G. Cadogan, 1979) and Horner (W.S. Wadsworth, 1977) reactions, in which an alkylidene-triphenylphosphorane (Y = PPh$_3^+$) or alkylidene phosphonate (Y = PO(OR)$_2$) is generated *in situ* by the treatment of a phosphonium salt or a phosphonate with strong base and then reacts with a carbonyl compound to yield an olefin. The generally good yields, the mild reaction conditions, the non-interference of ester and olefinic functions as well as the high degree of control over double bond position and stereochemistry (M. Schlosser, 1970; H. Bestmann, 1979) are main reasons for the widespread popularity of this type of olefin synthesis.

The reactivity of the Wittig reagents, alkylidene-triphenylphosphoranes, is determined by the substituents of the ylide carbon atom. If these have little effect on its carbanion character, the phosphoranes will be markedly nucleophilic and unstable towards water. They will react with carbonyl groups at low temperatures. If the alkylidene groups, however, bear electron withdrawing groups, the negative charge of the carbanion will be delocalized, the anion will be more stable to water, and the nucleophilicity of the carbon atom will decrease.

reactive towards:			
R–CHO	+ +	+	—
RR'CO	+ +	—	—
H$_2$O, O$_2$	+ +	±	—

Surprisingly carbonyl-substituted carbanions of phosphonates, in which the negative charge is delocalized over two oxygen atoms, are much more nucleophilic than the corresponding phosphoranes. This effect has first been observed by Horner, and has often been utilized in the synthesis of acylated olefins (R.D. Clark, 1975).

(68%)

The different behaviour of phosphoranes and phosphonates can be explained with the negative charge of the phosphonates in contrast to the electroneutral phosphoranes. If this charge is delocalized in an α-acylated phosphonate (a 5-center-6-electron system), the electron density and the nucleophilicity of the central carbon atom are high. In contrast phosphoranes with an electron withdrawing group at the carbanion are not reactive because the negative charge and nucleophilicity of the carbon atom are lowered (4-centre-4-electron system).

increasing nucleophilicity ⟶

In the reaction of a substituted ylide ($R^1CH=PPh_3$) with an aldehyde R^2CHO) a stereochemical problem arises. Much work has been carried out in order to achieve control of either *cis*- or *trans*-alkene formation. This work has been reviewed several times with always changing viewpoints (A. Maercker, 1965; L.D. Bergelson, 1964; M. Schlosser, 1970; H. Bestmann, 1979). Bergelson and Schemjakin demonstrated selective formation of *cis*-olefins in polar solvents or in the presence of Lewis bases, e.g. amines. In order to rationalize the observed stereoselectivity, they formulated a transient zwitterion in which both the bulky carbon substituents and the solvated phosphorus and oxygen ions are in *anti*-position. Cyclization of this zwitterion to an oxaphosphetane would yield a cyclic precursor of the olefin, in which both carbon substituents are *cis*. Schlosser formulated a similar mechanisms with zwitterions whose stereochemistry is dependent on the metallic counterion, e.g. lithium, in solution.

The simplest and most straightforward suggestions come from Bestmann's experiments in apolar solvents at low temperature. He states that in a kinetically controlled reaction between an ylide and a carbonyl compound the first step is the approach of the negatively charged carbon atom to the carbonyl carbon atom. This encounter should occur at an angle, which is close to 107°, if it is to form a C—C bond in which both partners are sp^3-hybridized (H.B. Bürgi, 1973). Bulky substituents (the quaternary phosphonium group and R^2 in the scheme) should be as far as possible from each other so that the formation of the carbon tetrahedrons is not disturbed. The *erythro*-P^+O^--betaine, which is presumably formed as an intermediate, is so short-lived, that it could not be detected even at -78 °C. It cyclizes immediately to an oxaphosphetane, in which both carbon substituents R^1 and R^2 are *cis*. Because this four-membered ring forms so quickly, Bestmann formulates the part of the Wittig reaction, in which the C—C bond is formed, as a "formal" [2 + 2]-cycloaddition. Originally the P—O bond is axial at the pentacovalent phosphorus (less hindered conformer). Breaking of the C—P bond occurs only, if this bond becomes axial by "pseudorotation" of the phosphorus substituents R. The C—P bond then breaks without simultaneous opening of the C—O bond. A P^+C^--betaine is formed. If R^1 is an electron donor and if the fixed ligands at the phosphorus are phenyl rings ($R=C_6H_5$), the lifetime of the betaine is very short and *cis*-alkenes are formed. If, however, R^1 is an electron acceptor (e.g. $COOCH_3$), the thermodynamically more stable conformational isomer is formed which yields the *trans*-olefin. If the R are cyclohexyl groups, the elimination of R_3PO is retarded and the proportion of the *trans*-isomer increases.

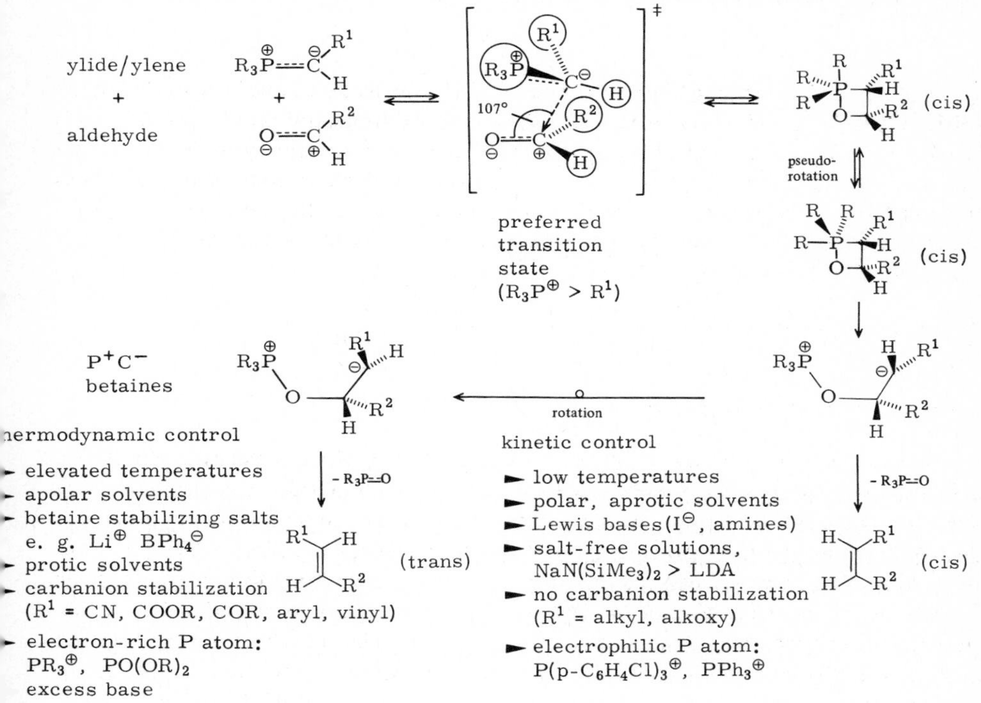

Another model considers the interactions of the phosphorus substituents, e.g. triphenyl, with the other substituents of the oxaphosphetane ring. The alkyl group R^1 forces the neighbouring phenyl ring out of the plane given by the phosphorus trigonal basis by roughly 50°. This leaves room for another alkyl group on the same side of the future oxaphosphetane ring. A *cis*-configuration is therefore favoured. If the phenyl groups are replaced by smaller groups, e.g. ethyl, even at -75 °C the *trans*-product is formed (M. Schlosser, 1982).

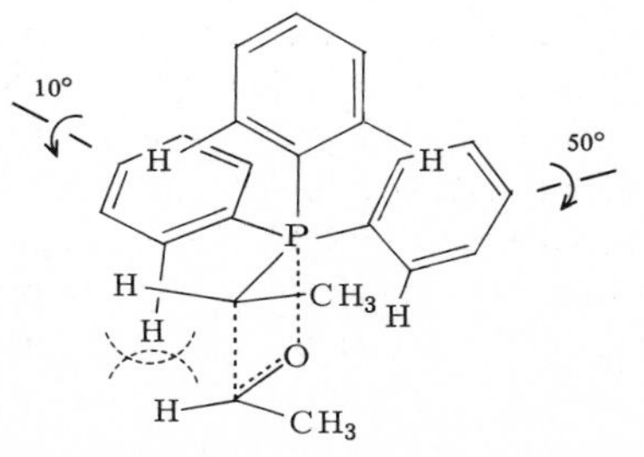

The consequences for olefin synthesis, especially polyenes, are the following:

(i) If alkyl groups are attached to the ylide carbon atom, *cis*-olefins are formed at low temperatures with stereoselectivity up to 98%. Sodium bis(trimethylsilyl)amide is a recommended base for this purpose.

(ii) Electron withdrawing groups at the ylide carbon atom give rise to *trans*-stereoselectivity. If the carbon atom is connected with a polyene, mixtures of *cis*- and *trans*-alkenes are formed. The *trans*-olefin is also stereoselectively produced when phosphonate diester α-carbanions are used, because the elimination of a phosphate ester anion is slow (W.S. Wadsworth, 1977).

Since is was discovered in 1953, the Wittig reaction has been applied in thousands of complex syntheses and a few typical or outstanding examples will now be discussed. A first general application was the synthesis of dienes and polyenes from aldehydes or ketones and allylic bromides. Before the event of the Wittig reaction, synthesis of such compounds in quantity was difficult. Nowadays carotenoids are produced in the 1,000 tons scale. Actually most of the knowledge of electronic factors in Wittig synthesis originates from industrial experiments on carotene synthesis and was first stated in the patent literature. One of the driving forces of these investigations was the need to replace the expensive, air- and moisture-sensitive phenyllithium base by less esoteric bases (H. Pommer, 1960, 1977). This was possible when the methylene hydrogen was acidified and the ylide was stabilized either by conjugated polyenes or carbonyl groups. In these syntheses mixtures of *cis* and *trans* olefination have been observed, but chemical or photochemical conversion to the natural and more stable *all-trans*-carotenoids is always possible (W. Reif, 1973). *In situ* generation of the carbonyl component by oxidative cleavage of the phosphorane leads to symmetrical olefins (H. Pommer, 1977).

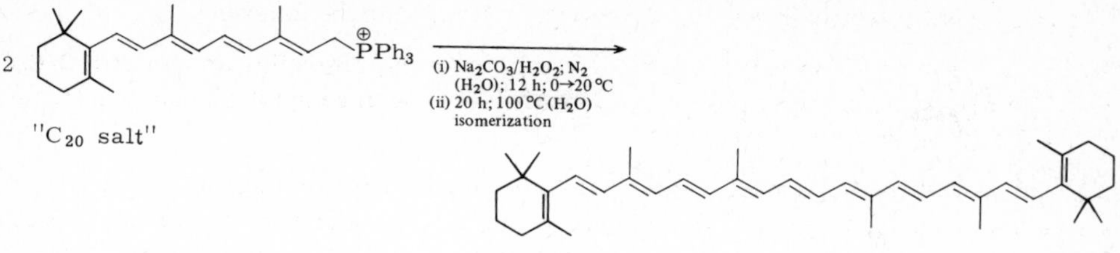

β-carotene
(80%)

Stereoselective syntheses of *cis*-olefins have been widely explored by Bestmann. An example of a pheromone synthesis is given below (H.J. Bestmann, 1976).

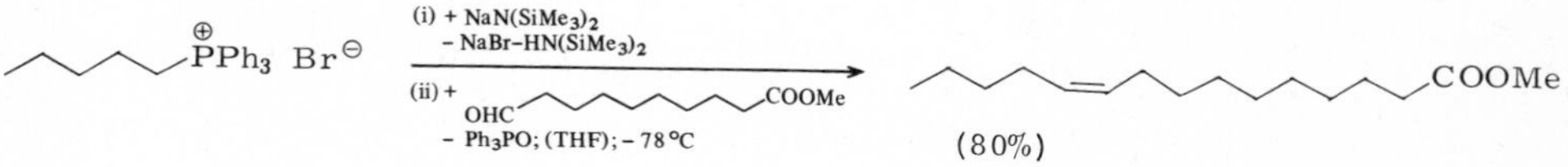

(80%)
> 98% (cis)

The synthesis of vitamin D₃ from a sensitive dienone was another early success of phosphorus ylide synthesis (H.H. Inhoffen, 1958 A). This Wittig reaction could be carried out without any isomerization of the diene. An excess of the ylide was needed presumably because the alkoxides formed from the hydroxy group in the educt removed some of the ylide.

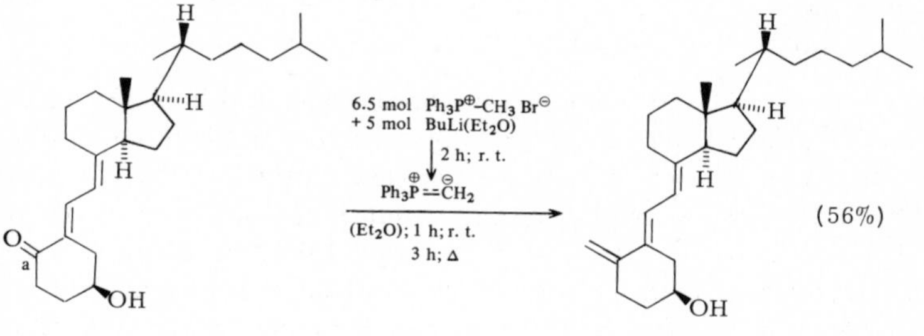

vitamin D₃

Later examples of the olefination of carbonyl compounds, which are extremely sensitive towards acid or base catalyzed rearrangements, have been given by G. Büchi and by R.B. Woodward.

The base catalyzed rearrangement of a monotosylated 1,2-diol on alumina, followed by immediate condensation of the sensitive ketone with methylenetriphenylphosphorane, gave the *exo*-methylene compound below (G. Büchi, 1966B).

Carbanions stabilized by phosphorus and acyl substituents have also been frequently used in sophisticated cyclization reactions under mild reaction conditions. Perhaps the most spectacular case is the formation of an ylide from the β-lactam given below using polymeric Hünig base (diisopropylaminomethylated polystyrene) for removal of protons. The phosphorus ylide in hot toluene then underwent an intramolecular Wittig reaction with an acetylthio group to yield the extremely acid-sensitive penicillin analogue (a "penem"; I. Ernest, 1979).

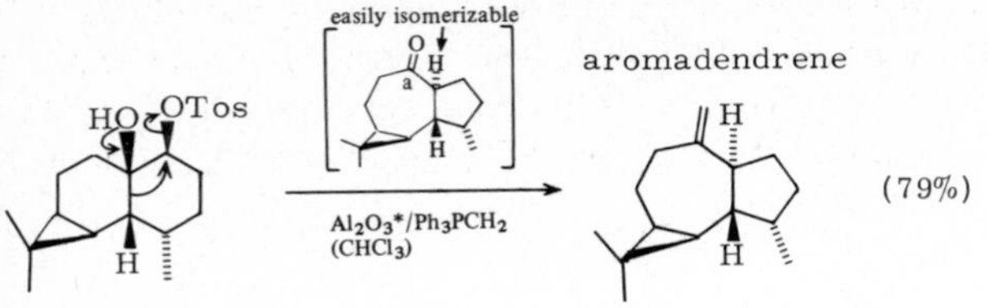

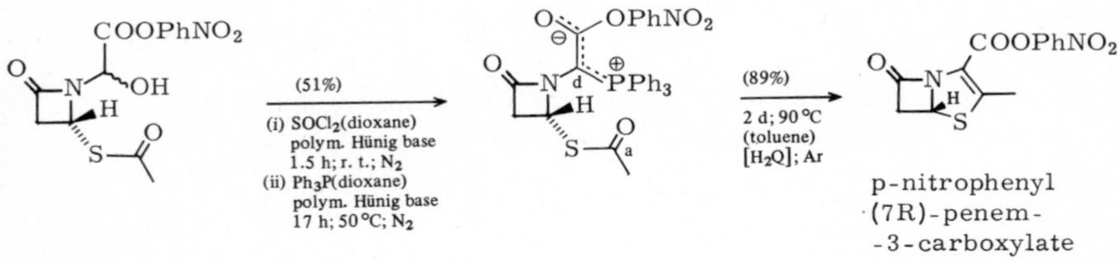

easily isomerizable

aromadendrene

(79%)

Al$_2$O$_3$*/Ph$_3$PCH$_2$
(CHCl$_3$)

(51%)

(i) SOCl$_2$(dioxane)
polym. Hünig base
1.5 h; r. t.; N$_2$
(ii) Ph$_3$P(dioxane)
polym. Hünig base
17 h; 50 °C; N$_2$

(89%)

2 d; 90 °C
(toluene)
[H$_2$Q]; Ar

p-nitrophenyl
(7R)-penem-
-3-carboxylate

In view of the synthetic importance of dicarbonyl compounds surprisingly little has been done, apart from carotene synthesis, on dialkenylation with Wittig reagents. However, from the few examples reported one may conclude, that no special problems are involved. Benzocyclobutanedione was converted by two equivalents of methoxycarbonylmethylidenetriphenylphosphorane to the corresponding diene in 85% yield (M.P. Cava, 1960), and o-phthalaldehyde reacted with a γ-butyrolactone-2-phosphonate (K.H. Büchel, 1965) to give o-phenylenebisalkene compounds (T. Minami, 1974). In polycycles where steric hindrance is important, the Wittig or Horner reactions may fail. An example is the tricyclic diketone shown below which could only be converted into mono-alkylidene derivatives (R.K. Hill, 1975).

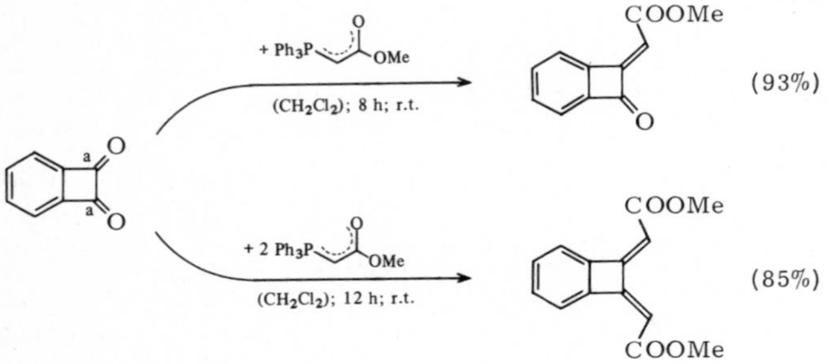

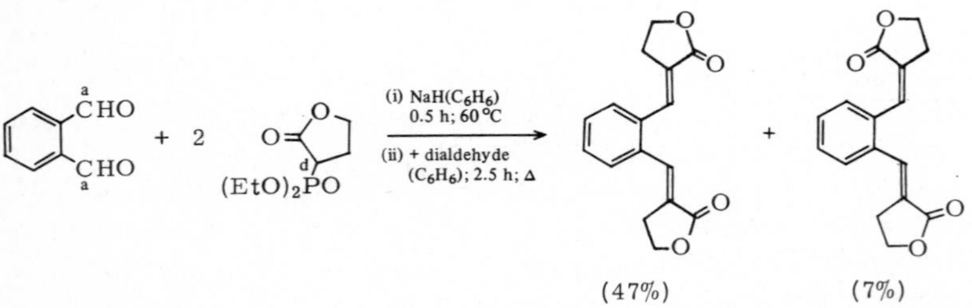

+ Ph$_3$P=CH-CO-OMe

(CH$_2$Cl$_2$); 8 h; r.t.

COOMe

(93%)

+ 2 Ph$_3$P=CH-CO-OMe

(CH$_2$Cl$_2$); 12 h; r.t.

COOMe

COOMe

(85%)

+ 2

(EtO)$_2$PO

(i) NaH(C$_6$H$_6$)
0.5 h; 60 °C

(ii) + dialdehyde
(C$_6$H$_6$); 2.5 h; Δ

(47%)

+

(7%)

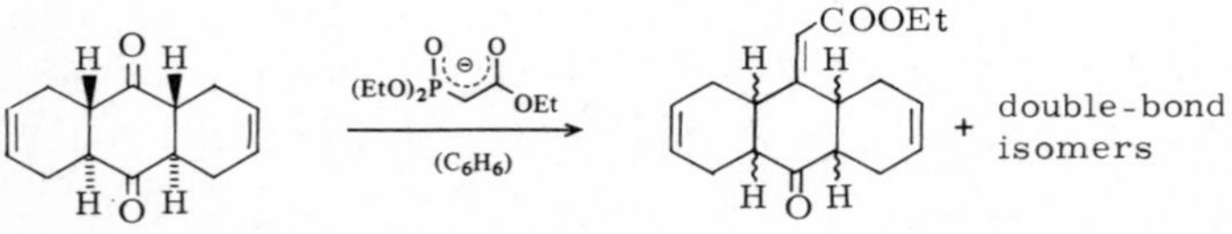

(EtO)$_2$P-CH$_2$-CO-OEt

(C$_6$H$_6$)

COOEt

+ double-bond
isomers

The phosphorus ylides of the Wittig reaction can be replaced by trimethylsilylmethyl-carbanions (Peterson reaction). These silylated carbanions add to carbonyl groups and can easily be eliminated with base to give olefins. The only by-products are volatile silanols. They are more easily removed than the phosphine oxides or phosphates of the more conventional Wittig or Horner reactions (D.J. Peterson, 1968).

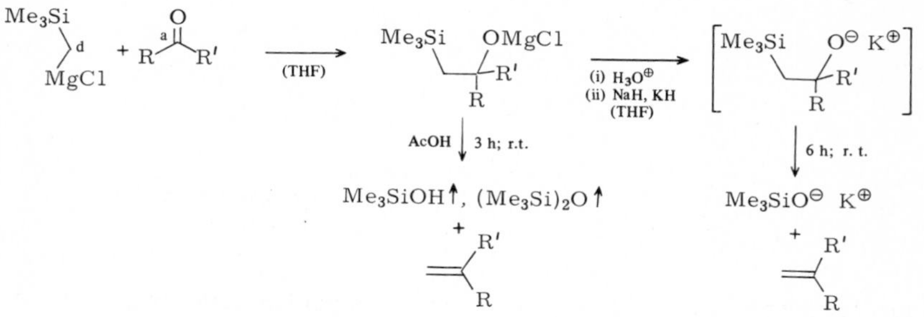

The Peterson reaction has two more advantages over the Wittig reaction: 1. it is some-times less vulnerable to sterical hindrance, and 2. groups, which are susceptible to nucleophilic substitution, are not attacked by silylated carbanions. The introduction of a methylene group into a sterically hindered ketone (R.K. Boeckman, Jr., 1973) and the syntheses of olefins with sulfur, selenium, silicon, or tin substituents (D. Seebach, 1973; B.T. Gröbel, 1974, 1977) illus-trate useful applications. The reaction is, however, more limited and time consuming than the Wittig reaction, since metallated silicon derivatives are difficult to synthesize and their reac-tions are rarely stereoselective (T.H. Chan, 1974).

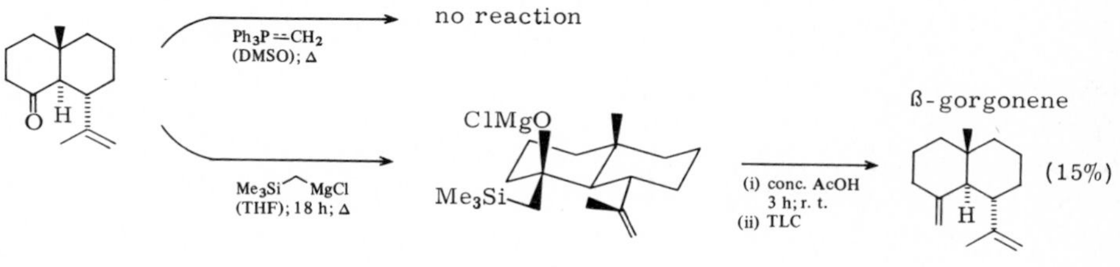

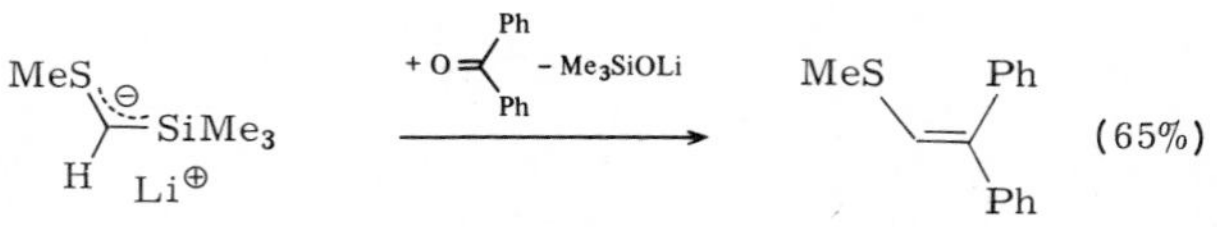

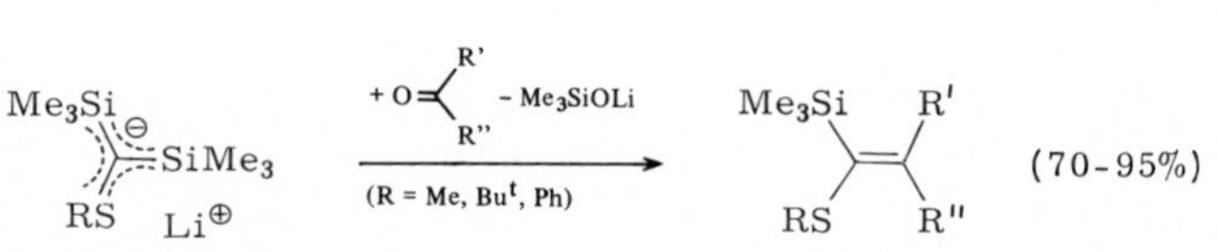

epoxide mixture of
gipsy moth pheromone

AcOOH/AcOH

(50%)

trans : cis = 1 : 1

Sulfur ylides (p. 7f.) cannot be used for the olefination of carbonyl compounds (R. Sowada, 1971; B.M. Trost, 1975A; C.R. Johnson, 1979). They are more useful in coupling reactions (p. 37f.) and the syntheses of epoxides (p. 42) and cyclopropanes (p. 69).

Syntheses of alkenes with three or four bulky substituents cannot be achieved with an ylide or by a direct coupling reaction. Sterical hindrance of substituents presumably does not allow the direct contact of polar or radical carbon synthons in the transition state. A generally applicable principle formulated by A. Eschenmoser indicates a possible solution to this problem: *If an intermolecular reaction is complex or slow, it is advisable to change the educt in such a way, that the critical bond formation can occur intramolecularly* (A. Eschenmoser, 1970).

The problem of the synthesis of highly substituted olefins from ketones according to this principle was solved by D.H.R. Barton. The ketones are first connected to azines by hydrazine and secondly treated with hydrogen sulfide to yield 1,3,4-thiadiazolidines. In this heterocycle the substituents of the prospective olefin are too far from each other to produce problems. Mild oxidation of the hydrazine nitrogens produces Δ^3-1,3,4-thiadiazolines.   The decisive step of carbon-carbon bond formation is achieved in a thermal reaction: a nitrogen molecule is cleaved off and the biradical formed recombines immediately since its two reactive centers are hold together by the sulfur atom. The thiirane (episulfide) can be finally desulfurized by phosphines or phosphites, and the desired olefin is formed. With very large substituents the 1,3,4-thiadiazolidines do not form with hydrazine. In such cases, however, direct thiadiazoline formation from thiones and diazo compounds is often possible, or a thermal reaction between alkylideneazinophosphoranes and thiones may be successful (D.H.R. Barton, 1972, 1974, 1975).

The only common synthons for alkynes are acetylide anions, which react as good nucleophiles with alkyl bromides (D.E. Ames, 1968) or carbonyl compounds (p. 58).

1.6 Alkanes, Alkenes, and Alkynes via Coupling Reactions

In the synthesis of molecules without functional groups the application of the usual "polar" synthetic reactions may be cumbersome, since the final elimination of hetero atoms can be difficult. Two solutions for this problem have been given in the previous sections, namely alkylation with nucleophilic carbanions and alkenylation with ylides. Another direct approach is to combine radical synthons in a non-polar reaction. Carbon radicals are, however, inherently short-lived and tend to undergo complex secondary reactions. Eschenmoser's principle (p. 34) again provides a way out. If one connects both carbon atoms via a metal atom which (i) forms and stabilizes the carbon radicals and (ii) can be easily eliminated, the intermolecular reaction is made intramolecular, and good yields may be obtained.

As stated above, intermolecular coupling reactions between carbon atoms are of limited use. In the classical Wurtz reaction two identical primary alkyl iodide molecules are reduced by sodium. n-Hectane ($C_{100}H_{202}$), for example, has been made by this method in 60% yield (G. Stallberg, 1956). The unsymmetrical coupling of two alkyl halides can be achieved via dialkylcuprates. The first halide, which may have a branched carbon chain, is lithiated and allowed to react with copper(I) salts. The resulting dialkylcuprate can then be coupled with alkyl or aryl iodides or bromides. Although the reaction probably involves radicals it is quite stereoselective and leads to inversion of chiral halides. For example, lithium diphenylcuprate reacts with (R)-2-bromobutane with 90% stereoselectivity to form (S)-2-phenylbutane (G.M. Whitesides, 1969).

$$2\ C_{50}H_{101}I \xrightarrow[\text{38 h; 154°C}]{\text{Na}} C_{100}H_{202} \qquad \begin{array}{c}\text{hectane}\\ \text{m. p. 115°C}\end{array}$$
$$(61\%)$$

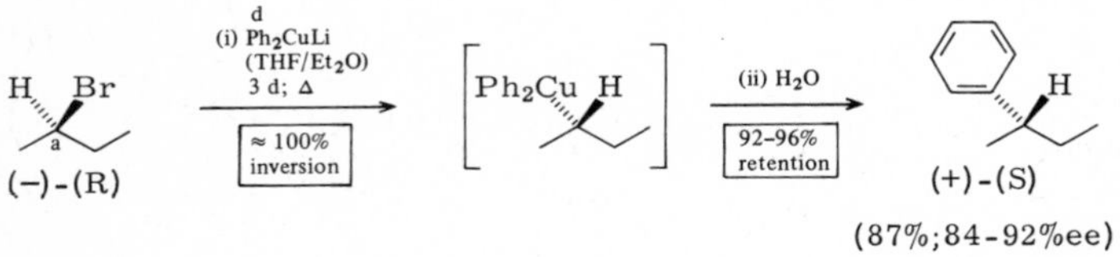

Trialkylboranes, which can be synthesized from olefins and diborane (p. 9), undergo alkyl coupling on oxidation with alkaline silver nitrate via short-lived silver organyls. Two out of three alkyl substituents are coupled in this reaction. Terminal olefins may be coupled by this reaction sequence in 40 - 80% yield. With non-terminal olefins yields drop to 30 - 50% (H.C. Brown, 1972C, 1975).

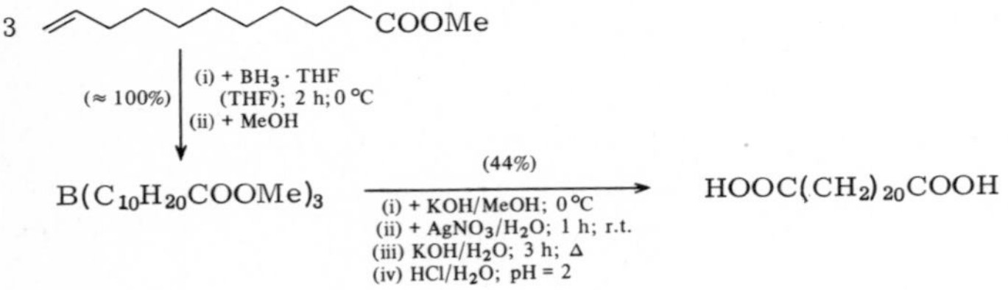

Cis-olefins or *cis, trans*-dienes can be obtained from alkynes in similar reaction sequences. The alkyne is first hydroborated and then treated with alkaline iodine. If the other substituents on boron are alkyl groups, a *cis*-olefin is formed (G. Zweifel, 1967). If they are *cis*-alkenyls, a *cis, trans*-diene results. The reactions are thought to be iodine-assisted migrations of the *cis*-alkenyl group followed by *trans*-deiodoboronation (G. Zweifel, 1968). *Trans, trans*-dienes are made from haloalkynes and alkynes. These compounds are added one after the other to thexylborane. The alkenyl(1-haloalkenyl)thexylboranes are converted with sodium methoxide into *trans, trans*-dienes (E. Negishi, 1973). The thexyl group does not migrate.

(cis)-alkenes:

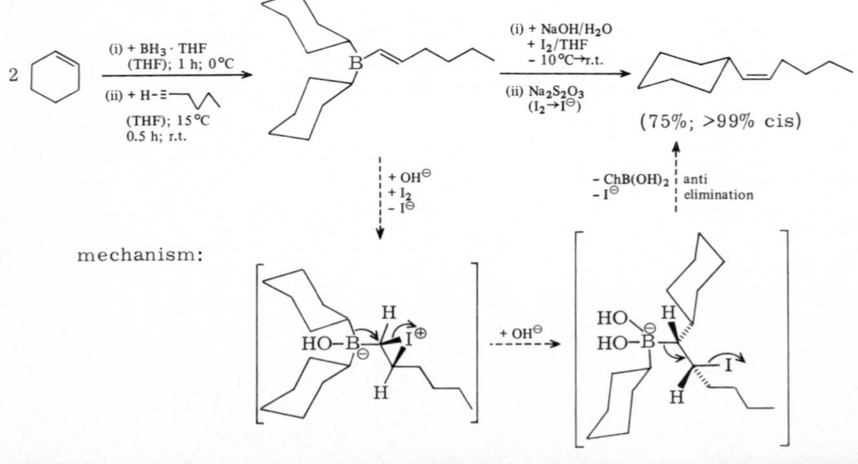

conjugated (trans, cis)-dienes:

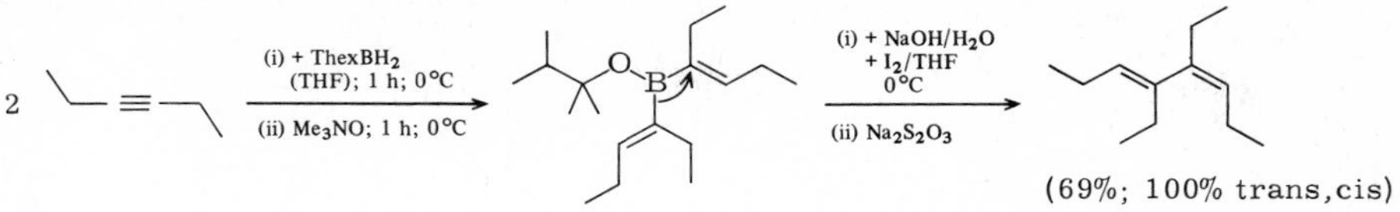

(69%; 100% trans,cis)

(trans)-alkenes:

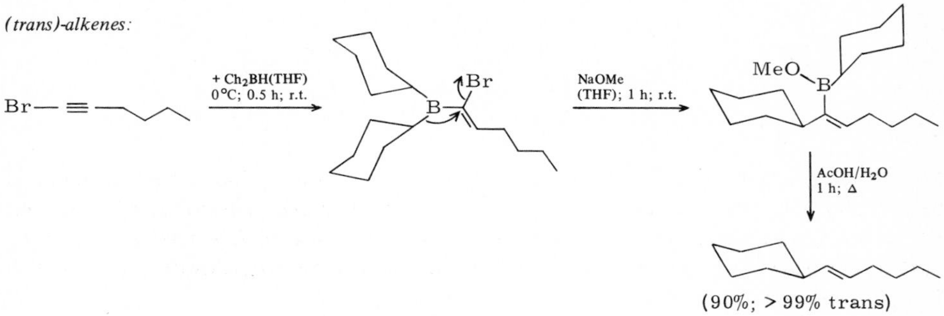

(90%; > 99% trans)

conjugated (trans, trans)-dienes:

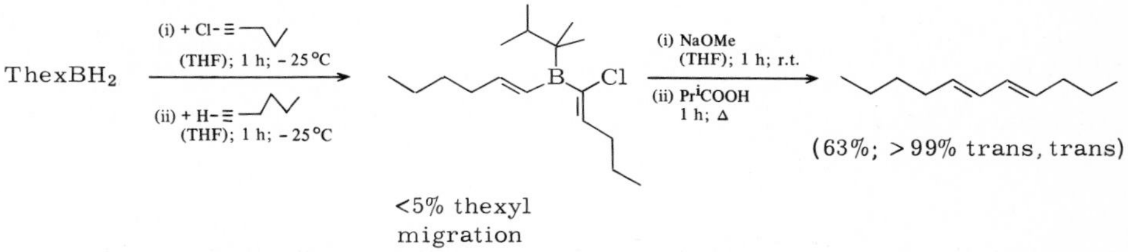

(63%; > 99% trans, trans)

<5% thexyl
migration

Two efficient syntheses of strained cyclophanes indicate the synthetic potential of allyl or benzyl sulfide intermediates, in which the combined nucleophilicity and redox activity of the sulfur atom can be used. The dibenzylic sulfides from xylylene dihalides and -dithiols can be methylated with dimethoxycarbenium tetrafluoroborate (H. Meerwein, 1960; R.F. Borch, 1968, 1969; from trimethyl orthoformate and BF₃, 3:4). The sulfonium salts are deprotonated and rearrange to methyl sulfides (Stevens rearrangement). Repeated methylation and Hofmann elimination yields double bonds (R.H. Mitchell, 1974).

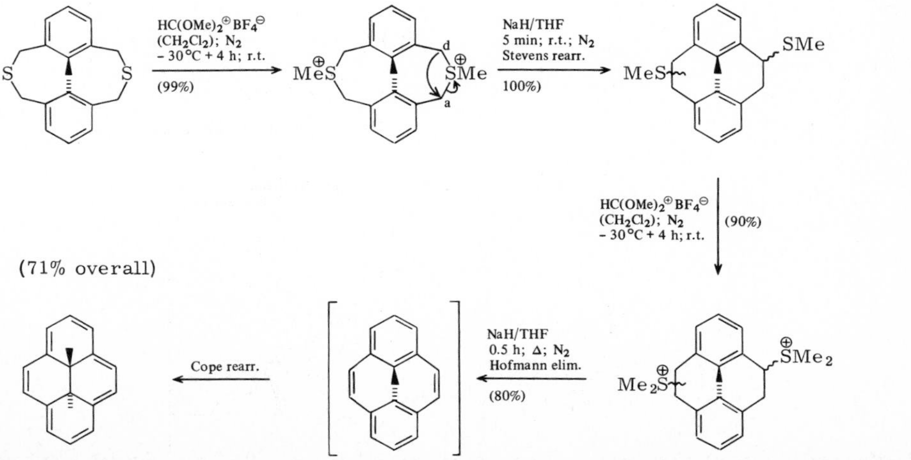

(71% overall)

Treatment of dibenzylic sulfides with triethylphosphite and UV-light also led to cyclophanes in high yield (H.A. Staab, 1979).

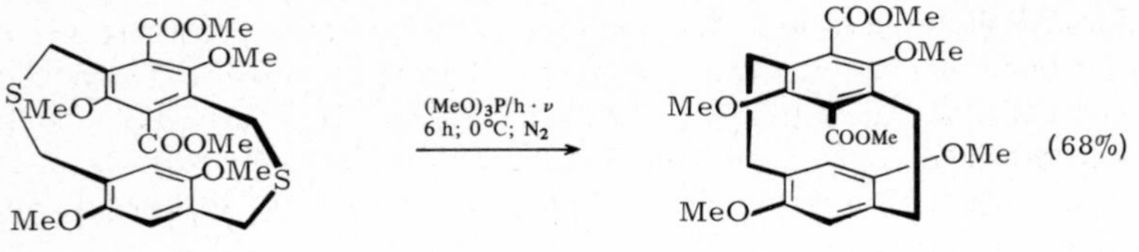

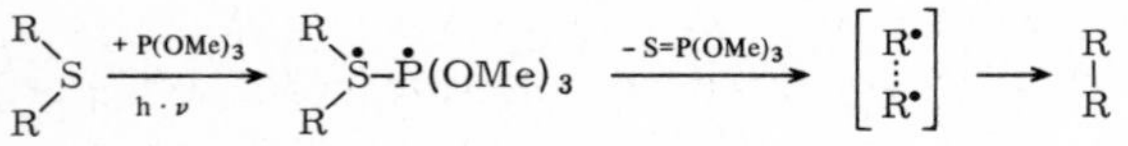

Diallylsulfonium salts undergo intramolecular allylic rearrangement with strong bases to yield 1,5-dienes after reductive desulfurization. The straight-chain 1,5-dienes may be obtained by double sulfur extrusion with concomitant allylic rearrangements from diallyl disulfides. The first step is achieved with phosphines or phosphites, the second with benzyne. This procedure is especially suitable for the synthesis of acid sensitive olefins and has been used in oligoisoprene synthesis (G.M. Blackburn, 1969).

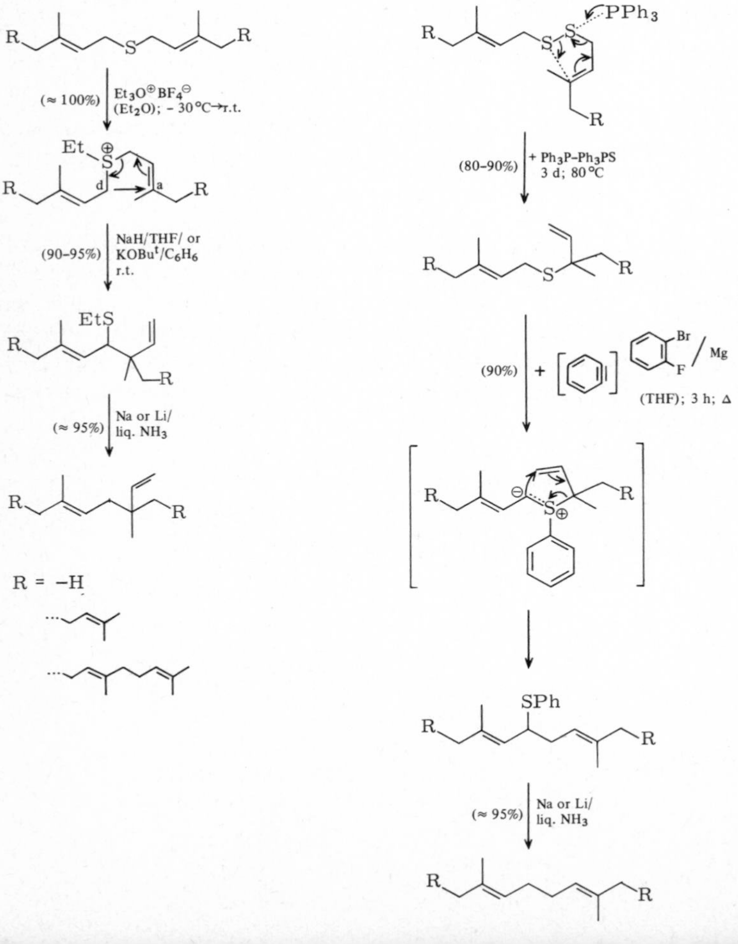

Alkenes or alkynes, which are connected with electron-rich metal atoms, e.g. Fe(0) or Ni(0), by σ- or π-bonds, are highly activated toward electropositive reagents such as protons or oxidants. If the same ligands are bound to metals in higher oxidation states, e.g. Hg(II), Cu(I) or Pd(II), the reverse situation exists. These metals withdraw electron density from the carbon-carbon multiple bonds and activate them toward attack by nucleophiles and reducing agents. Therefore, the reactivity of the carbon ligand as well as the stability of the whole metal complex changes drastically, if one oxidizes or reduces the metal in it. Since the electron density on the metal also depends on its extra ligands, e.g. phosphines or carbon monoxide, one can prepare compounds with carbon-metal bonds of "adjustable" strength and induce various rearrangements within the carbon ligands.

Terminal acetylenes react with copper(II) salts or with copper(I) salts in the presence of oxygen to form bisacetylides, which undergo alkyne coupling (Glaser coupling) in aqueous ammonium solutions or hot pyridine (G. Eglington, 1963). This reaction has been used for the synthesis of polyenes and of macrocycles (F. Sondheimer, 1963). Unsymmetrical dialkynes can be made, if one of the alkynes is submitted to an "umpolung" (p. 17) by bromination and is then allowed to react with a copper(I) acetylide (Cadiot-Chodkiewicz coupling; G. Eglington, 1963).

Glaser coupling:

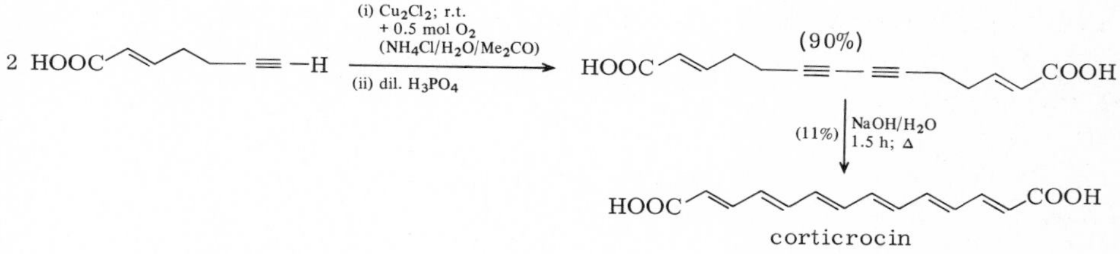

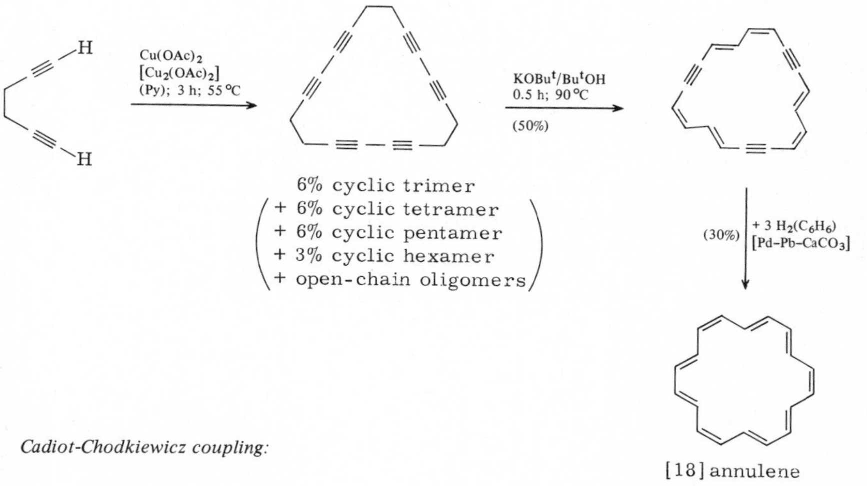

corticrocin

6% cyclic trimer
+ 6% cyclic tetramer
+ 6% cyclic pentamer
+ 3% cyclic hexamer
+ open-chain oligomers

[18] annulene

Cadiot-Chodkiewicz coupling:

(75%)

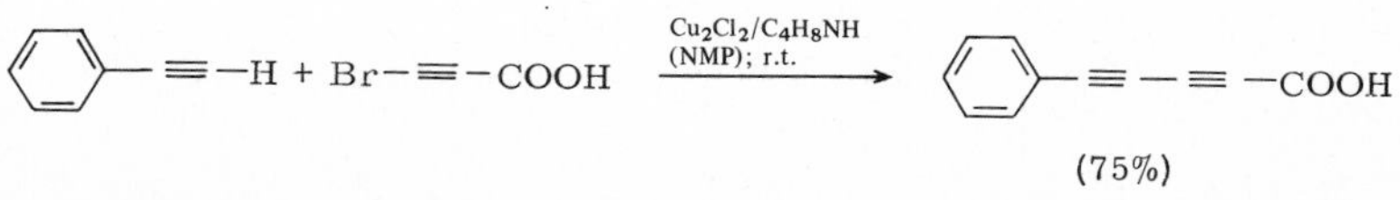

Reductive coupling of carbonyl compounds to yield olefins is achieved with titanium (0), which is freshly prepared by reduction of titanium(III) salts with $LiAlH_4$ or with potassium. The removal of two carbonyl oxygen atoms is driven by TiO_2 formation. Yields are often excellent even with sensitive or highly hindered olefins. (J.E. McMurry, 1974, 1976A,B).

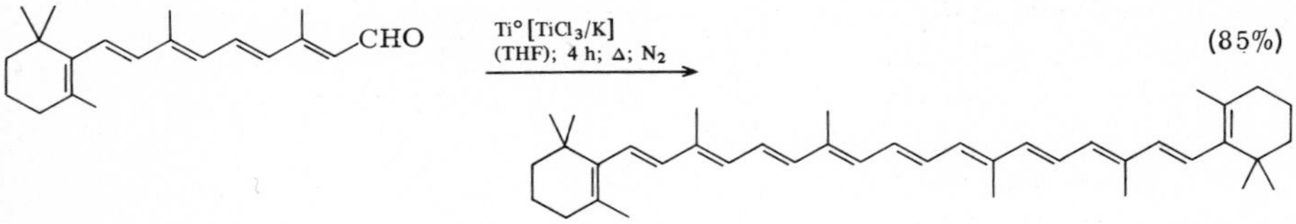

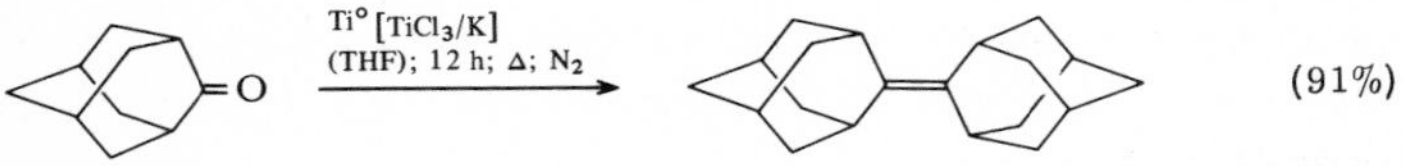

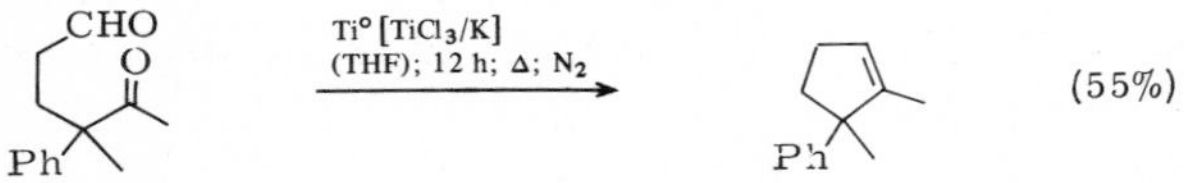

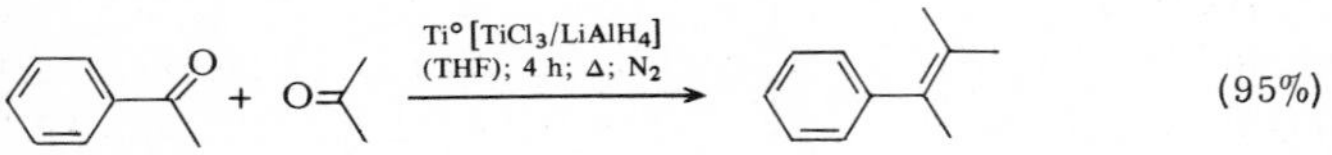

Nickel(0) forms a π-complex with three butadiene molecules at low temperature. This complex rearranges spontaneously at 0 °C to afford a bisallylic system, from which a large number of interesting olefins can be obtained. The scheme given below and the example of the synthesis of the odorous compound muscone (R. Baker, 1972, 1974; A.P. Kozikowski, 1976) indicate the variability of such rearrangements (P. Heimbach, 1970). Nowadays many rather complicated cycloolefins are synthesized on a large scale by such reactions and should be kept in mind as possible starting materials, e.g. after ozonolysis.

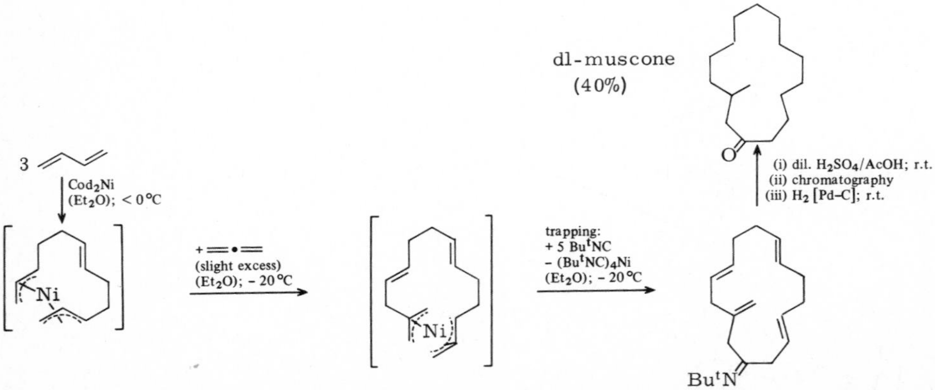

Nickel-allyl complexes prepared from Ni(CO)$_4$ and allyl bromides are useful for the olefination of alkyl bromides and iodides (E.J. Corey, 1967 B; A.P. Kozikowski, 1976). The reaction has also been extended to the synthesis of macrocycles (E.J. Corey, 1967 C, 1972A).

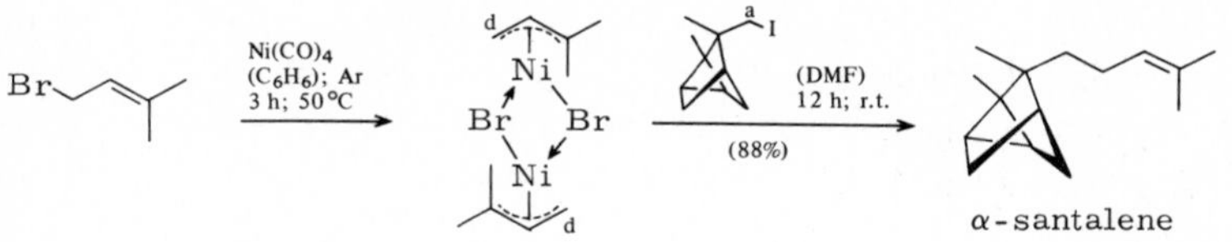

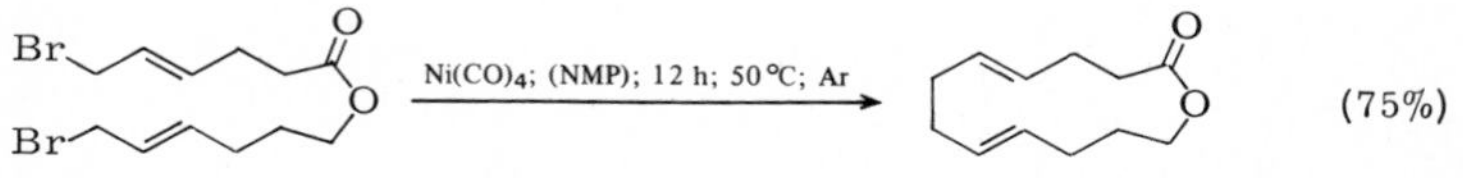

α-santalene

(75%)

1.7 Alcohols and Epoxides

Alcohols can be synthesized by the addition of carbanions to carbonyl compounds (W.C. Still, 1976) or epoxides. Both types of reactions often produce chiral centres, and stereoselectivity is an important aspect of these reactions.

The alkyl addition to monosubstituted epoxides is regio- and stereoselective. It occurs at the less substituted carbon atom. Lithium dialkylcuprates are especially suitable (G.H. Posner, 1975). They react much faster with aldehydes and epoxides, than with ketones and esters. The conversion of complex aldehydes and ketones to secondary and tertiary alcohols is preferably done with zinc organyls (Reformatsky reaction). These reagents are less reactive and more selective than the corresponding Grignard reagents, from which they are often obtained by reaction with anhydrous zinc chloride. A particular efficient method for the alkylation of carbonyl compounds uses the alkyl bromide directly. In this process both reagents are poured over a column containing heated zinc granules (J.F. Ruppert, 1976). With cyclic ketones the Reformatsky reaction often is highly stereoselective (Y. Mazur, 1960).

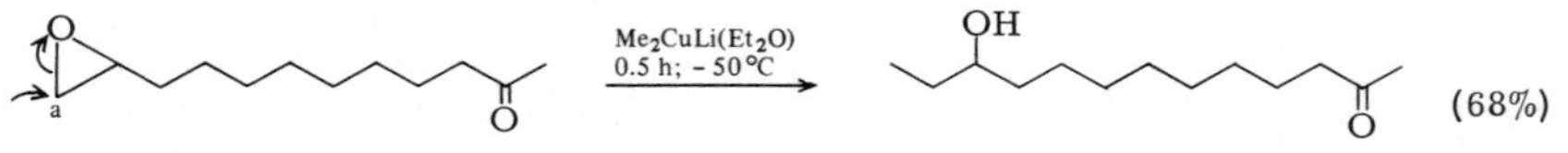

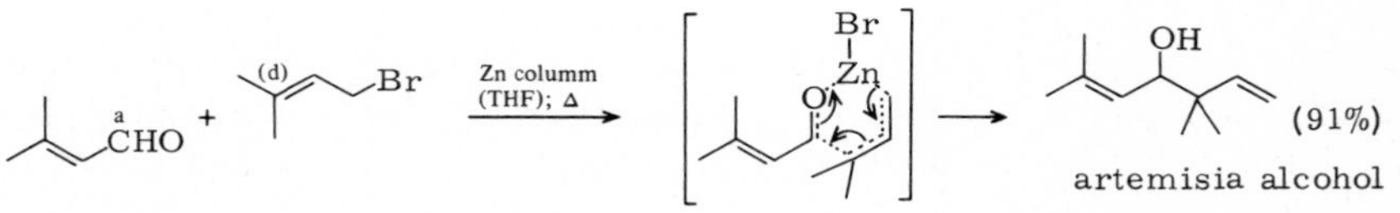

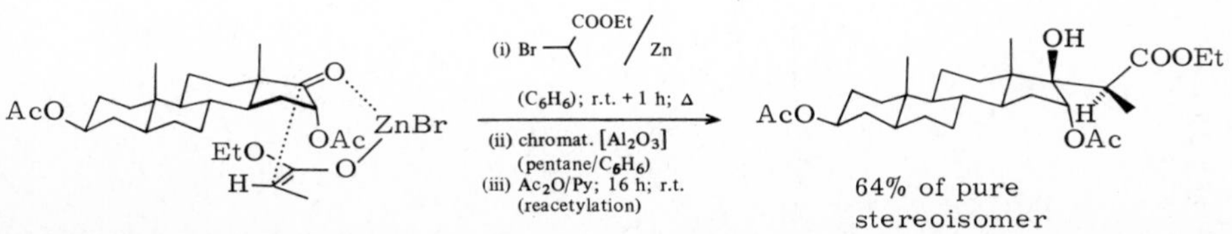

artemisia alcohol

64% of pure
stereoisomer

It is also possible to convert carbonyl groups into oxirane rings with certain carbenoid synthons. The classical Darzens reaction, which involves addition of anions of α-chloroacetic esters, has recently been largely replaced by the addition of sulfonium ylides (R. Sowada, 1971; C.R. Johnson, 1979).

An interesting aspect of this reaction is the contrasting stereoselective behaviour of the dimethylsulfonium and dimethyloxosulfonium methylides in reactions with cyclic ketones (E.J. Corey, 1963 B, 1965 A; C.E. Cook, 1968). The small, reactive dimethylsulfonium ylide prefers axial attack, but with the larger, less reactive oxosulfonium ylide only the thermo-dynamically favored equatorial addition is observed.

1.8 Aldehydes, Ketones, and Carboxylic Acids

The most general synthetic route to ketones uses the reaction of carboxylic acids (or their derivatives) or nitriles with organometallic compounds (M.J. Jorgenson, 1970). Lithium car-boxylates react with organolithium compounds to give stable *gem*-diolates, which are decom-posed by water to the ketone (R.G. Riley, 1974). Formation of tertiary alcohols, which results from the reaction of the ketone with the organometallic compound, can generally be avoided, if the acid derivative is always in excess relative to the lithium organyl. Primary, secondary and even tertiary lithium alkanides can be used with generally good success, with lithium alke-nyls yields are less satisfactory. Chiral ketones may partly epimerize, if the chiral center is in the α-position (P.D. Bartlett, 1935). This problem can be avoided, if lithium alkanides or lithium dialkylcuprates (from Grignard reagents and copper(I) salts) are applied and the work-up is done quickly at low temperature (G.H. Posner, 1975).

(i) Mg (Et$_2$O); N$_2$
(ii) + ZnCl$_2$–MgCl$_2$

(Et$_2$O); Δ

(52%)

partial racemization

+ EtLi (Et$_2$O)
0.5 h; Δ

H$_2$C
0°C

(78%)

no racemization

Bu$_2^t$CuMgI, Ph$_2$CuLi

(Et$_2$O); N$_2$; 0°C

(60%)   (85%)

no epimerization

Olefins yield aldehydes, if they are treated with carbon monoxide and hydrogen at elevated temperature and high pressure. The best catalysts for this procedure, which is mostly exploited in industry, are rhodium-carbonyls and -phosphines. A related carbonylation procedure for laboratory use takes advantage of the nucleophilicity of carbon monoxide ligands in negatively charged carbonylferrates. The commercial reagent of choice is disodium tetracarbonylferrate (J.P. Collman, 1972, 1975A, 1978; A.P. Kozikowski, 1976). It can be first alkylated or acylated with the corresponding halides or anhydrides to form a nucleophilic acyl synthon bound to Fe(1-). This can be trapped by several procedures, e.g. by protonation to yield aldehydes or by ethene to form ethyl ketones in high yield. Higher alkenes are unsuitable (M.P. Cooke, 1975). Other ketones can be obtained by alkylation of the acylferrate(1-) with alkyl iodides. Chiral bromides and tosylates react with total inversion. (S)-2-Octyl tosylate, for example, yielded 99% optically pure (R)-3-methyl-2-nonanone, when it was treated with disodium tetracarbonylferrate and methyl iodide (J.P. Collman, 1972).

X = Cl, OCOR   Fe(0) intermediates   + AcOH / r.t. → R–CHO

+ Na$_2$Fe(CO)$_4$ + L
(THF); 1 h; r.t.; N$_2$

[R–Fe$^{\ominus}$(CO)$_3$L]

+ R'I
1 d; r.t.

R–X   X = Br, I, OTos

C$_2$H$_4$

L = CO, C$_2$H$_4$, Ph$_3$P,
NMP, HMPTA

aldehydes:

Cl ⌇ Br
+ Na$_2$Fe(CO)$_4$
1 bar CO
(THF); 1 h; r.t.

+ AcOH

Cl ⌇ CHO   (82%)

+ Na$_2$Fe(CO)$_4$
(THF); 1 h; r.t.; Ar

+ AcOH

HOOC ⌇ CHO   (81%)

ethyl ketones via ethylene trapping:

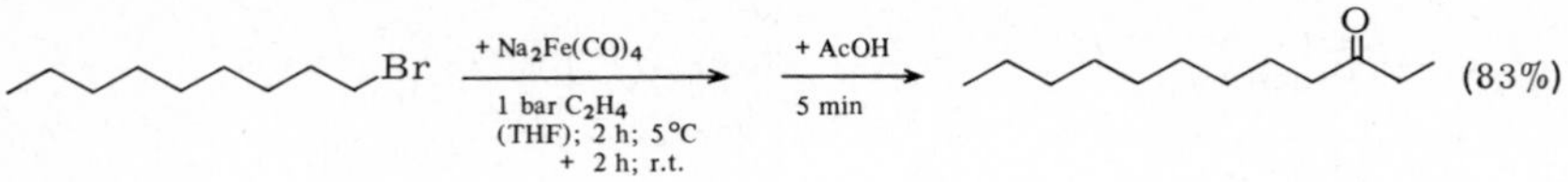

ketones:

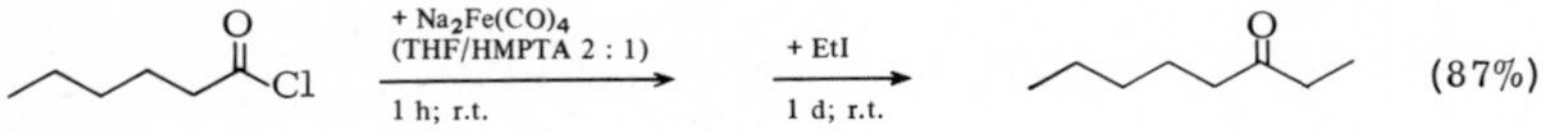

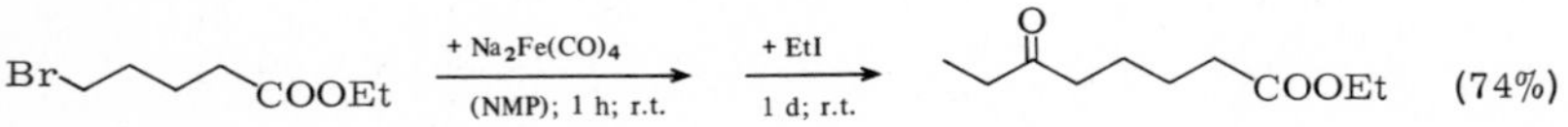

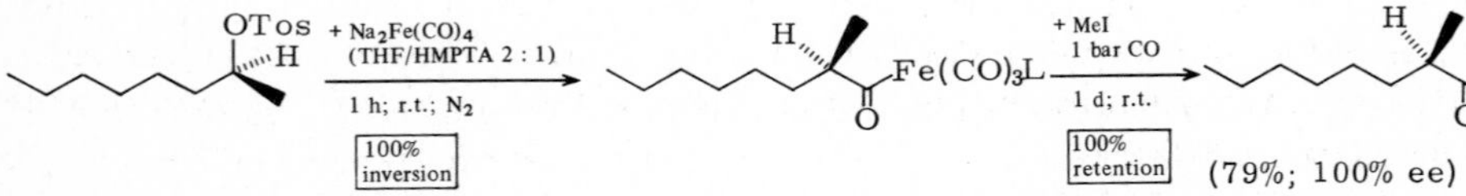

Trialkylboranes are useful reagents for β-alkylation of α,β-unsaturated carbonyl compounds. Only one alkyl group is regioselectively transferred to the β-position. Although this reaction can produce a large variety of aldehydes, it is an alkylation procedure and does belong to section 1.4 (see p. 21). A real carbonylation of boranes, however, is also known (H.C. Brown, 1972C, 1975). Carbon monoxide adds to trialkylboranes smoothly (see p. 9). The alkyl substituents migrate stepwise under appropriate conditions from boron to the carbonyl carbon atom. At each step the reaction can be halted with trapping reagents and by use of slowly migrating alkyl groups (e.g. thexyl, norbornyl, cyclooctane-1,5-diyl). In the presence of trialkoxyalanates the primarily formed acylborane is immediately reduced, and aldehydes are obtained by oxidative cleavage. If a thexylborane is treated with water and 70 bar CO at 50 °C and thereafter with alkaline H_2O_2, a ketone results. The final migration product is trapped with 1,2-ethanediol at high temperatures as the B-alkylboradioxolane, and tertiary alcohols are formed with alkaline H_2O_2.

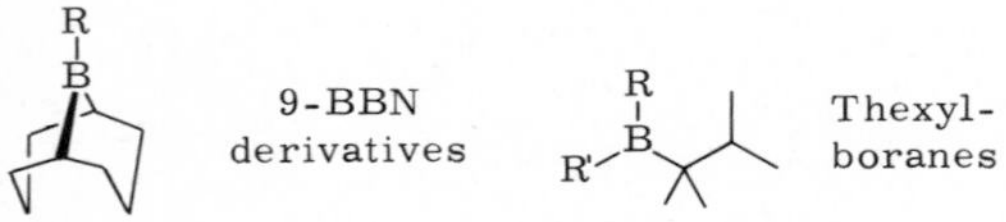

aldehydes:

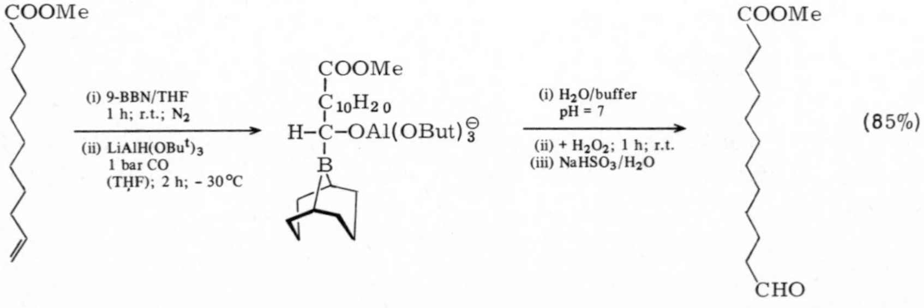

ketones:

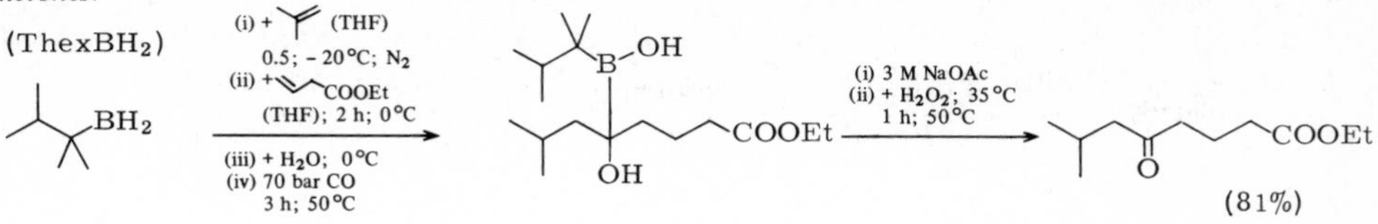

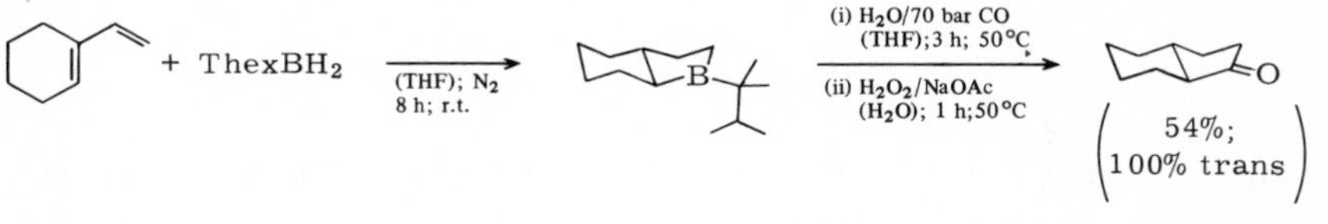

The synthesis of aldehydes from alkyl bromides and α-carbanions of cyclic imidoesters (p. 22) as well as formation of ketones from aldehydes via alkylation of dithiane carbanions (p. 21) have been mentioned earlier. Steglich (R. Lohmar, 1980) developed a related procedure to connect two halides to a heterocycle providing a masked carbonyl group. The 2-phenyloxazolinone formed from hippuric acid is dialkylated, e.g. with geranyl bromide. Hydrolysis and treatment with lead tetraacetate leads to symmetric ketones. Wittig reactions of carbonyl compounds with alkoxymethylenephosphoranes can give aldehydes in good yield (W. Nagata, 1967).

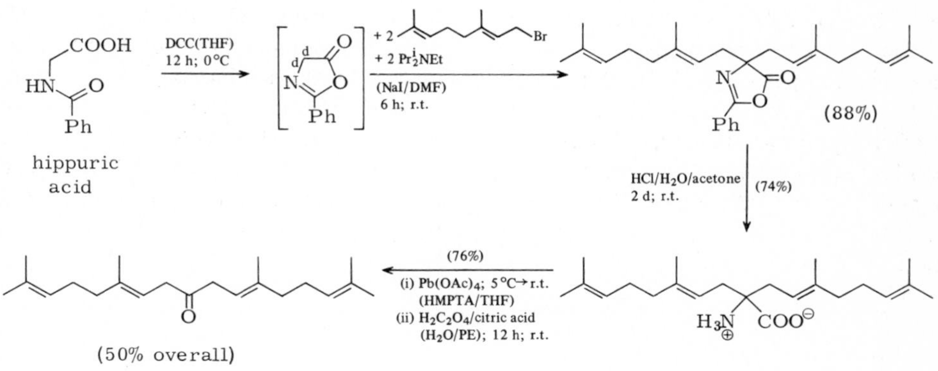

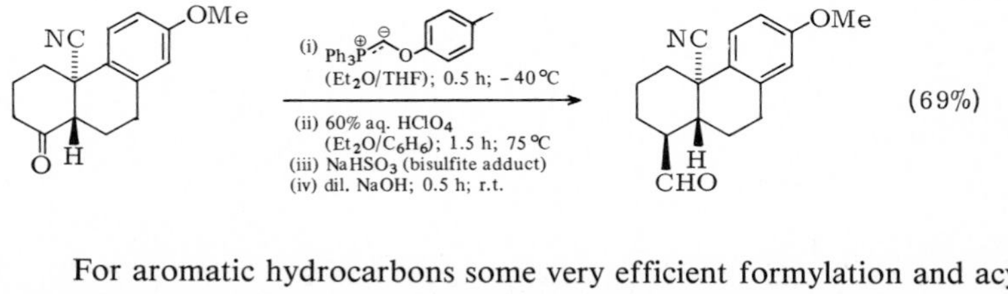

For aromatic hydrocarbons some very efficient formylation and acylation procedures are known (e.g. Friedel-Crafts, Vilsmeier, Gattermann-Koch). They are treated in introductory text books.

Carboxyl and nitrile groups are usually introduced in synthesis with commercial carboxylic acid derivatives, nitriles, or cyanide anion. Carbanions can be carboxylated with carbon dioxide (H.F. Ebel, 1970) or dialkyl carbonate (J. Schmidlin, 1957).

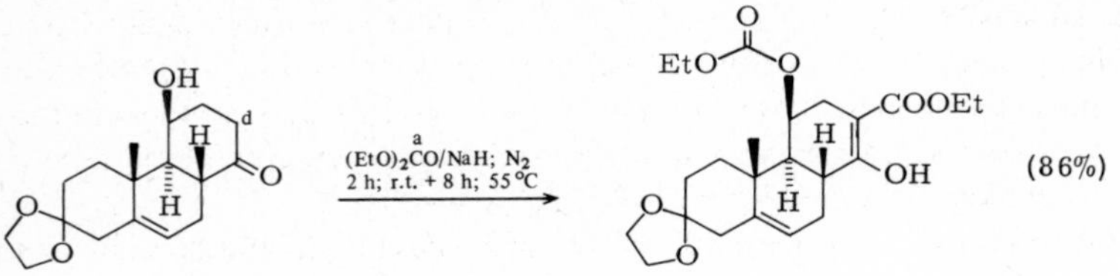

(86%)

Ketones can be converted to nitriles with an additional carbon atom with the aid of the TosMIC reagent (J.R. Bull, 1975; O.H. Oldenziel, 1973).

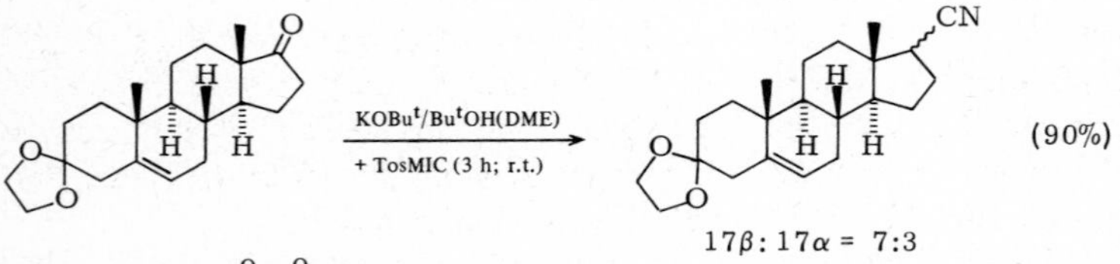

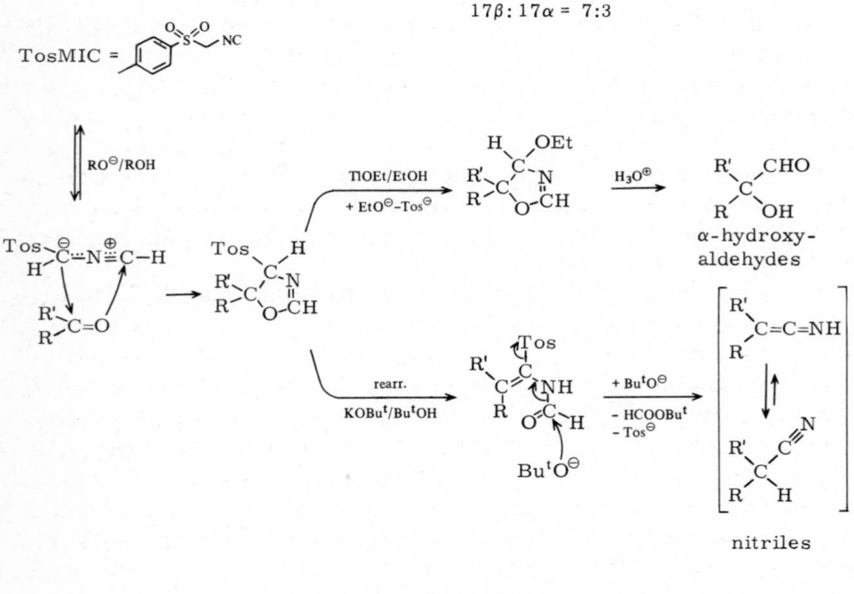

(90%)

$17\beta : 17\alpha = 7:3$

TosMIC =

α-hydroxy-aldehydes

nitriles

1.9 1,2-Difunctional Compounds

The reactions described so far can be considered as alkylation, alkenylation, or alkynylation reactions. In principle all polar reactions in syntheses, which produce monofunctional carbon compounds, proceed in the same way: a carbanion reacts with an electropositive carbon atom, and the activating groups (e.g. metals, boron, phosphorus) of the carbanion are lost in the work-up procedures. We now turn to reactions, in which the hetero atoms of both the acceptor and donor synthons are kept in a difunctional reaction product.

The choice of synthons for the syntheses of difunctional target molecules depends first of all on the number of carbon atoms, which separate the functional groups. 1,2-Difunctional products can only be obtained directly (without using additional oxidation reactions) from an a^1- and a d^1-synthon or two r^1-synthons (r = radical reaction site). 1,3-Difunctional compounds are directly available only from combinations of a^1- or a^2-synthons with d^2- or d^1-synthons, respectively. For syntheses of 1,4-difunctional compounds one needs a^1-, a^2-, or a^3-synthons together with d^3-, d^2-, or d^1-synthons. 1,5-Difunctional molecules come usually from a^2- or a^3-synthons together with d^3- or d^2-synthons. 1,6-Difunctional target molecules are available from cleavage of six-membered carbocycles (see section 1.14). For target molecules, in which the functional groups are separated by more than four non-functional carbon atoms, one has to devise individual syntheses with difunctional educts.

Before we start with a systematic discussion of the syntheses of difunctional molecules, we have to point out a formal difficulty. A carbon-carbon multiple bond is, of course, considered as one functional group. With these groups, however, it is not clear, which of the two carbon atoms has to be named as the functional one. A 1,3-diene, for example, could be considered as a 1,2-, 1,3-, or 1,4-difunctional compound. An α,β-unsaturated ketone has a 1,2- as well as a 1,3-difunctional structure. We adhere to useful, although arbitrary conventions. Dienes and polyenes are separated out as a special case. α,β-Unsaturated alcohols, ketones, etc. are considered as 1,3-difunctional. We call a carbon compound 1,2-difunctional only, if two neighbouring carbon atoms bear hetero atoms.

The most general methods for the syntheses of 1,2-difunctional molecules are based on the oxidation of carbon-carbon multiple bonds (p. 107) and the opening of oxiranes by hetero atoms (p. 113ff.). There exist, however, also a few useful reactions in which an a^1- and a d^1-synthon or two r^1-synthons are combined. The classical polar reaction is the addition of cyanide anion to carbonyl groups, which leads to α-hydroxynitriles (cyanohydrins). It is used, for example, in Strecker's synthesis of amino acids and in the homologization of monosaccharides. The α-hydroxy group of a nitrile can be easily substituted by various nucleophiles, the nitrile can be solvolyzed or reduced. Therefore a large variety of terminal difunctional molecules with one additional carbon atom can be made. Equally versatile are α-methylsulfinyl ketones (H.G. Hauthal, 1971; T. Durst, 1979), which are available from acid chlorides or esters and the dimsyl anion. Carbanions of these compounds can also be used for the synthesis of 1,4-dicarbonyl compounds (p. 60f.).

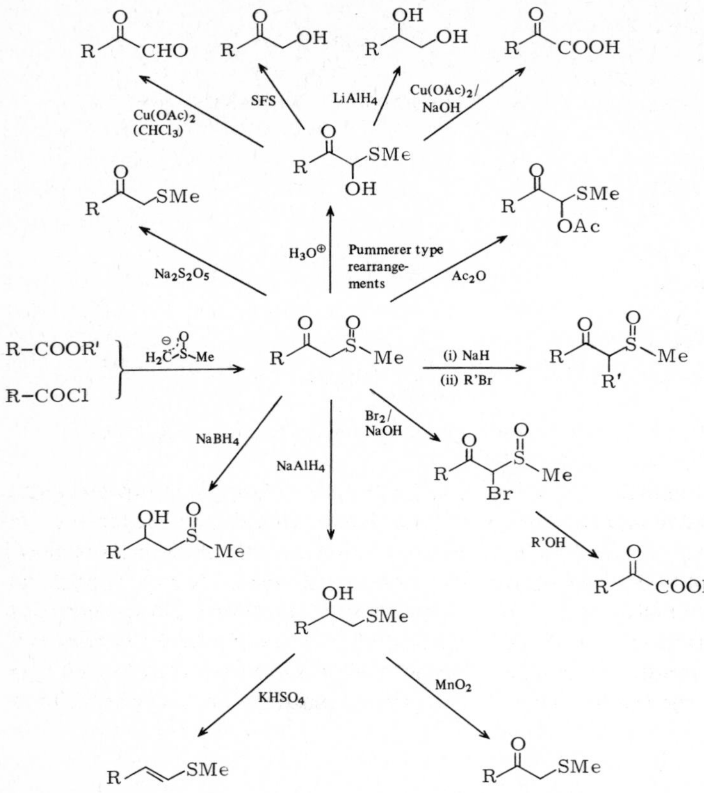

After the "umpolung" of an aldehyde group by conversion to a 1,3-dithian-2-ide anion (p. 17) it can be combined with a carbonyl group (D. Seebach, 1969, 1979; B.-T. Gröbel, 1977 B). Analogous reagents are tosylmethyl isocyanide (TosMIC), which can be applied in the nucleophilic formylation of ketones (O.H. Oldenziel, 1974), and dichloromethyl lithium (G. Köbrich, 1969; P. Blumbergs, 1972; H. Taguchi, 1973).

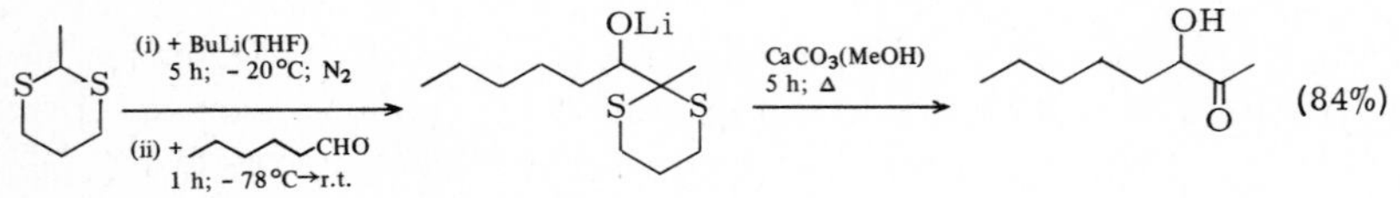

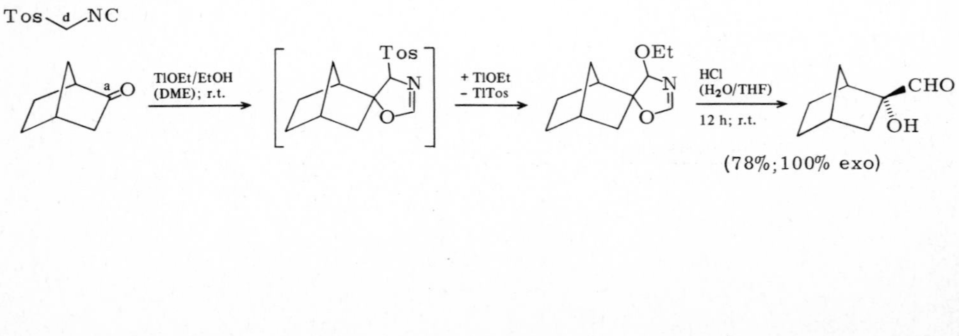

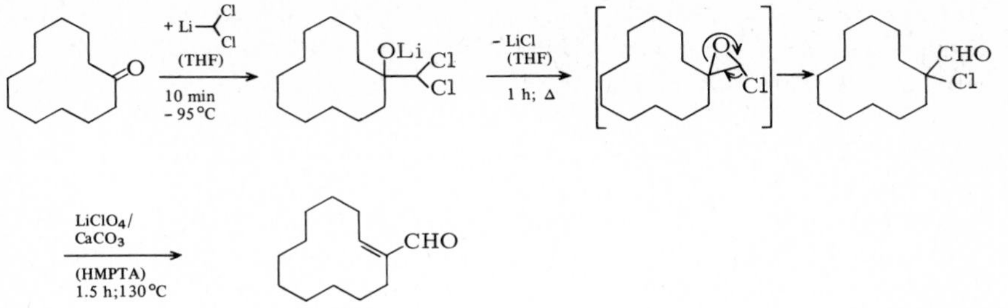

(81%)

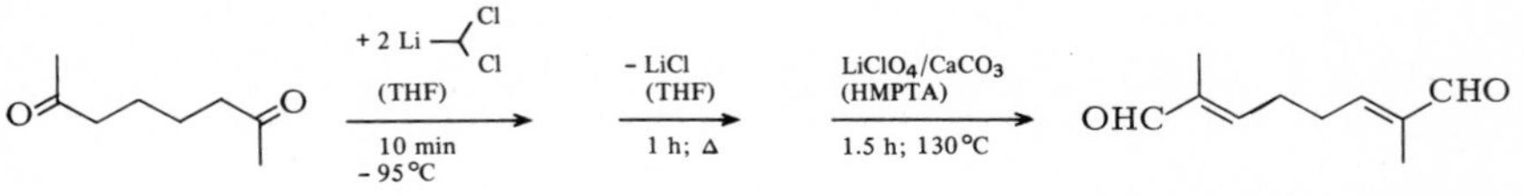

(68%; >99% trans, trans)

Terminal alkyne anions are popular reagents for the acyl anion synthons (RCH_2CO^-). If this nucleophile is added to aldehydes or ketones, the triple bond remains. This can be converted to an alkynemercury(II) complex with mercuric salts and is hydrated with water or acids to form ketones (M.M.T. Khan, 1974). The more substituted carbon atom of the alkynes is converted preferentially into a carbonyl group. Highly substituted α-hydroxyketones are available by this method (J.A. Katzenellenbogen, 1973). Acetylene itself can react with two molecules of an aldehyde or a ketone (V. Jäger, 1977). Hydration then leads to 1,4-dihydroxy-2-butanones. The 1,4-diols tend to condense to tetrahydrofuran derivatives in the presence of acids.

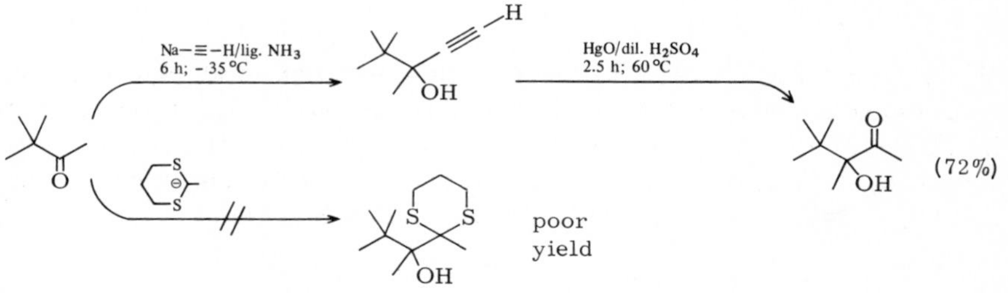

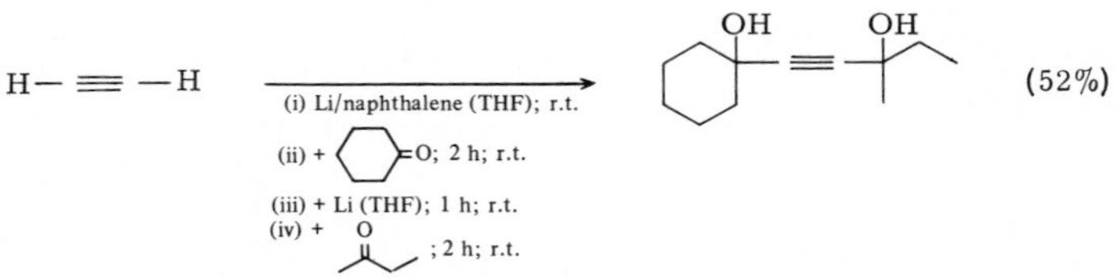

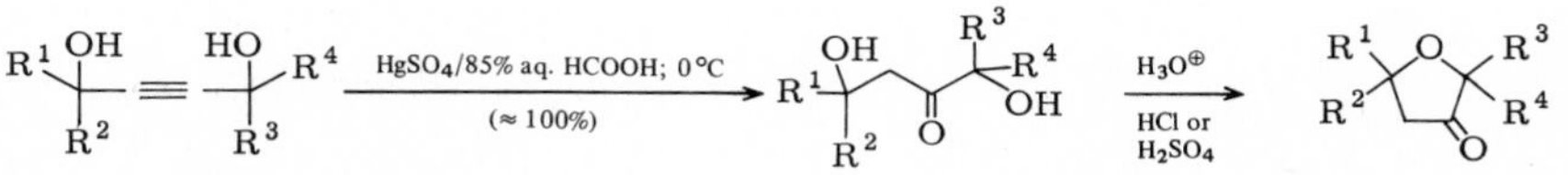

(72%)

poor yield

(52%)

The most important r^1-synthons are obtained from reduction of carbonyl compounds by metals. Corresponding synthetic reactions are the pinacol and acyloin coupling reactions. In the pinacol reduction carbonyl groups are reduced by titanium(II), magnesium, or zinc to ketyl radicals, which combine to give 1,2-diols (S. Tyrlik, 1973; T. Mukaiyama, 1973). The reaction is particularly successful in intramolecular formation of cyclic diols (E.J. Corey, 1976). The same holds true for the reductive coupling of two ester groups to form acyloins (= α-hydroxy ketones; S.M. McElvain, 1948; J.J. Bloomfield, 1976). The yields are generally better, if the alkoxide anions and endiolate dianions are trapped as silyl ethers. Undesirable side-reactions caused by those strong bases, e.g. cleavage, aldol type reactions, and dehydrations, are thus prevented (K. Rühlmann, 1971). 1,2-Dicarboxylic esters yield 1,2-bis(trimethylsiloxy)-cyclobutenes, which can be opened thermally. This has been used in a procedure for ring expansion of carbocycles (T. Mori, 1969).

pinacol coupling:

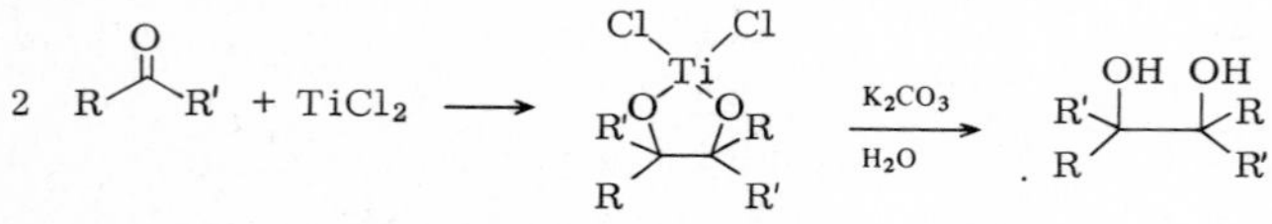

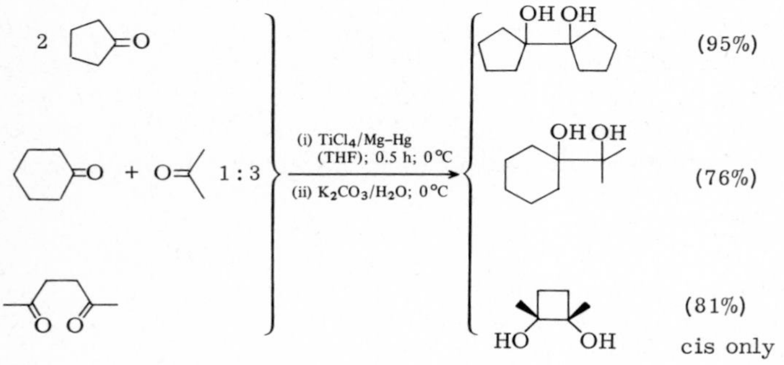

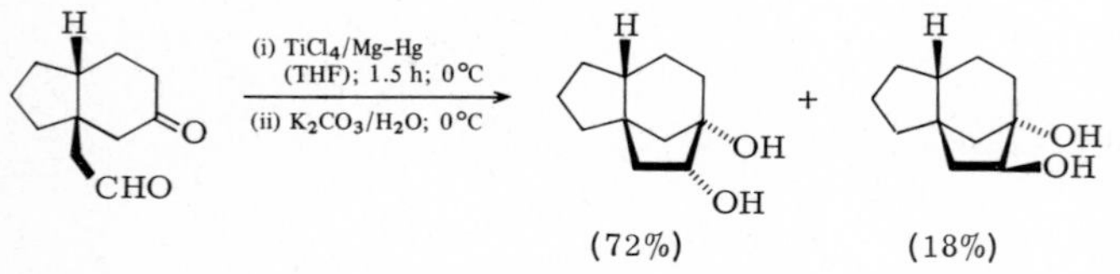

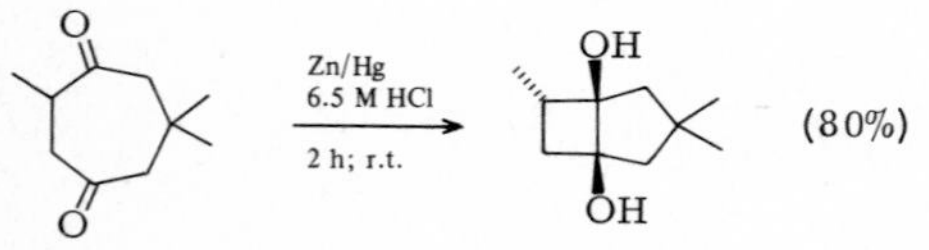

acyloin coupling

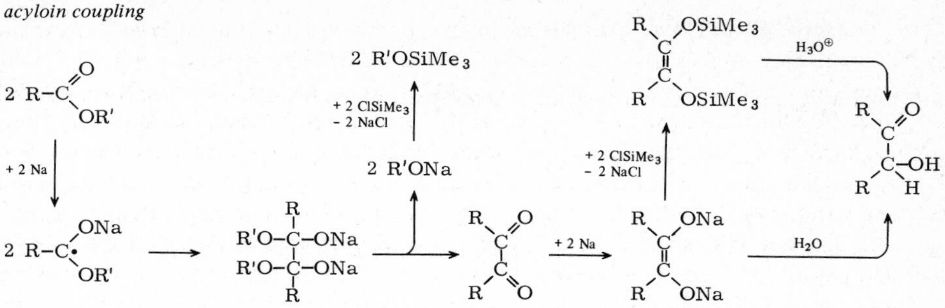

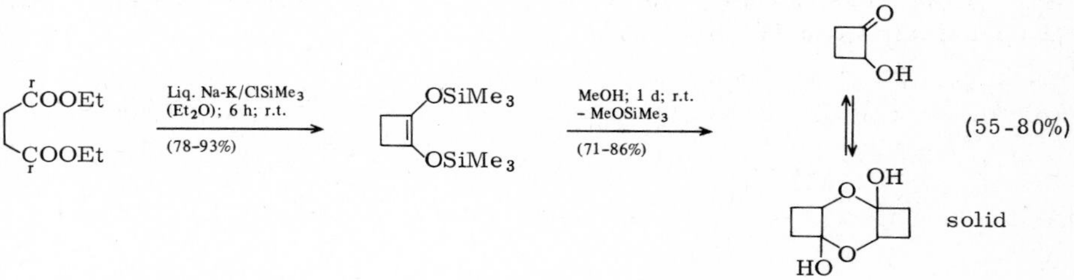

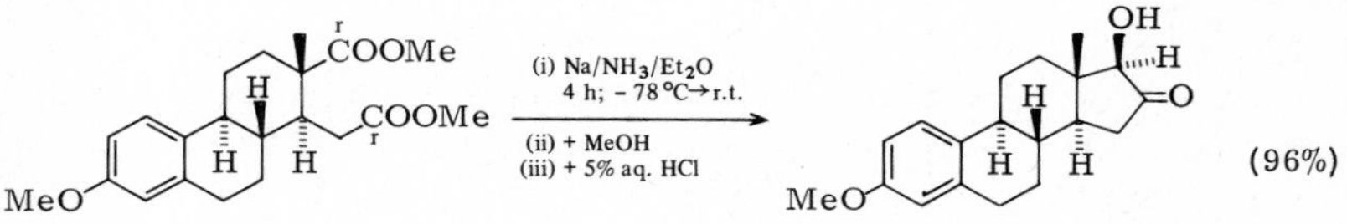

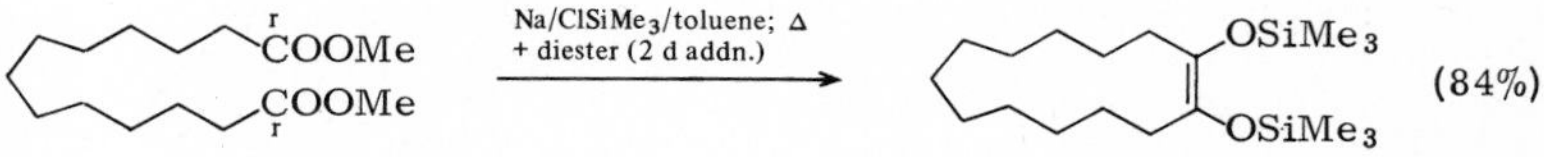

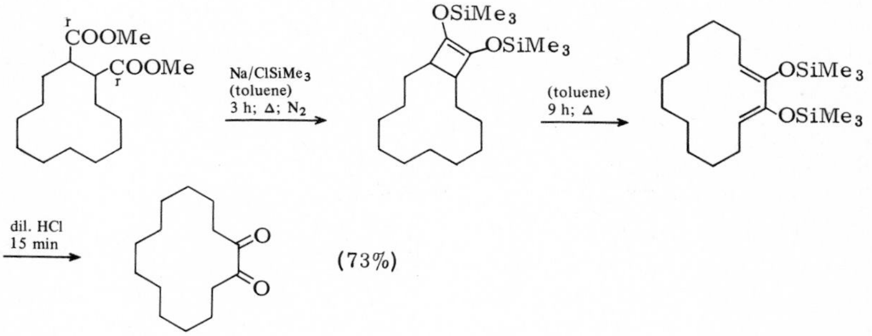

$$2 \ R-\overset{O}{\underset{OR'}{C}} \quad \xrightarrow{+ 2 \ Na} \quad 2 \ R-\overset{ONa}{\underset{OR'}{\overset{\bullet}{C}}} \quad \longrightarrow \quad \begin{matrix} R'O-\overset{R}{\underset{}{C}}-ONa \\ R'O-\underset{R}{\overset{}{C}}-ONa \end{matrix}$$

$$\xrightarrow{\begin{matrix} + 2 \ ClSiMe_3 \\ - 2 \ NaCl \end{matrix}} \quad 2 \ R'OSiMe_3$$

$$2 \ R'ONa \quad \longrightarrow \quad \overset{R}{\underset{R}{\overset{}{C}}}\overset{O}{\underset{O}{\overset{}{C}}} \quad \xrightarrow{+ 2 \ Na} \quad \begin{matrix} R-\overset{ONa}{\overset{\parallel}{C}} \\ R-\underset{ONa}{\overset{\parallel}{C}} \end{matrix} \xrightarrow{\begin{matrix} + 2 \ ClSiMe_3 \\ - 2 \ NaCl \end{matrix}} \begin{matrix} R \quad OSiMe_3 \\ \overset{\parallel}{C} \\ R \quad OSiMe_3 \end{matrix}$$

$$\xrightarrow{H_3O^\oplus} \quad \begin{matrix} R \\ \overset{}{C}=O \\ R-\overset{}{C}-OH \\ H \end{matrix} \xleftarrow{H_2O}$$

$$\overset{r}{\underset{r}{\overset{COOEt}{\underset{COOEt}{|}}}} \quad \xrightarrow[\substack{(Et_2O); \ 6 \ h; \ r.t. \\ (78-93\%)}]{Liq. \ Na-K/ClSiMe_3} \quad [\text{OSiMe}_3 \ \text{cyclobutene}] \quad \xrightarrow[\substack{- \ MeOSiMe_3 \\ (71-86\%)}]{MeOH; \ 1 \ d; \ r.t.} \quad (55-80\%)$$

solid

$$\xrightarrow[\substack{(i) \ Na/NH_3/Et_2O \\ 4 \ h; \ -78°C \to r.t. \\ (ii) \ + \ MeOH \\ (iii) \ + \ 5\% \ aq. \ HCl}]{} \quad (96\%)$$

$$\xrightarrow[\substack{+ \ diester \ (2 \ d \ addn.)}]{Na/ClSiMe_3/toluene; \ \Delta} \quad (84\%)$$

$$\xrightarrow[\substack{(toluene) \\ 3 \ h; \ \Delta; \ N_2}]{Na/ClSiMe_3} \quad \xrightarrow[\substack{(toluene) \\ 9 \ h; \ \Delta}]{}$$

$$\xrightarrow[\substack{15 \ min}]{dil. \ HCl} \quad (73\%)$$

1.10 1,3-Difunctional Compounds

Aldol additions and ester condensations have always been and still are the most popular reactions for the formation of carbon-carbon bonds (A.T. Nielsen, 1968). The carbonyl group acts as an a^1-synthon, the enol or enolate as a d^2-synthon. Both reactions will be treated together here, and arguments, which are given for aldol additions, are also valid for ester condensations. Many famous name reactions belong to this category*). The products of aldol additions may be either β-hydroxy carbonyl compounds or, after dehydration, α,β-unsaturated carbonyl compounds.

The usual base or acid catalyzed aldol addition or ester condensation reactions can only be applied as a useful synthetic reaction, if both carbonyl components are identical. Otherwise complicated mixtures of products are formed. If two different aldehydes or esters are to be combined, it is essential that one of the components is transformed quantitatively into an enol whereas the other component remains as a carbonyl compound in the reaction mixture.

We begin with the discussion of intramolecular reactions. An example of a regioselective Dieckmann condensation (J.P. Schaefer, 1967) used an educt with two ester groups, of which only one could form an enolate. Regioselectivity was dictated by the structure of the educt.

A regioselective aldol condensation described by Büchi succeeds for sterical reasons (G. Büchi, 1968). If one treats the dialdehyde given below with acid, both possible enols are probably formed in a reversible reaction. Only compound **A**, however, is found as a product, since in **B** the interaction between the enol and ester groups which are in the same plane hinders the cyclization. Büchi used acid catalysis instead of the usual base catalysis. This is often advisable, when sterical hindrance may be important. It works, because the addition of a proton or a Lewis acid to a carbonyl oxygen acidifies the neighbouring CH-bonds.

---

* Knoevenagel, Dieckmann, Stobbe, and Reformatsky reactions to name just a few.

An example of an intermolecular aldol type condensation, which works only under acidic catalysis is the Knoevenagel condensation of a sterically hindered aldehyde group in a formyl-porphyrin with a malonic ester (J.-H. Fuhrhop, 1976). Self-condensations of the components do not occur, because the ester groups of malonic esters are not electrophilic enough, and because the porphyrin-carboxaldehyde cannot form enolates.

The selective intermolecular addition of two different ketones or aldehydes can sometimes be achieved without protection of the enol, because different carbonyl compounds behave differently. For example, attempts to condense acetaldehyde with benzophenone fail. Only self-condensation of acetaldehyde is observed, because the carbonyl group of benzophenone is not sufficiently electrophilic. With acetone instead of benzophenone only β-hydroxyketones are formed in good yield, if the aldehyde is slowly added to the basic ketone solution. Aldols are not produced. This result can be generalized in the following way: aldehydes have more reactive carbonyl groups than ketones, but enolates from ketones have a more nucleophilic carbon atom than enolates from aldehydes (G. Wittig, 1968).

Electrophilicity of the formyl group decreases, if it is converted with cyclohexylamine into an cyclohexylimino group. α-CH groups, however, are still acidic enough in these Schiff bases to be deprotonated by LDA at -78 °C. The resulting lithium salts are stable at low temperatures and do not undergo proton-metal exchange with added ketones or aldehydes. Therefore one has a highly nucleophilic carbanion, an electrophilic carbonyl compound, and an almost unreactive imino group in the reaction mixture. Addition of the enamine anion to the carbonyl group is the only reaction (G. Wittig, 1968).

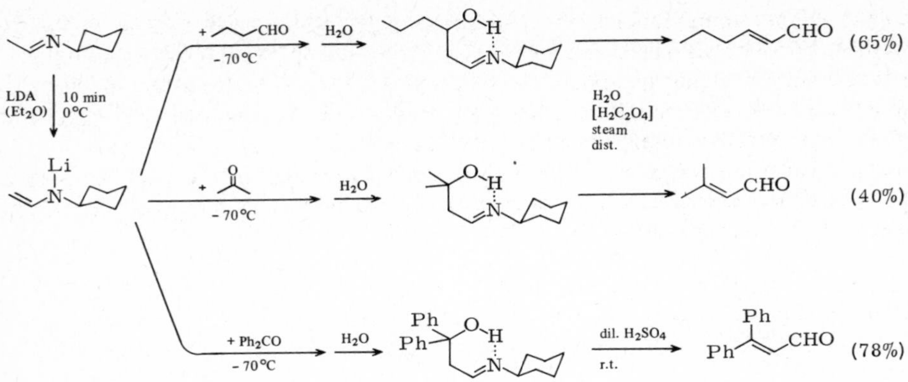

There also exists an acid-catalyzed regioselective condensation of the aldol type, namely the Mannich reaction (B. Reichert, 1959; H. Hellmann, 1960; see also p. 263f.). The condensation of secondary amines with aldehydes yields immonium salts, which react with ketones to give 3-amino ketones (=Mannich bases). Ketones with two enolizable CH_2-groupings may form 1,5-diamino-3-pentanones, but monosubstitution products can always be obtained in high yield. Unsymmetrical ketones react preferentially at the most highly substituted carbon atom. Sterical hindrance can reverse this regioselectivity. Thermal elimination of amines leads to the α,β-unsaturated ketone. Another efficient pathway to vinyl ketones starts with the addition of terminal alkynes to immonium salts. On mercury(II) catalyzed hydration the product is converted to the Mannich base (H. Smith, 1964).

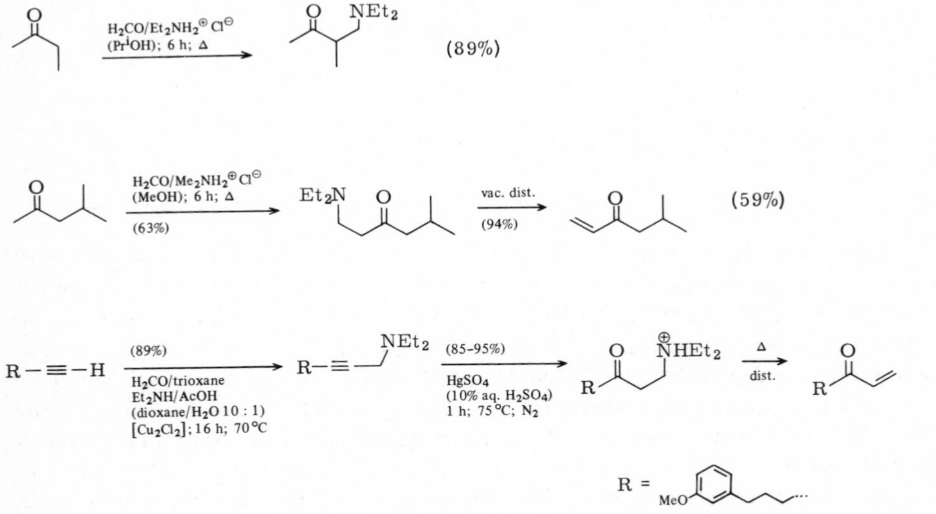

Low molecular mass enol esters (e.g. acetates; H.O. House, 1965) or enol ethers (e.g. silyl ethers; H.O. House, 1969) of ketones can be synthesized regioselectively and/or separated by distillation. Treatment with lithium alkyls converts them into the corresponding lithium enolates without isomerization of the double bond (H.O. House, 1965; G. Stork, 1968 C; I. Kuwajima, 1975). It is also possible to produce the "kinetic" lithium enolate (p. 11) directly at low temperature in almost quantitative yield (G. Stork, 1974 A; I. Kuwajima, 1976). The lithium enolates may then be reacted with an aldehyde or ketone again without metal exchange. A single adduct is often formed in such "directed" aldol additions. The Lewis acid catalyzed aldol addition of O-silyl-enols and enol acetates with aldehydes is equally regioselective. Even phenylacetaldehyde, which forms enolates very easily, reacts only as a carbonyl component (T. Mukaiyama, 1974).

A classical way to achieve regioselectivity in an $(a^1 + d^2)$-reaction is to start with α-carbanions of carboxylic acid derivatives and electrophilic ketones. Most successful are condensations with 1,3-dicarbonyl carbanions, e.g. with malonic acid derivatives, since they can be produced at low pH, where ketones do not enolize. Succinic acid derivatives can also be deprotonated and added to ketones (Stobbe condensation). In the first example given below a Dieckmann condensation on a nitrile follows a Stobbe condensation, and selectivity is dictated by the tricyclic educt: neither the nitrile group nor the ketone is enolizable (W.S. Johnson, 1945, 1947).

In a sophisticated variation of the Knoevenagel condensation ("Panizzi") methyl 3,3-dimethoxypropanoate (from ketene and dimethoxymethenium tetrafluoroborate; D.J. Crosby, 1962) is used as a d^2-reagent. Because only one carbonyl group activates the methylene group, a strong base with no nucleophilic properties (p. 10) has to be used. A sodium-sand mixture, which presumably reacts to form silicate anions in the heat, was chosen by Büchi (G. Büchi, 1973). In the α-deprotonated ester the negative charge destroys the electrophilicity of the neighbouring ester group, and no self-condensation occurs. Non-enolizable formic acid ester, on the other hand, adds smoothly to form the 2-formyl substituted product. Asymmetric aldol type additions may be performed with chiral esters (J.D. Morrison, 1971).

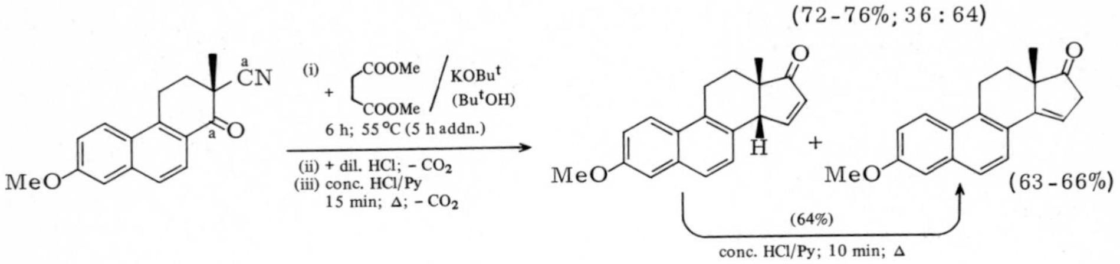

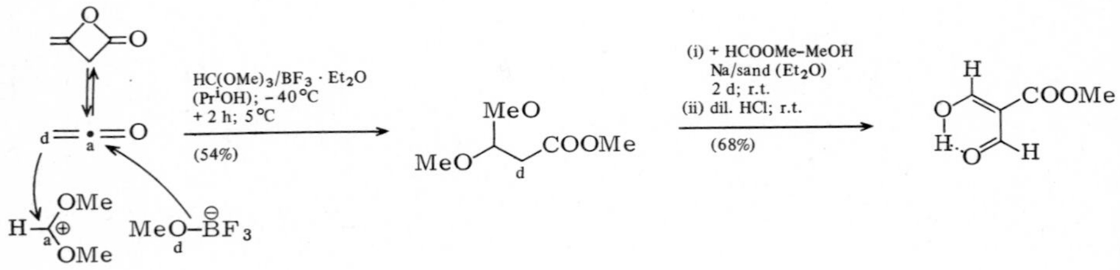

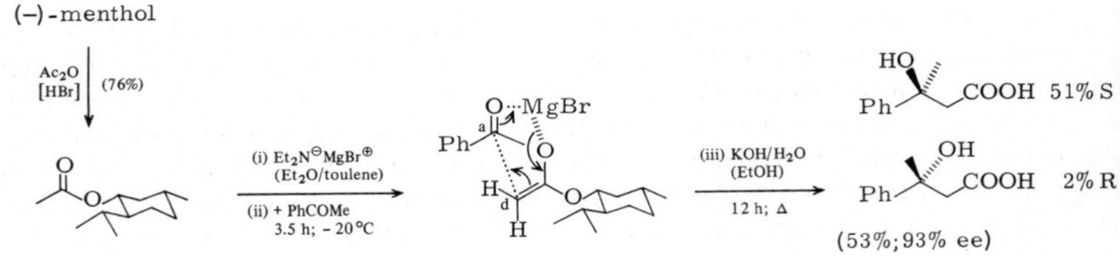

α-Hydrogen atoms of carboxylic acid derivatives may be substituted by bromine and then be transformed to bromozinc groups. These metallated acid derivatives do not undergo self-addition, but can be added selectively to aldehydes and ketones (Reformatsky reaction, p. 41). The analogous cadmium organyls have similar properties, but are less frequently employed in synthesis.

The high nucleophilicity of sulfur atoms is preserved, even if it is bound to electron withdrawing carbonyl groups. Thiocarboxylates, for example, substitute bromine, e.g. of α-bromo ketones. In the presence of bases the α-acylthio ketones deprotonate and rearrange to episulfides. After desulfurization with triphenylphosphine 1,3-diketones are formed in good yield. Thiolactams react in the same way and A. Eschenmoser (1970) has used this sequence in his vitamin B_{12} synthesis (p. 237).

We terminate the description of aldol type reactions with recent developments concerning regio- and stereoselective syntheses. Such procedures imply the use of pure stereoisomers of enolates (see p. 12). C.H. Heathcock (W.A. Kleschick, 1977) added the *trans*-enolate* of a ketone with one bulky substituent (see p. 13) to benzaldehyde and obtained the pure *erythro* ketol in excellent yield. This result was explained in terms of a six-centre transition state. The carbon chain of the enolate remains *trans* and the lithium counterion of the enolate binds both oxygen atoms in the reacting array with chair conformation. The rearrangement of the *axial* methyl group to the more stable *equatorial* position does not occur, because the reaction is carried out at -72 °C. The same effect of complete kinetic control may be achieved, if the enolates are formed by a bulky base and trapped with dialkylboron triflates $R_2BOSO_2CF_3$, at low temperatures (see p. 13). The dialkylboron enolates add stereoselectively to carbonyl groups, since the boron atom again tends to bind both 1,3-oxygen atoms in a pericyclic process and to keep them in a *syn*-conformation. The boron is finally removed by oxidation with a peroxomolybdenum complex and an *erythro*-ketol is formed (D.A. Evans, 1979).

J. Mulzer (1979, 1980) succeeded to produce *threo*-3-hydroxycarboxylic acids in 95% diastereomeric purity. The starting materials were carboxylate dianions, which are readily prepared from carboxylic acids with LDA in THF, and aldehydes with a bulky substituent. At low temperatures *threo*-products are formed in high yields. This fact has been traced back to a pure electronic effect, namely electron donation from the dianion HOMO into the aldehyde LUMO. No positive effect of chelation was observed that could explain the stereoselectivity. It is thought, that an exceptionally strong HOMO-LUMO interaction is caused by the presence of two electron donating oxygen atoms in the enolate. These mechanistic arguments are, however, only preliminary explanations of recent results. They will certainly be modified.

---

* See footnote on p. 12.

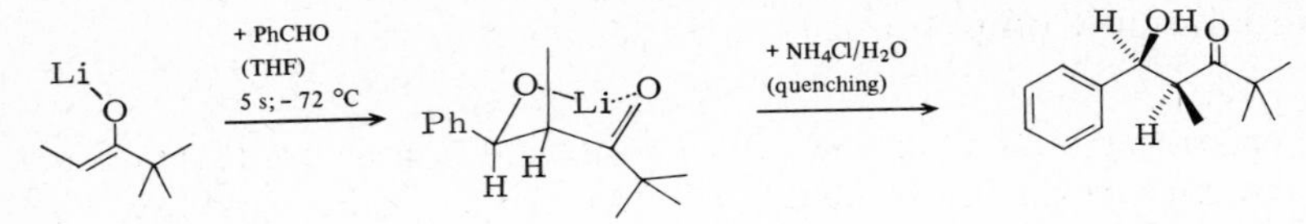

(trans)≡(Z)   (78%; 100% erythro*)

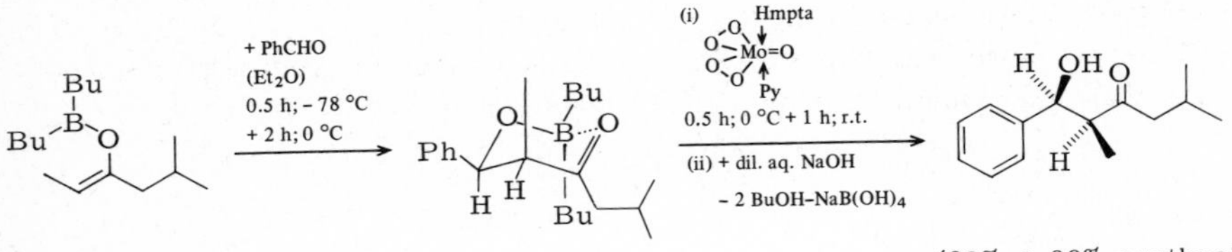

(trans)≡(Z)   (82%; > 99% erythro*)

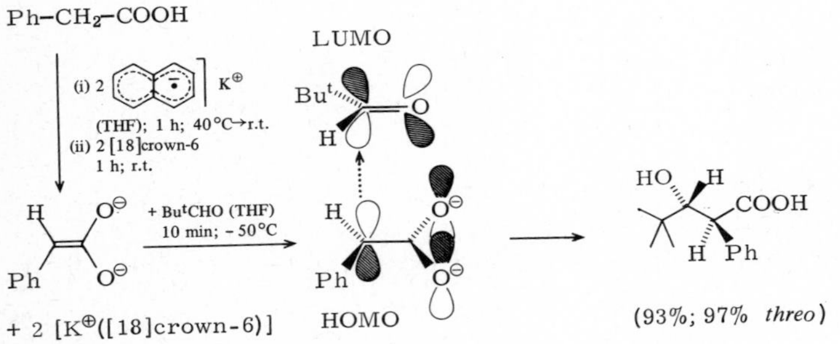

Ph–CH₂–COOH

(93%; 97% *threo*)

*erythro = (R,R) + (S,S)
 threo = (R,S) + (S,R)

Finally (d² + a¹)-additions of 1-alkenyl and 1-alkynyl anions to carbonyl groups should be mentioned. Examples are the addition of vinylmagnesium bromide to ketones, e.g. in the first step of Torgov's steroid synthesis (I.N. Nazarov, 1957), and the famous alkynylation of estrone by Inhoffen, which made available non-digestible oral contraceptives (H.H. Inhoffen, 1939).

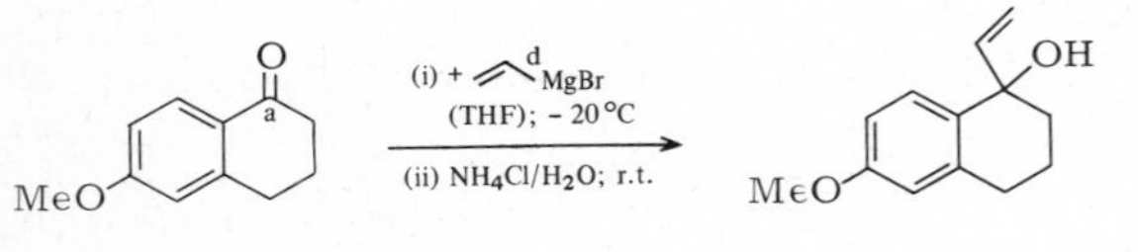

(86%)

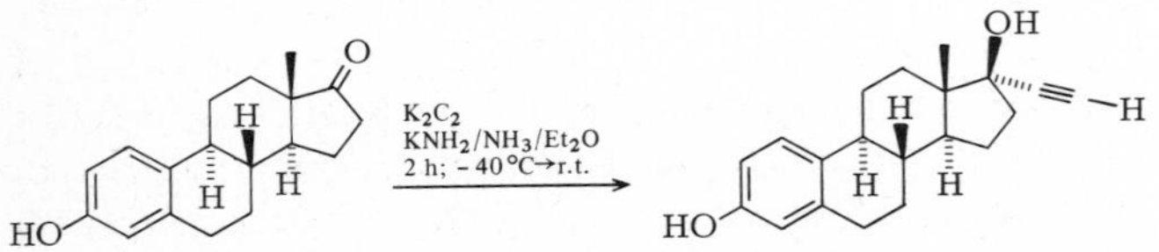

estrone   17α-ethynylestradiol

(90%)

1.11 1,4-Difunctional Compounds

A classical reaction leading to 1,4-difunctional compounds is the nucleophilic substitution of the bromine of α-bromo carbonyl compounds (a^2-synthons) with enolate type anions (d^2-synthons). Regio- and stereoselectivities, which can be achieved by an appropiate choice of the enol component, are similar to those described in the previous section. Just one example of a highly functionalized product (W.L. Meyer, 1963) is given.

2,3-Dibromopropene was used in an efficient thermodynamically controlled alkylation. Hydrolysis of the vinyl bromide yielded a 1,4-diketone (S.C. Welch, 1979).

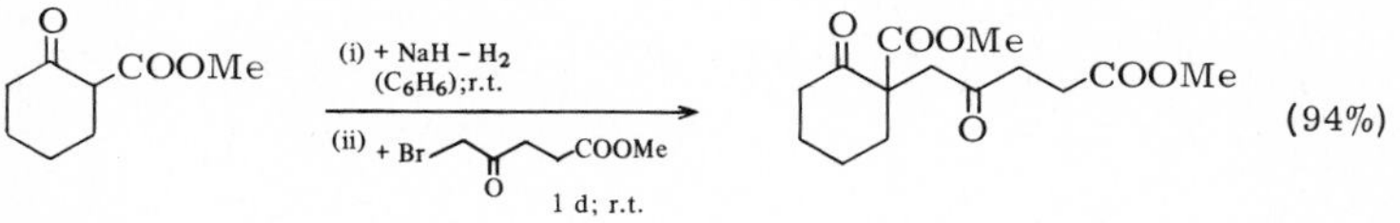

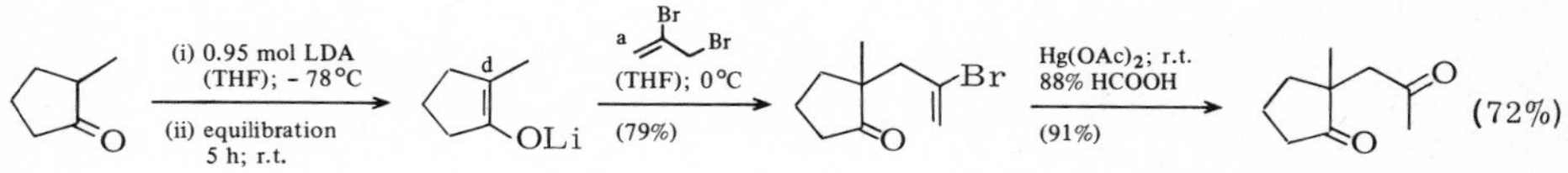

Epoxides provide another useful a^2-synthon. Nucleophilic ring opening with dianions of carboxylic acids (P.L. Creger, 1972) leads to γ-hydroxy carboxylic acids or γ-lactones. Addition of imidoester anions to epoxides yields γ-hydroxyaldehyde derivatives after reduction (H.W. Adickes, 1969).

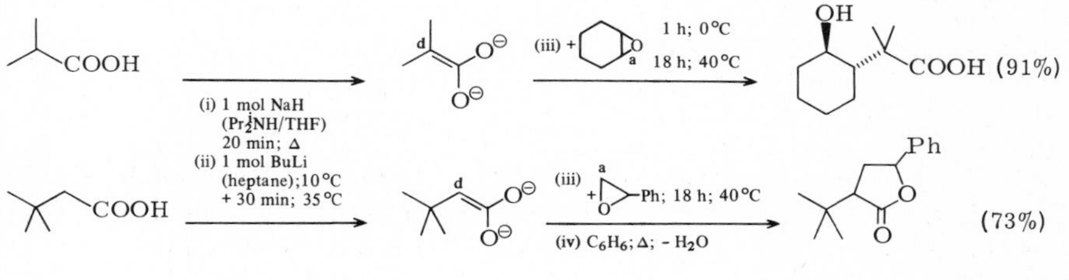

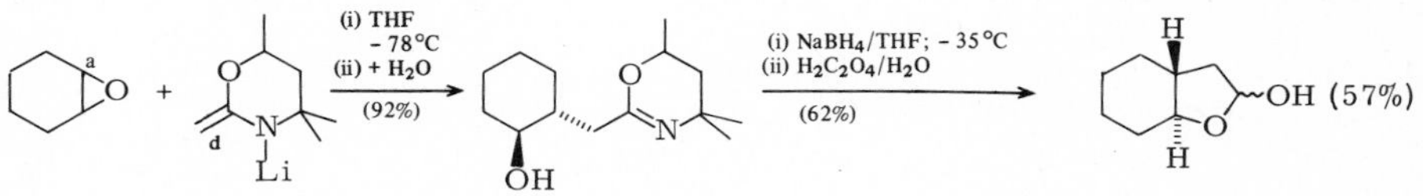

2-Methylsulfinyl enolates are more recently developed d^2-reagents. They are readily prepared from carboxylic esters and dimsyl anion. Methanesulfenic acid can be eliminated thermally after the condensation has taken place. An example is found in Bartlett's Brefeldin synthesis (P.A. Bartlett, 1978).

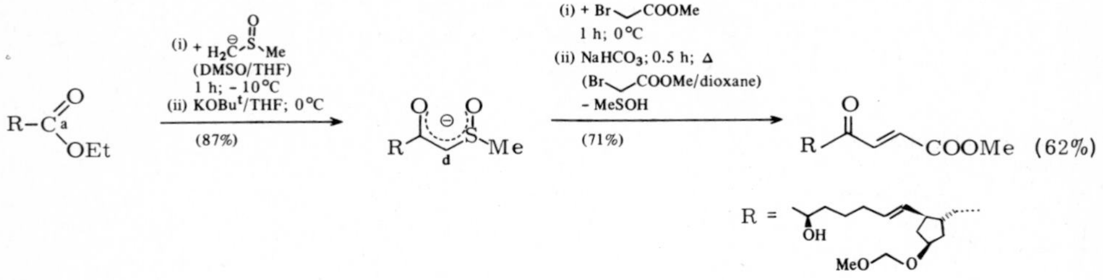

Treatment of O-silyl enols with silver oxide leads to radical coupling via silver enolates. If the carbon atom bears no substituents, two such r^2-synthons recombine to symmetrical 1,4-dicarbonyl compounds in good yield (Y. Ito, 1975).

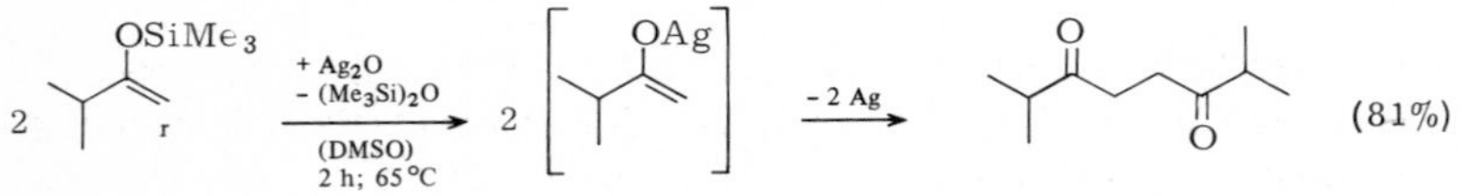

The second important pathway to 1,4-disubstituted compounds is provided by Michael additions of d^1-synthons to electron deficient carbon-carbon double bonds (a^3-synthons). The addition of nitroalkane anions to α,β-unsaturated carbonyl compounds is a popular route to 1,4-dicarbonyl compounds (M.A. Adams, 1979), because the CH—NO$_2$-group is transformed into an oxo group, if treated first with strong base and afterwards with strong acid (Nef reaction; W.E. Noland, 1955). Other examples of regio- and stereoselective Michael additions of d^1-reagents are given below (I. Felner, 1967; H.H. Inhoffen, 1958 B; B.-T. Gröbel, 1977B).

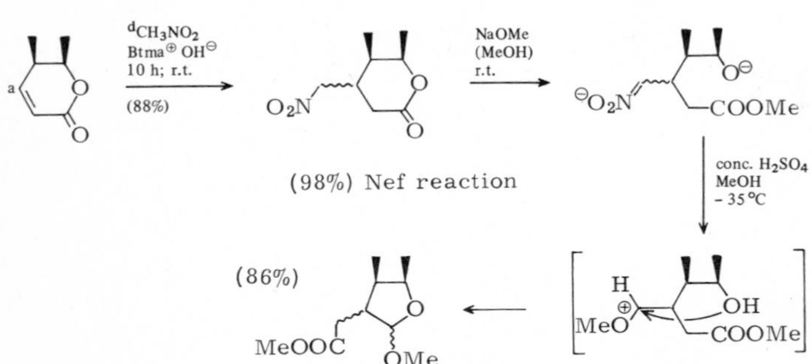

Nef reaction:

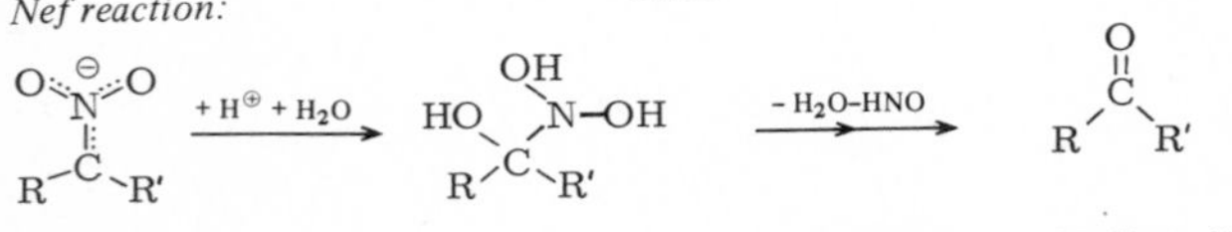

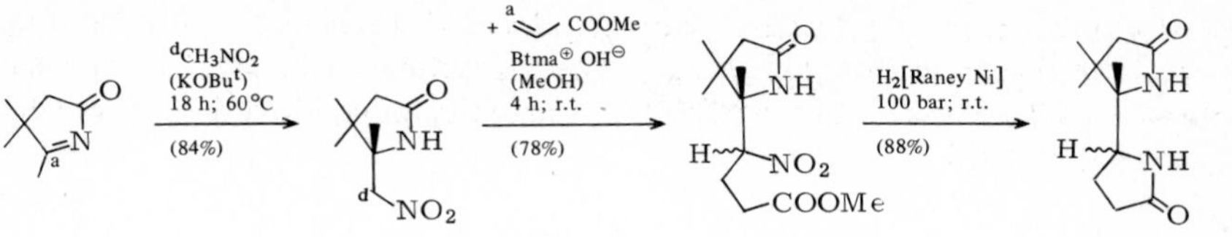

erythro/threo 11 : 9

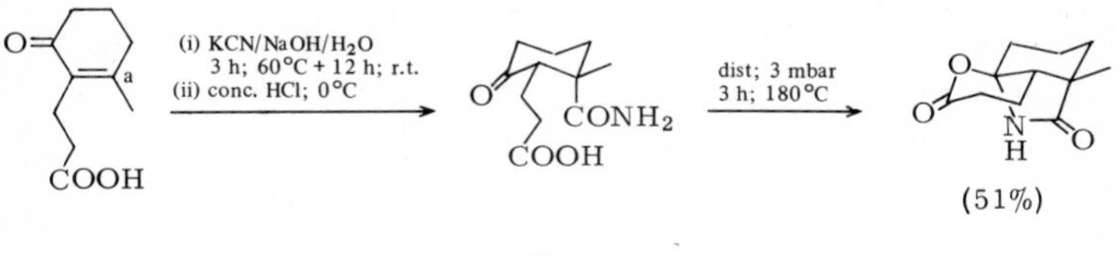

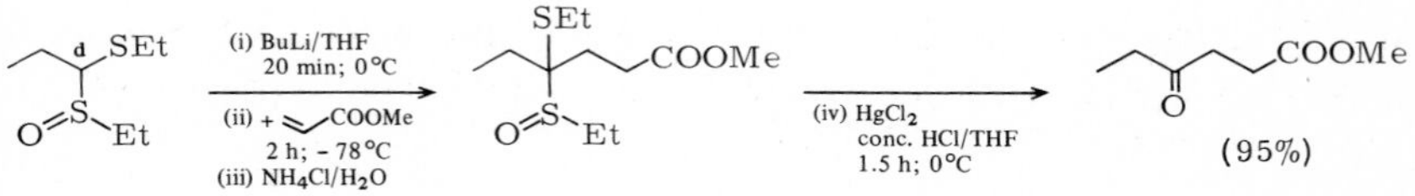

(51%)

(95%)

The addition of acetylides to oxiranes yields 3-alkyn-1-ols (F. Sondheimer, 1950; M.A. Adams, 1979; R.M. Carlson, 1974, 1975; K. Mori, 1976). The acetylene dianion and two a^1-synthons can also be used. 1,4-Diols with a carbon triple bond in between are formed from two carbonyl compounds (V. Jäger, 1977, see p. 49). The triple bond can be either converted to a *cis*- or *trans*-configured double bond (M.A. Adams, 1979) or be hydrated to give a ketone (see p. 54 and p. 118).

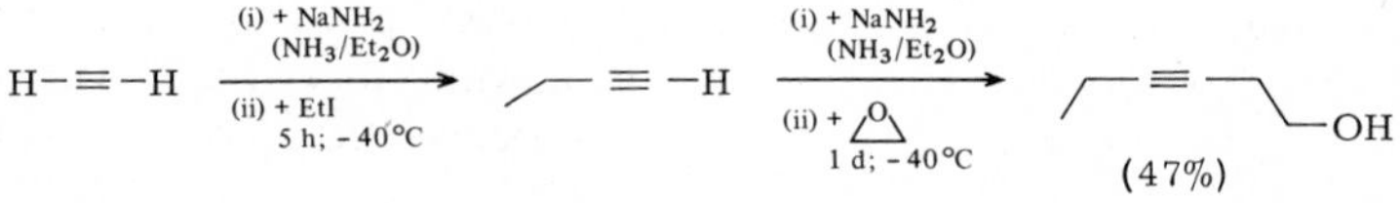

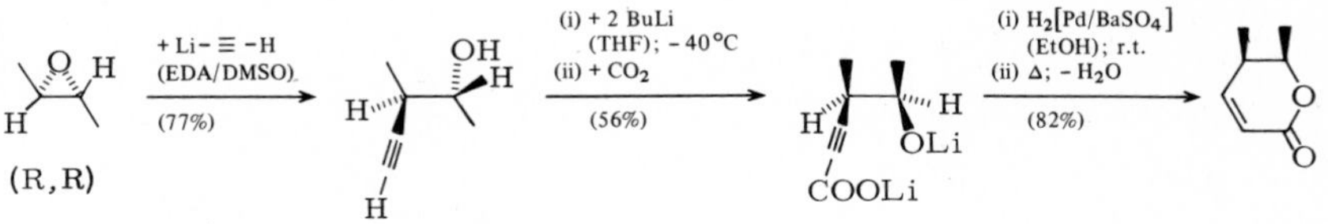

(47%)

(R,R)   (77%)   (56%)   (82%)

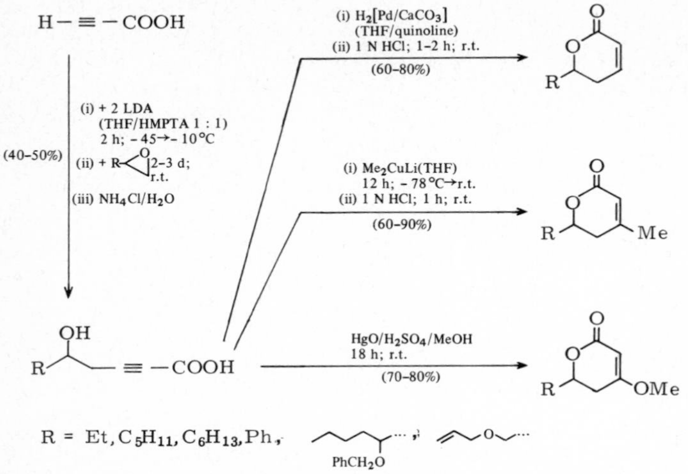

H— ≡ —COOH

(i) H₂[Pd/CaCO₃]
   (THF/quinoline)
(ii) 1 N HCl; 1–2 h; r.t.
(60–80%)

(40–50%)

(i) + 2 LDA
   (THF/HMPTA 1 : 1)
   2 h; – 45→– 10 °C
(ii) + R—◁ O 2–3 d;
           r.t.
(iii) NH₄Cl/H₂O

(i) Me₂CuLi(THF)
   12 h; – 78 °C→r.t.
(ii) 1 N HCl; 1 h; r.t.
(60–90%)

OH
R⟍⟍— ≡ —COOH

HgO/H₂SO₄/MeOH
18 h; r.t.
(70–80%)

R = Et, C₅H₁₁, C₆H₁₃, Ph,   ⟍⟍⟍⟍ᵣⁱ,   ⟍⟍O⟍⟍
                              PhCH₂O

The last group of reactions uses ring opening of carbonyl or 1-hydroxyalkyl substituted cyclopropanes, which operate as a⁴-synthons. d⁰-Synthons, e.g. hydroxide or halides, yield 1,4-disubstituted products (E. Wenkert, 1970 A). (1-Hydroxyalkyl)- and (1-haloalkyl)-cyclopropanes are rearranged to homoallylic halides, e.g. in Julia's method of terpene synthesis (M. Julia, 1961, 1974; S.F. Brady, 1968; J.P. McCormick, 1975).

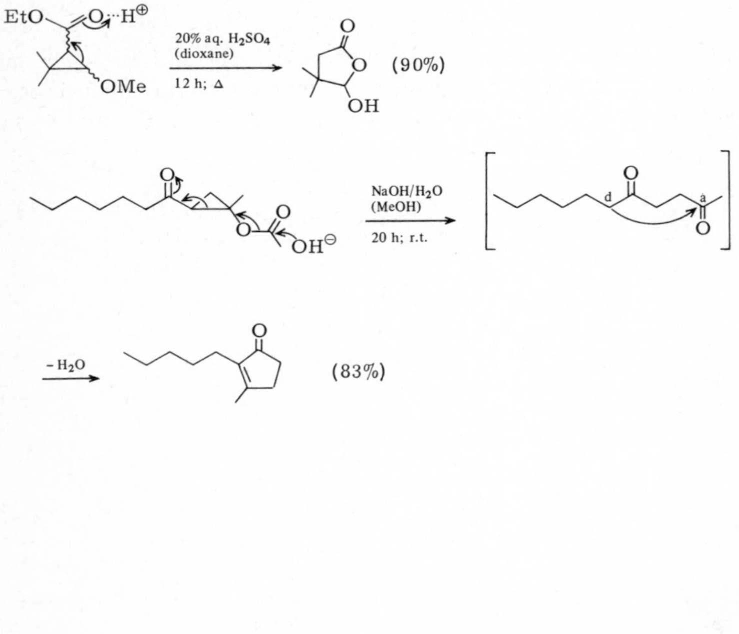

EtO⟍⟍O⟍⋯H⊕

⟍◁
    ''OMe

20% aq. H₂SO₄
(dioxane)
12 h; Δ

(90%)

OH

NaOH/H₂O
(MeOH)
20 h; r.t.

– H₂O

(83%)

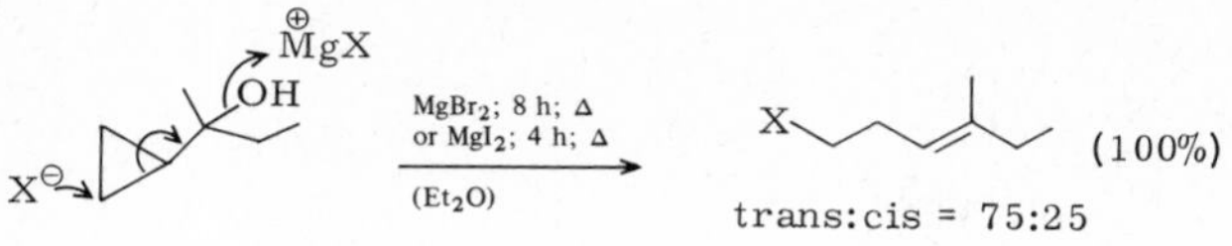

$$X \rightarrow \sim\sim\sim \quad (100\%)$$

$$trans : cis = 75 : 25$$

Julia's terpenoid synthesis

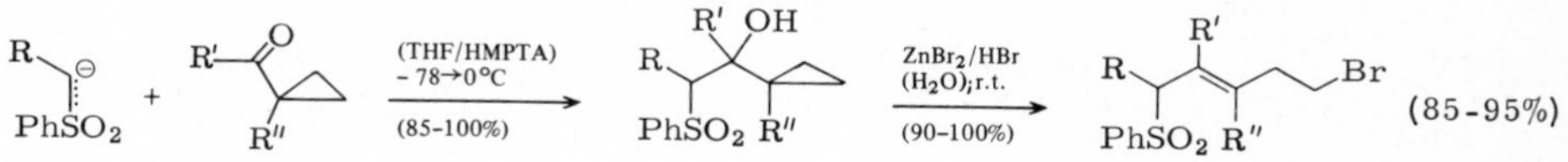

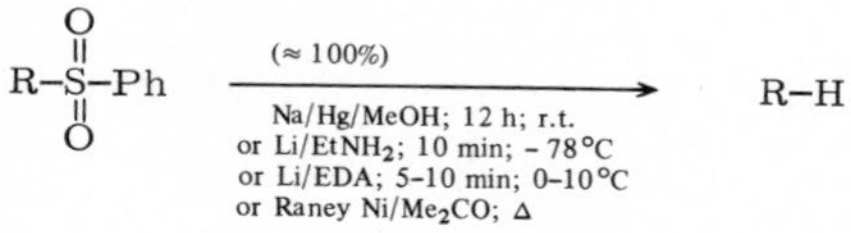

$$(85-95\%)$$

$$R{-}H$$

Heterosubstituted cyclopropanes can be synthesized from appropiate olefins and carbenes. Since cyclopropane resembles olefins in its reactivity and is thus an electron-rich carbocycle (p. 69ff.), it forms complexes with Lewis acids, e.g. $TiCl_4$, and is thereby destabilized. This effect is even more pronounced in cyclopropanone ketals. If one of the alcohols forming the ketal is a silanol, the ketal is stable and distillable. The O—Si-bond is cleaved by $TiCl_4$ and a d^3-reagent is formed. This reacts with a ^{1}-reagents, e.g. aldehydes or ketals. Various 4-substituted carboxylic esters are available from 1-alkoxy-1-siloxycyclopropanes in this way (E. Nakamura, 1977). If one starts with 1-bromo-2-methoxycyclopropanes, the bromine can be selectively substituted by lithium. Subsequent treatment of this reagent with carbonyl compounds yields (2-methoxycyclopropyl)methanols, which can be transformed to β,γ-unsaturated aldehydes (E.J. Corey, 1975B).

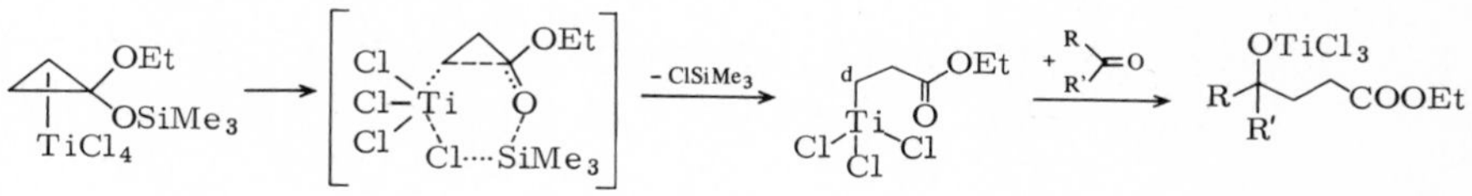

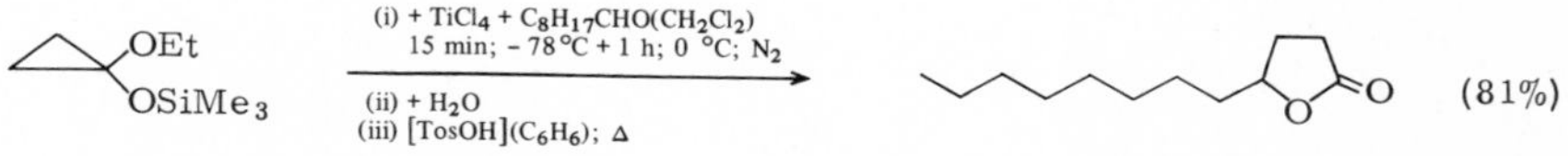

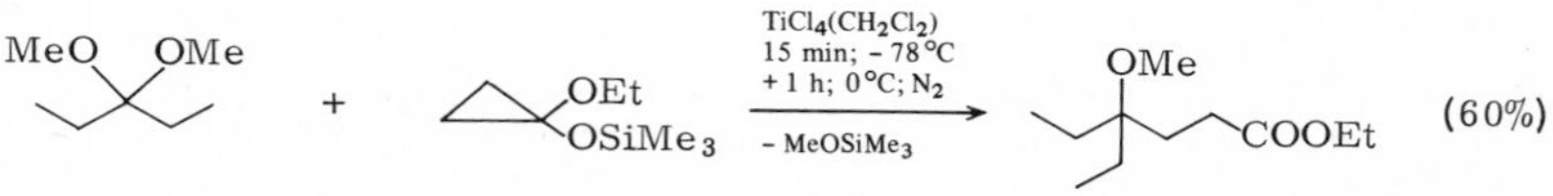

$$(81\%)$$

$$(60\%)$$

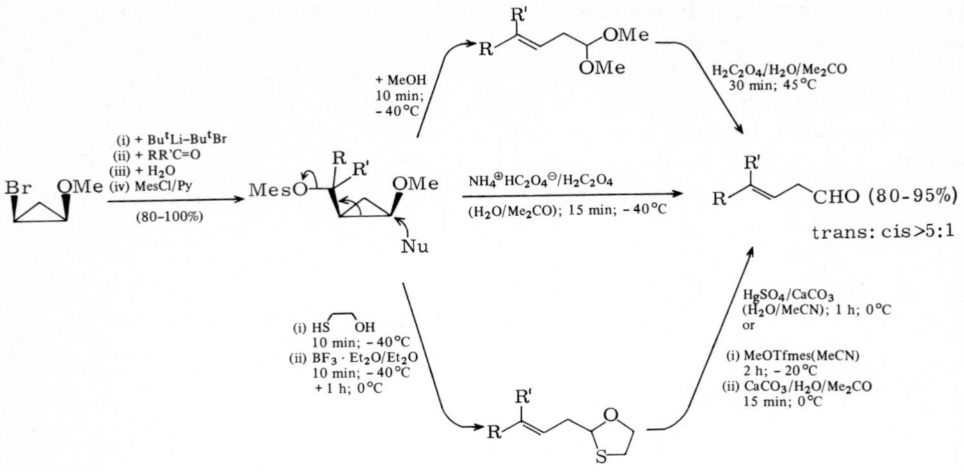

Some other methods to synthesize 1,4-dicarbonyl compounds *via* cyclopropane or cyclobutanone derivatives are given in sections 1.13.1 and 1.13.2.

1.12 1,5-Difunctional Compounds

The Michael reaction is of central importance here. This reaction is a vinylogous aldol addition, and most facts, which have been discussed in section 1.10, also apply here: the reaction is catalyzed by acids and by bases, and it may be made regioselective by the choice of appropriate enol derivatives. Stereoselectivity is also observed in reactions with cyclic educts. An important difference to the aldol addition is, that the Michael addition is usually less prone to sterical hindrance. This is evidenced by the two examples given below, in which cyclic 1,3-diketones add to α,β-unsaturated carbonyl compounds (K. Hiroi, 1975; H. Smith, 1964).

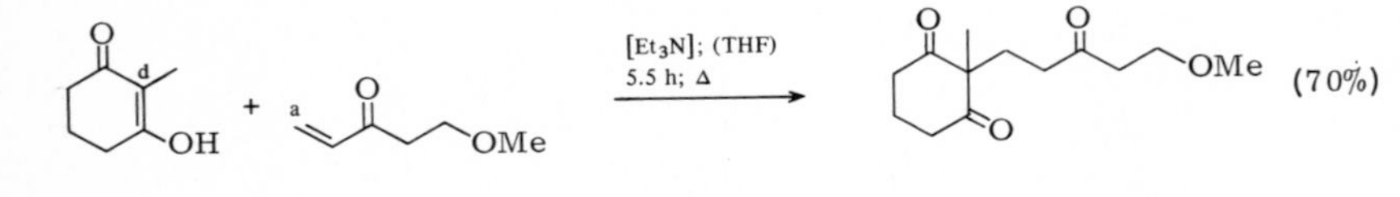

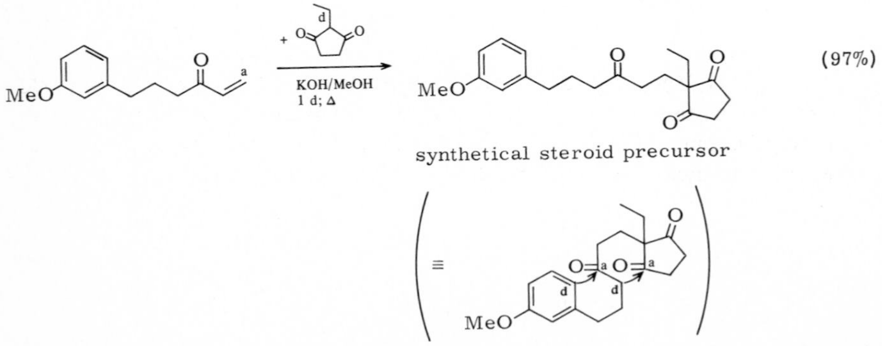

synthetical steroid precursor

   Torgov introduced an important variation of the Michael addition: allylic alcohols are used as vinylogous a³-synthons and 1,3-dioxo compounds as d²-reagents (S.N. Ananchenko, 1962, 1963; H. Smith, 1964; C. Rufer, 1967). Mild reaction conditions have been successful in the addition of 1,3-dioxo compounds to vinyl ketones. Potassium fluoride can act as weakly basic, non-nucleophilic catalyst in such Michael additions under essentially non-acidic and non-basic conditions (Y. Kitahara, 1964).

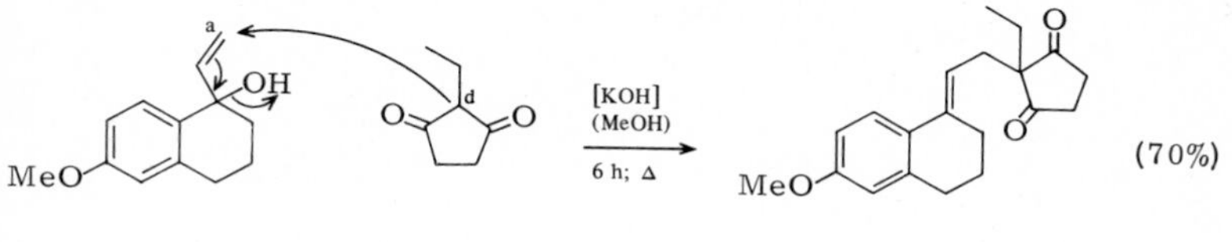

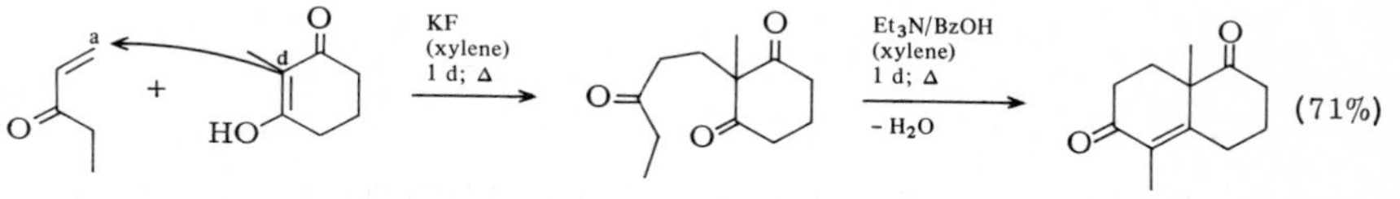

   The addition of large enolate synthons to cyclohexenone derivatives via Michael addition leads to equatorial substitution. If the cyclohexenone conformation is fixed, e.g. as in decalones or steroids, the addition is highly stereoselective. This is also the case with the δ-addition to conjugated dienones (Y. Abe, 1956). Large substituents at C-4 of cyclic a³-synthons direct incoming carbanions to the *trans*-position at C-3 (A.R. Battersby, 1960). The thermodynamically most stable products are formed in these cases, because the addition of 1,3-dioxo compounds to activated double bonds is essentially reversible.

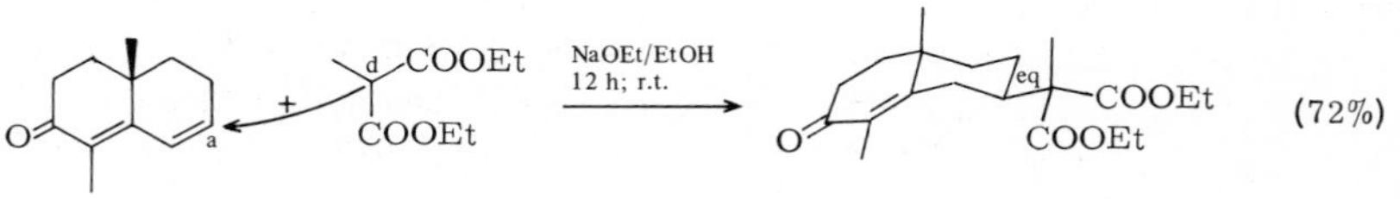

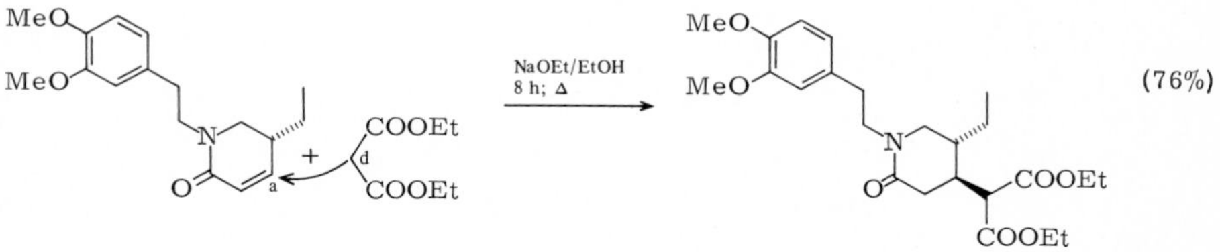

   Michael addition of enolates to vinyl ketones yields saturated 1,5-diketones, which may be again enolized and may undergo a further aldol addition. This sequence is important in the synthesis of polycyclic compounds and is called Robinson anellation (*anellus*, lat. small ring). Some representative examples are summarized below (W.S. Johnson, 1956; J.A. Marshall,

1968; R. Zurflüh, 1968; R.M. Coates, 1968). The regio- and stereoselectivities which are indicated can all be understood, if one considers the arguments given in the discussions of aldol and Michael addition reactions as well as the relative reactivities of enolates (sections 1.1 and 1.10).

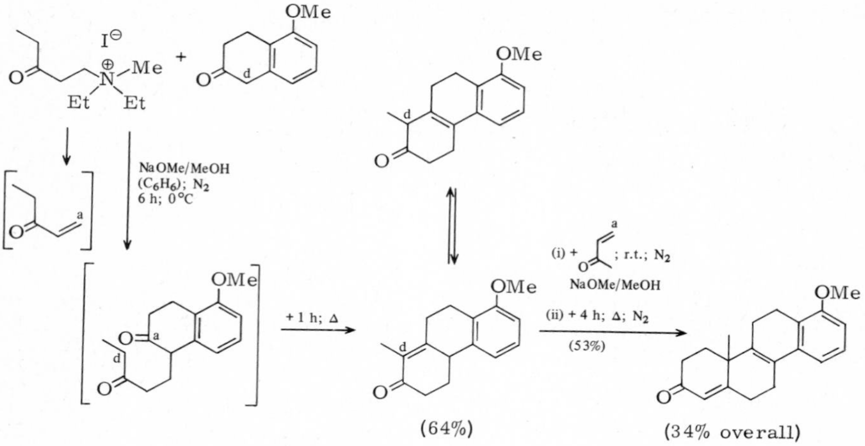

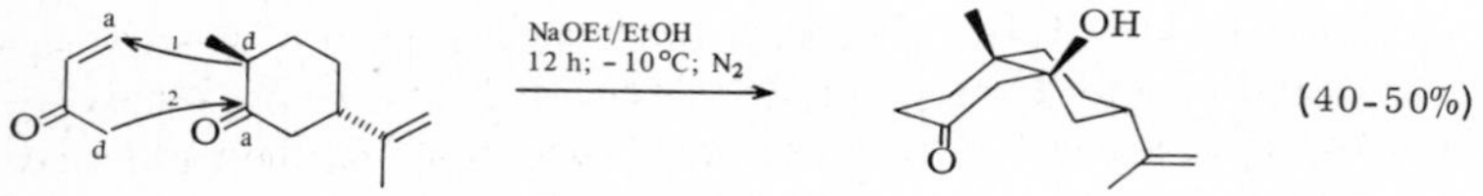

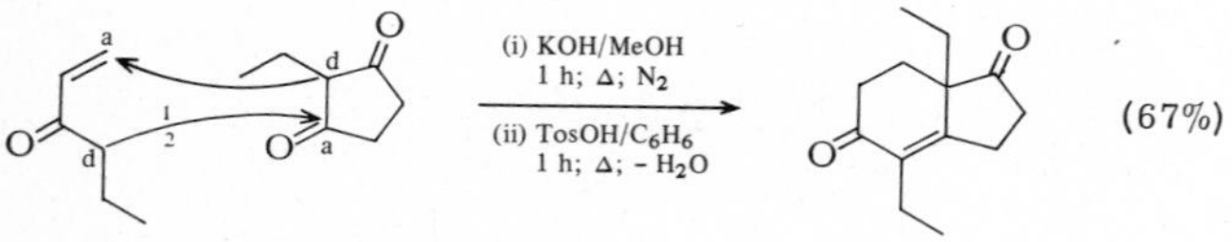

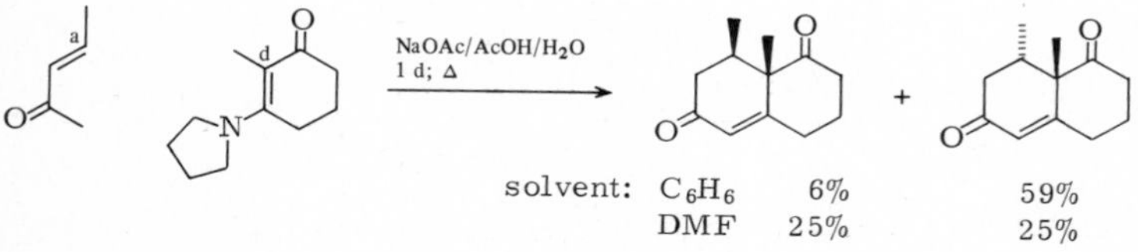

If a Michael reaction uses an unsymmetrical ketone with two CH-groups of similar acidity, the enol or enolate is first prepared in pure form (p. 9ff.). To avoid equilibration one has to work at low temperatures. The reaction may then become slow, and it is advisable to further activate the carbon-carbon double bond. This may be achieved by the introduction of an extra electron-withdrawing silyl substituent at C-2 of an a^3-synthon. Treatment of the Michael adduct with base removes the silicon, and may lead as well to an aldol addition (G. Stork, 1973, 1974 B; R.K. Boeckman, Jr., 1974).

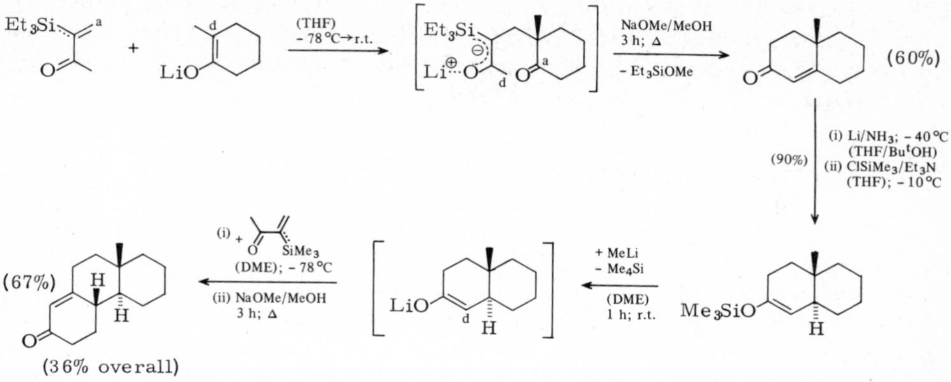

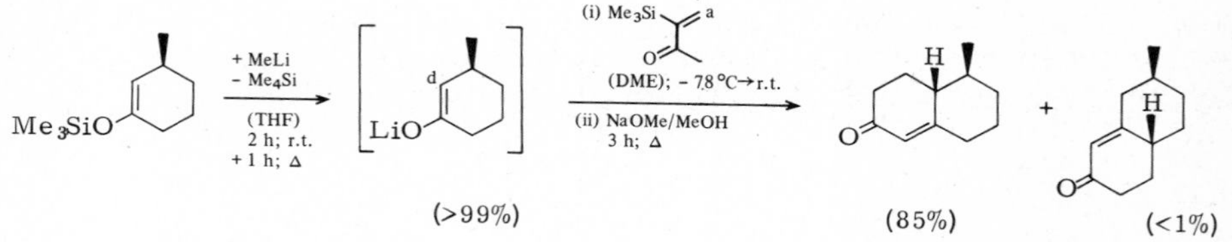

The synthesis of spiro compounds from ketones and methoxyethynyl propenyl ketone exemplifies some regioselectivities of the Michael addition. The electrophilic triple bond is attacked first, next comes the 1-propenyl group. The conjugated keto group is usually least reactive. The ethynyl starting material has been obtained from the addition of the methoxyethynyl anion to the carbonyl group of crotonaldehyde (G. Stork, 1962 B, 1964A).

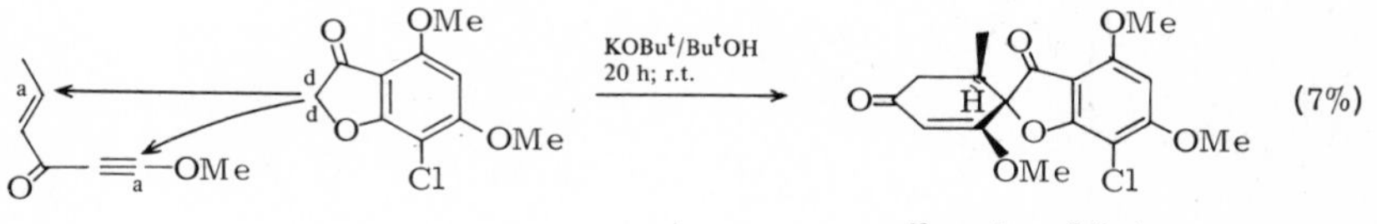

dl-griseofulvin

1.13 Carbocycles

In the previous sections various synthetic methods for cyclic alkanes and alkenes have already been discussed. In the vast majority of cases, these procedures were analogous to those used in the synthesis of acyclic carbon compounds. The only difference was that both the electron acceptor and donor or two radical moieties were present in a single molecule, so that intramolecular ring formation was achieved. Further reactions of this type are discussed in later sections, but will not be considered in this paragraph. Here we shall introduce reactions, by which some cycloalkanes can be obtained directly using specific coupling reactions of two open-chain educts. These special ring forming reactions are difficult to systematize under more general headings. They can, however, often be accomplished in particularly high yield and stereoselectivity. They also provide useful intermediates of general synthetic applicability, and are therefore discussed separately. We shall also indicate some useful ring cleavage reactions of strained cyclopropane and cyclobutane derivatives. Fragmentation and cleavage reactions of larger rings are discussed in section 1.14.

1.13.1 Cyclopropane and Cyclopropene Derivatives

The majority of preparative methods which have been used for obtaining cyclopropane derivatives involve carbene addition to an olefinic bond. If acetylenes are used in the reaction, cyclopropenes are obtained. Heteroatom-substituted or vinyl cyclopropanes come from alkenyl bromides or enol acetates (A. de Meijere, 1979; E.J. Corey, 1975 B; E. Wenkert, 1970 A). The carbenes needed for cyclopropane syntheses can be obtained *in situ* by α-elimination of hydrogen halides with strong bases (R. Köster, 1971; E.J. Corey, 1975 B), by copper catalyzed decomposition of diazo compounds (E. Wenkert, 1970 A; S.D. Burke, 1979; N.J. Turro, 1966), or by reductive elimination of iodine from *gem*-diiodides (J. Nishimura, 1969; D. Wendisch, 1971; J.M. Denis, 1972; H.E. Simmons, 1973; C. Girard, 1974).

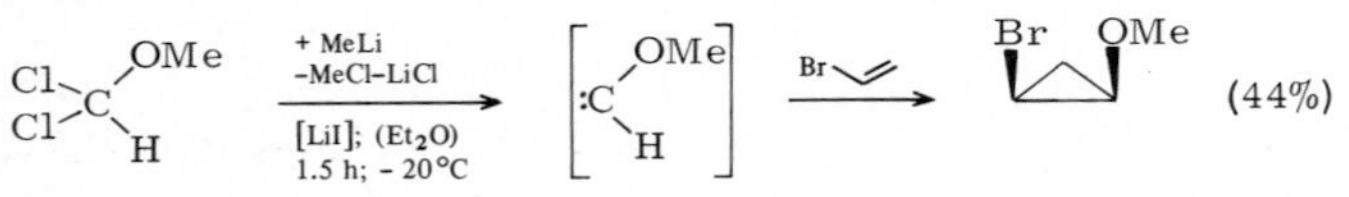

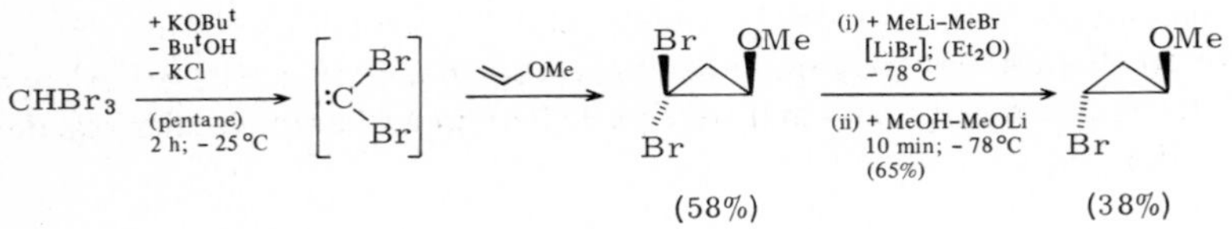

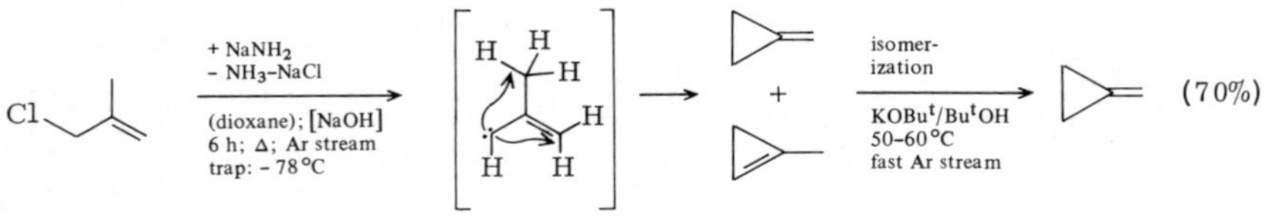

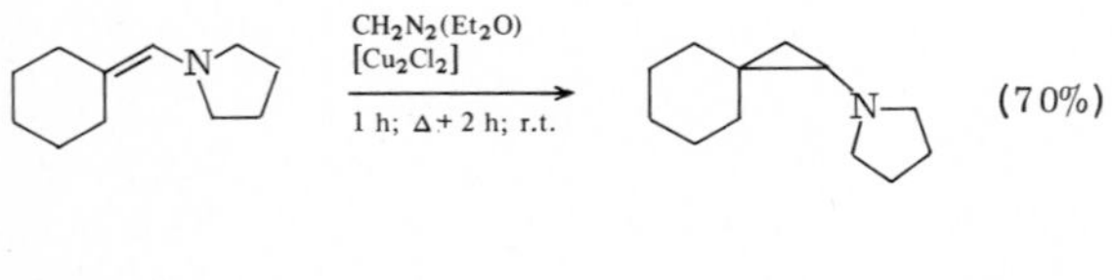

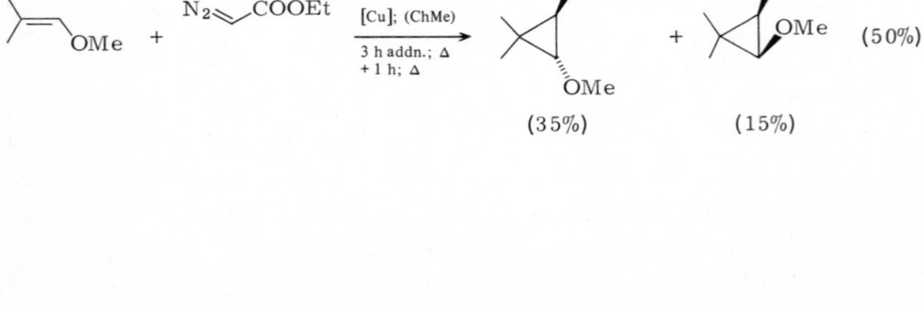

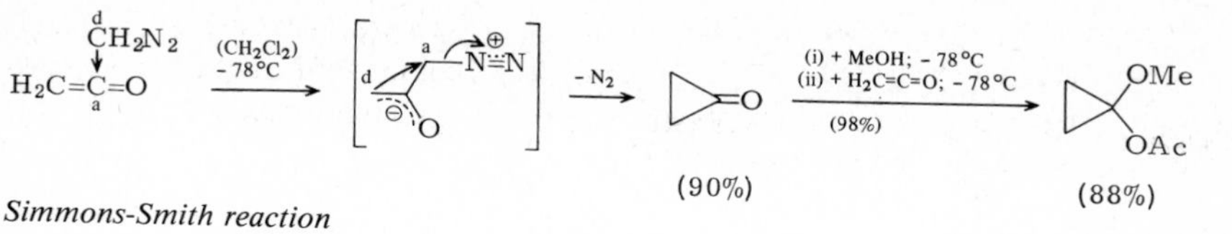

Simmons-Smith reaction

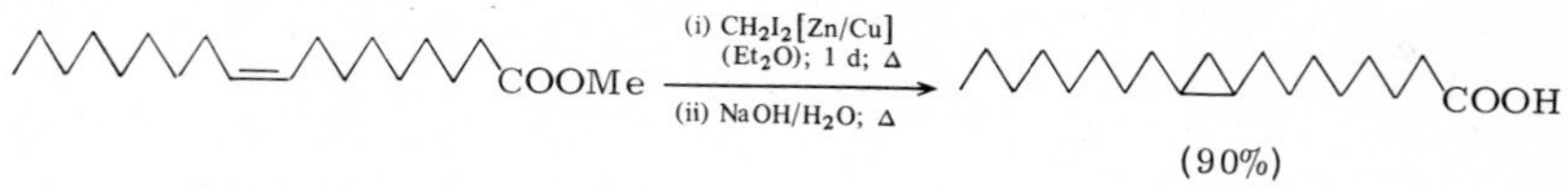

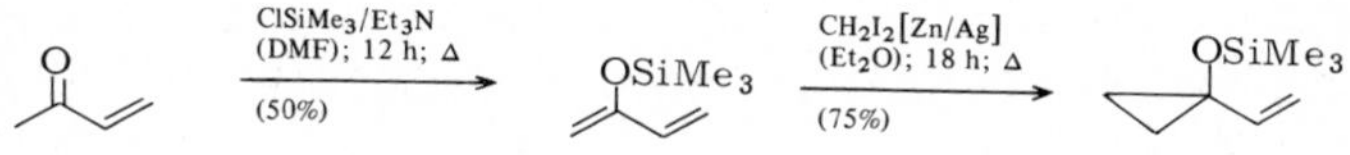

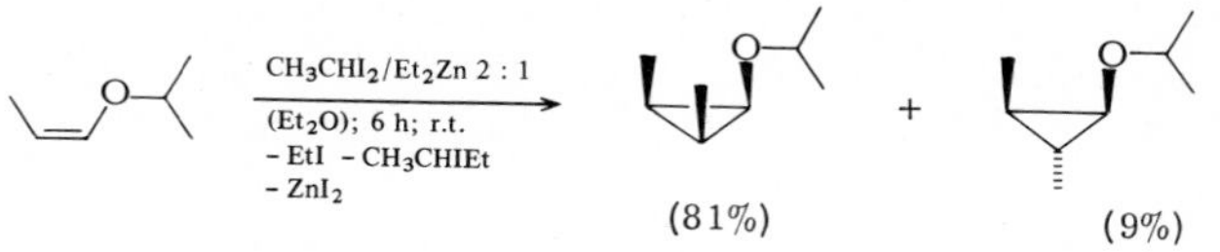

Another widely used route to cyclopropanes involves the addition of sulfonium ylides to α,β-unsaturated carbonyl compounds (S.R. Landor, 1967; R. Sowada, 1971; C.R. Johnson, 1973B, 1979; B.M. Trost, 1975 A). Non-activated double bonds are not attacked. Sterical hindrance is of little importance in these reactions because the C—S bond is extraordinarily long and the sulfur atom is relatively far away from the ylide carbon bound to it. The last example shown below illustrates control of regio- and stereoselectivity in an intramolecular reaction (R.S. Matthews, 1971).

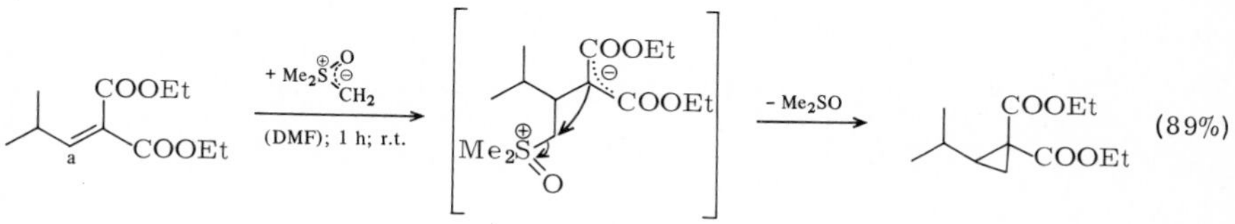

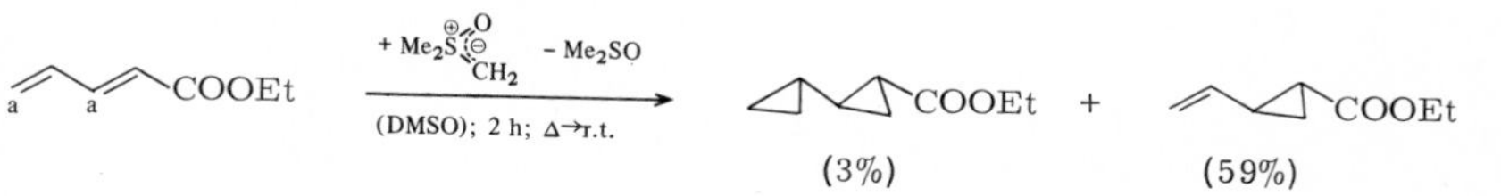

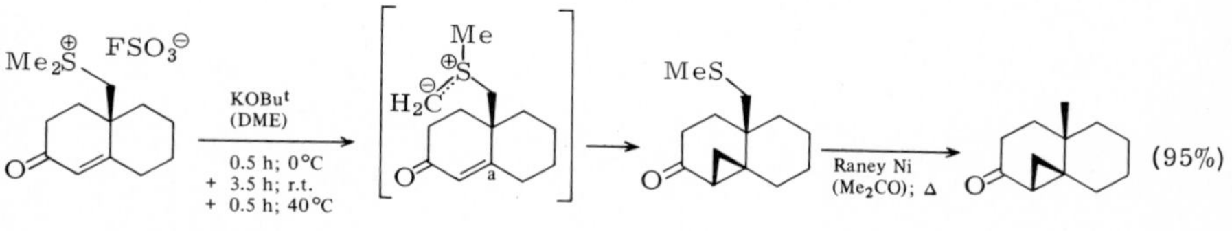

The growing importance of cyclopropane derivatives (A. de Meijere, 1979), as synthetic intermediates originates in the unique, olefin-like properties of this carbocycle. Cyclopropane derivatives with one or two activating groups are easily opened (see. p. 62f.). Some of these reactions are highly regio- and stereoselective (E. Wenkert, 1970 A, B; E.J. Corey, 1956 A, B, 1975; see p. 64). Many appropriately substituted cyclopropane derivatives yield 1,4-difunctional compounds under mild nucleophilic or reductive reaction conditions. Such compounds are especially useful in syntheses of cyclopentenone derivatives and of heterocycles (see also section 1.13.3).

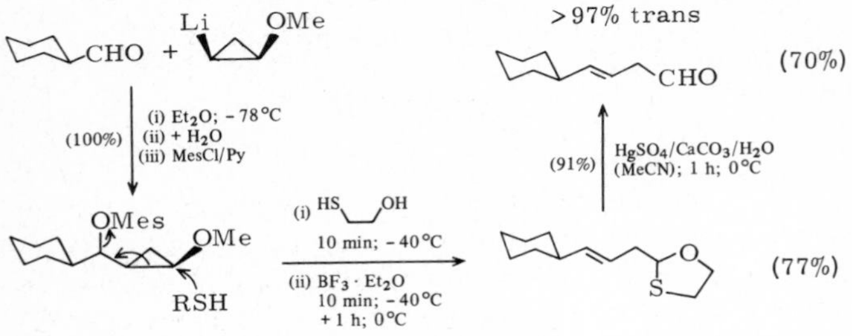

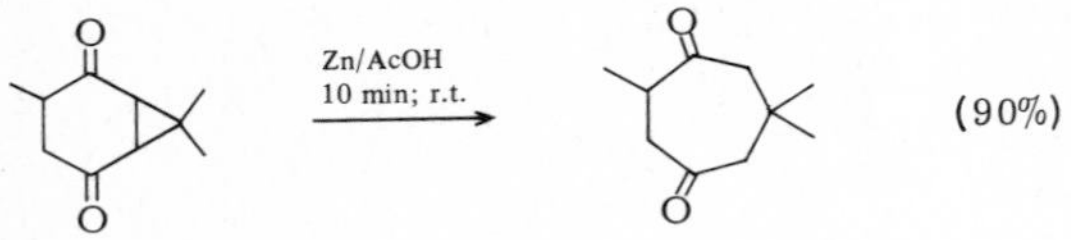

1-Siloxy-1-vinylcyclopropanes rearrange thermally to give 1-siloxycyclopentene derivatives or may be hydrolyzed to cyclobutanones by acids (J.M. Conia, 1975; B.M. Trost, 1973; G.C. Girard, 1974). The cyclopropane ring may also be opened by base-catalyzed hydrolysis of the silyl ethers to form ethyl ketones.

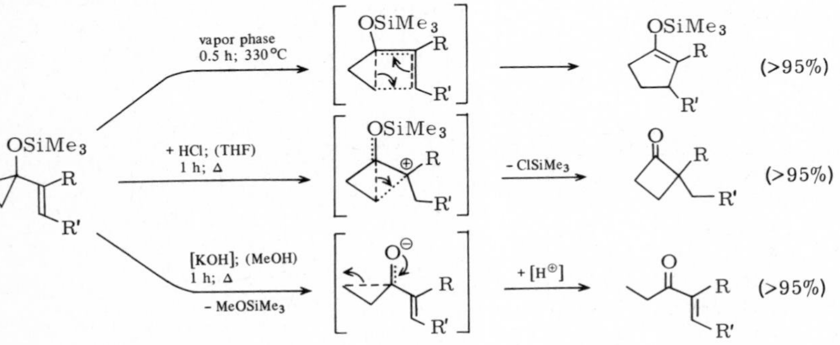

(Bromomethyl)- or (hydroxymethyl)cyclopropane derivatives undergo acid-catalyzed homoallylic rearrangements to yield *trans*-olefins (J.P. McCormick, 1975; S.F. Brady, 1968; M. Julia, 1974). This rearrangement is the basis of Julia's terpene synthesis (see. p. 63).

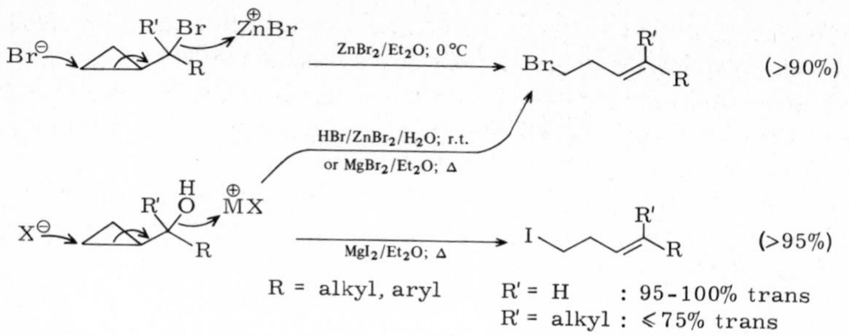

R = alkyl, aryl   R' = H   : 95-100% trans
R' = alkyl : ⩽75% trans

If a bromomethyl- or vinyl-substituted cyclopropane carbon atom bears a hydroxy group, the homoallylic rearrangement leads preferentially to cyclobutanone derivatives (J. Salaun, 1974). Addition of amines to cyclopropanone (N.J. Turro, 1966) yields β-lactams after successive treatment with tert-butyl hypochlorite and silver(I) salts (H.H. Wasserman, 1975). For intramolecular cyclopropane formation see section 1.16.

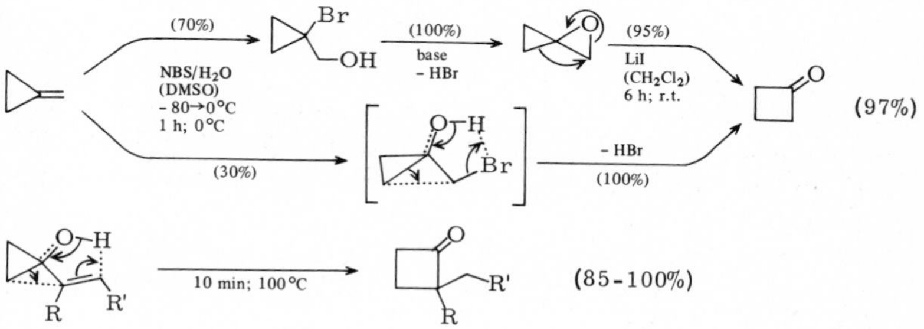

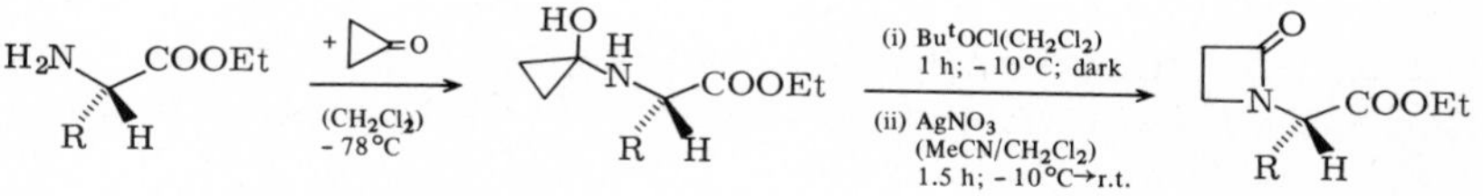

R	amino acid	yield
H	Gly	33%
Me	Ala	47%
Pr^i, Bu^i	Val, Leu	65%
PhCH$_2$	Phe	70%

1.13.2 Cyclobutane Derivatives

Only relatively few examples of interesting target molecules containing C$_4$ rings are known. These include caryophyllene (E.J. Corey, 1963 A, 1964) and cubane (J.C. Barborak, 1966). The photochemical [2 + 2]-cycloaddition applied by Corey yielded mainly the *trans*-fused isomer, but isomerization with base leads via enolate to formation of the more stable *cis*-fused ring system.

Within the cubane synthesis the initially produced cyclobutadiene moiety (see p. 292) is only stable as an iron(0) complex (M. Avram, 1964; G.F. Emerson, 1965; M.P. Cava, 1967). When this complex is destroyed by oxidation with cerium(IV) in the presence of a "dienophilic" quinone derivative, the cycloaddition takes place immediately. Irradiation leads to a further cyclobutane ring closure. The cubane synthesis also exemplifies another general approach to cyclobutane derivatives. This starts with cyclopentanone or cyclohexane-dione derivatives which are brominated and treated with strong base. A Favorskii rearrangement then leads to ring contraction.

Ketones and aldehydes have been cyclized with diphenylsulfonium cyclopropylide followed by Li⁺-catalyzed rearrangement to produce cyclobutanone derivatives (B.M. Trost, 1973). These compounds react on heating with *tert*-butoxybis(dimethylamino)methane to give vinylogous amides (H. Bredereck, 1963, 1965, 1968). Reaction with propylene bis(thiotosyl-ate) in buffered ethanol leads to replacement of the dimethylaminomethylene unit by the propylenedithio unit. The yield of this reaction sequence is often close to quantitative and correlates with the molecular size of the carbonyl educt because small vinylogous amides are much more sensitive to decomposition than are larger ones. The resulting spirobicyclic systems may then be opened by nucleophiles, and the 1,3-dithiane can be hydrolyzed. 1,4-Dicarbonyl compounds are formed. The versatility of the dithiane group as an acyl anion equivalent allows various modifications of this reaction scheme. Condensation reactions of the 1,4-dicarbonyl functions lead to cyclopentenone rings. Thus a method for spiroanellation of such ring on a carbonyl containing ring is given (B.M. Trost, 1975B).

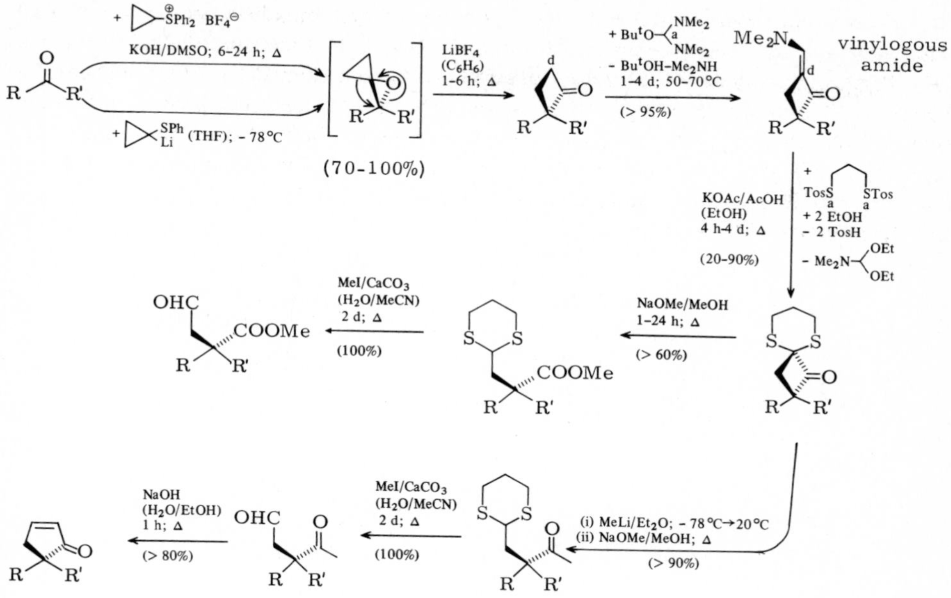

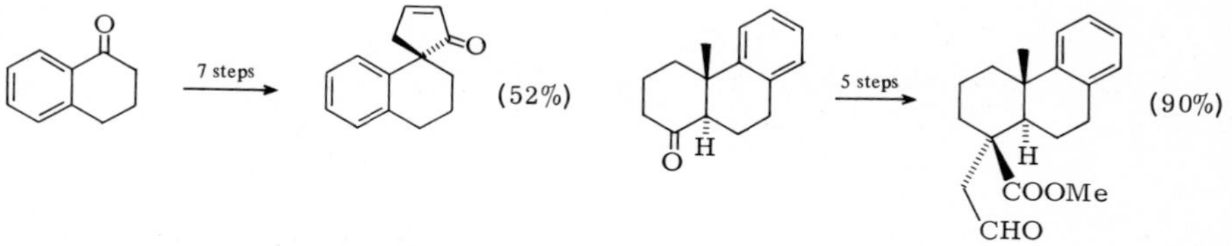

Highly strained polycycloalkanes consisting of several interconnected cyclopropane or cyclobutane rings undergo thermal and hydrogenolytic ring opening with great ease (D. Kaufmann, 1979). Those bonds which release the largest amount of strain energy are cleaved selectively.

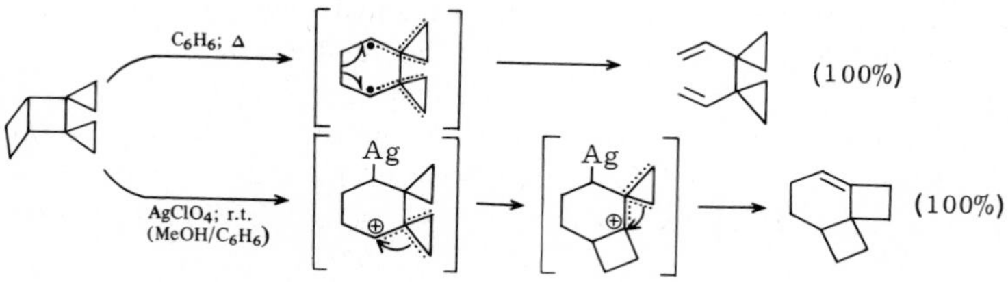

Simple cyclobutanes do not readily undergo such reactions, but cyclobutenes do. Benzocyclobutene derivatives tend to open to give extremely reactive dienes, namely *ortho*-quinodimethanes (examples of syntheses see on p. 253, 254, and 269). Benzocyclobutenes and related compounds are obtained by high-temperature elimination reactions of bicyclic benzene derivatives such as 3-isochromanone (C.W. Spangler, 1973, 1976, 1977), or more conveniently in the laboratory, by Diels-Alder reactions (R.P. Thummel, 1974) or by cyclizations of silylated acetylenes with 1,5-hexadiynes in the presence of (cyclopentadienyl)dicarbonylcobalt (W.G. Aalbersberg, 1975; R.P. Thummel, 1980).

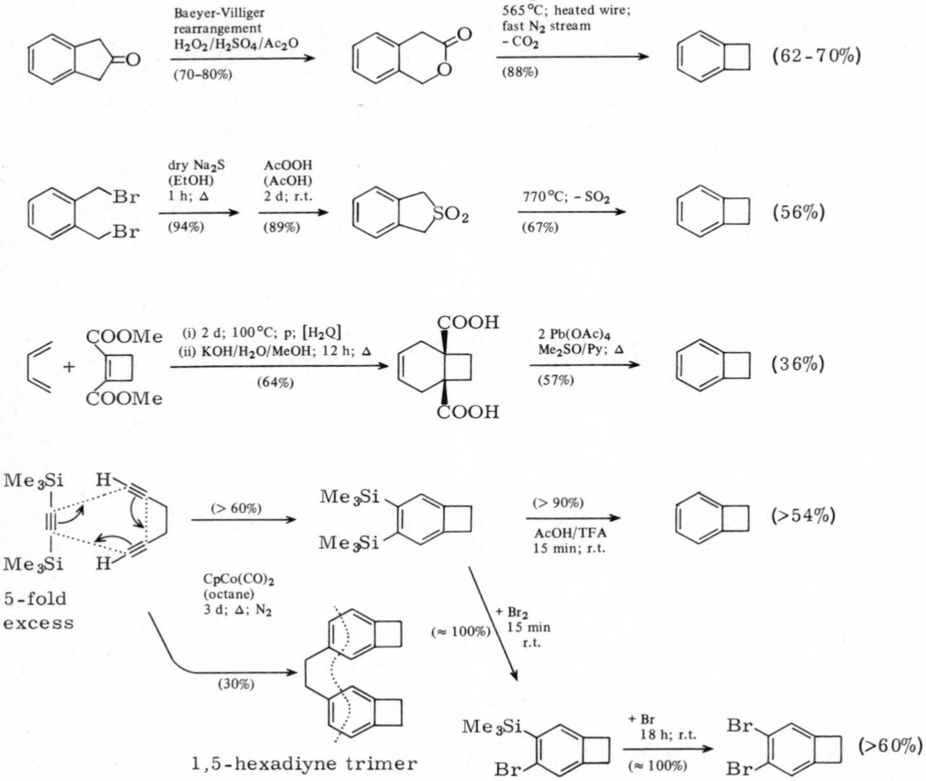

1,5-hexadiyne trimer

The thermal ring opening of 1,2-bis(trimethylsiloxy) cyclobutenes (from acyloin condensation of 1,2-dicarboxylic esters) was used in ring expansion prodecures (see p. 51).

1.13.3 Cyclopentane Derivatives

Cyclopentane reagents used in synthesis are usually derived from cyclopentanone (R.A. Ellison, 1973). "Classically" they are made by base-catalyzed intramolecular aldol or ester condensations (see also p. 52). An important example is 2-methylcyclopentane-1,3-dione. It is synthesized by intramolecular acylation of diethyl propionylsuccinate dianion followed by saponification and decarboxylation. This cyclization only worked with potassium t-butoxide in boiling xylene (R. Bucourt, 1965). Faster routes to this diketone start with succinic acid or its

anhydride. A Friedel-Crafts acylation with 2-acetoxy-2-butene in nitrobenzene or with pro-
pionylchloride in nitromethane leads to acylated adducts, which are deacylated in aqueous
acids (V.J. Grenda, 1967; L.E. Schick, 1969). A new promising route to substituted
cyclopent-2-enones makes use of intermediate 5-nitro-1,3-diones (D. Seebach, 1977).

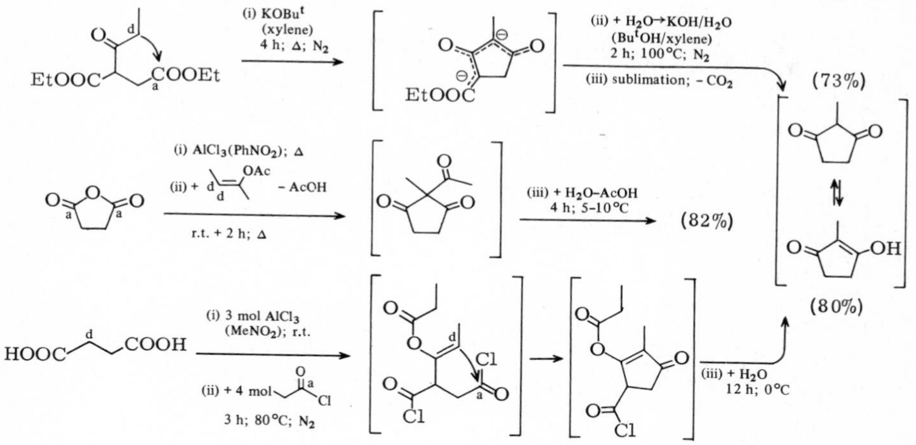

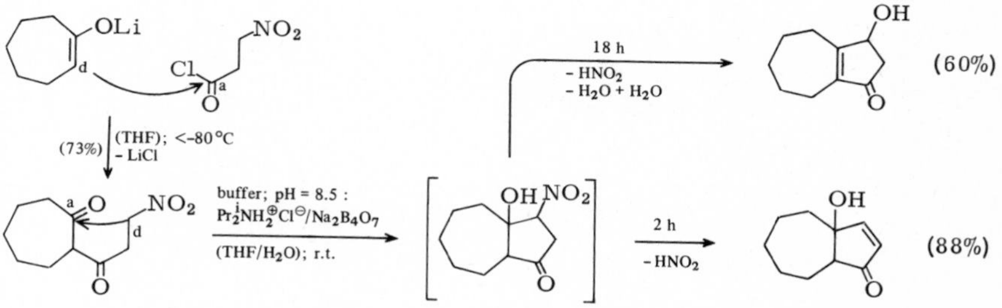

Cyclopentene-1-carboxaldehydes are obtained from cyclohexene precursors by the sequence cyclohexene → cyclohexane-1,2-diol → open-chain dialdehyde → cyclopentane aldol.
The main advantage of this ring contraction procedure is, that the regio-and stereoselectivity
of the Diels-Alder synthesis of cyclohexene derivatives can be transferred to cyclopentane
synthesis (G. Stork, 1953; G. Büchi, 1968).

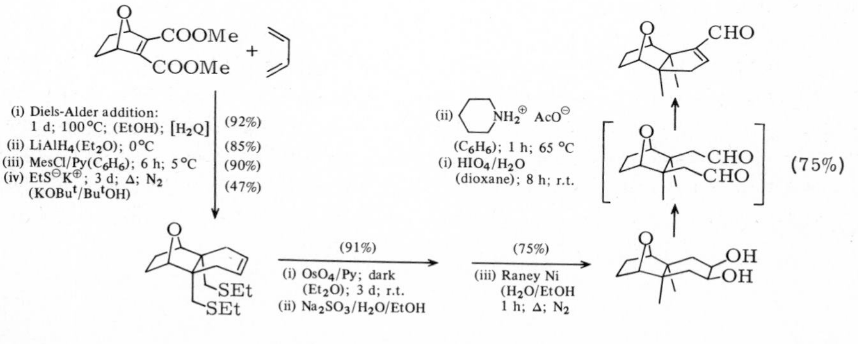

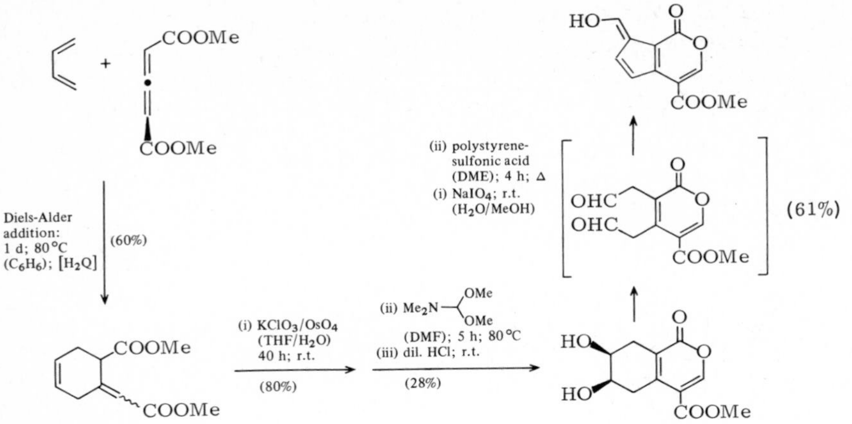

A now classical synthesis of the cyclopentanone ring in steroids together with the introduction of the angular methyl group was introduced by W.S. Johnson (1957). The α-methylene group of a cyclohexanone precursor is protected by aldol condensation with an aromatic aldehyde. The tertiary α'-carbon atom is methylated with methyl iodide, and oxidative cleavage with alkaline H_2O_2 converts the 2-(arylmethylene)cyclohexanone into a 1,6-dioic acid. Dieckmann condensation of the corresponding diester, saponification, and decarboxylation finally yields a cyclopentanone ring.

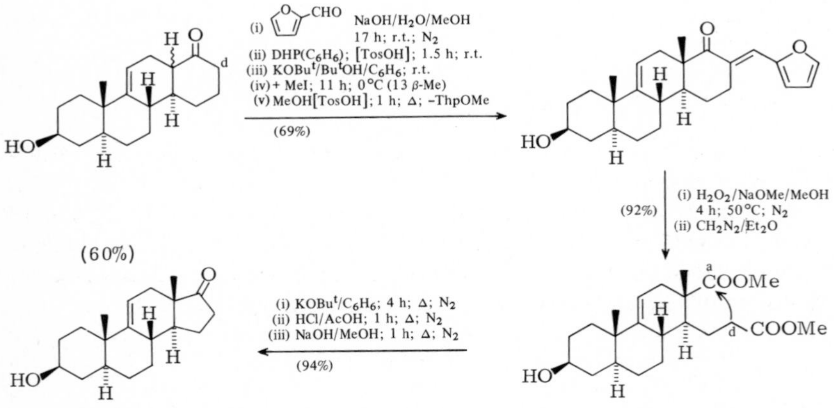

Some straightforward, efficient cyclopentanellation procedures were developed recently. Addition of a malonic ester anion to a cyclopropane-1,1-dicarboxylic ester followed by a Dieckmann condensation (S. Danishefsky, 1974) or addition of β-ketoester anions to a (1-phenylthiocyclopropyl)phosphonium cation followed by intramolecular Wittig reaction (J.P. Marino, 1975) produced cyclopentanones. Another procedure starts with a [2+2]-cycloaddition of dichloroketene to alkenes followed by regioselective ring expansion with diazomethane. The resulting 2,2-dichlorocyclopentanones can be converted to a large variety of cyclopentane derivatives (A.E. Greene, 1979; J.-P. Deprés, 1980).

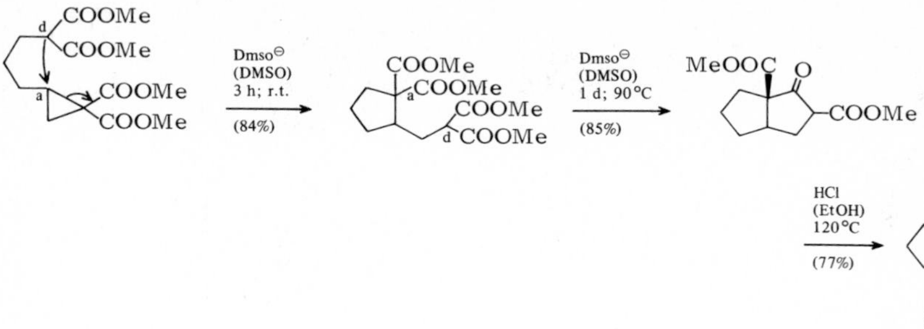

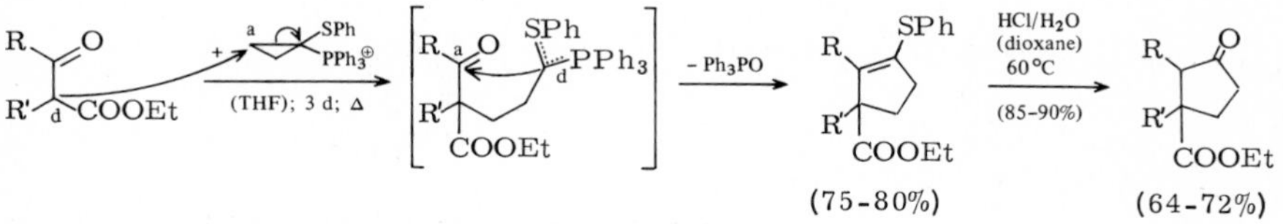

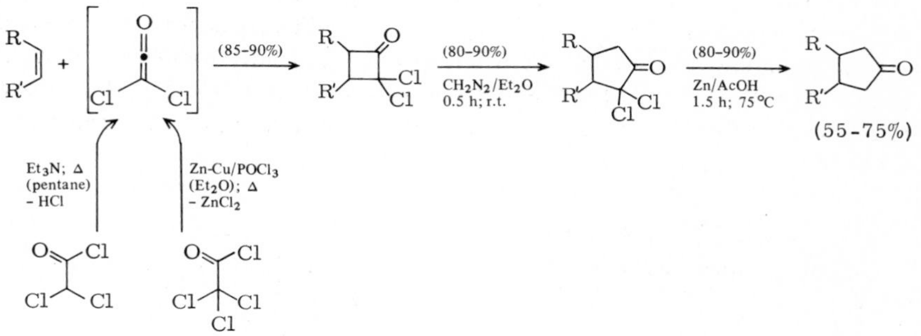

Cyclopentanones may also be synthesized from α,β-unsaturated ketones and diiodomethane. The ketone is converted to the O-silyl enol, and carbene is added to the enol double bond using the Simmons-Smith reaction (see p. 69f.). Thermal rearrangement of the resulting 1-siloxy-1-vinylcyclopropane and acid-catalyzed hydrolysis of the silyl enol ether leads to cyclopentanones in excellent yields (C. Girard, 1974). Very high temepratures, however, are needed, and this obviously limits the generality of this rearrangement reaction.

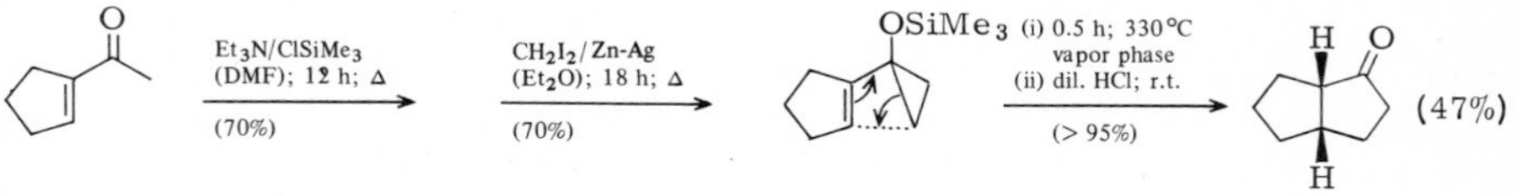

Another useful route to cyclopentanes is the ring contraction of 2-bromocyclohexanones by a Favorskii rearrangement to give cyclopentanecarboxylic acids. If α,β-dibromoketones are used, ring opening of the intermediate cyclopropanone leads selectively to β,γ-unsaturated carboxylic acids (S.A. Achmad, 1963, 1965; J. Wolinsky, 1965).

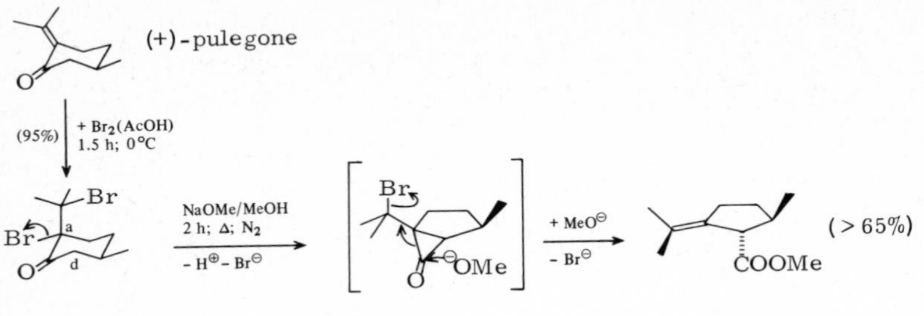

(+)-pulegone
(95%)
+ Br2(AcOH)
1.5 h; 0°C
Br
Br
a
d
NaOMe/MeOH
2 h; Δ; N2
- H⊕- Br⊖
Br
OMe
+ MeO⊖
- Br⊖
COOMe
(> 65%)

1.13.4 Cyclohexane and Cyclohexene Derivatives

Thermal [2+4]-cycloaddition of an olefin, preferably with electron withdrawing substituents, to a diene yields cyclohexene derivatives in high yields (W. Carruthers, 1978). High temperatures must be used with electron-deficient dienes. This ''Diels-Alder'' reaction is stereoselective: the olefin adds suprafacially to the diene, and so the configuration of both components is retained. With cyclic dienes oligocyclic products are formed, which have preferentially the *endo*-configuration. The Diels-Alder reaction is one of the most powerful synthetic reactions of all since its application can result *simultaneously* in an increase of (i) the number of rings, (ii) the number of asymmetric centres, and (iii) the number of functional groups (E.J. Corey, 1967A). A few examples illustrate the wide range of applications of this classical electrocyclization reaction. The first examples demonstrate the effect of substituents on regio- and stereoselectivity (G. Büchi, 1966B; R.B. Woodward, 1952). All reaction schemes show the synthesis of complex polyfunctional molecules from relatively simple, symmetric precursors (G. Büchi, 1968; D.A. Evans, 1972; R.A. Ruden, 1974), which would be difficult to synthesize by other means.

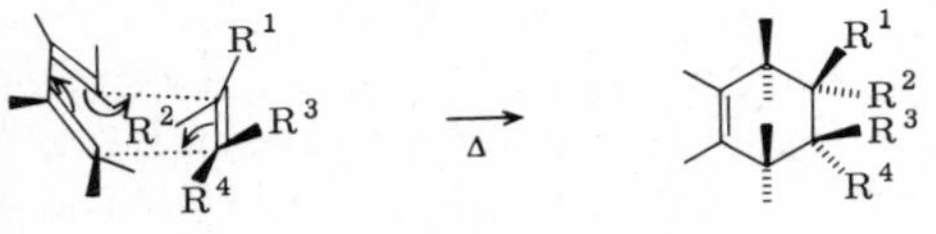

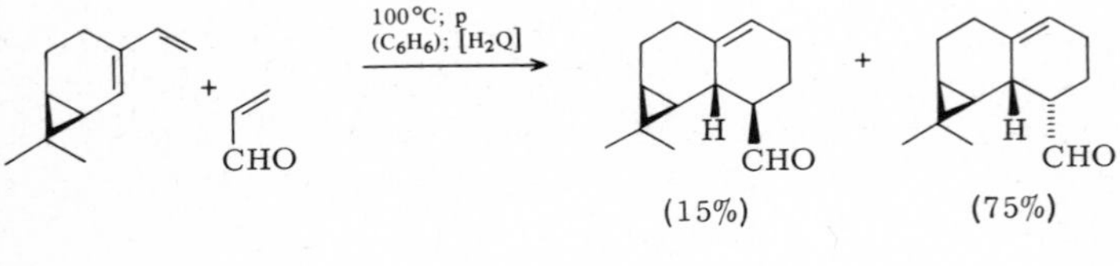

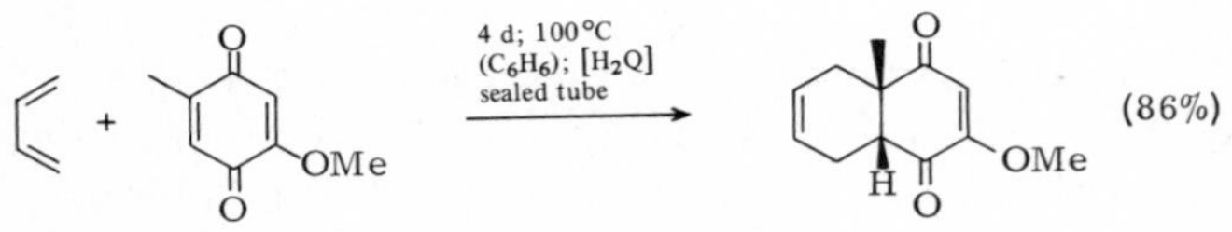

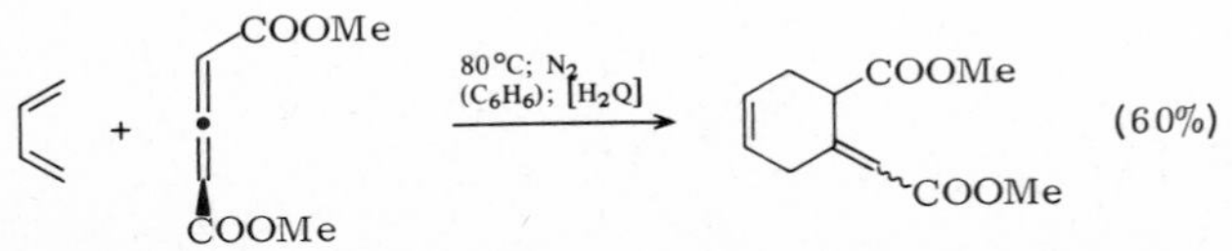

R1
R2
R3
R4
Δ
R1
R2
R3
R4

100°C; p
(C6H6); [H2Q]
CHO
H
CHO
H
CHO
(15%)
(75%)

O
OMe
O
4 d; 100°C
(C6H6); [H2Q]
sealed tube
O
H
O
OMe
(86%)

COOMe
COOMe
80°C; N2
(C6H6); [H2Q]
COOMe
COOMe
(60%)

A major difficulty with the Diels-Alder reaction is its sensitivity to sterical hindrance. Tri- and tetrasubstituted olefins or dienes with bulky substituents at the terminal carbons react only very slowly. Therefore bicyclic compounds with angular substituents are often obtained in low yields, and polar reactions are more suitable for such target molecules, e.g. steroids. There exists, however, one recent report on the reaction of a tetrasubstituted alkene with a 1,1-disubstituted diene to produce a cyclohexene intermediate containing three contiguous quaternary carbon atoms (S. Danishefsky, 1979). This reaction was assisted by large polarity differences between the electron rich diene and the electron deficient ene component.

Several substituted cyclohexane derivatives may also be obtained by the reduction of a benzenoid precursor. Partial reduction of resorcinol, for example, and subsequent methylation yields 2-methylcyclohexane-1,3-dione, which is frequently used in steroid synthesis (M.S. Newman, 1960; see also p. 64f.). From lithium-ammonia reduction of alkoxybenzenes 1-alkoxy-1,4-cyclohexadienes are obtained (E.J. Corey, 1968 D).

Intramolecular condensation reactions to generate six-membered carbocycles are mentioned in section 1.12, the polyene cyclization in section 1.15.

1.14 1,6-Difunctional Compounds

Cyclohexene derivatives can be oxidatively cleaved under mild conditions to give 1,6-dicarbonyl compounds. The synthetic importance of the Diels-Alder reaction described above originates to some extent from this fact, and therefore this oxidation reaction is discussed in this part of the book.

The most common procedure is ozonolysis at -78 °C (P.S. Bailey, 1978) in methanol or methylene chloride in the presence of dimethyl sulfide or pyridine, which reduce the intermediate ozonides to aldehydes. Unsubstituted cyclohexene derivatives give 1,6-dialdehydes, enol ethers or esters yield carboxylic acid derivatives. Oxygen-substituted $C=C$ bonds in cyclohexene derivatives, which may also be obtained by Birch reduction of alkoxyarenes (see p. 95ff.), are often more rapidly oxidized than non-substituted bonds (E.J. Corey, 1968 D; G. Stork, 1968 A,B). Catechol derivatives may also be directly cleaved to afford conjugated hexadienedioic acid derivatives (R.B. Woodward, 1963). Highly regioselective cleavage of the more electron-rich double bond is achieved in the ozonization of dienes (W. Knöll, 1975).

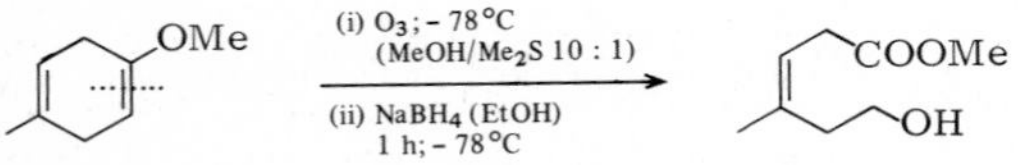

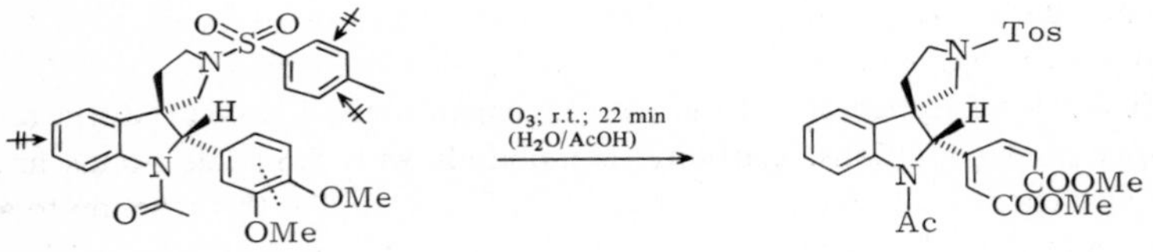

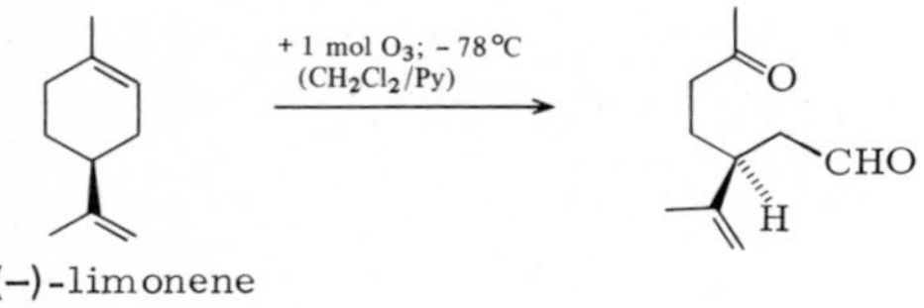

Another method to achieve selectivity in oxidative splitting of $C=C$ bonds to carbonyl groups is controlled epoxidation followed by periodate cleavage (J.P. Nagarkatti, 1973).

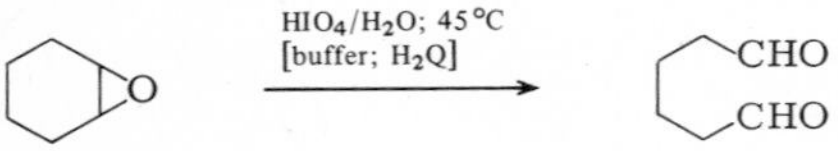

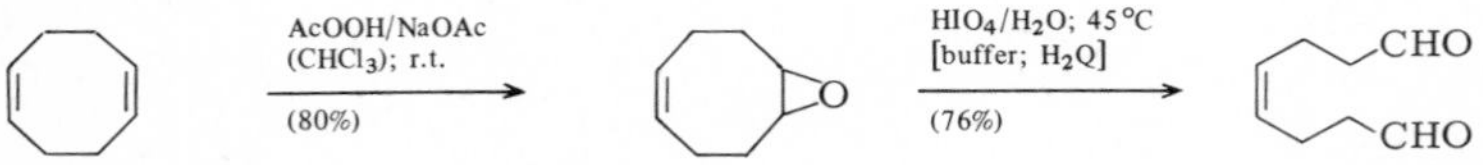

6-Keto acids are obtained by acylation of cyclopentanone enamines (see p. 13f.) with acid chlorides and subsequent base-catalyzed *retro*-aldol cleavage (S. Hünig, 1960).

Most of the synthetic reactions leading to substituted carbon compounds can be reversed. *Retro*-aldol or *retro*-Diels-Alder reactions, for example, are frequently used in the degradative fragmentation of complex molecules to give simpler fragments. In synthesis, such procedures are of limited value. There are, however, some noteworthy fragmentations of easily accessible polycycles with *n* rings, and those lead to useful products possessing *n*-1 rings, but the same total number of carbon atoms.

A 2-cyclohexenone derivative can be transformed into the corresponding epoxy tosylhydrazone by sequential treatment with peracid and tosylhydrazine. The elimination of nitrogen and *p*-toluenesulfinate and fragmentation after rearrangement to the 3-tosylazo allylic alcohol may occur under mild conditions. Carbonyl compounds with 5,6-triple bonds are formed in high yields (J. Schreiber, 1967; M. Tanabe, 1967). If one applies this reaction to a 9,10-epoxy-1-decalone, a ten-membered 5-cyclodecyn-1-one ring is formed (D. Felix, 1971). This product is an important intermediate in the perfume industry and has been used on a large scale. For this purpose Eschenmoser developed a synthesis in which the readily removed styrene was split off instead of a sulfinic acid. Thus a 1-amino-2-phenylaziridine hydrazone was used instead of a tosylhydrazone (D. Felix, 1968).

Similar fragmentations to produce 5-cyclodecen-1-ones and 1,6-cyclodecadienes have employed 1-tosyloxy-4a-decalols and 5-mesyloxy-1-decalyl boranes as educts. The ring-fusing carbon-carbon bond was smoothly cleaved and new π-bonds were thereby formed in the macrocycle (P.S. Wharton, 1965; J.A. Marshall, 1966). The mechanism of these reactions is probably E2, and the positions of the leaving groups determine the stereochemistry of the olefinic product.

four-bond fragmentation

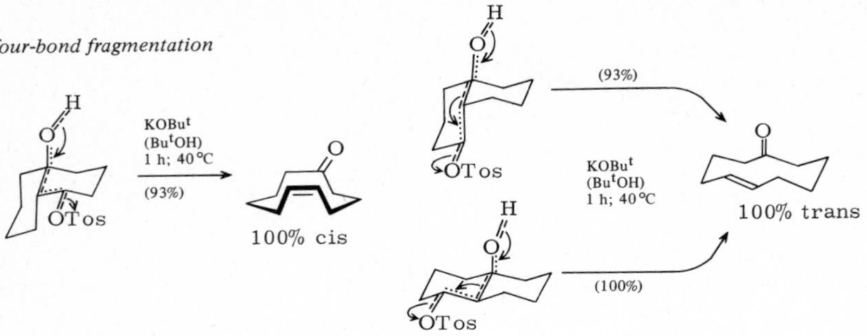

100% cis

(93%)

(100%)

100% trans

five-bond fragmentation

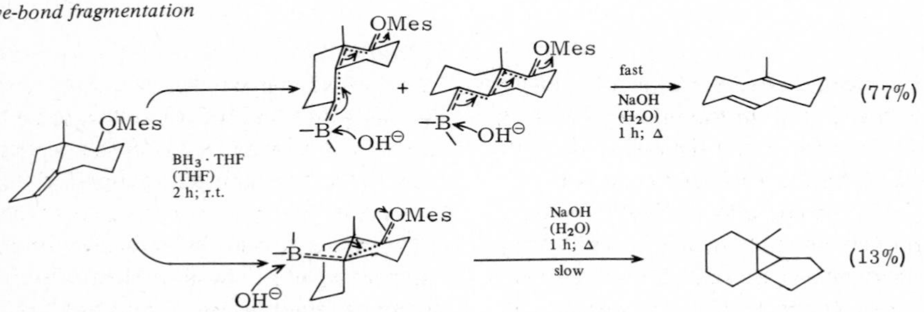

(77%)

(13%)

All these fragmentation reactions occur readily, when the C—X bond and the breaking C—C bond have the *trans* antiparallel arrangement. Olefin formation can proceed along a concerted pathway. The stereochemistry of the olefinic product is determined by the orientation of groups in the cyclic precursor. If the *trans* antiparallel arrangement of the breaking bonds has been attained, the relative orientation of the hydrogen atoms in the precursor is retained in the olefin.

Conventional synthetic schemes to produce 1,6-disubstituted products, e.g. reaction of a^3- with d^3-synthons, are largely unsuccessful.

1.15 Acid Catalyzed Cyclizations

Open-chain 1,5-polyenes (e.g. squalene) and some oxygenated derivatives are the biochemical precursors of cyclic terpenoids (e.g. steroids, carotenoids). The enzymic cyclization of squalene 2,3-oxide, which has one chiral carbon atom, to produce lanosterol introduces seven chiral centres in one totally stereoselective reaction. As a result, organic chemists have tried to ascertain, whether squalene or related olefinic systems could be induced to undergo similar stereoselective cyclizations in the absence of enzymes (W.S. Johnson, 1968, 1976).

A simple acid-catalyzed cyclization transforms ψ-ionone into α-ionone (W. Kimel, 1957, 1958). Further treatment with protic acids transforms the α-ionone to the thermodynamically more stable β-ionone.

pseudoionone
(ψ-ionone)

α-ionone

β-ionone

Lewis acid: (i) BF$_3$/C$_6$H$_6$; 10 min; 0→10 °C
(ii) neutralization: dil. NaOH; < 10 °C

92 : 8   (70%)   (6%)

protic acid: H$_2$SO$_4$/AcOH 7 : 3; 20 min; 10 °C
+ 20 min; r. t.

10 : 90   (7%)   (65%)

$-A = -\overset{\ominus}{B}F_3, -H$

The achiral triene chain of (*all-trans-*)-3-demethyl-farnesic ester as well as its (6-*cis*-)-isomer cyclize in the presence of acids to give the decalol derivative with four chiral centres whose relative configuration is well defined (P.A. Stadler, 1957; A. Eschenmoser, 1959; W.S. Johnson, 1968, 1976). A monocyclic diene is formed as an intermediate (G. Stork, 1955). With more complicated 1,5-polyenes, such as squalene, oily mixtures of various cyclization products are obtained. The 18,19-glycol of squalene 2,3-oxide, however, cyclized in modest yield with picric acid catalysis to give a complex tetracyclic natural product with nine chiral centres. Picric acid acts as a protic acid of medium strength whose conjugated base is non-nucleophilic. Such acids activate oxygen functions selectively (K.B. Sharpless, 1970).

The early Eschenmoser-Stork results indicated, that stereoselective cyclizations may be achieved, if monocyclic olefins with 1,5-polyene side chains are used as substrates in acid treatment. This assumption has now been justified by many syntheses of polycyclic systems. A typical example synthesis is given with the last reaction. The cyclization of a trideca-3,7-dien-11-ynyl cyclopentenol leads in 70% yield to a 17-acetyl A-norsteroid with correct stereochemistry at all ring junctions. Ozonolysis of ring A and aldol condensation gave *dl*-progesterone (M.B. Gravestock, 1978; see p. 252f.).

H$_2$SO$_4$/
HCOOH/
H$_2$O
6 h; r.t.

(60-70%)

2, 4, 6-trinitrophenol
(MeNO$_2$); 1 d; r.t.

(erythro-18,19-glycol)

7% dl-malabaricanediol
7% dl-18,19-epimalabaricanediol

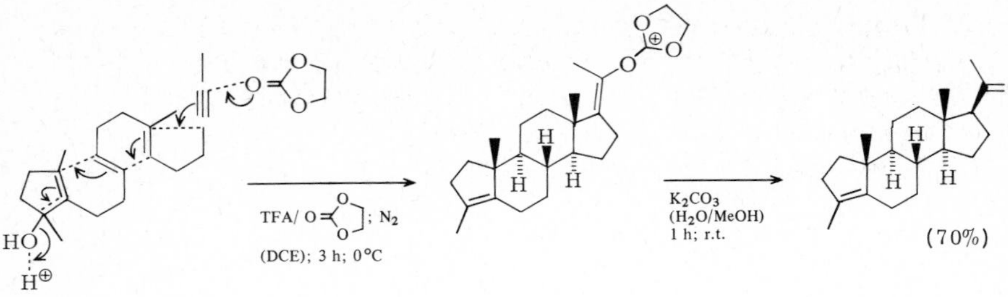

1.16 Bridged Carbocycles

In polycyclic hydrocarbon skeletons the adjacent rings may not only have one common bond ("fused ring systems"). Bridges, which consist of one or more carbon atoms, may connect two tertiary ring carbon atoms which are termed bridgeheads. Bridged structures are found in many natural products from plants (terpenes, alkaloids) and transpose the three-dimensionality of the carbon tetrahedron into larger structures. Whereas fused cyclic systems are more or less planar, often with kinks from *cis*-linked rings, bridged systems may approach the space filling of a sphere or a hemisphere.

The syntheses of bridged carbon compounds start from monocyclic or fused polycyclic compounds with appropriate donor and acceptor centres. We start again with the most simple and straightforward synthetic reaction. This is a Diels-Alder type combination of dienes and alkenes in which the diene or both components are cyclic. As with fused ring systems (see p. 78 f.) this reaction is again stereoselective. The double bond of the cyclohexene boat, which is formed in the reaction, and the substituents on the carbon atoms of the educt olefin lie on the same side (= *endo*) in the major product. Sterically strained cyclopropene rings are particularly reactive as "ene"-components (M.L. Deem, 1972). With electron-deficient dienes reactions are still selective but they give low yields and require high temperatures (H.E. Zimmerman, 1969A; R.C. Cookson, 1956). Benzyne (G. Wittig, 1956; O.L. Chapman, 1973, 1975; R.H. Levin, 1978; R.W. Hoffmann, 1967) has also been used as "ene"-component. Examples of intramolecular Diels-Alder reactions are given on pp. 253f., 269, and 298.

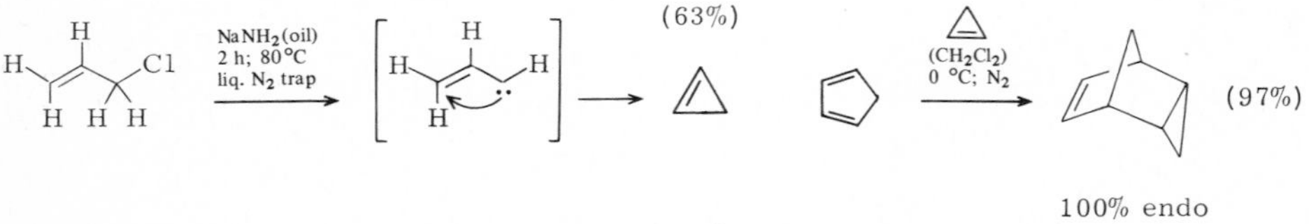

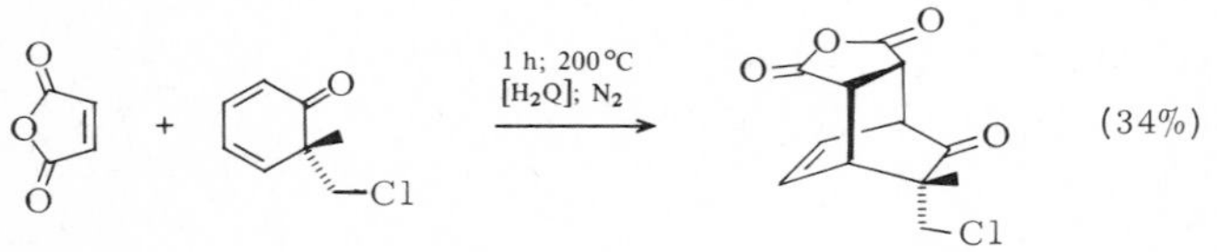

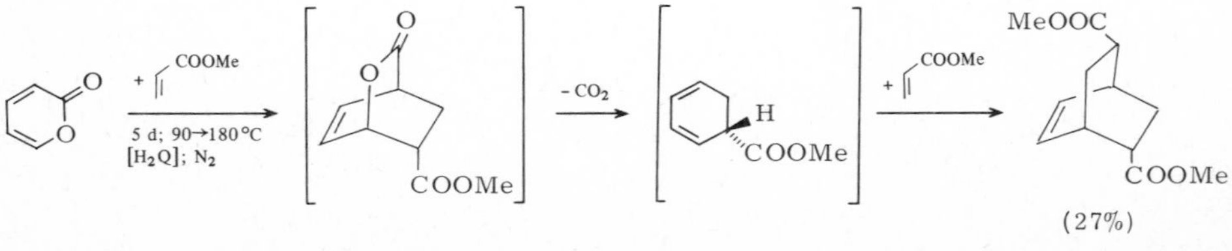

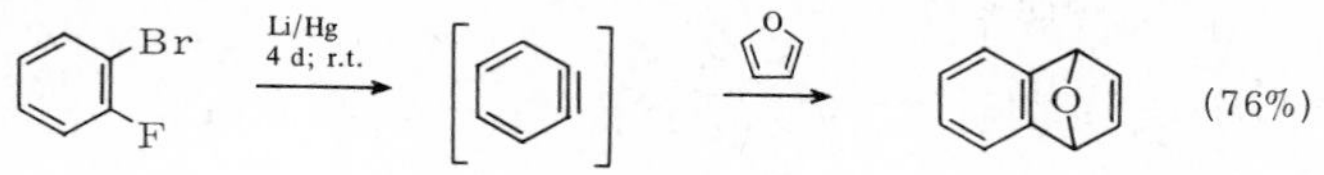

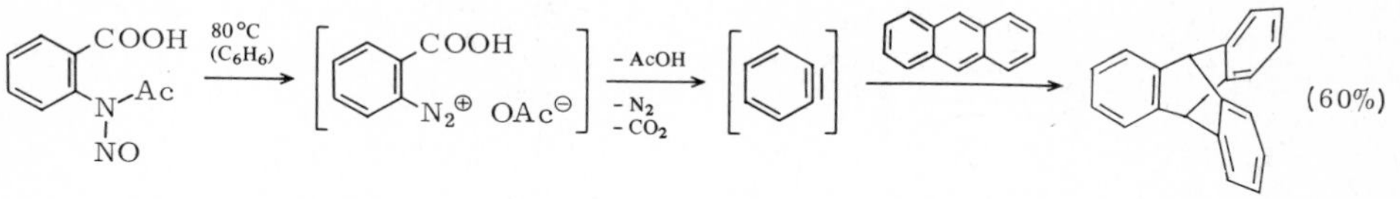

(76%)

(60%)

Intramolecular reactions between donor and acceptor centres in fused ring systems provide a general route to bridged polycyclic systems. The *cis*-decalone mesylate given below contains two d^2-centres adjacent to the carbonyl function and one a^1-centre. Treatment of this compound with base leads to reversible to enolate formation, and the C-3 carbanion substitutes the mesylate on C-7 (J. Gauthier, 1967; A. Bélanger, 1968).

In an intramolecular aldol condensation of a diketone many products are conceivable, since four different enols can be made. Five- and six-membered rings, however, will be formed preferentially. Kinetic or thermodynamic control or different acid-base catalysts may also induce selectivity. In the Lewis acid-catalyzed aldol condensation given below, the more substituted enol is formed preferentially (E.J. Corey, 1963 B, 1965B).

Another synthesis of a bridged hydrocarbon takes advantage of high electron release from the *para*-position of phenolate anions, which may be used to transform the phenol moiety into a substituted cross-conjugated cyclohexadienone system (S. Masamune, 1961, 1964).

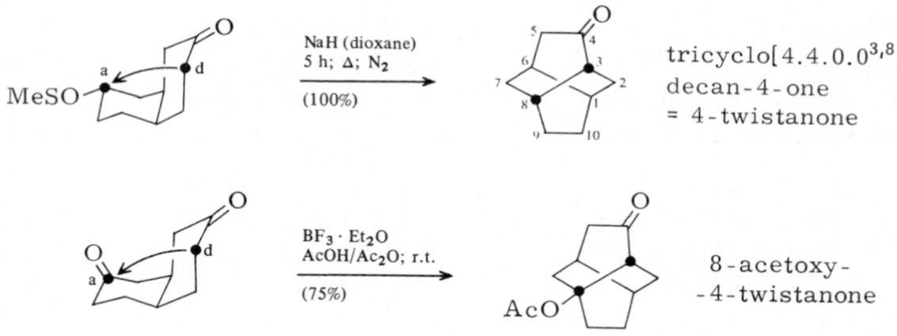

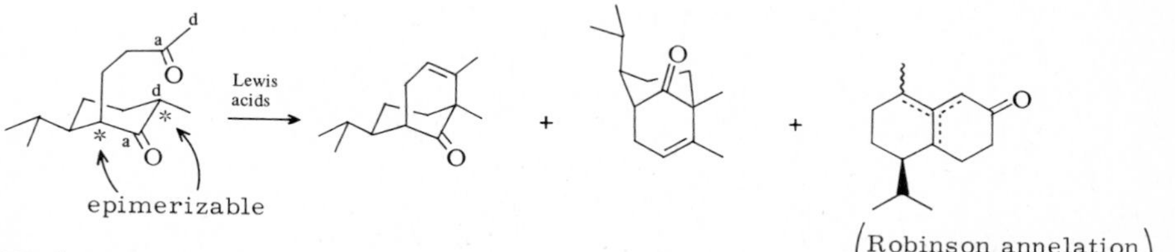

tricyclo[4.4.0.0^{3,8}]-
decan-4-one
= 4-twistanone

8-acetoxy-
-4-twistanone

epimerizable

BF$_3$(CH$_2$Cl$_2$); 16 h; r. t. :   (32%)   (8%)
SnCl$_4$(C$_6$H$_6$); 45 min; Δ :   (28%)   (28%)

(Robinson annelation products)

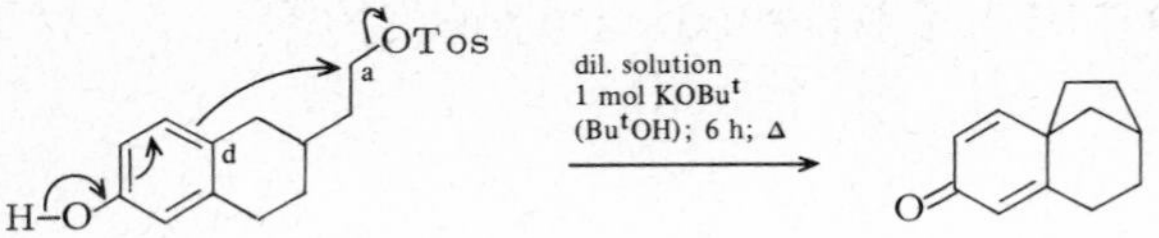

(90%)

As final examples, the intramolecular cyclopropane formation from cycloolefins with diazo groups (S.D. Burke, 1979), intramolecular cyclobutane formation by photochemical cycloaddition (p. 72, 269f., 295f.), and intramolecular Diels-Alder reactions (p. 139f., 298) are mentioned. The application of these three cycloaddition reactions has led to an enormous variety of exotic polycycles (E.J. Corey, 1967A).

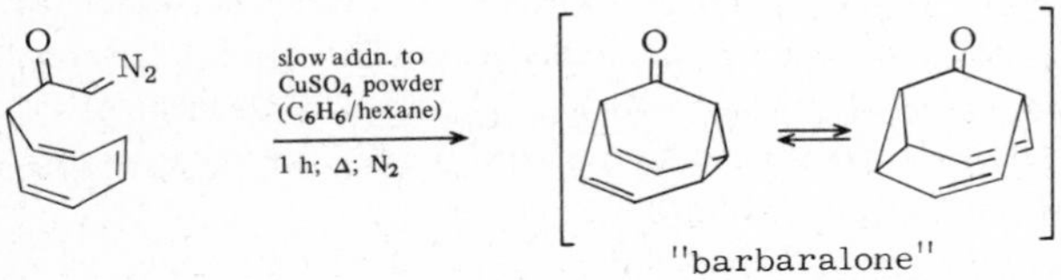

"barbaralone"   (30%)

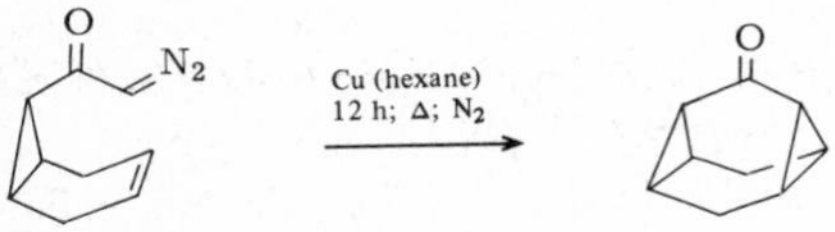

"triasteranone"   (73%)

2 Selective Functional Group Interconversions (FGI)

Introduction and Summary

- *Functional group interconversions (FGI) are mainly employed to achieve one of the following synthetic objectives:*
    - *change of the oxidation number of specific carbon atoms (e.g. reduction of carbonyl compounds, oxidation of alcohols)*
    - *introduction, removal, or substitution of hetero atoms (e.g. bromination or hydroxylation; olefins from 1,2-diols; exchange of nitrogen for oxygen)*
    - *connection of monomers or cyclization by formation of new $C—X—C$ or $C—X=C$ bonds (e.g. esterification, amidation, formation of Schiff bases and heterocycles)*
    - *reversible protection of reactive sites to allow regioselective synthetic reactions (e.g. peptide and nucleic acid syntheses).*

- *Regioselectivity in FGIs is dominated by the pattern of substituents and by steric effects in the substrate as well as by the choice of appropriate reagents.*

    Examples:

    - *for steric effects: aluminum hydrides bind to double bonds of internal olefins much slower than to terminal ones. Tetrasubstituted double bonds cannot be hydrogenated catalytically.*
    - *for electronic effects: MnO_2 oxidizes allylic alcohols, but does usually not affect saturated alcohols.*

- *Functional group selectivity is often easy to achieve in reduction and condensation reactions since several highly selective reagents for reduction and for protection of functional groups are available.*

    Examples:

    Metal hydrides reduce preferably polar double bonds, whereas catalytic hydrogenation is somewhat selective for non-polar double bonds. Selective protection of amino groups in amino acids.

- *The enantioselective introduction of chiral centres into an achiral molecule can nowadays be achieved most easily using chiral hydrides.*

    Example:

    Wilkinson type catalysts.

- *Heterocycle syntheses are often possible from difunctional open-chain precursors, including olefins as 1,2-difunctional reagents, and an appropiate nucleophile or electrophile containing one or more hetero atoms. The choice of the open-chain precursor is usually dictated by the longest carbon chain within the heterocycle to be synthesized.*

Examples:

Pyrroles from 1,4-dicarbonyl compounds and ammonia; isoxazolines from olefins and nitrile oxides.

Construction of a carbon skeleton seldom yields the target molecule directly. Almost inevitably some of the functional groups are in the wrong oxidation state, contain the wrong hetero atoms, or are undesired derivatives. The functional groups have then to be manipulated. Since most target (or intermediate) molecules contain several functional groups, these interconversions have to be selective. Selectivity may arise from the specific electronic structure of the functional group (e.g. C=C double bonds are more nucleophilic than C=O double bonds) or from its steric environment (e.g. a sterically hindered ketone does not react with bulky boranes). In this part of the book selected FGIs which are synthetically "realistic" will be discussed. "Realistic" means that good control of single reactions in complicated molecules has been achieved.

2.1 Reduction

Of all synthetic operations hydrogenation of organic molecules is probably the most highly developed. Knowledge of selectivities is comparably far advanced.

Common reducing agents are hydrogen in the presence of metallic or complex catalysts (e.g. Ni, Pd, Pt, Ru, Rh), hydrides (e.g. alanes, boranes, $LiAlH_4$, $NaBH_4$), reducing metals (e.g. Li, Na, Mg, Ca, Zn), and low-valent compounds of nitrogen (e.g. N_2H_4, N_2H_2), phosphorus (e.g. triethyl phosphite, triphenylphosphine), and sulfur (e.g. $HO\text{-}CH_2\text{-}SO_2Na$ = SFS, sodium dithionite = $Na_2S_2O_4$).

Catalytic hydrogenation is mostly used to convert C≡C triple bonds into C=C double bonds and alkenes into alkanes or to replace allylic or benzylic hetero atoms by hydrogen (H. Kropf, 1980). Simple theory postulates *cis-* or *syn-*addition of hydrogen to the C—C triple or double bond with heterogeneous (R. L. Augustine, 1965, 1968, 1976; P. N. Rylander, 1979) and homogeneous (A. J. Birch, 1976) catalysts. Sulfur functions can be removed with reducing metals, e.g. with Raney nickel (G. R. Pettit, 1962 A). Heteroaromatic systems may be reduced with the aid of ruthenium on carbon.

Hydrides are available in many molecular sizes and possessing different reactivities. *$LiAlH_4$* reduces most unsaturated groups except alkenes and alkynes. *$NaBH_4$* is less reactive and reduces only aldehydes and ketones, but usually no carboxylic acids or esters (N.G. Gaylord, 1956; A. Hajós, 1979).

The conversion of carboxylic acid derivatives (halides, esters and lactones, tertiary amides and lactams, nitriles) into aldehydes can be achieved with bulky *aluminum hydrides* (e.g. DIBAL = diisobutylaluminum hydride, lithium trialkoxyalanates). Simple addition of three equivalents of an alcohol to $LiAlH_4$ in THF solution produces those deactivated and selective reagents, e.g. lithium triisopropoxyalanate, $LiAlH(OPr^i)_3$ (J. Malek, 1972).

Diborane or alkylboranes are used for reduction of alkenes and alkynes via hydroboration (see p. 9) followed by hydrolysis of the borane with acetic acid (H.C. Brown, 1975).

Halides and tosylates are reduced by *LiAlH₄*, if S_N2 displacement is easy (N.G. Gaylord, 1956; A. Hajós, 1966, 1979; W.G. Brown, 1951). *Tin hydrides* are especially reactive to these substrates (e.g. Bu₃SnH; W.T. Brady, 1970).

The use of *reducing metals* nowadays is mainly restricted to acyloin and pinacol coupling reactions (see p. 50f.) and Birch reductions of arenes (A.A. Akhrem, 1972; see p. 95f.) and activated C—C multiple bonds (see p. 95ff.).

Low-valent nitrogen and phosphorus compounds are used to remove hetero atoms from organic compounds. Important examples are the Wolff-Kishner type reduction of ketones to hydrocarbons (R.L. Augustine, 1968; D. Todd, 1948; R.O. Hutchins, 1973B) and Barton's olefin synthesis (p. 34f.) both using *hydrazine derivatives*.

Table 1 gives a broad summary of the reactions of the common classes of reducing agents. In the following sections some typical examples of synthetically useful reductions (in the educt order given on the table) together with some more sophisticated methods of stereoselective hydrogenations will be discussed.

Table 1. Reactivity of reducing agents towards functional groups (adapted from: J. B. Hendrickson, D. J. Cram, and G. S. Hammond, Organic Chemistry, 3rd ed., McGraw-Hill *1970*, with modifications).

educt	product	H₂ with catalyst	NaBH₄	LiAlH₄	AlH(OR)₃⁻ or R₂AlH	B₂H₆ or (R₂BH)₂	Li, Na	other reagents
$\begin{matrix} A & C \\ B & D \end{matrix}$ (alkene)	$\begin{matrix} A & C \\ H & H \\ B & D \end{matrix}$	+++ S	–	–	+++ S	+++ S	(+)	N₂H₂ = diimine (extremely mild) chiral Wilkinson type catalysts S
A—≡—B	$\begin{matrix} A & B \\ H & H \end{matrix}$	++– S	–	–	++– S	++– S	–	
	$\begin{matrix} A & H \\ H & B \end{matrix}$	–	–	–	–	–	+++ S	
⬡—OR	⬡—OR	–	–	–	–	–	+++ S	
R–CH₂–X (i) R₂CH–X (i)	R–CH₃ R₂CH₂	+++	–	+++	–	–	+++	Bu₃SnH
R₃C–X (i) Ar–X (i)	R₃CH Ar–H	+++	–	–	–	–	+++	Bu₃SnH
ROH, ROR	R–H	–	–	–	–	–	–	convert to olefin or tosylate

Table 1. (Continued)

educt	product	H_2 with catalyst	$NaBH_4$	$LiAlH_4$	$AlH(OR)_3^{\ominus}$ or R_2AlH	B_2H_6 or $(R_2BH)_2$	Li, Na	other reagents
Ar–C–Y (ii)	Ar–C–H	+++	–	–	–	–	+++	
>C–C< (O)	H C–C OH	+++	–	+++	(+)	+++	+++	
R–SH, R–S–R	R–H	+++ Raney Ni	–	–	–	–	+++	Raney nickel without H_2
$R–NO_2$	$R–NH_2$	+++	–	(+)(iii)	–	–	+++	Sn^{2+}, Ti^{2+} etc.
R–CHO, $R_2C=O$	$R–CH_3$, R_2CH_2	(+–)	(+–)	(+–)	(+–)	(+–)	(+–)	(i) N_2H_4 (ii) KOBut/DMSO o[r] (i) TosNHNH₂ (ii) $NaBH_4$ or (i) HSC_2H_4SH (ii) Raney Ni or Clemmensen red.
	$R–CH_2OH$, $R_2CH–OH$	(+)	+++ S	+++	+++ S	+++	+++	
$R_2C=NOH$	$R_2CH–NH_2$	(+)	–	+++	–	–	+++	Sn, Zn, Ti^{2+}
R–COOH	$R–CH_2OH$	–	–	+++	–	+++	–	
R–COOR'	$R–CH_2OH$	(+)	–	+++	–	(+)	(+)	
R–COCl	R–CHO	(+–)	(+–)	(+–)	+++	–	–	Bu_3SnH, $Na_2Fe(CO)_4$
$R–CONR'_2$	R–CHO	–	+++	(+–)	+++	–	–	DIBAL
$R–CONR'_2$	$R–CH_2–NR'_2$	–	–	+++	–	+++	–	$NaBH_4 + CoCl_2$ or RCOOH (i) $Et_3O^+BF_4^-$ (ii) $NaBH_4$
R–C≡N	$R–CH_2NH_2$	+++	–	+++	+++	+++	+++	DIBAL

+++ Synthetically useful reaction
++– Rate of secondary reactions can be kept comparatively small
(+) Slow reaction or complex product mixture
(+–) Reaction does not stop at or does not reach the desired oxidation state
– No reaction

S Regio- and/or stereoselectivity has been achieve[d] in syntheses of complex molecules
(i) X = Cl, Br, I, OMes, OTos
(ii) Y = OH, OR, NR_2
(iii) complex mixtures, if R = aryl

If it is necessary to reduce one group in a given molecule without affecting any other unprotected reducible group, the following reactivity orders for "ease of reduction" toward catalytic hydrogenation, LiAlH₄, and diborane may serve as a guideline.

reactivity	catalytic hydrogenation	complex hydrides	boranes
high	$-C\equiv C-$ $>C=C<$ $-COCl$ $-C\equiv N$	$-COCl$ $-CHO$ $>C=O$ $>C=N\diagdown$	$-C\equiv C-$ $>C=C<$ $-COOH$ $-CONR_2$ $-C\equiv N$
medium	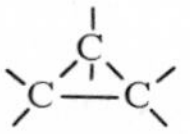 $-CHO$ $>C=O$ $-NO_2$ $Ar\!-\!\overset{\mid}{\underset{\mid}{C}}\!-\!OR$ 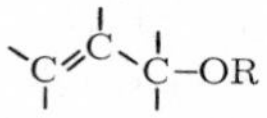 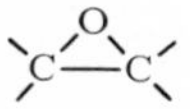 $\diagdown C\!\!\overset{}{\diagdown}\!\overset{\mid}{C}\!-\!OR$ (epoxide, $C\!-\!O\!-\!C$) $>C\!-\!X$ (aziridine, $C\!-\!N\!-\!C$) heteroarenes	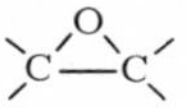 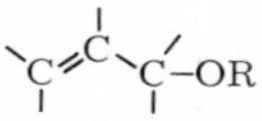 (anhydride, $\diagup C\!-\!O\!-\!C\diagdown$) $\diagup C\!\overset{O}{\diagdown}\!\overset{\mid}{C}\!-\!OR$ $Ar\!-\!\overset{\mid}{\underset{\mid}{C}}\!-\!OR$ $-COOR$ $-CONR_2$ $-C\equiv N$ $>C\!-\!X$ $-NO_2$ $-COO^{\ominus}$	$-CHO$ $>C=O$ $-CONR_2$ $-C\equiv N$ 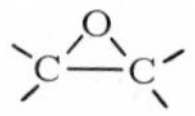 (epoxide, $C\!-\!O\!-\!C$)
low	$-COOR$ $-CONR_2$ arenes	pyridines	$-COOR$ heteroarenes
very low	$-COOH$ $-COO^{\ominus}$	$-C\equiv C-$ $>C=C<$ arenes	arenes

2.1.1 Hydrogenation of Carbon-Carbon Multiple Bonds and Cyclopropane Rings

The carbon-carbon triple bond is the most readily hydrogenatable functional group with several reagents. Its full or partial hydrogenation can take place in the presence of nearly all other functionalities. Most important is its selective conversion into *cis*-double bonds. The *cis*-hydrogenation is readily accomplished using diisobutylaluminum hydride (DIBAL; E. Winterfeldt, 1975; W.J. Gensler, 1963) or hydrogen (B.W. Baker, 1955; F. Sondheimer, 1962) with palladium catalysts (e.g. 1-2% Pd on $BaSO_4$ or $CaCO_3$, poisoned by quinoline and/or lead(II) acetate; H. Lindlar, 1973).

The less hindered *trans*-olefins may be obtained by reduction with lithium or sodium metal in liquid ammonia or amine solvents (Birch reduction). This reagent, however, attacks most polar functional groups (except for carboxylic acids; R.E.A. Dear, 1963; J. Fried, 1968), and their protection is necessary (see section 2.6).

Terminal alkynes are only reduced in the presence of proton donors, e.g. ammonium sulfate, because the acetylide anion does not take up further electrons. If, however, an internal $C{\equiv}C$ triple bond is to be hydrogenated without any reduction of terminal, it is advisable to add sodium amide to the alkyne solution first. On catalytic hydrogenation the less hindered triple bonds are reduced first (N.A. Dobson, 1955, 1961).

The reduction of medium-size cycloalkynes, however, always yields considerable amounts of the less strained *cis*-cycloalkenes (A.C. Cope, 1960 A; M. Svoboda, 1965). Cyclodecyne, for example, is reduced almost exclusively to *cis*-cyclodecene.

In polycyclic systems the Birch reduction of $C{=}C$ double bonds is also highly stereoselective, e.g. in the synthesis of the thermodynamically favored *trans*-fused steroidal skeletons (see p. 96 and p. 251).

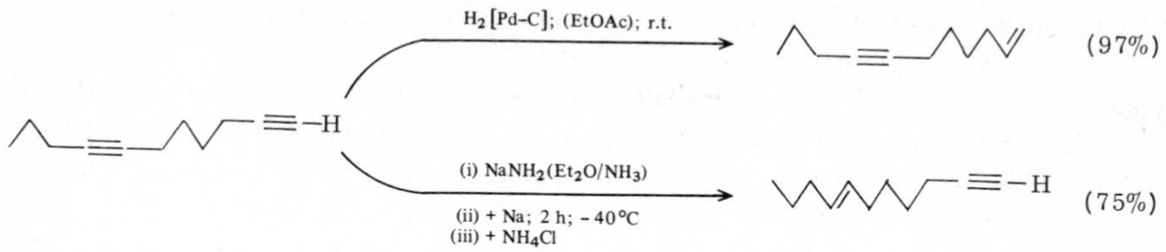

H$_2$ [Pd–C]; (EtOAc); r.t. ⟶ (97%)

(i) NaNH$_2$ (Et$_2$O/NH$_3$)
(ii) + Na; 2 h; –40 °C
(iii) + NH$_4$Cl ⟶ (75%)

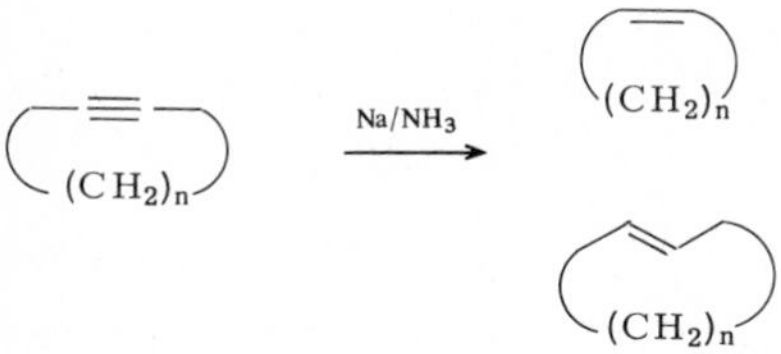

	n = 7	n = 8	n = 9	n = 10
	19%	94%	47%	9%
	71%	2%	53%	38%

Na/NH$_3$ ⟶

Carbon-carbon double bonds are usually reduced using hydrogen and a heterogeneous catalyst. The activity of hydrogenation catalysts decreases in the order Pd > Rh > Pt > Ni > Ru. Catalysts other than Pd are especially chosen to minimize migration of hydrogen, e.g. if one wants to deuterate a C=C double bond (Ni, Ru), or if hydrogenolysis of sensitive groups is to be prevented (Rh). The ease of hydrogenation is inversely proportional to the number and size of substituents at the C—C multiple bond (W.F. Newhall, 1958). Hydrogenation of tetrasubstituted double bonds is strongly retarded and occurs only, if the double bond shifts to a less hindered position in the presence of catalysts. It may also happen, however, that a di- or trisubstituted double bond isomerizes to a tetrasubstituted double bond, which resists to reduction. Such isomerizations are particularly fast in the presence of protic acids (D.H.R. Barton, 1956).

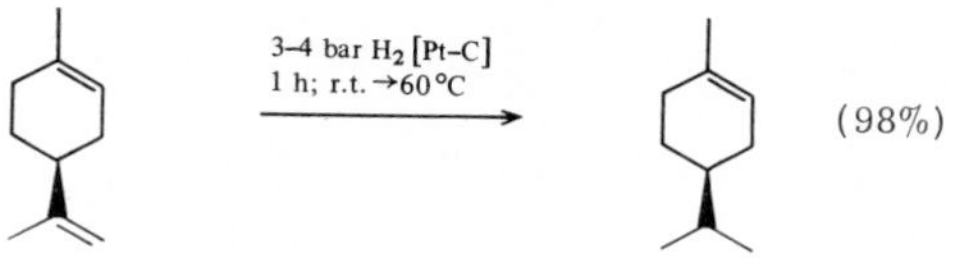

3–4 bar H$_2$ [Pt–C]
1 h; r.t. →60 °C ⟶ (98%)

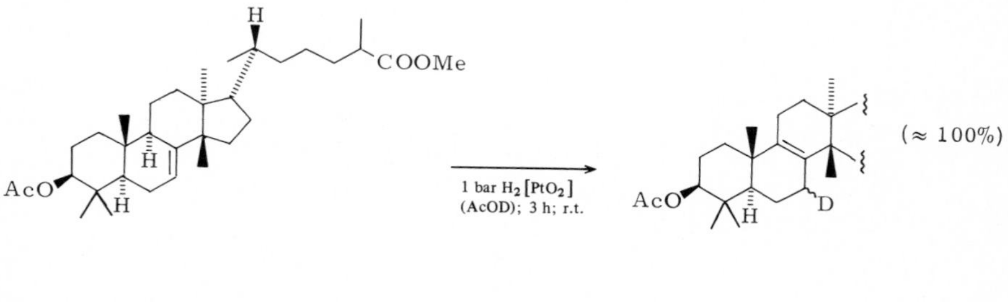

1 bar H$_2$ [PtO$_2$]
(AcOD); 3 h; r.t. ⟶ (≈ 100%)

The commonly accepted mechanism of heterogeneously catalyzed hydrogenation involves activation of both the hydrogen and the C—C multiple bond adsorbed on the metal surface. First one hydrogen atom is transferred to the least hindered position of the multiple bond to give a half-hydrogenated adsorbed species. This reaction is fully reversible and accounts for double-bond shifts and *cis-trans* isomerizations as well as for hydrogen scrambling, when deuterium is used for reduction. Under conditions of high hydrogen availability (high pressure, rapid agitation) another hydrogen atom adds to the same side of the half-hydrogenated bond as the first (R.L. Augustine, 1976). If the hydrogenation proceeds more slowly, this *syn* or *cis* selectivity is largely lost.

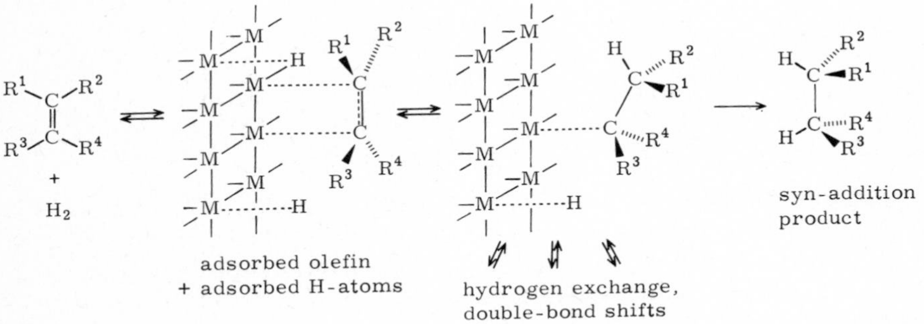

adsorbed olefin
+ adsorbed H-atoms

hydrogen exchange,
double-bond shifts

syn-addition
product

Considering the properties of the substrates, the highest stereoselectivities in *cis*-hydrogenations are observed with chiral cycloalkenes of rigid conformation. The hydrogenation of the tricyclic system given below, for example, led selectively to the α-hydrogenated *trans*-fused product because the aromatic ring keeps the methyl groups rigidly in the axial position (G. Stork, 1962).

Cis-olefins may also be obtained very selectively by reduction with diimine (N_2H_2, S. Hünig, 1965; C.E. Miller, 1965). The reagent can be used at low temperatures and has been employed in the selective reduction of C=C double bonds, e.g. in the presence of a sensitive peroxidic function (W. Adam, 1978).

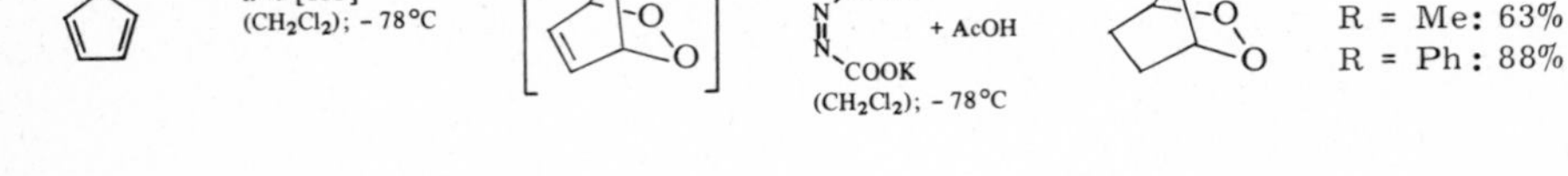

Asymmetric hydrogenation has been achieved with dissolved Wilkinson type catalysts (A.J. Birch, 1976). If, for example, the chiral ligand (—)-(R,R)-1,2-ethanediylbis[(2-methoxyphenyl)phenylphosphine] is reacted with a 1,5-cyclooctadiene-rhodium(I) complex, a chiral rhodium(I) complex forms, which catalyzes hydrogenation via a chiral rhodium hydride. The polar C=C double bonds of α-acylamino acrylic acids were hydrogenated in excellent optical yields with this catalyst. Wilkinson catalysts are also useful for regioselective hydrogenations of less substituted C=C double bonds. They are much more discriminating for stereochemical hindrance than are heterogeneous catalysts (W.S. Knowles, 1975; B.D. Vineyard, 1977; D. Valentine, Jr., 1978).

The Birch reductions of C=C double bonds with alkali metals in liquid ammonia or amines obey other rules than do the catalytic hydrogenations (D. Caine, 1976). In these reactions regio- and stereoselectivities are mainly determined by the stabilities of the intermediate carbanions. If one reduces, for example, the α,β-unsaturated decalone below with lithium, a dianion is formed, whereof three different conformations (A), (B), and (C) are conceivable. Conformation (A) is the most stable, because repulsion disfavors the *cis*-decalin system (B) and in (C) the conjugation of the dianion is interrupted. Thus, protonation yields the *trans*-decalone system (G. Stork, 1964B).

Similar rules hold true for the Birch reduction of substituted benzene rings to give 1,4-dihydro derivatives (A.A. Akhrem, 1972; A.J. Birch, 1972). In the first step an anion radical is formed, which is selectively protonated at the site of highest electron density. The resulting pentadienyl radical is further reduced to the corresponding anion and selectively protonated *para* to the first proton. Alkoxy- and dialkylaminosubstituted benzene rings are reduced in this way to produce the corresponding 2,5- or 3,6-dihydro derivatives, whereas benzoic acid derivatives are hydrogenated at 1,4-positions. Enol ethers from alkoxyarenes may be converted to ketones by acidic hydrolysis.

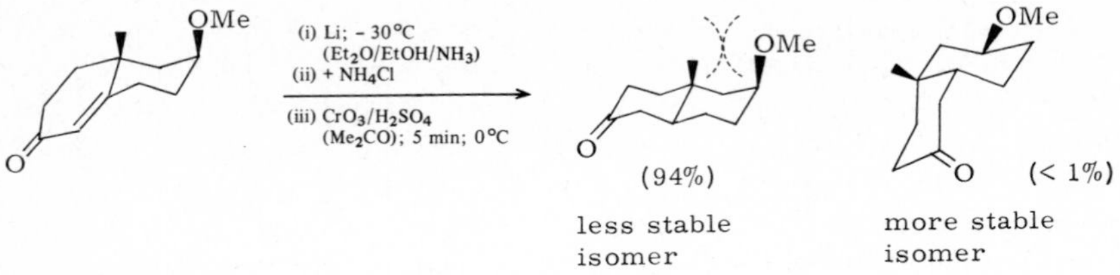

(94%)   less stable isomer

(< 1%)   more stable isomer

transition states:

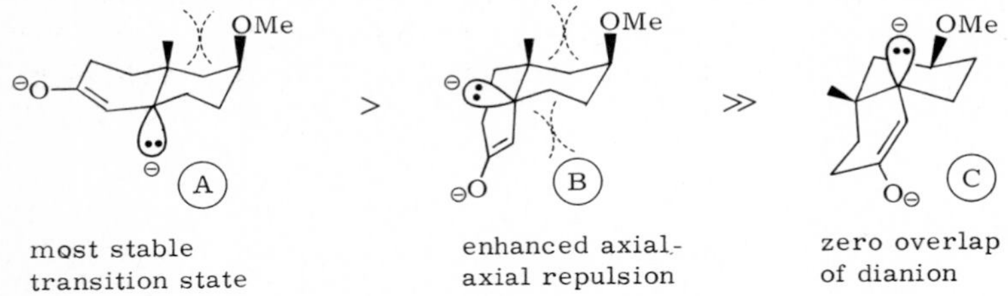

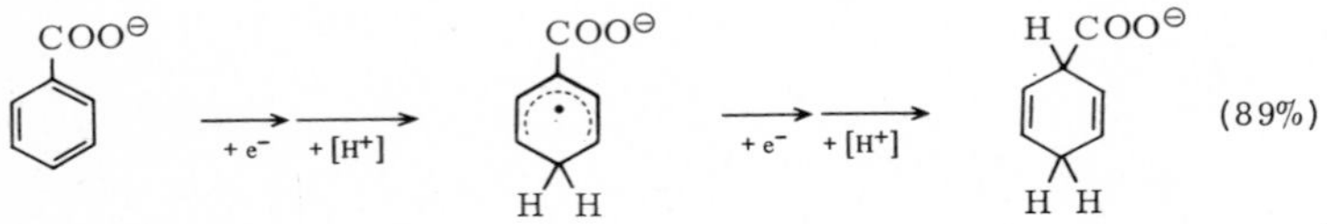

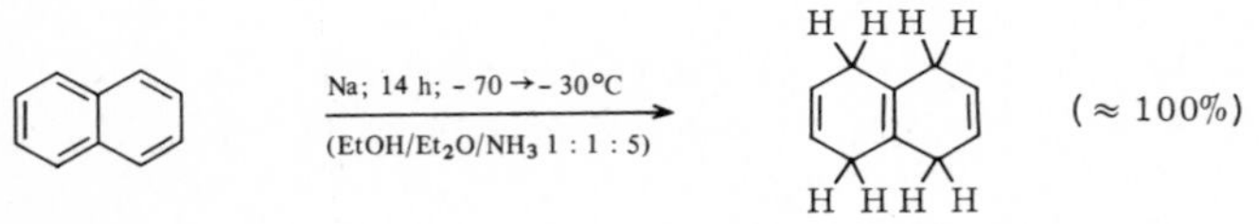

A > B ≫ C

most stable transition state

enhanced axial-axial repulsion

zero overlap of dianion

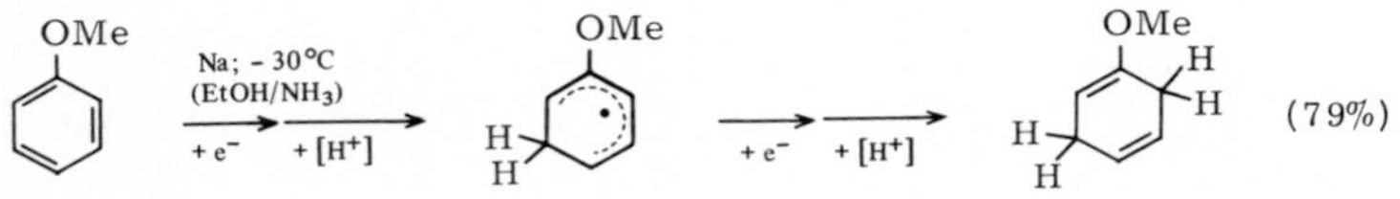

(79%)

(89%)

(≈ 100%)

Selective reduction of a benzene ring (W. Grimme, 1970) or a C=C double bond (J.E. Cole, 1962) in the presence of protected carbonyl groups (acetals or enol ethers) has been achieved by Birch reduction. Selective reduction of the C=C double bond of an α,β-unsaturated ketone in the presence of a benzene ring is also possible in aprotic solution, because the benzene ring is reduced only very slowly in the absence of a proton donor (D. Caine, 1976).

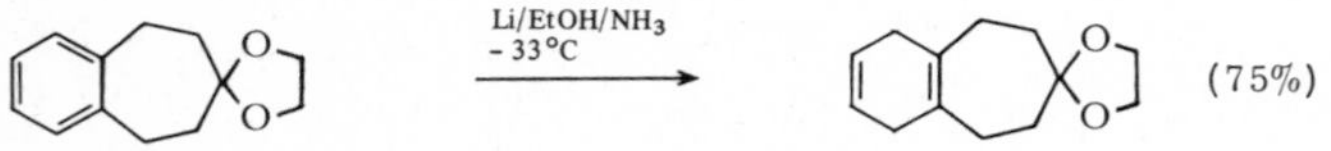

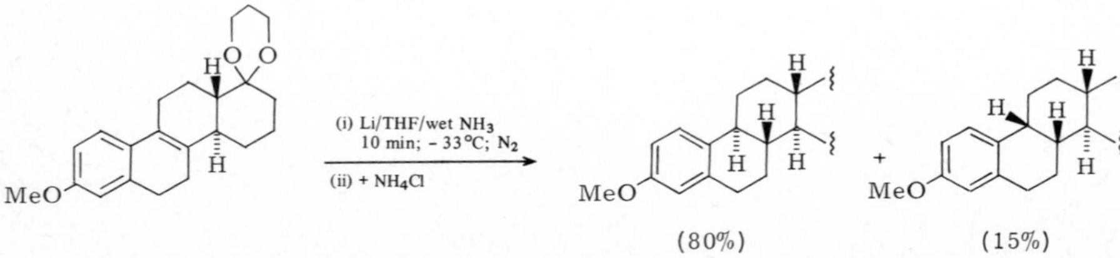

(75%)

(80%)   +   (15%)

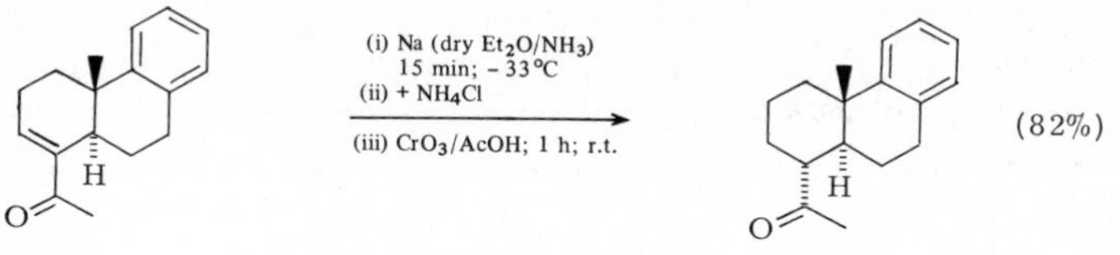

(i) Na (dry Et$_2$O/NH$_3$)
15 min; – 33 °C
(ii) + NH$_4$Cl
(iii) CrO$_3$/AcOH; 1 h; r.t.

(82%)

Cyclopropane rings are opened hydrogenolytically, e.g., over platinum on platinum dioxide (Adam's catalyst) in acetic acid at 2 - 4 bars hydrogen pressure. The bond, which is best accessible to the catalyst and most activated by conjugated substituents, is cleaved selectively (W.J. Irwin, 1968; R.L. Augustine, 1976). Synthetically this reaction is useful as a means to hydromethylate C=C double bonds via carbenoid addition (see p. 68f.; Z. Majerski, 1968; C.W. Woodworth, 1968).

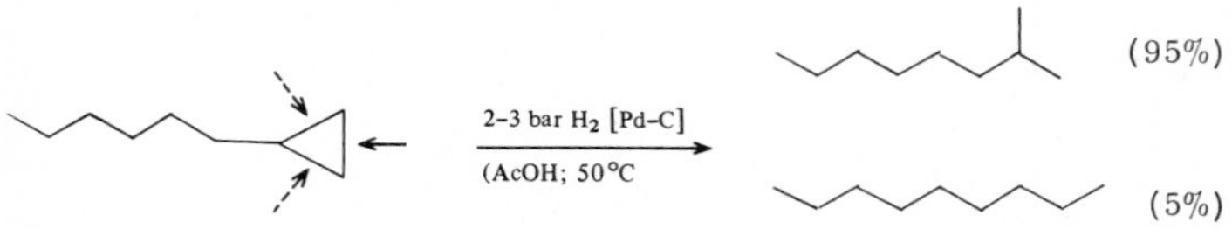

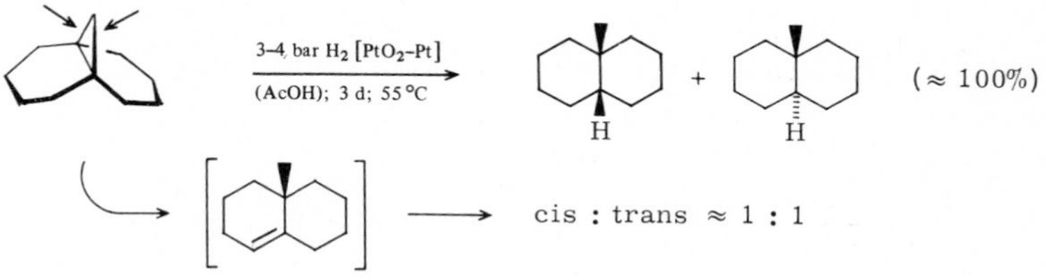

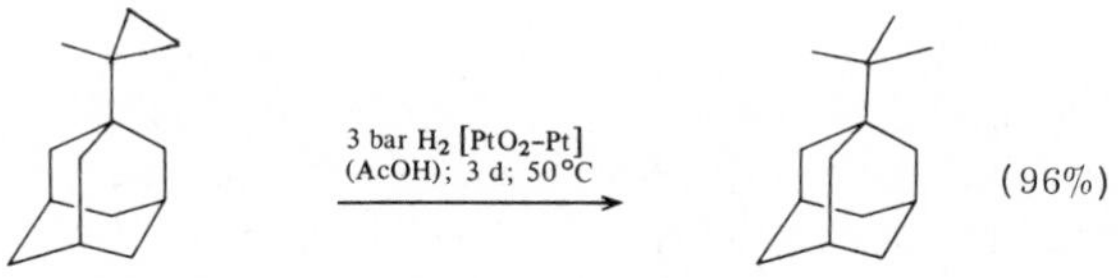

2-3 bar H$_2$ [Pd–C]
(AcOH; 50 °C)

(95%)

(5%)

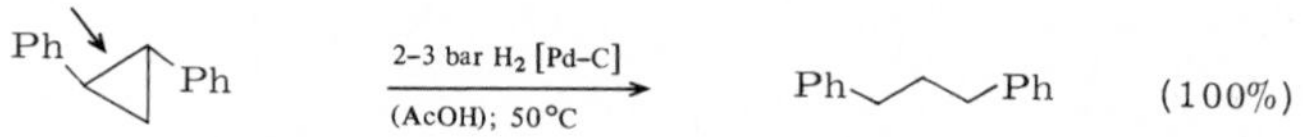

2-3 bar H$_2$ [Pd–C]
(AcOH); 50 °C

(100%)

3-4 bar H$_2$ [PtO$_2$–Pt]
(AcOH); 3 d; 55 °C

(≈ 100%)

cis : trans ≈ 1 : 1

3 bar H$_2$ [PtO$_2$–Pt]
(AcOH); 3 d; 50 °C

(96%)

2.1.2 Reduction of Aldehydes, Ketones, and Carboxylic Acid Derivatives

These polar functional groups are mostly reduced to the corresponding alcohols with hydride reagents (A. Hajós, 1966, 1979). The general selectivities are indicated in table 1 (p. 89f.) and a few specific examples will be given here.

Hydroborates with an electron-withdrawing cyano group reduce protonated C=O and C=N double bonds selectively. Aldehydes and ketones are therefore only reduced below pH 3, whereas imines are hydrogenated already at pH 6 (see p. 103). The bulky salt tetrabutylammonium cyanoborate reduces aldehydes much faster than ketones in slightly acidic HMPTA solutions (R.O. Hutchins, 1973 A). Hydrotris(alkylthio)borates show the same selectivity (Y. Maki, 1977). Bulky aluminum hydrides, e.g. diisobutylaluminum hydride (DIBAL), convert esters or amides into the corresponding aldehydes or hemiacetals (E.J. Corey, 1975A). Efficient conversion of esters into ethers is possible with $NaBH_4$-BF_3 reagent, if the alcohol component is tertiary (G.R. Pettit, 1962 B). α,β-Unsaturated carbonyl compounds are reduced with alane (AlH_3, from 3 $LiAlH_4$ + $AlCl_3$; D.C. Wigfield, 1973; W.G. Dauben, 1973) or with $NaBH_4$ in the presence of cerium(III) ions (J.-L. Luche, 1978) to give allylic alcohols. $LiAlH_4$ itself tends to reduce the C=C double bond.

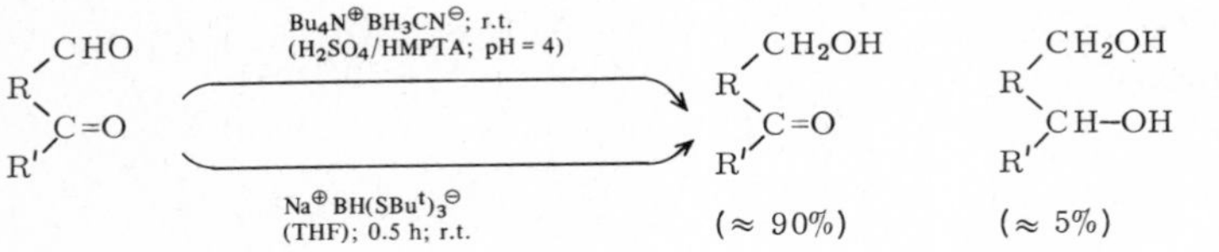

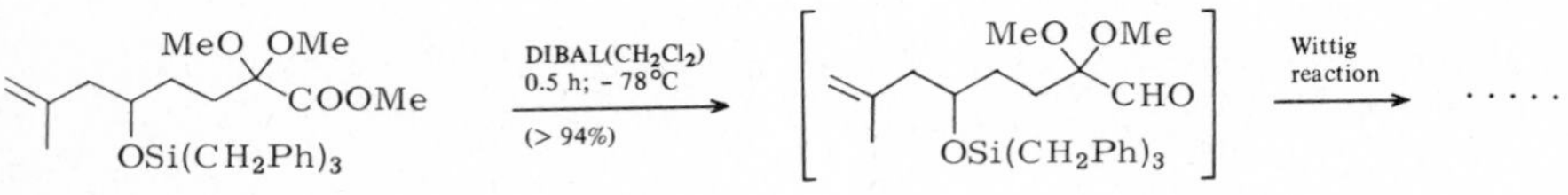

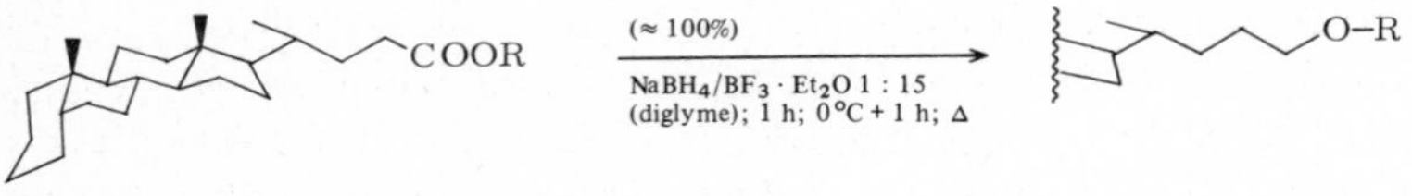

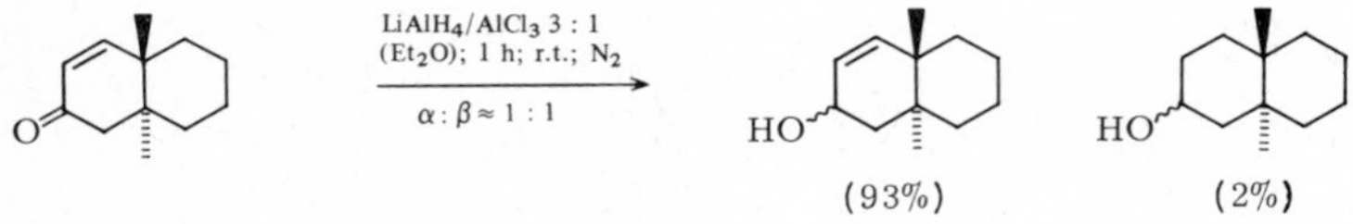

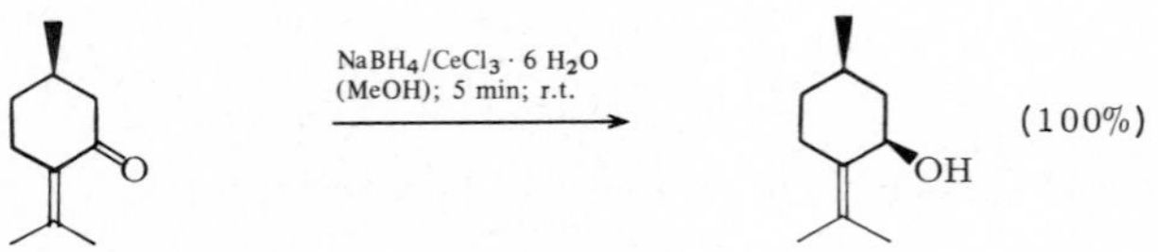

An important aspect of the reduction of carbonyl compounds is the stereoselectivity of these reactions. With open-chain compounds such selectivity is difficult to achieve, although there exist some preferences. D.J. Cram's rule (1952), for example, predicts the steric course of hydride and carbanion addition to open-chain aldehydes and ketones containing a chiral α-

carbon atom with three substituents of different size (l, m, and s). The rule has been substantiated and revised several times (J.D. Morrison, 1971). A version by M. Chérest *et al.* (1968) explains the observed diastereomeric product ratios with a sterically favored approach of the hydride donor or carbanion *anti* to the large group (l). In the preferred transition state the small carbonyl oxygen is near to the medium-size group (m), the bulky alkyl group R near to the small group (s). But the extent of asymmetric induction is usually unsatisfactory, and therefore such reactions are of very limited value in complex organic syntheses.

preferred staggered
transition state

	three product	
R	Ch	Ph
Me	62%	74%
Et	67%	76%
Pr^i	80%	83%
Bu^t	62%	98%

Synthetically useful stereoselective reductions have been possible with cyclic carbonyl compounds of rigid conformation. Reduction of substituted cyclohexanone and cyclopentanone rings by hydrides of moderate activity, e.g. $NaBH_4$ (J.-L. Luche, 1978), leads to alcohols via hydride addition to the less hindered side of the carbonyl group. Hydrides with bulky substituents are especially useful for such regio- and stereoselective reductions, e.g. lithium hydrotri-*t*-butoxyaluminate (C.H. Kuo, 1968) or lithium hydrotrialkylborates (from trialkylboranes + hydroaluminates; H.C. Brown, 1972 B; S. Krishnamurthy, 1976).

$NaBH_4/CeCl_3 \cdot 6 H_2O$
(MeOH); 5 min; r.t.

(99%)

$Li^{\oplus} AlH(OBu^t)_3^{\ominus}$
(THF); 16 h; r.t.

(100%)

17α 17β
($\approx 90\%$)   ($\approx 10\%$)

$Li^{\oplus} AlH(OMe)_3^{\ominus}$ +

$\xrightarrow{\Delta}$ $Al(OMe)_3$ + $Li^{\oplus} BHSia_3^{\ominus}$

(i) $Li^{\oplus}BHSia_3^{\ominus}$(THF); 2 h; – 78 °C
(ii) $H_2O_2/OH^{\ominus}$(Et$_2$O/H$_2$O); 1 h; r.t.

(> 98%; 99.6% trans)

Another possibility for asymmetric reduction is the use of chiral alcohols in Meerwein-Ponndorf reductions (e.g. R,R,R-isobornyloxymagnesium bromide; H. Gerlach, 1966) or of chiral complex hydrides derived from LiAlH$_4$ and chiral alcohols, e.g. quinine (O. Červinka, 1973), N-methylephedrine (I. Jacquet, 1974), or 1,4-bis(dimethylamino)butanediol (D. Seebach, 1974). But stereoselectivities are mostly below 50%*. At the present time attempts to form chiral alcohols from ketones are less successful than the asymmetric reduction of C=C double bonds via hydroboration or hydrogenation with Wilkinson type catalysts (G. Zweifel, 1963; H.B. Kagan, 1978; see p. 95).

reducing agent:	example:	R : S
	$\xrightarrow[\text{(60\%; 30\%ee)}]{\text{(C}_6\text{H}_6\text{); 2 h; r.t.}}$	35 : 65
	$\xrightarrow[\text{(100\%; 23\%ee)}]{\text{(Et}_2\text{O); 4 h; }\Delta}$	61 : 39
	$\xrightarrow[\text{(> 90\%; 89\%ee)}]{\text{(Et}_2\text{O); 2 h; 0 °C}}$	94 : 6
	$\xrightarrow[\text{(92\%; 47\%ee)}]{\text{(Et}_2\text{O); 4 h; r.t.}}$	73 : 27
	$\xrightarrow[\substack{\text{(ii) H}_2\text{O}_2\text{/OH}^\ominus\text{; r.t.}\\ \text{(Et}_2\text{O/H}_2\text{O)}}]{\substack{\text{(i) (THF); 5 h; } -78\text{ °C}\\ +16\text{ h; 0 °C}}}$ 98% (R,R)	—

The direct four-electron reduction of the carbonyl group to the methylene group is possible directly with zinc (Clemmensen reductions; E. Vedejs, 1975) or, more conveniently, *via* the corresponding hydrazones or thioacetals (F. Asinger, 1970). The drastic conditions required in the original Wolff-Kishner reduction of hydrazones (alkali hydroxide, above 200 °C;

---

* percent stereoselectivity = % stereisomer A — % stereoisomer B in the product mixture A + B

J.E. Cole, 1962; potassium *t*-butoxide in boiling toluene; M.F. Grundon, 1963) can be avoided, if potassium *t*-butoxide in DMSO is used (D.J. Cram, 1962). Another possibility is the use of hydrazone derivatives bearing good leaving groups, e.g. tosyl. They may eliminate molecular nitrogen on reduction with hydrides (R.O. Hutchins, 1973 B, 1975; E.J. Taylor, 1976). Hydrazones of α,β-usaturated carbonyl compounds yield rearranged olefins. Desulfurization of thioacetals and thioketals occurs with Raney nickel (G.R. Pettit, 1962 A; F. Asinger, 1970). Several examples are known for the direct reduction of the keto group the methylene group in α,β-unsaturated ketones (D.H.R. Barton, 1968).

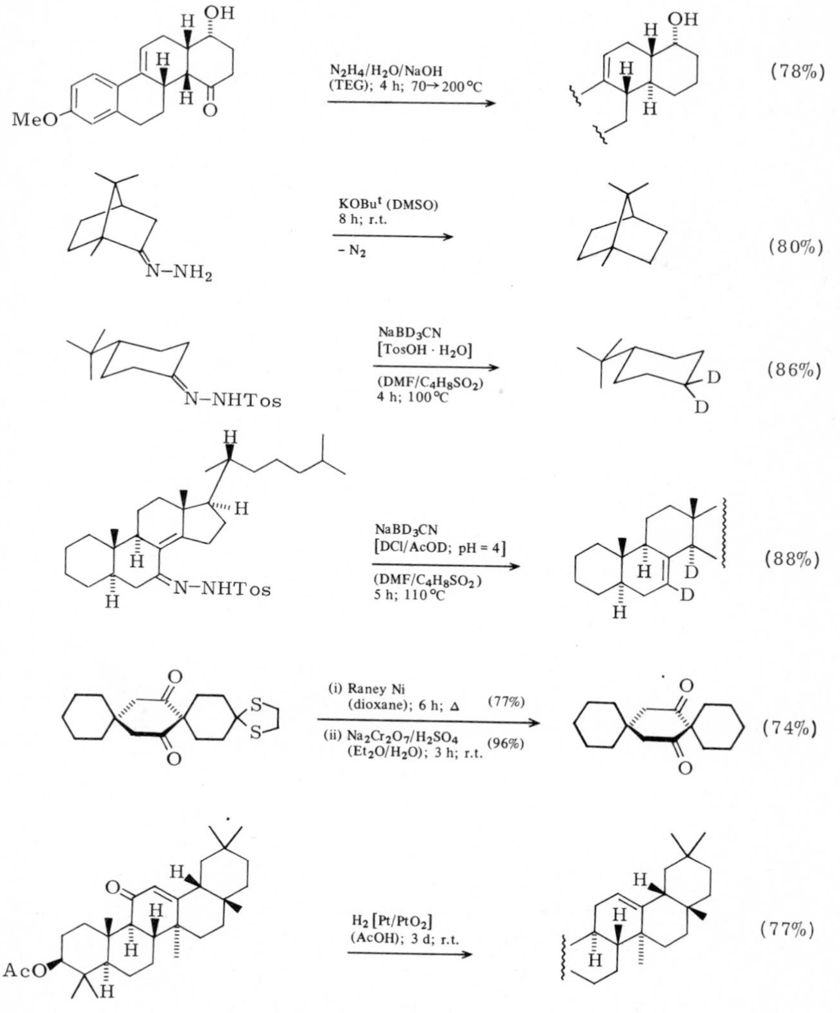

2.1.3 Reduction of Nitrogen Compounds

Amides can be reduced to amines with $LiAlH_4$, although the reaction proceeds slower than the reduction of most other functional groups (W.A. Harrison, 1961; W.A. Ayer, 1968), which have to be re-oxidized afterwards if desired. Diborane is also useful and does not attack ester groups, but C=C double bonds (H.C. Brown, 1964). Sodium tetrahydroborate reduces amides only in the presence of acidic catalysts, e.g. $CoCl_2$ (prim. and sec. amides only; T. Satoh, 1969) or carboxylic acids (N. Umino, 1976). Secondary and tertiary amides are O-alkylated with Meerwein's reagent ($Et_3O^+BF_4^-$), and the resulting carbenium ions are reduced in high yield with $NaBH_4$ in ethanol (R.F. Borch, 1968). In all these cases the C—N linkages remain intact after reduction. Cleavage into amines and alcohols (from the reduction of the acyl moiety) occurs only occasionally. Esters, in contrast, are almost always cleaved on reduction because alkoxide ions are easily cleaved from the intermediate hemiacetal anions. In amides the carbonyl oxygen bound to boron or aluminum is removed much more easily than an amide anion. Nitriles are converted into aldehydes by several reducing agents, e.g. DIBAL, complex hydrides, or catalytic hydrogenation (E. Winterfeldt, 1975).

R–$CONH_2$ —[$NaBH_4/CoCl_2$ 1 : 2, (H_2O); 1 h; r.t.]→ R–CH_2NH_2 (≈ 70%)

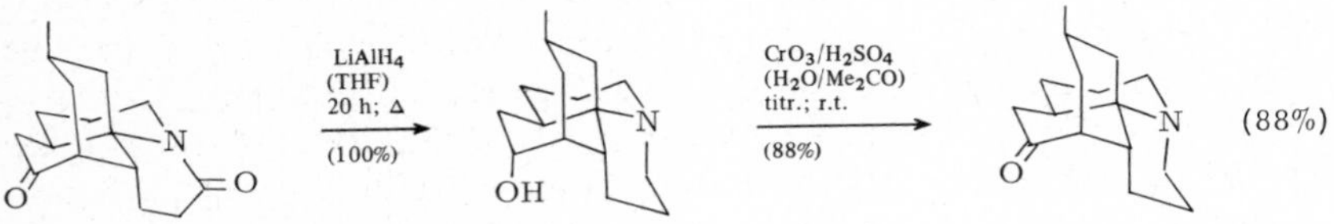

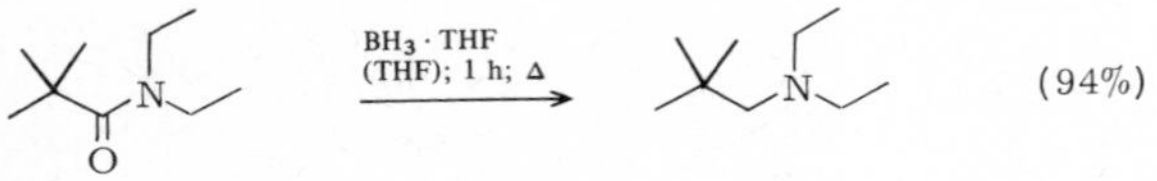

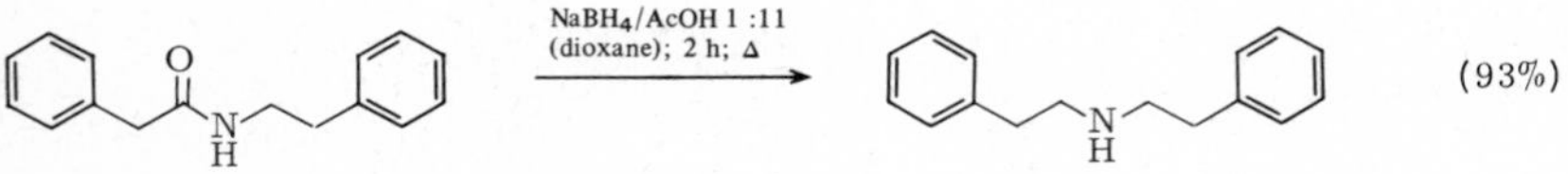

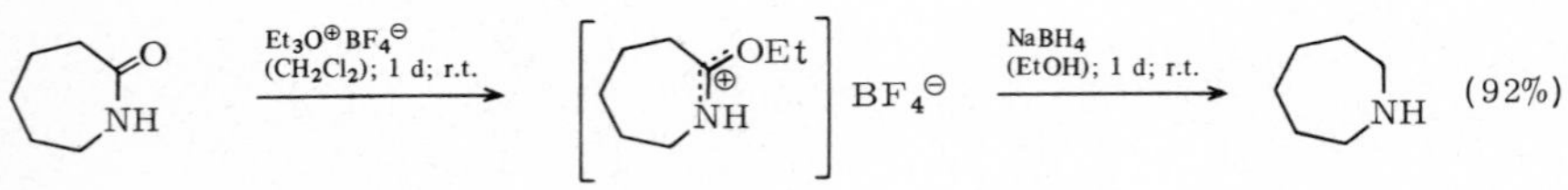

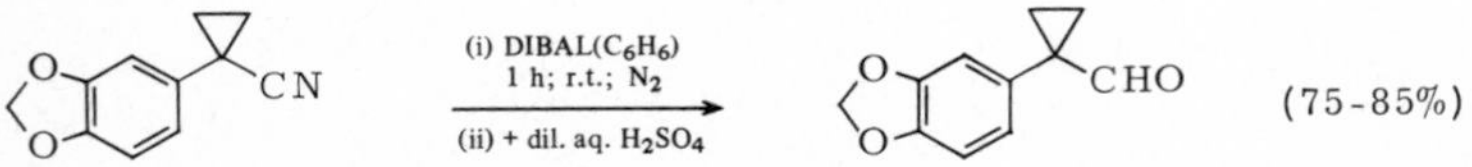

Ketones may also be converted into amines, if they are first reacted with ammonium salts or methoxyamine and then reduced with sodium trihydrocyanoborate at pH 7, where carbonyl groups are not attacked (M.-H. Boutigue, 1973).

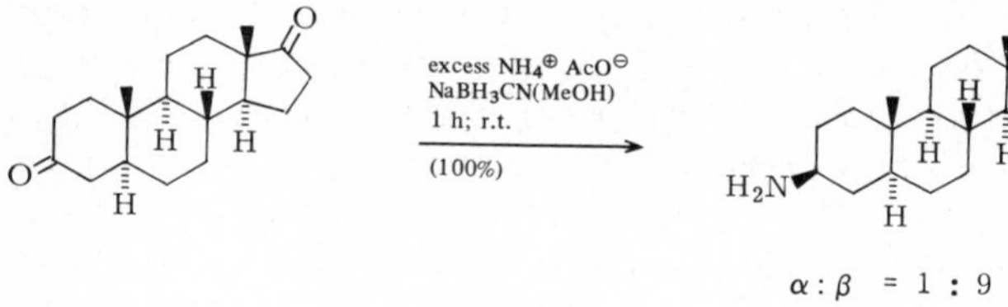

$$\alpha : \beta = 1 : 9$$

Nitro groups are efficiently reduced with hydrogen over Raney nickel catalyst (I. Felner, 1967), with hydrides, or with metals.

As mentioned above, nonaromatic C=N double bonds are easily reduced by hydrides (G. Stork, 1963). AlH_3 reduces activated pyridine rings (M. Ferles, 1970). Mild and selective reduction of pyridine rings, e.g. with $NaBH_4$ or dithionite, neccessitates quaternization or protonation of the nitrogen atom and/or electron withdrawing substituents on a carbon atom (U. Eisner, 1972; S.F. Dyke, 1972). Catalytic hydrogenation of pyridine rings also works best with 3,5-disubstituted electron-deficient pyridinium cations, but further reduction to tetra- and hexahydropyridines is a side-reaction in all reductions (U. Eisner, 1972; M. Ferles, 1970; R.M. Acheson, 1976A). The complete hydrogenation of neutral pyridine rings occurs in catalytic hydrogenation, e.g. with rhodium on carbon or Adam's catalyst (Pt/PtO$_2$) (M. Freifelder, 1962, 1963; G.M. Coppola, 1978).

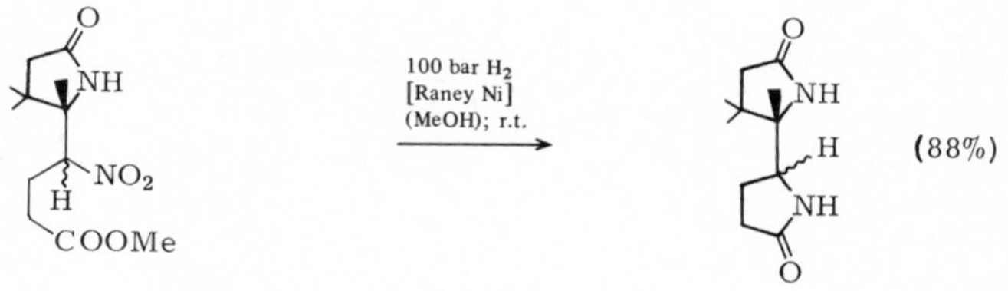

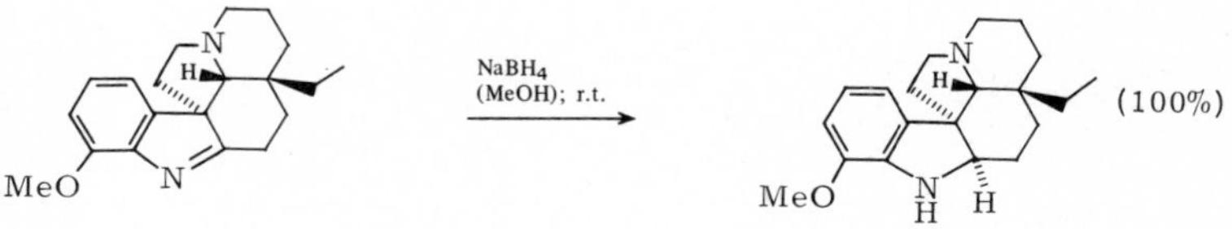

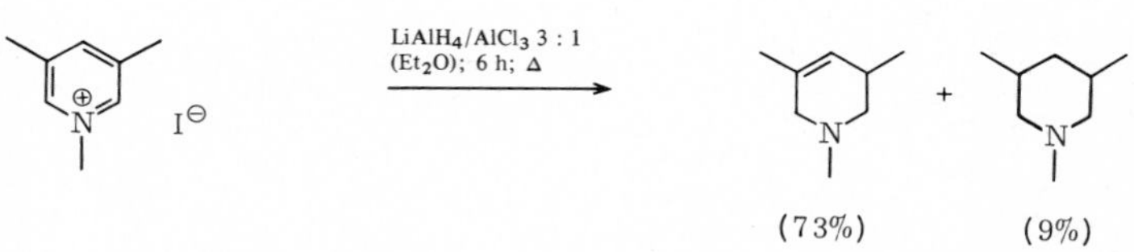

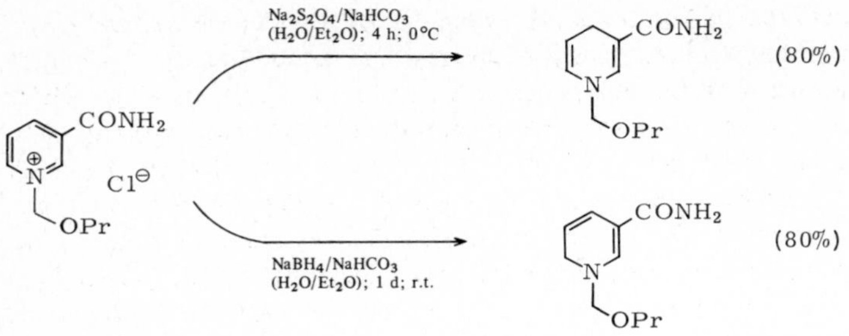

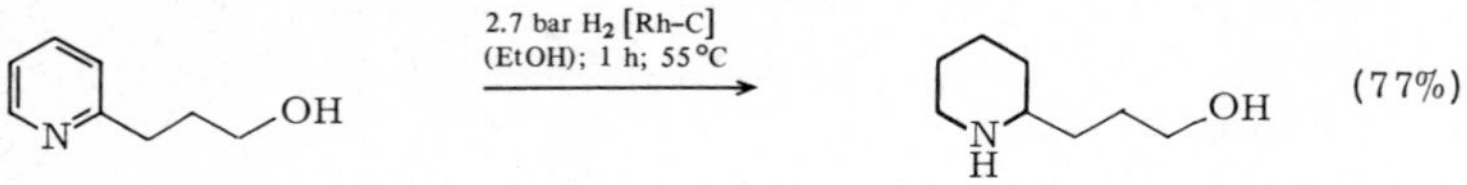

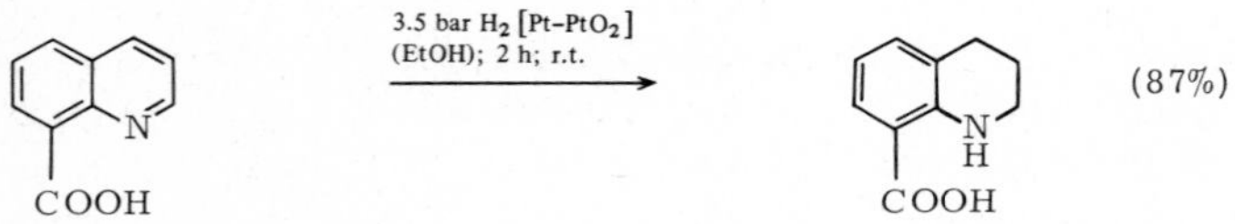

2.1.4 Reductive Cleavage of Carbon-Heteroatom Bonds

Single-bond cleavage with molecular hydrogen is termed hydrogenolysis. Palladium is the best catalyst for this purpose, platinum is not useful. Desulfurizations are most efficiently performed with Raney nickel (with or without hydrogen; G.R. Pettit, 1962 A) or with alkali metals in liquid ammonia or amines. The scheme below summarizes some classes of compounds most susceptible to hydrogenolysis.

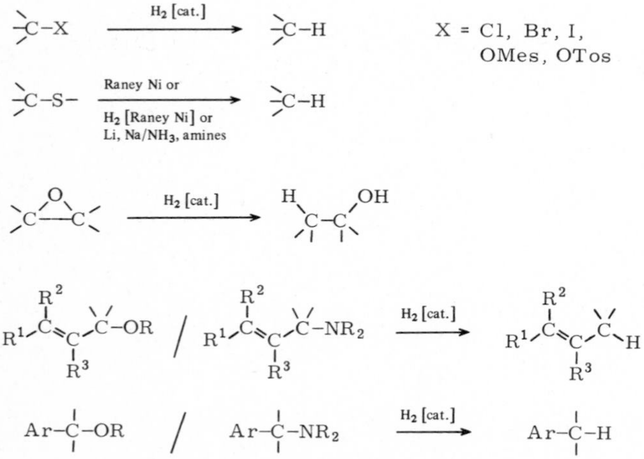

The hydrogenolysis of cyclopropane rings (C—C bond cleavage) has been described on p. 97). In syntheses of complex molecules reductive cleavage of alcohols, epoxides, and enol ethers of β-keto esters are the most important examples, and some selectivity rules will be given. Primary alcohols are converted into tosylates much faster than secondary alcohols. The tosylate group is substituted by hydrogen upon treatment with LiAlH$_4$ (W. Zorbach, 1961). Epoxides are also easily opened by LiAlH$_4$. The hydride ion attacks the less hindered carbon atom of the epoxide (H.B. Henbest, 1956). The reduction of sterically hindered enol ethers of β-keto esters with lithium in ammonia leads to the α,β-unsaturated ester and subsequently to the saturated ester in reasonable yields (R.M. Coates, 1970). Tributyltin hydride reduces halides to hydrocarbons stereoselectively in a free-radical chain reaction (L.W. Menapace, 1964) and reacts only slowly with C=O and C=C double bonds (W.T. Brady, 1970; H.G. Kuivila, 1968).

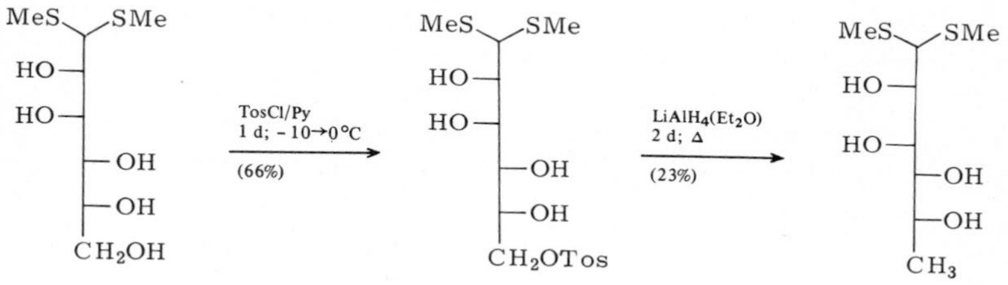

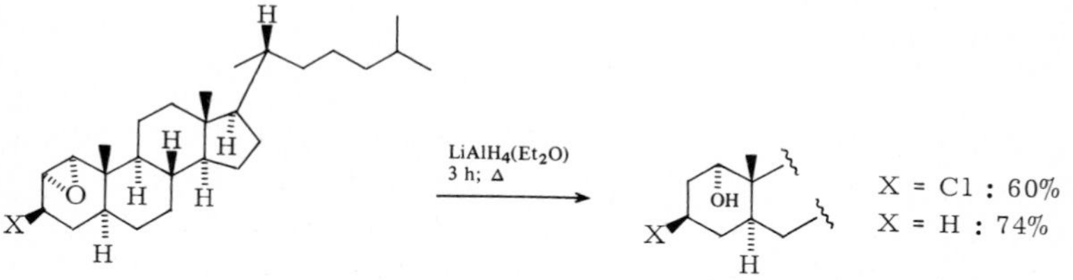

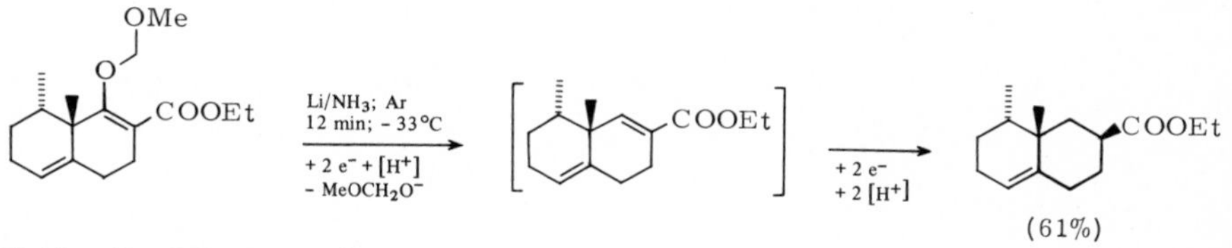

$$R\!-\!X + Bu_3Sn^{\bullet} \longrightarrow R^{\bullet} + Bu_3SnX$$

$$R^{\bullet} + Bu_3SnH \longrightarrow R\!-\!H + Bu_3Sn^{\bullet}$$

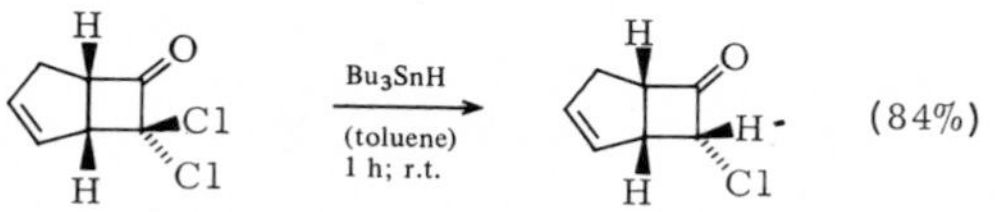

2.2 Oxidation

The standard redox potentials of inorganic oxidants used in organic synthesis are generally around or above $+1.0$ V. Organic substrates do not have such high potentials. The E_0 values for the CH_4/CH_3OH and C_2H_6/C_2H_5OH couples are at $+0.59$ V and 0.52 V, respectively. The oxidation of alcohols and aldehydes corresponds to E_o values around 0.0 V (W.M. Latimer, 1952). Therefore all applied oxidants are, in thermodynamic terms, able to oxidize or to dehydrogenate all hydrocarbons and all oxidizable functional groups of organic molecules.

   Specifity in synthetic oxidation procedures is only possible, if one C—H, C—C, or C—X bond reacts much faster than all other bonds. Therefore it depends much more on the structure of the organic substrate than on the oxidant used, and large excesses of oxidants should always be avoided. The rationale for the use of a large variety of oxidizing reagents in organic syntheses is largely based on empirical results with specific oxidations of specific substrates (R.L. Augustine, 1969, 1971; K.B. Wiberg, 1965; W.S. Trahanovsky, 1973, 1978; D. Arndt, 1975). It has been shown frequently, that presumably "general" rules previously developed for the use of different oxidants depend on the nature of the substrates. Nevertheless, such rules nowadays constitute the only basis for the choice of an oxidant, and some of them will therefore be delineated within this section. A first guideline of preferential uses of oxidants may be deduced from table 2. Very few of the oxidants, however, are selective, and other easily oxidizable groups have generally to be protected (see section 2.6).

Table 2. Some selective oxidation reactions.

Non-activated carbon atoms	
$\geq$C–H $\longrightarrow$ $\geq$C–OH	microorganisms; enzymes
$\longrightarrow$ $\geq$C–Cl	$ArICl_2$
$\geq$C–C$\leq$ (H H) $\longrightarrow$ $\geq$C=C$\leq$	$Ar-\overset{O}{\overset{\|}{C}}-Ar'/h\cdot\nu$ } "remote oxidation"

Activated non-funtional carbon atoms	
	SeO_2; $Hg(OAc)_2$
	SeO_2; $CrO_2(OBu^t)_2$; CrO_3Py_2; K_2CrO_4
	$O_2/KOBu^t/(EtO)_3P$; $LDA/[Mo^{VI}O(O_2)_2(Hmpta)(Py)]$
	SeO_2
	$Hg(OAc)_2$
	(i) $LDA/PhSeBr$ (ii) H_2O_2, $NaIO_4$; SeO_2; quinones (DDQ, TCQ)

Z = carbonyl, vinyl, aryl

Table 2. (Continued)

Carbon atoms of C=C double bonds		
$>C=C<$ → epoxide $>C\!-\!C<$ (O bridge)		(i) halohydrin (ii) $OH^{\ominus}$ peracids : less hindered epoxide
→ $>C\!-\!C<$ with OH, OH anti		(i) peracids (ii) $OH^{\ominus}$; (i) $AgOAc/I_2/C_6H_6$ (ii) $OH^{\ominus}$, $LiAlH_4$
→ $>C\!-\!C<$ with OH, OH syn		(i) $AgOAc$, $Cu(OAc)_2/I_2/H_2O/AcOH$ or $Tl(OAc)_3/H_2O/AcOH$ (ii) $OH^{\ominus}$, $LiAlH_4$; OsO_4Py_2 : less hindered syn-diol
→ $>C\!-\!C<$ with OH, X anti		$HOCl$, $HOBr$; NCS, NBS/H_2O; (i) peracids (ii) HX
→ $>C\!-\!C<$ with OH, X syn		CrO_2Cl_2
→ $>C\!-\!C<$ with X, X anti		Cl_2, Br_2, I_2
$R^2\overset{R^1}{\underset{R^3}{C}}=\overset{R^4}{C}$ → $R^2\overset{R^1}{\underset{R^3}{C}}-\overset{R^4}{C}=O$		$TTN = Tl(NO_3)_3$; $TTFA = Tl(OTfac)_3$
$>C=C<$ → $\overset{|}{\underset{H}{C}}-\overset{|}{\underset{OH}{C}}$ syn		(i) B_2H_6 (ii) $H_2O_2/OH^{\ominus}$

Carbon atoms of C≡C triple bonds		
$R-C\equiv C-H$ → $R\!-\!CH_2\!-\!CHO$		(i) R_2BH (ii) $H_2O_2/OH^{\ominus}$
→ $R\!-\!CO\!-\!CH_3$		Hg^{2+}/H^+
$R-C\equiv C-R'$ → $R\!-\!CH(OMe)\!-\!CO\!-\!R'$		$Tl^{3+}/MeOH$
→ $R\!-\!CO\!-\!CO\!-\!R'$		$KMnO_4$, RuO_4
$R-C\equiv C-H$ → $R-COOH$		Tl^{3+}, $KMnO_4$, RuO_4

Oxygen-containing functional groups		
$\overset{H}{\underset{OH}{>C}}$ → $>C=O$		CrO_3/Py, CrO_3/H_2SO_4 $Me_2CO/Al(OR)_3$ (Oppenauer ox.) $DMSO + DCC$, Ac_2O, SO_3, P_4O_{10}, $COCl_2$
$R-CH_2OH$ → $R-CHO$		Ag_2CO_3, $Pb(OAc)_4$, $CrO_2(OBu^t)_2$
→ $R-COOH$		many strong oxidants, AgO

Table 2. (Continued)

	RuO$_4$
	MnO$_2$, DDQ
	Cu(OAc)$_2$

Oxidative rearrangement or cleavage of ketones

	Tl^{3+}/H$_2$O, MeOH
	peracids
	(i) NH$_2$OH (ii) H$^+$ (Beckmann rearr.)
	CrO$_3$, KMnO$_4$; (i) SeO$_2$ (ii) HIO$_4$

2.2.1 Oxidation of Non-functional Carbon Atoms

Chromic acid, potassium permanganate, and chlorine attack alkanes under vigorous conditions. The relative rates of chromic acid oxidation of primary, secondary, and tertiary C—H bonds, for example, are approximately 1 : 100 : 10000 (F. Mareš, 1961; J. Roček, 1957, 1959). Oxidations of complex saturated hydrocarbons with external oxidants are, however, of limited synthetic use because complicated mixtures of products are generally obtained in low yield. A classic attempt in this field is the oxidative degradation of the side chain on C-17 of steroids (e.g. cholesterol, stigmasterol). Such a degradation would convert inexpensive steroids into precious steroidal hormones. Simple chromic acid oxidation led to a useful androstenolone derivative in 7-8% overall yield from cholesterol (Schering process, 1957, 1977; L.F. Fieser, 1959). Recently fluorination of steroids has been shown to be quite selective for tertiary C—H bonds (D.H.R. Barton, 1976). This reaction, however, has not been further exploited. If the oxidizing agent, e.g. (dichloroiodo)benzene, is covalently bound to the steroid, yields of regio- and stereoselective chlorination may be as high as 60% (R. Breslow, 1977; see page 257). This "remote oxidation" approach has also not been tested on a large scale so far. A less esoteric method is biological oxidation with the aid of specialized microorganisms. Such biotechnological procedures allow highly regio- and stereoselective hydroxylations of complex organic molecules even on an industrial scale. The enormous data collection of K. Kieslich (1976) and the introductory texts by G. Bonse (1978) or R.A. Johnson (1978) should be consulted for the scope of selective biological transformations of organic substrates.

cholesterol

Schering process

(i) Ac$_2$O (DCE); 5 h; Δ
(ii) PyBr$^+$ Br$_3^-$ (DCE); 12 h; −15 °C   } protection
(iii) CrO$_3$/H$_2$SO$_4$/AcOH (DCE/SiO$_2$);
       12 h; 12 °C + 12 h; r.t.   } oxidation
(iv) Zn/AcOH/H$_2$O; 9 h; r.t.   } deblocking
(v) removal of carboxylic acids
(vi) H$_2$NNHCONH$_2$/AcOH (EtOH); 2 h; Δ   } purification
(vii) CH$_3$COCOOH (NaOAc/H$_2$O/AcOH);
       10 min; Δ

androstenolone acetate
(7-8%)

(i) F$_2$/NaOTFac/TFA
   (CFCl$_3$/CH$_2$Cl$_2$)
   [inh.: PhNO$_2$]
   0.5 h; −25 °C

(ii) Zn/AcOH/H$_2$O; Δ

17α-fluoro :   40%
17α,25-difluoro : 20%

Conventional regioselective oxidations of C—H bonds require neighboring activating groups. Very common in synthesis is the oxidation of allylic methylene or methyl groups to yield allylic alcohols or α,β-unsaturated carbonyl compounds. The most successful reagents are derivatives of selenium(IV) and chromium(VI). Microorganisms are also used but will not be discussed here. The primary reaction is the electrophilic attack of the highly electropositive metal atoms upon the less hindered side of the C=C double bond. Selenium dioxide is added to olefins in an "ene" reaction to yield allylseleninic acids, which rearrange to allyl selenoxylates (D. Arigoni, 1973; K.B. Sharpless, 1972, 1977). These are then cleaved thermally or hydrolytically. The original allylic methyl, methylene, or methine group is oxidized by this sequence to give a carbonyl or alcohol group. The oxidation proceeds regioselectively *trans* to the largest substituent of the C=C double bond (K.B. Sharpless, 1972; D. Arigoni, 1973; reviews: E.N. Trachtenberg, 1969; H.J. Reich, 1978; N. Rabjohn, 1949, 1976).

"ene" reaction

allylseleninic acid

2,3-sigmatropic rearrangement

selenoxylic acid allyl ester

solvolysis (H$_2$O, EtOH)

Δ
− H$_2$O − Se

oxidation (MnO$_2$, Cr^{VI})

SeO$_2$/H$_2$O
(dioxane); 3 h; Δ

(57%)

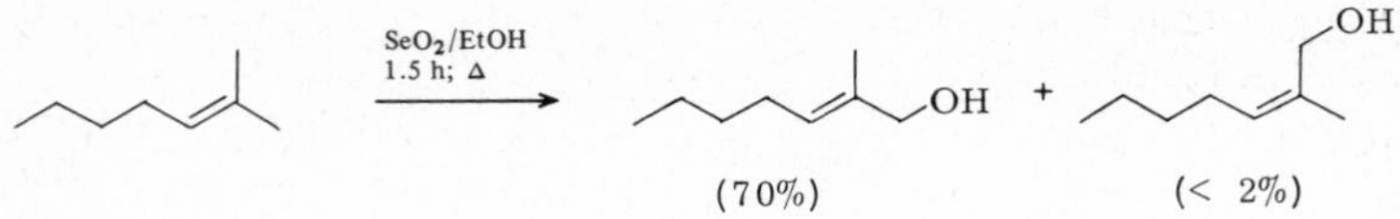

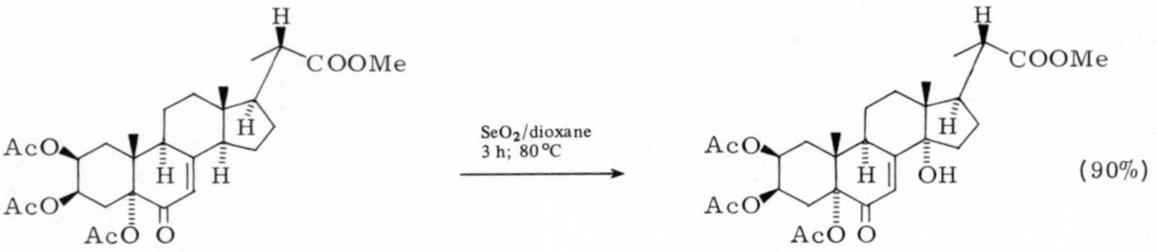

(70%)   (< 2%)

(90%)

The allylic oxidation with derivatives of chromic acid (K.B. Wiberg, 1965; H.G. Bosche, 1975), e.g. di-*t*-butyl chromate (K. Fujita, 1961), CrO_3/pyridine (W.G. Dauben, 1969), or K_2CrO_4 (C.Y. Cuilleron, 1970), gives similar products in occasionally very high yields.

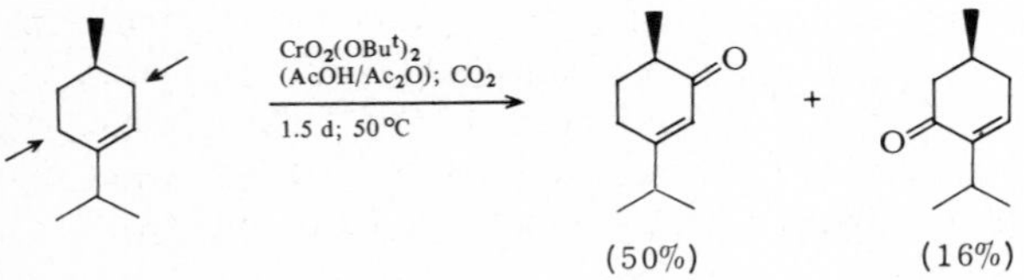

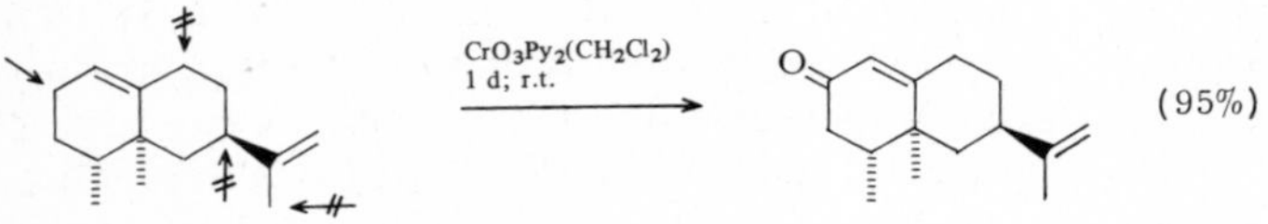

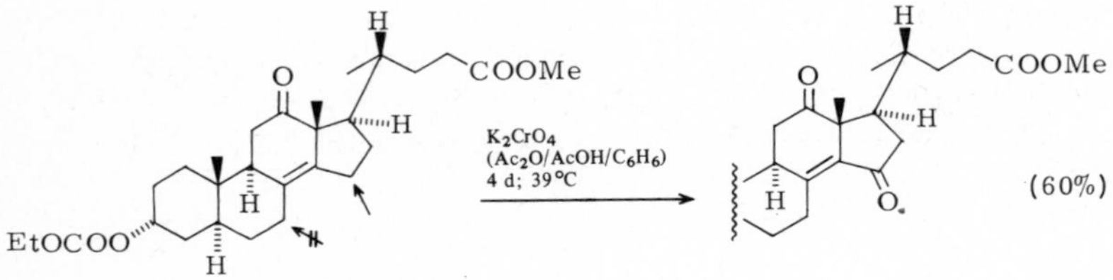

(50%)   (16%)

(95%)

(60%)

Alkyl groups attached to aromatic rings are oxidized more readily than the ring in alkaline media. Complete oxidation to benzoic acids usually occurs with nonspecific oxidants such as $KMnO_4$, but activated tertiary carbon atoms can be oxidized to the corresponding alcohols (R. Stewart, 1965; D. Arndt, 1975). With mercury(II) acetate, allylic and benzylic oxidations are also possible. It is most widely used in the mild dehydrogenation of tertiary amines to give enamines or heteroarenes (M. Shamma, 1970; H. Arzoumanian, 1971; A. Friedrich, 1975).

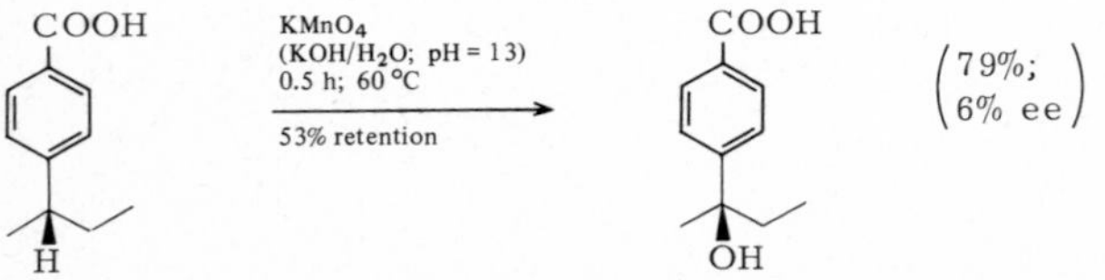

$\left(\begin{array}{l}79\%; \\ 6\% \text{ ee}\end{array}\right)$

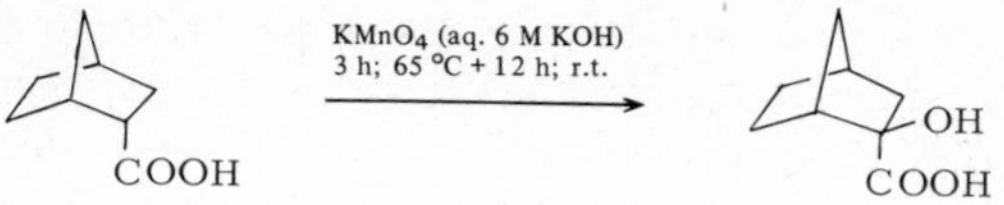

(67%)

(80%)

(99%)

The α-oxidation of carbonyl compounds may be performed by addition of molecular oxygen to enolate anions and subsequent reduction of the hydroperoxy group, e.g. with triethyl phosphite (E.J. Bailey, 1962; J.N. Gardner, 1968 A,B). If the initially formed α-hydroperoxide possesses another enolizable α-proton, dehydration to the 1,2-dione occurs spontaneously, and further oxidation to complex product mixtures is usually observed.

free-radical chain oxygenation:

(46%)

(73%)

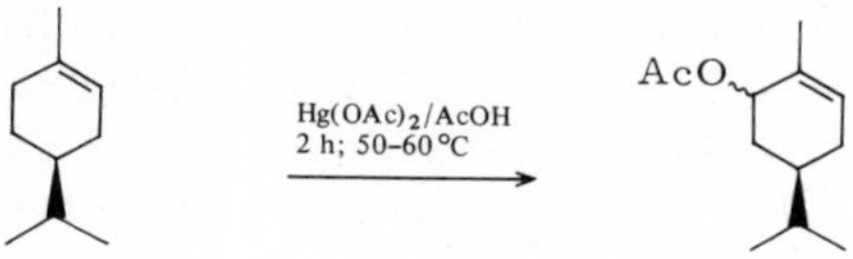

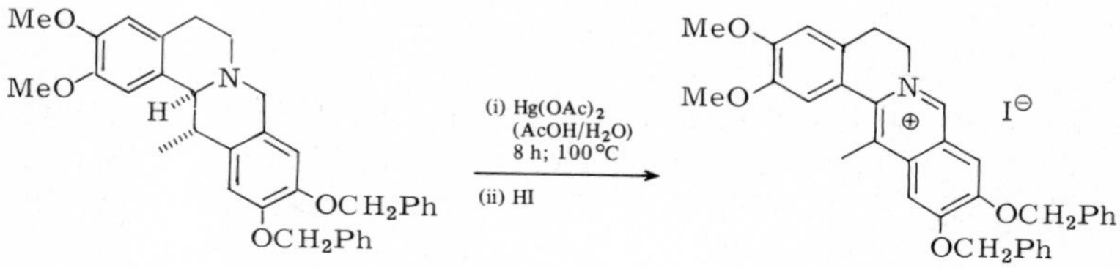

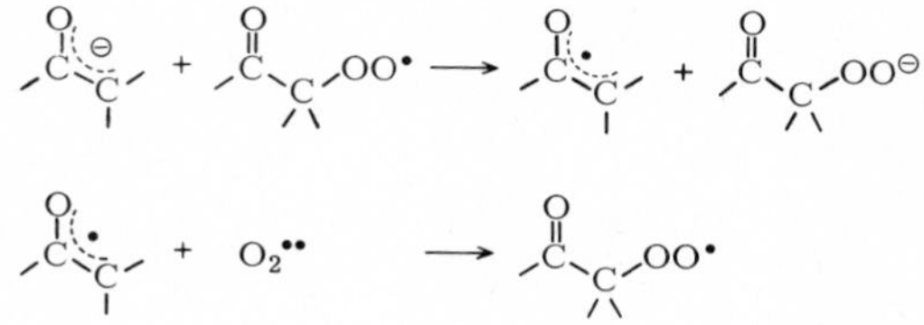

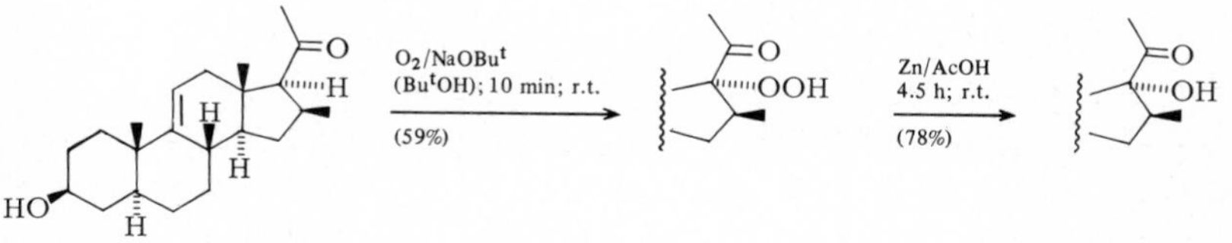

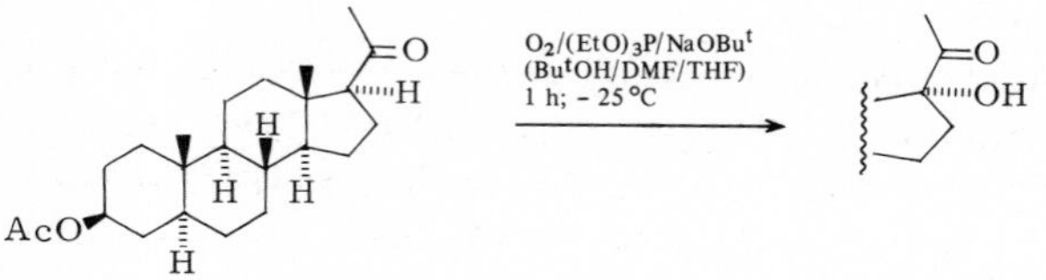

E. Vedejs (1978) developed a general method for the sterically controlled electrophilic α-hydroxylation of enolates. This uses a bulky molybdenum(VI) peroxide complex, MoO(O₂)₂(HMPTA)(Py), which is rather stable and can be stored below 0 °C. If this peroxide is added to the enolate in THF solution (base: e.g. LDA) at low temperatures, one O—O bond is broken, and a molybdyl ester is formed. Excess peroxide is quenched with sodium sulfite after the reaction has occurred, and the molybdyl ester is cleaved to give the α-hydroxy carbonyl compound in high yield. This reaction was used in the regio- and stereoselective α-hydroxylation of a terpenoid aldehyde (S.P. Tanis, 1979). Oxygen addition to cyclic enolates occurs preferentially at the less hindered side of the ring.

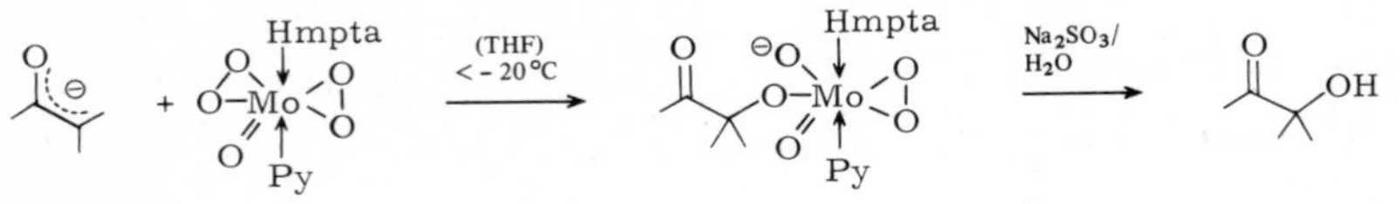

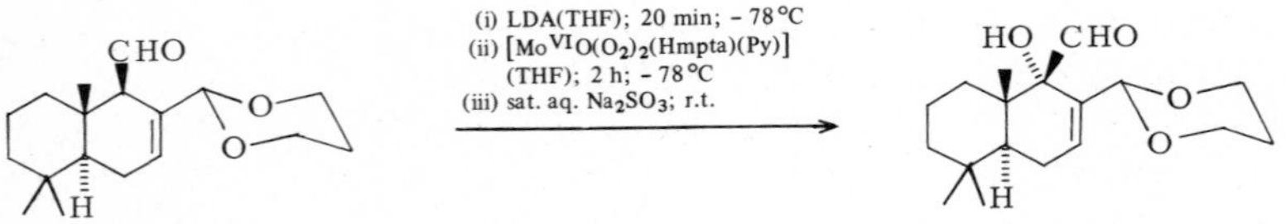

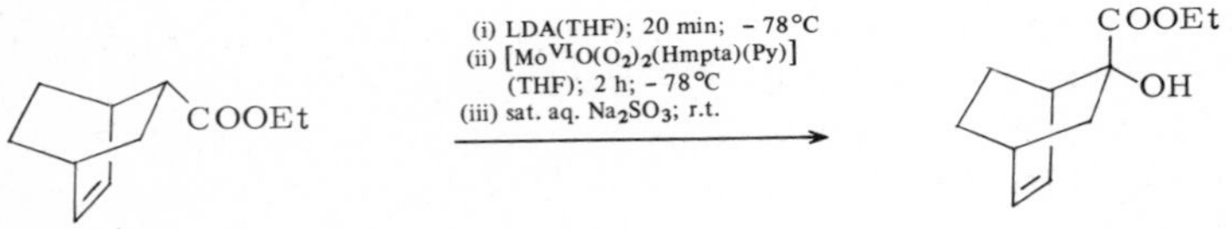

SeO₂ oxidizes ketones and aldehydes to 1,2-dioxo compounds (N. Rabjohn, 1949, 1976; N. Trachtenberg, 1969). Some systems give α,β-unsaturated carbonyl compounds as by-products or even as the major product, especially if the β-carbon site is activated (K. Wiesner, 1958). The reaction sequence is similar to that described for the allylic oxidation of olefins with SeO₂, except that the C=O double bond is attacked instead of a C=C double bond. For the dehydrogenation of C—C single bonds activated by carbonyl, vinyl, or aryl groups various other reagents have also been used frequently (see p. 125f.), e.g. phenylselenium bromide or quinones.

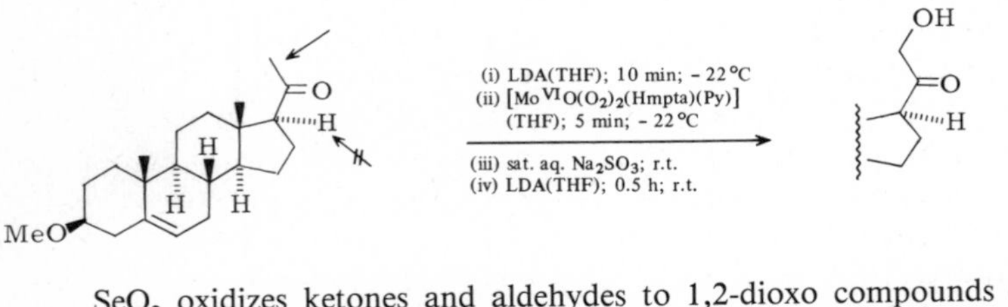

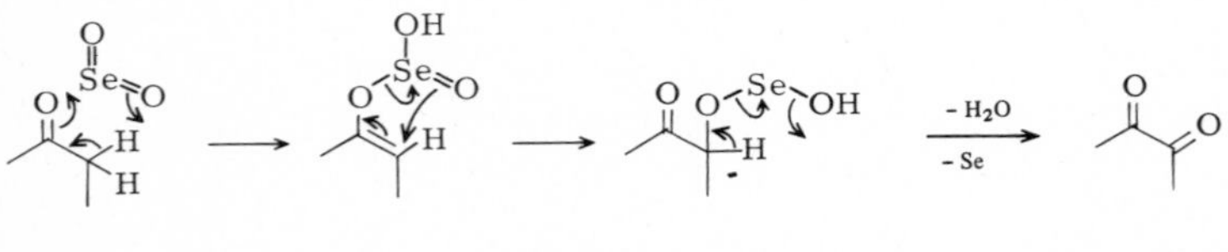

$$R\text{-}C\underset{O\text{-}O\text{-}H}{\overset{O}{\diagdown}} \quad + \quad {>}C{=}C{<} \quad \longrightarrow \quad R\text{-}C\underset{OH}{\overset{O}{\diagdown}} \quad + \quad {>}C\overset{O}{\underset{\diagup\diagdown}{}}C{<}$$

2.2.2 Oxidation of Carbon Atoms in Carbon-Carbon Multiple Bonds

Oxidation of olefins and dienes provides the classic means for syntheses of 1,2- and 1,4-difunctional carbon compounds. The related cleavage of cyclohexene rings to produce 1,6-dioxo compounds has already been discussed in section 1.14. Many regio- and stereoselective oxidations have been developed within the enormously productive field of steroid syntheses. Our examples for regio- and stereoselective C=C double bond oxidations as well as the examples for C=C double bond cleavages (see p. 80f.) are largely selected from this area.

Peracids or mixtures of hydrogen peroxide and formic acid have been frequently used to transform olefins into epoxides, alcohols, or *trans*-glycols. The reactivities of the peracids parallel the acidities of the corresponding acids (D. Swern, 1953, 1970; S.N. Lewis, 1969; B. Plesničar, 1978). In cycloalkenes the less hindered side of the C=C double bond is epoxidized selectively (H.C. Brown, 1970). If two different C=C double bonds are present within a molecule, the bond with the higher electron density reacts faster (electron-releasing alkyl groups, no electron-withdrawing groups; W. Hückel, 1955; R.B. Woodward, 1958; W. Knöll, 1975).

Epoxide opening with nucleophiles occurs at the less substituted carbon atom of the oxirane ring. Catalytic hydrogenolysis yields the more substituted alcohol. The scheme below contains also an example for *trans*-dibromination of a C=C double bond followed by dehydrobromination with strong base for overall conversion into a conjugated diene. The bicyclic tetraene then isomerizes spontaneously to the aromatic 1,6-oxido[10]annulene (E. Vogel, 1964).

increasing reactivity of peracids

R =	CH_3			H		COOH	CF_3
pK$_a$(RCOOH):	4.8	4.2	3.9	3.8	3.4	2.9	< 0

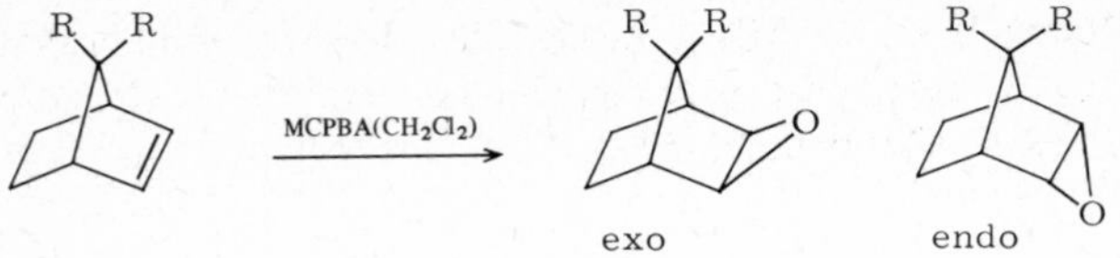

R = H: 20 min; r.t.   99%   < 1%

R = CH₃: 1 d; r.t.   < 10%   90%

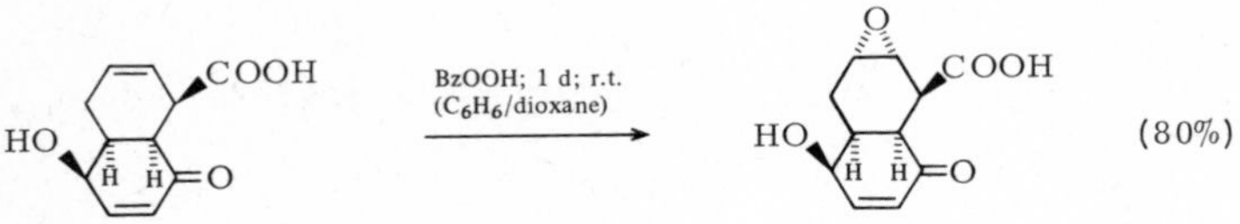

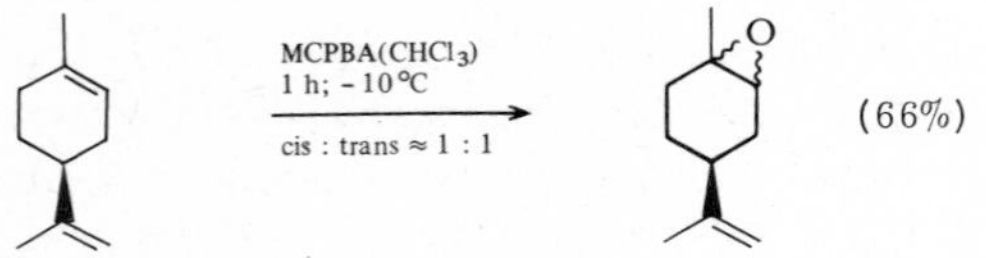

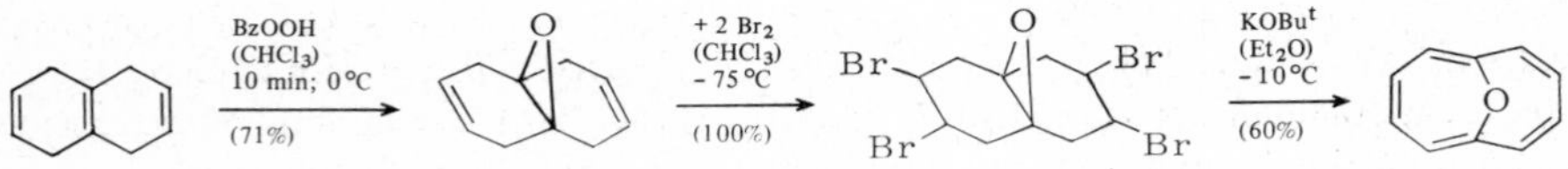

(80%)

(66%)

Conjugated dienes are often converted into 2,3-unsaturated 1,4-diols by successive treatment with peracids and hydroxides (R.B. Woodward, 1957).

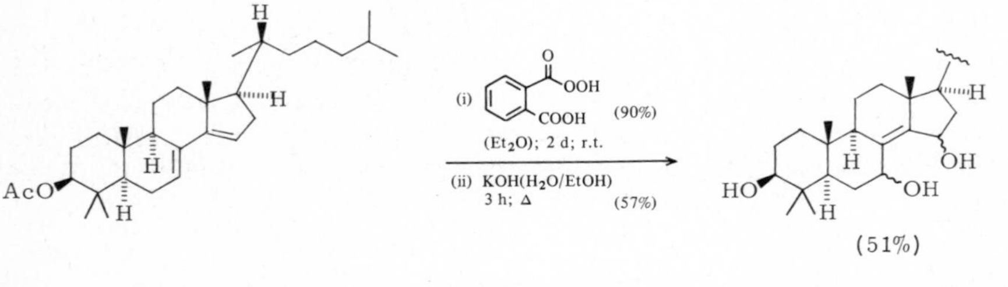

As has been shown above, in cyclic systems the epoxy group is introduced on the less hindered side, and its hydrolysis leads to *trans*-diols. Another dihydroxylation procedure works with acyl hypoiodites, which are prepared *in situ* from iodine and silver(I) carboxylates in dry aprotic solvents. First O-acyl iodohydrins are formed. From these an iodide anion is removed by Ag⁺, and the slightly nucleophilic carboxyl oxygen adds to the carbocation to form a *cis*-fused cyclic dioxolenium cation, which is opened by carboxylate to give *trans*-diol derivatives (Prévost reaction; C.V. Wilson, 1957; R. Criegee, 1979). The selectivity and the stereochemistry depend strongly on solvent, metal ions, substrate, and temperature, but *diaxial* glycols are obtainable in good yield (R.C. Cambie, 1977).

Prévost reaction:

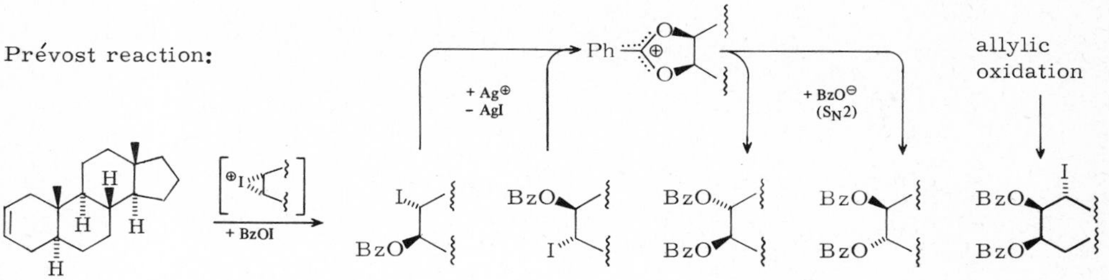

AgOBz/I$_2$ (C$_6$H$_6$); 1 h; Δ	6%	–	21%	46%	12%
„ / „ (C$_6$H$_6$); 3 d; r.t.	10%	–	–	51%	33%
„ / „ (CCl$_4$); 3 d; r.t.	9%	–	–	60%	24%
TlOBz/I$_2$ (C$_6$H$_6$); 1 h; Δ	–	65%	–	–	–
			(eq,eq)	(ax,ax)	

Stereoselective *cis*-dihydroxylation of the more hindered side of cycloalkenes is achieved with silver(I) or copper(II) acetates and iodine in wet acetic acid (Woodward glycolization; J.B. Siddall, 1966; L. Mangoni, 1973; R. Criegee, 1979) or with thallium(III) acetate via organothallium intermediates (E. Glotter, 1976). In these reactions the intermediate dioxolenium cation is supposed to be opened hydrolytically, not by S$_N$2 reaction.

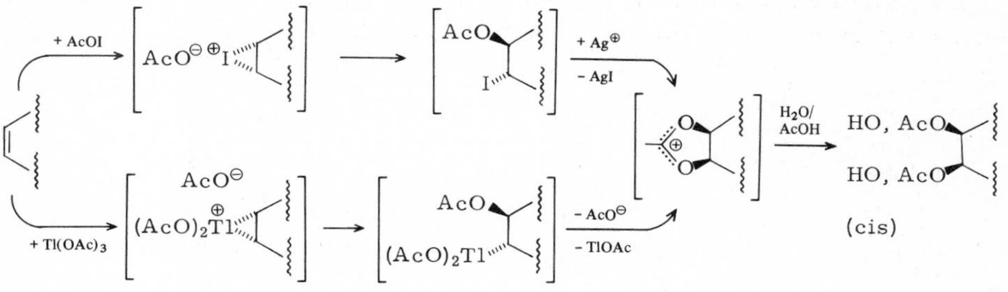

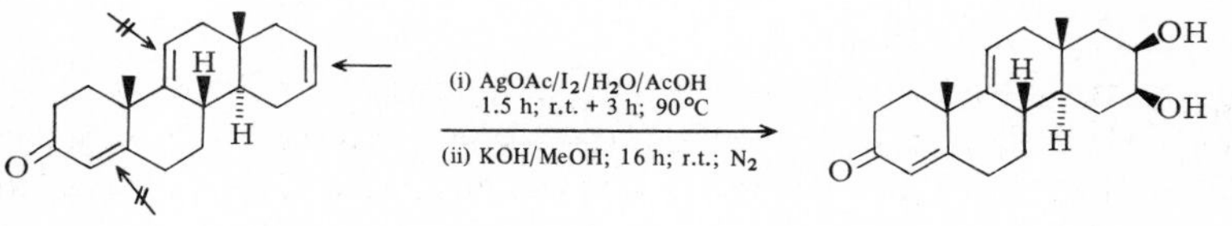

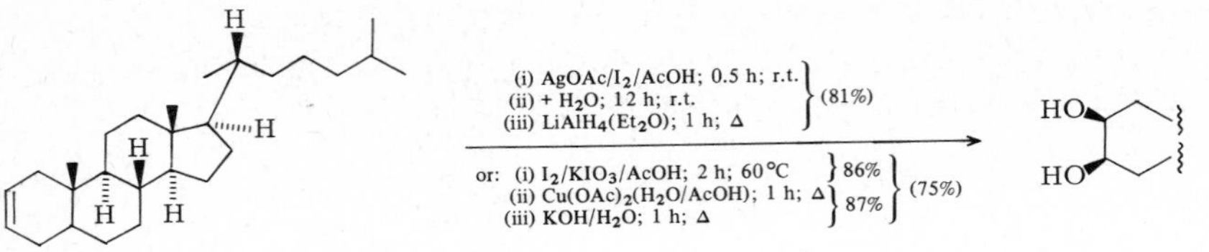

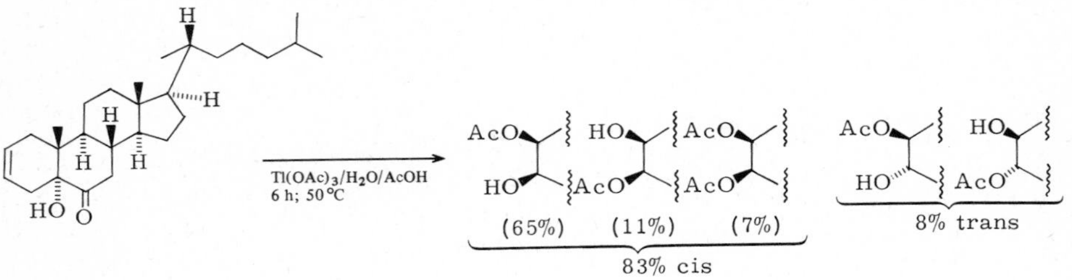

(i) AgOAc/I₂/AcOH; 0.5 h; r.t.
(ii) + H₂O; 12 h; r.t.   } (81%)
(iii) LiAlH₄(Et₂O); 1 h; Δ

or: (i) I₂/KIO₃/AcOH; 2 h; 60°C } 86%
(ii) Cu(OAc)₂(H₂O/AcOH); 1 h; Δ } 87% } (75%)
(iii) KOH/H₂O; 1 h; Δ

Tl(OAc)₃/H₂O/AcOH
6 h; 50°C

(65%)   (11%)   (7%)   8% trans

83% cis

The less hindered side of C=C double bonds is best *cis*-dihydroxylated with osmium tetroxide. Cyclic osmate diesters are isolable, often crystalline intermediates. The very expensive and very toxic OsO₄ may be used in catalytic amounts with KClO₄, H₂O₂, or N-oxides (R. Criegee, 1979) to regenerate the OsO₄ *in situ*. OsO₄ and especially its bis(pyridine) complex are bulky reagents and attack more accessible and more electron-rich C=C double bonds stereo- and regioselectively (S. Bernstein, 1956; E.J. Corey, 1964B). The glycolization with KMnO₄ is largely limited to molecules with no other oxidizable groups (R. Criegee, 1979).

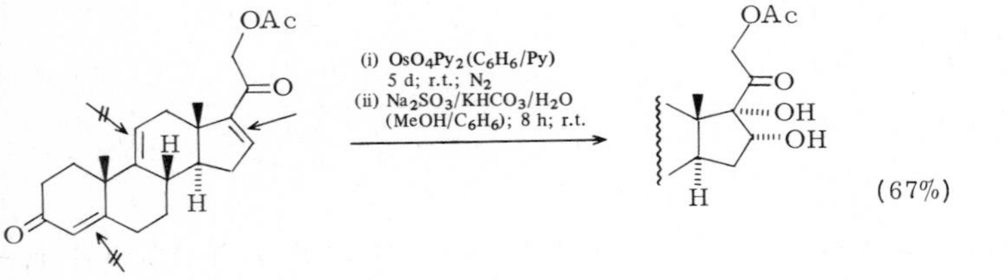

(i) OsO₄Py₂(C₆H₆/Py)
5 d; r.t.; N₂
(ii) Na₂SO₃/KHCO₃/H₂O
(MeOH/C₆H₆); 8 h; r.t.

(67%)

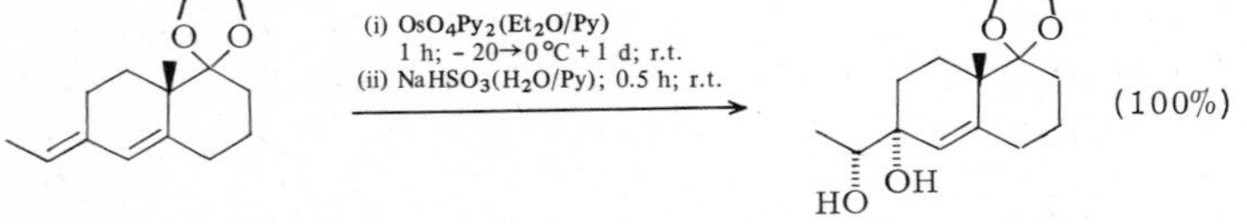

(i) OsO₄Py₂(Et₂O/Py)
1 h; – 20→0°C + 1 d; r.t.
(ii) NaHSO₃(H₂O/Py); 0.5 h; r.t.

(100%)

Thallium(III) acetate reacts with alkenes to give 1,2-diol derivatatives (see p. 115) while thallium(III) nitrate leads mostly to rearranged carbonyl compounds via organothallium compounds (E.C. Taylor, 1970, 1976; R.J. Ouellette, 1973; W. Rotermund, 1975; R. Criegee, 1979). Very useful reactions in complex syntheses have been those with olefins and ketones (see p. 121f.) containing conjugated aromatic substituents, e.g. porphyrins (G.W. Kenner, 1973; K.M. Smith, 1975).

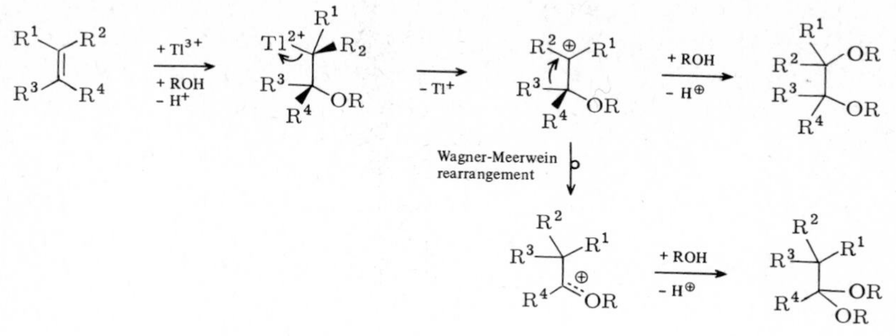

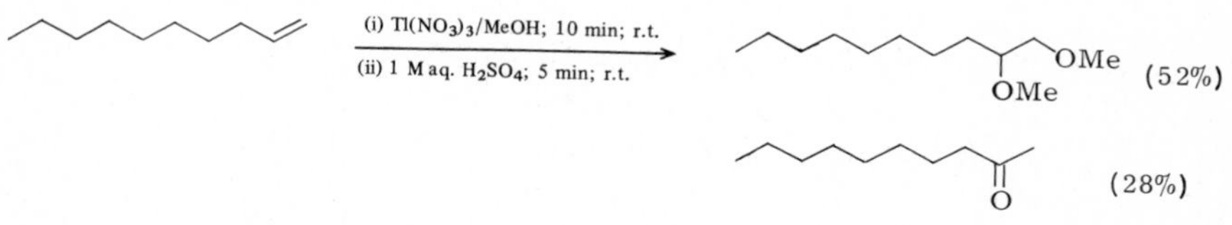

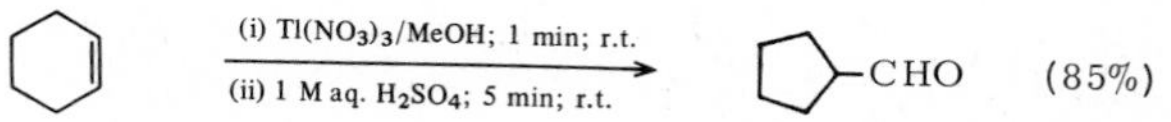

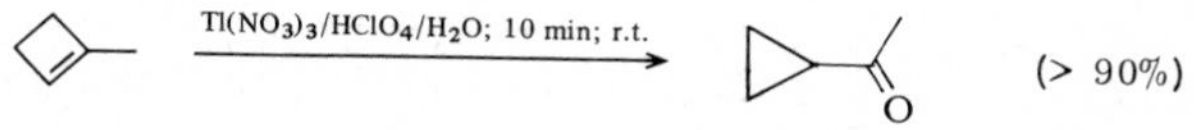

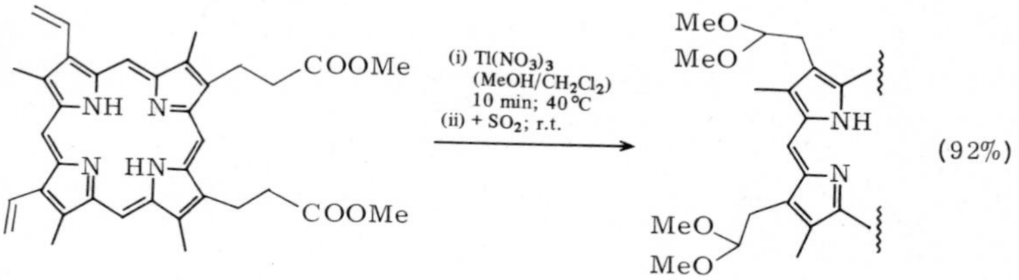

Regio- and stereoselective monohydroxylations of C=C double bonds are achieved in high yield by successive hydroboration and oxidative cleavage of the borane with H_2O_2 (G. Zweifel, 1963; H.C. Brown, 1975). This procedure is much more suitable for laboratory use than a similar one with organoaluminum compounds first discovered by K. Ziegler (1960). Boranes will add to the least hindered site of the C=C double bond, and oxidative cleavage proceeds with retention of configuration (H. Fuhrer, 1970). Upon reaction of thexylborane with dienes cyclic boranes may be formed, which yield diols on oxidative cleavage. Treatment of the cyclic boranes with acids leads to hydrogenation of the less hindered carbon atom (H.C. Brown, 1972 A).

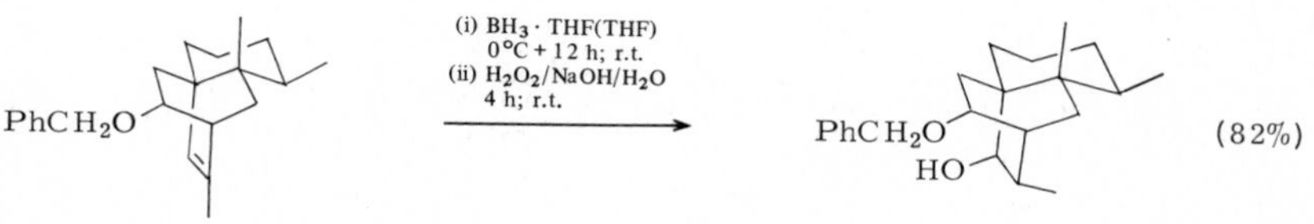

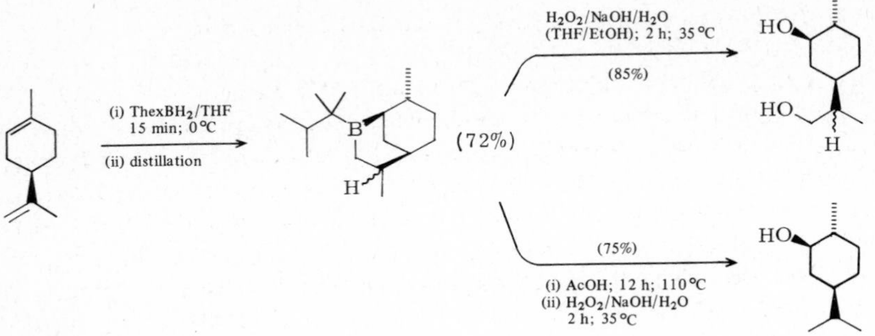

C≡C triple bonds are "hydrated" to yield carbonyl groups in the presence of mercury (II) ions (see p. 49) or by successive treatment with boranes and H_2O_2. The first procedure gives preferentially the most highly substituted ketone, the latter the complementary compound with high selectivity (T.W. Gibson, 1969).

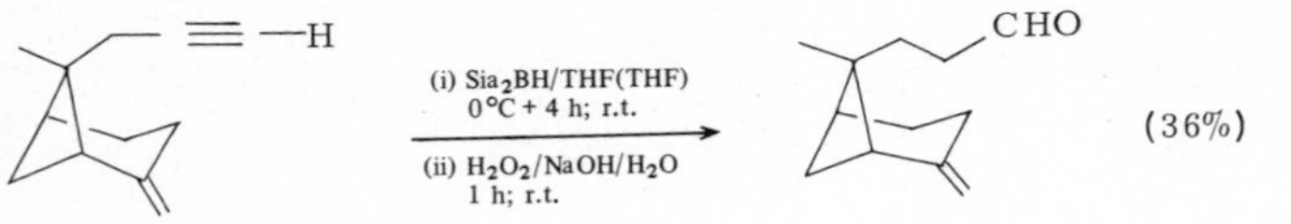

Internal alkynes are oxidized to acyloins by thallium(III) in acidic solution (A. McKillop, 1973; G.W. Rotermund, 1975) or to 1,2-diketones by permanganate or by *in situ* generated ruthenium tetroxide (D.G. Lee, 1969, 1973; H. Gopal, 1971). Terminal alkynes undergo oxidative degradation to carboxylic acids with loss of the terminal carbon atom with these oxidants.

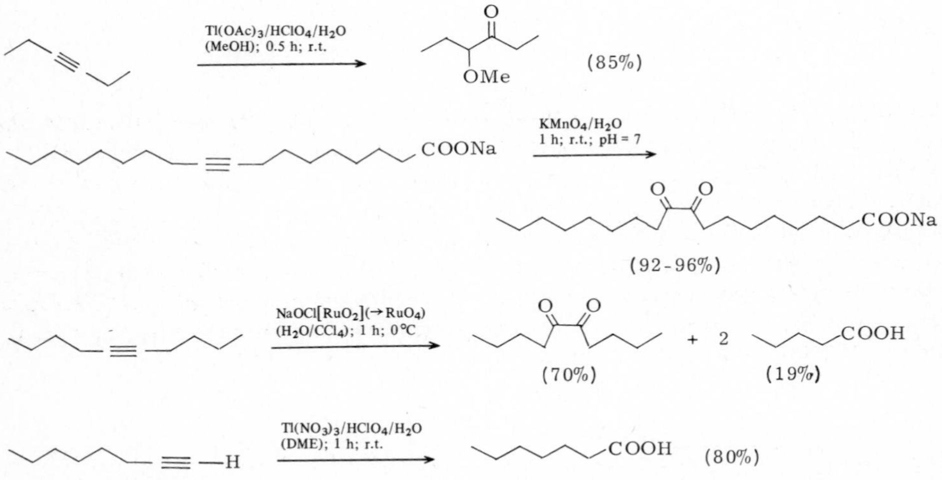

2.2.3 Oxidation of Alcohols to Aldehydes, Ketones, and Carboxylic Acids

The oxidation of primary alcohols to give aldehydes requires mild oxidants and careful control of reaction conditions. The CrO_3/pyridine complex has occasionally been used, but further oxidation to carboxylic acids is generally a problem. Di-*t*-butyl chromate (H.G. Bosche, 1975), MnO_2 (D. Arndt, 1975), or Ag_2CO_3 (D. Manegold, 1975) in aprotic solvents, lead tetraacetate in pyridine (G.W. Rotermund, 1975), copper(II) acetate (W.G. Nigh, 1973), and DDQ (D. Walker, 1967; H.H. Stechl, 1975) are more reliable oxidants. MnO_2 and high-potential quinones oxidize allylic alcohols with high selectivity, whereas copper(II) acetate has often been used for the synthesis of sensitive α-keto aldehydes. The widely used Moffatt-Pfitzner oxidation works with *in situ* formed adducts of dimethyl sulfoxide with dehydrating agents, e.g. DCC, Ac_2O, SO_3, P_4O_{10}, or $COCl_2$ (K.E. Pfitzner, 1965; A.H. Fenselau, 1966; J.G. Moffatt, 1971; D. Martin, 1971). A classical procedure is the Oppenauer oxidation with ketones and aluminum alkoxide catalysts (C. Djerassi, 1951; H. Lehmann, 1975). All of these reagents also oxidize secondary alcohols to ketones but do not attack $C=C$ double bonds or activated $C-H$ bonds.

$CH_3(CH_2)_{14}CH_2OH$ $\xrightarrow[\text{(ii) } H_2C_2O_4/H_2O]{\text{(i) } CrO_2(OBu^t)_2 \ \text{(PE); 14 d; r.t.}}$ $CH_3(CH_2)_{14}CHO$ (95%)

$\xrightarrow{MnO_2\text{(hexane); 0.5 h; 0°C}}$ (> 90%)

$\xrightarrow{MnO_2(Et_2O)\text{; 1 h; r.t.}}$ (64%)

$\xrightarrow{Ag_2CO_3(C_6H_6)\text{; 12 h; } \Delta}$ (95%)

$\xrightarrow{Cu(OAc)_2/O_2 \ \text{(MeOH); 15 min; r.t.}}$ (80-95%)

$\xrightarrow{ortho\text{-TCQ} \ (CCl_4)\text{; 15 h; r.t.}}$ (100%)

Moffatt-Pfitzner oxidation:

$Me_2SO + DCC + H^{\oplus}$

$\xrightarrow[\text{(ii) DnpNHNH}_2/\text{dil. aq. } H_2SO_4]{\text{(i) DCC/DMSO} \ [H_3PO_4]\text{; 20 h; r.t.}}$ Dnp-NH-N=   (61%)

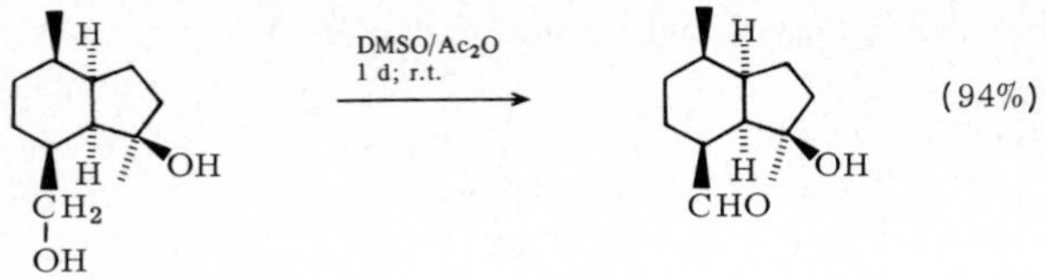

(94%)

Oppenauer oxidation:

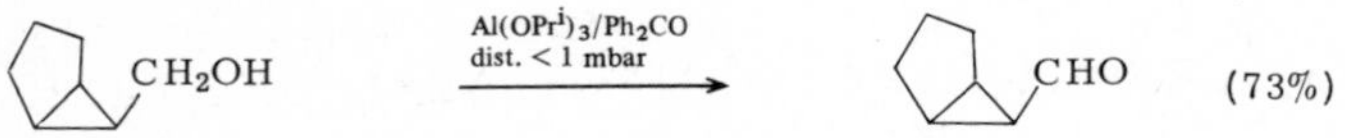

$Al(OPr^i)_3/Ph_2CO$
dist. < 1 mbar

(73%)

The conversion of primary alcohols and aldehydes into carboxylic acids is generally possible with all strong oxidants. Silver(II) oxide in THF/water is particularly useful as a neutral oxidant (E.J. Corey, 1968 A). The direct conversion of primary alcohols into carboxylic esters is achieved with MnO_2 in the presence of hydrogen cyanide and alcohols (E.J. Corey, 1968 A,D). The remarkably smooth oxidation of ethers to esters by ruthenium tetroxide has been employed quite often (D.G. Lee, 1973). Dibutyl ether affords butyl butanoate, and tetrahydrofuran yields butyrolactone almost quantitatively. More complex educts also give acceptable yields (M.E. Wolff, 1963).

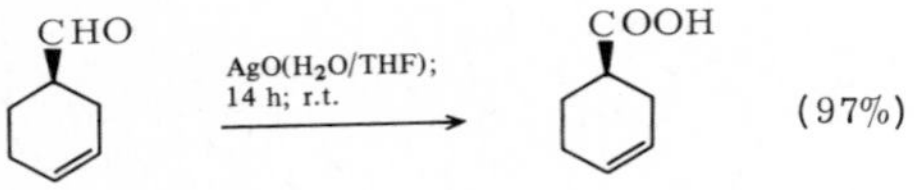

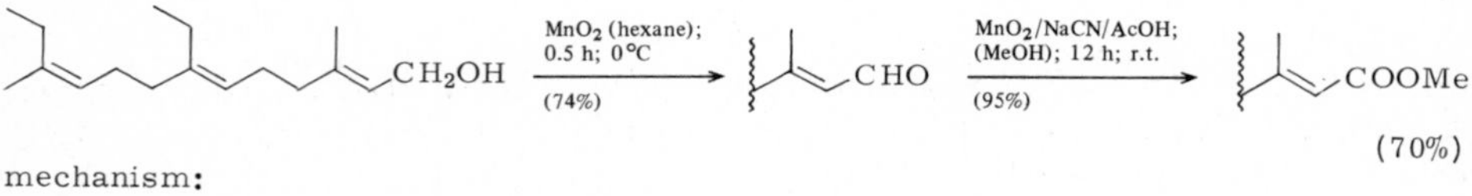

$AgO(H_2O/THF)$;
14 h; r.t.

(97%)

MnO_2 (hexane);
0.5 h; 0°C

(74%)

$MnO_2/NaCN/AcOH$;
(MeOH); 12 h; r.t.

(95%)

(70%)

mechanism:

$$R\text{–}CHO + HCN \rightleftharpoons R\text{–}\overset{H}{\underset{CN}{C}}\text{–}OH \xrightarrow[+ \ MeOH]{MnO_2} R\text{–}\overset{HO}{\underset{CN}{C}}\text{–}OMe \xrightarrow{-\ HCN} R\text{–}COOMe$$

RuO_4; (CCl₄);
3 d; r.t.

(42%)

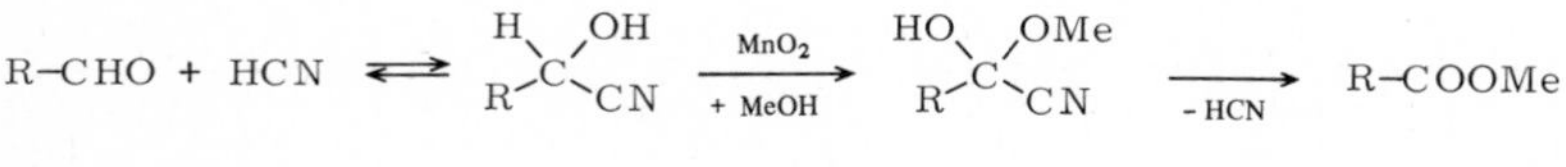

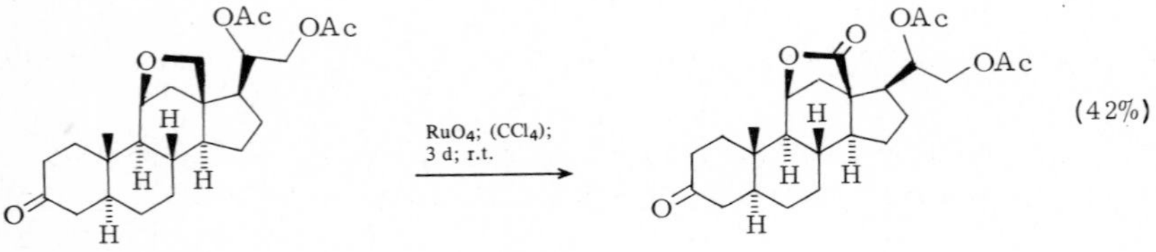

Selective oxidation of secondary alcohols to ketones is usually performed with CrO_3/H_2SO_4 1:1 in acetone (Jones' reagent) or with CrO_3Py_2 (Collin's reagent) in the presence of acid-sensitive groups (H.G. Bosche, 1975; C. Djerassi, 1956; W.S. Allen, 1954). As mentioned above, α,β-unsaturated secondary alcohols are selectively oxidized by MnO_2 (D.G. Lee, 1969; D. Arndt, 1975) or by DDQ (D. Walker, 1967; H.H. Stechl, 1975).

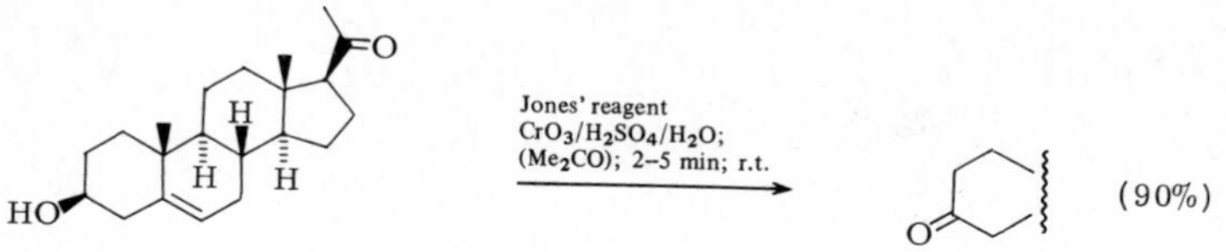

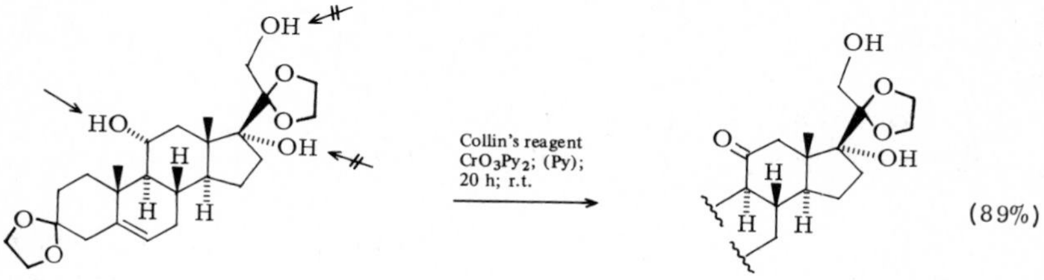

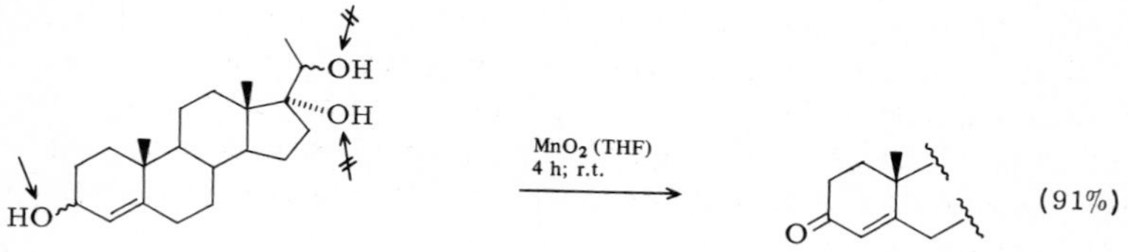

Tertiary alcohols are usually degraded unselectively by strong oxidants. Anhydrous chromium trioxide leads to oxidative ring opening of tertiary cycloalkanols (L.F. Fieser, 1948).

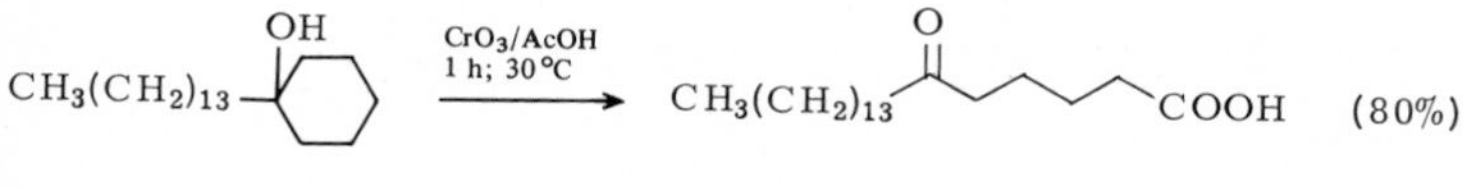

2.2.4 Oxidative Rearrangement and Cleavage of Ketones and Aldehydes

Ketones are rearranged oxidatively by reaction of the corresponding enols with thallium(III), e.g. to yield pyrroleacetic acids from acetyl pyrroles (G.W. Kenner, 1973 B; W. Rotermund, 1975).

Peracids convert ketones into esters in high yield. The peracid adds to the carbonyl group, and one carbon substituent migrates to the positively polarized peroxy oxygen (C.H. Hassall, 1957; W.D. Emmons, 1955; B. Plesničar, 1978). The migration tendency follows the

order *t*-alkyl > *s*-alkyl > benzyl > phenyl > *n*-alkyl > methyl, if no steric effects are coun-
teracting. This Baeyer-Villiger oxidation is particularly useful with bridged polycyclic ketones.
From these compounds lactones are formed, in which the new ester bond replaces a C—C
single bond with complete retention of configuration. Thus, upon hydrolysis of the lactone, a
cis-disubstituted ring results (J. Meinwald, 1960; R.R. Sauers, 1961). Aldehydes are converted
into sensitive formate esters by selenous peracid (F. Nakatsubo, 1970). Ketones may also be
converted into amides or lactams via the similar Beckmann rearrangement of ketoximes
(W.Z. Heldt, 1960).

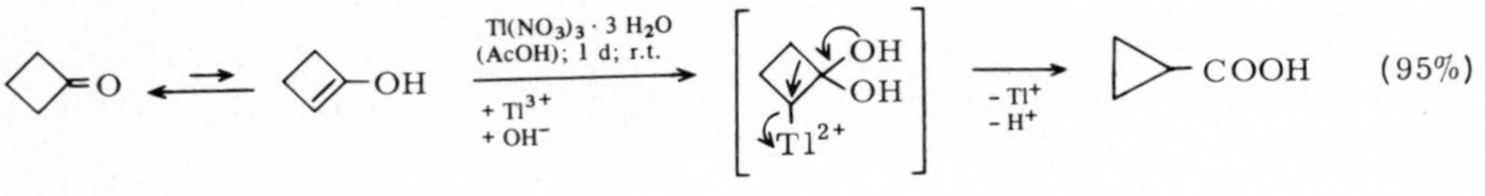

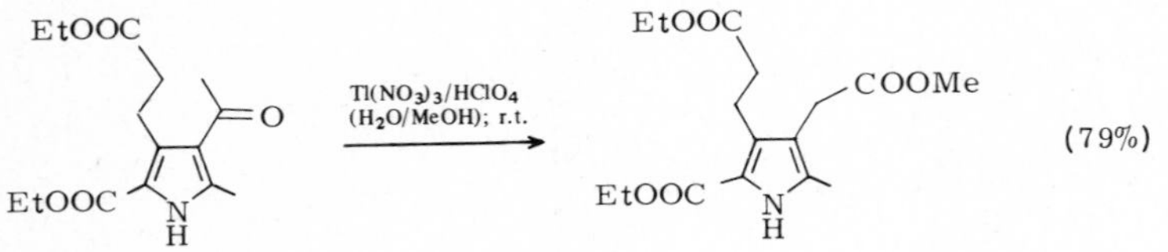

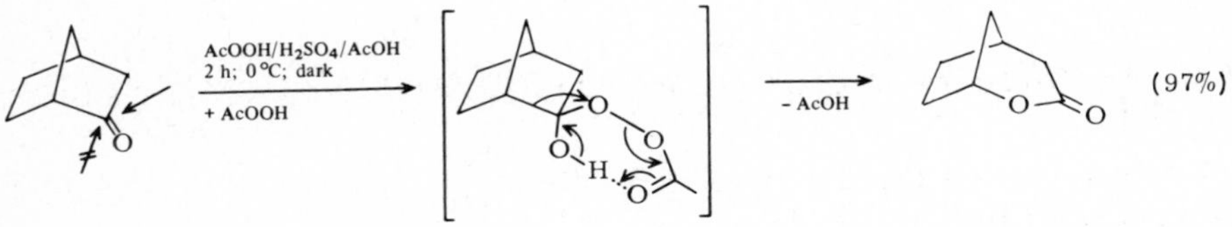

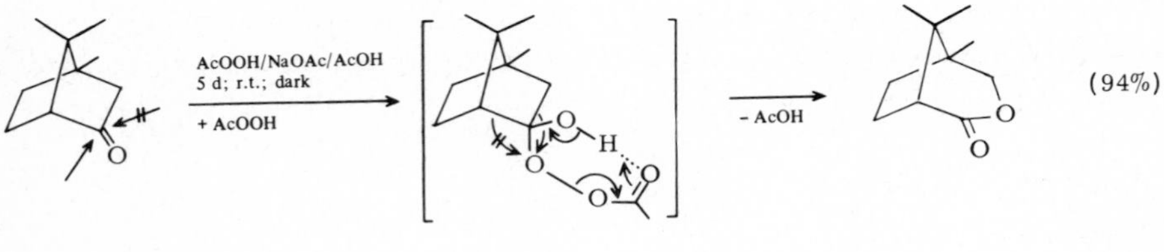

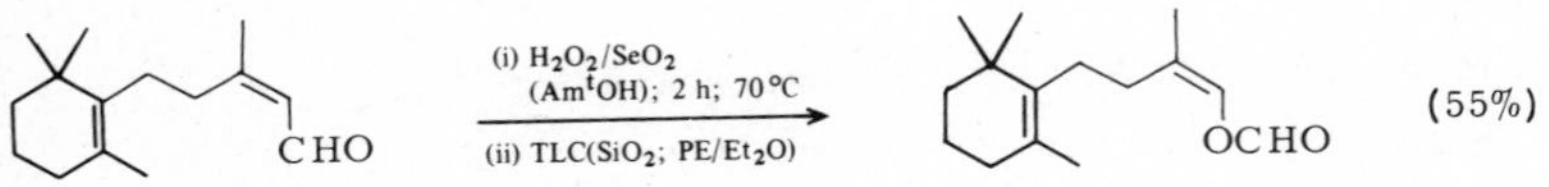

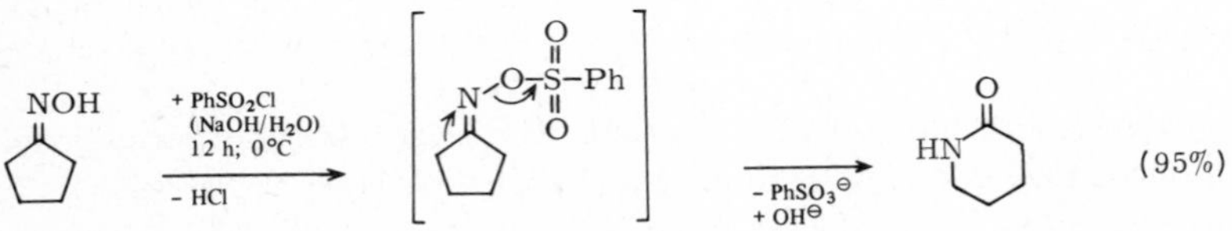

Saturated cyclic ketones are cleaved by CrO_3, by $KMnO_4$, or by successive treatment with SeO_2 and periodate to give open-chain α, ω -dicarboxylic or ω -acyl carboxylic acids in good yield (K.B. Wiberg, 1965; D. Arndt, 1975). This reaction should be taken into consideration together with the ozonolysis of cycloalkenes (p. 80) and the *retro*-aldol cleavage reaction of Hünig (p. 81) for the synthesis of difunctional compounds with distances of five or more bonds between the functional groups.

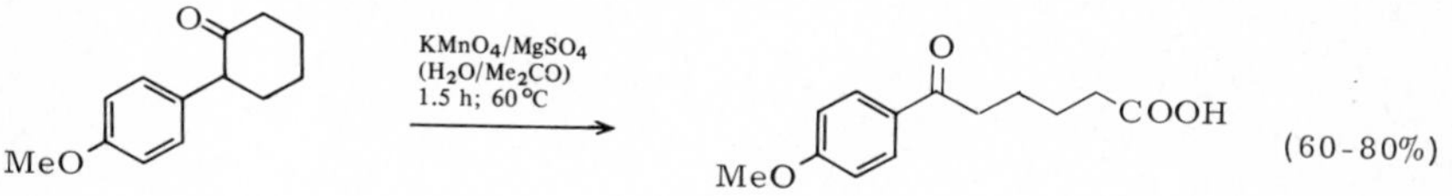

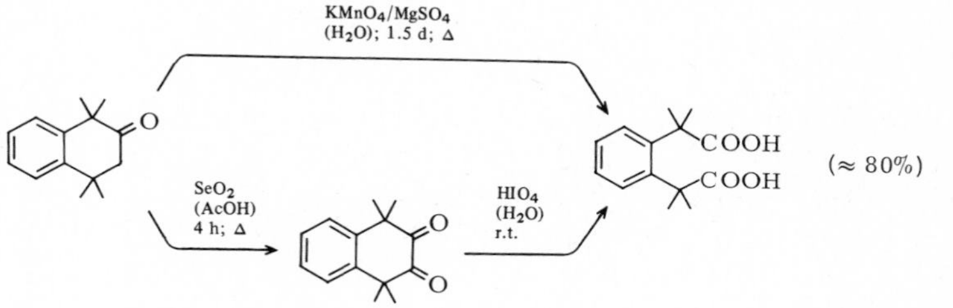

2.3 Olefin Syntheses by Dehydrogenation and Other Elimination Reactions

Classic procedures for $C{=}C$ double bond formation involve β-elimination of two vicinal substituents X and Y from a C—C single bond.

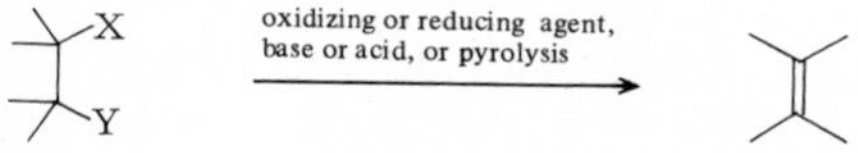

X, Y = H, OH, $OH_2^{\oplus}$, OMes, OTos, Cl, Br, I,

   $NMe_3^{\oplus}$, $SMe_2^{\oplus}$, OOCR, etc.

Table 3 summarizes some selective elimination reactions of synthetic interest.

Table 3. Selective elimination reactions.

● *Dehydrogenations (= oxidative β-eliminations)*

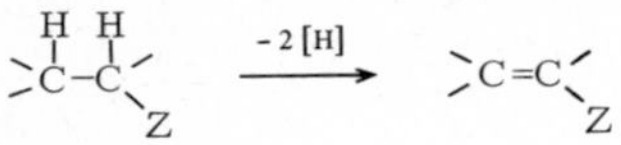

Z = carbonyl, vinyl, aryl SeO_2; PhSeBr; quinones
Z = NR_2 (tertiary amines) $Hg(OAc)_2$

● *Acid-, base-, or heat-induced β-eliminations*

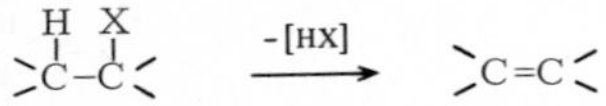

X = Cl/Br/I   strong bases; $Li^{\oplus}$/DMF
X = $OH/OH_2^{\oplus}$ strong acids; DMSO/Δ
    OH→O–S–OMes: mesyl sulfite CH_3SO_2Cl/SO_2/weak base

X = $NMe_3^{\oplus}$, $SMe_2^{\oplus}$, $\overset{\oplus}{N}Me_2$ base/Δ
                                   $\underset{O^{\ominus}}{|}$

X = RCOO (ester) $\Delta: -RCOOH$ } ester pyrolysis
X = RCSO (thionoester) $\Delta: -RCOSH$ }

X = ROCSO (thionocarbonate) $\Delta: -ROH$ $-OCS$ } Chugaev
X = RSCSO (xanthate) $\Delta: -RSH$ $-OCS$ } reaction

● *Reductive β-eliminations*

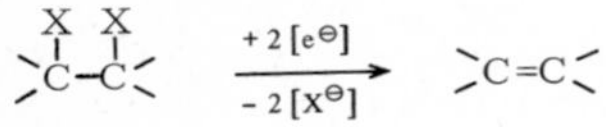

X, X = Cl/Br/I   Mg; Zn; NaI/acetone
                                      PhLi; $Na^{\oplus}Dmso^{\ominus}$

X, X = $\diagup^{O}\diagdown$ Zn/NaI
X, X = $\diagup^{S}\diagdown$ R_3P; $(RO)_3P$ (see section 1.5)
X, X = OH   (i) $Im_2C=S$ (ii) $(RO)_3P/\Delta$

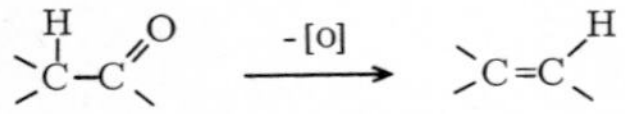

(i) TosNHNH₂ (ii) RLi

● *Degradation of carboxylic acid derivatives*

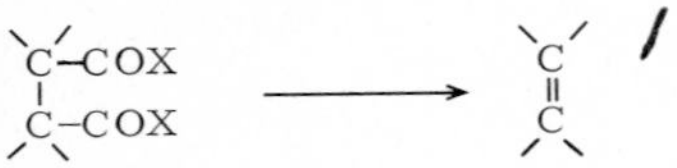

X = OH: $Pb(OAc)_4$/Py
                                      X = $OOBu^t$: $h \cdot \nu$

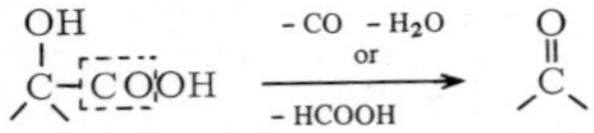

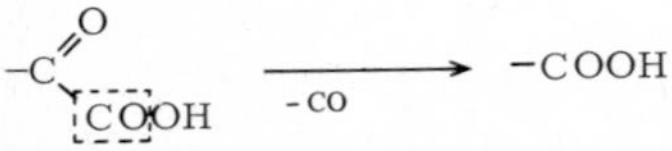

$H^{\oplus}/\Delta$ }
                                                          } decarbonylation
$H^{\oplus}/\Delta$ }

Table 3. (Continued)

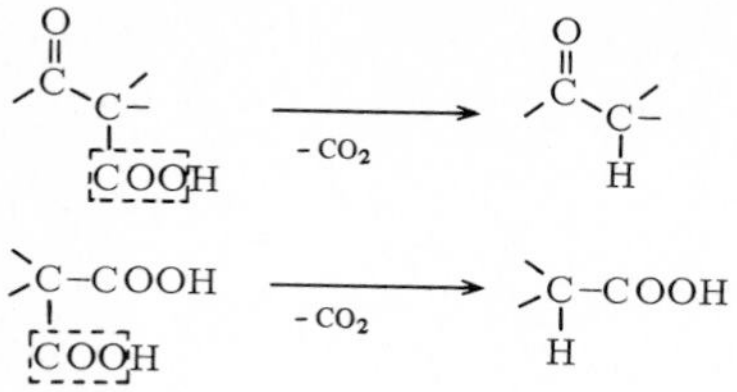

	$CO_3^{2\ominus}$, $H^{\oplus}/\Delta$	$\left.\begin{array}{l} \\ \\ \\ \\ \end{array}\right\}$ decarboxylation
	$H^{\oplus}/\Delta$	
	(i) RCOOEt + N_2H_4 (ii) HNO_2, RONO (iii) Δ or (i) RCOCl + $N_3^{\ominus}$ (ii) Δ (see next section)	
	$NaOCl/OH^{\ominus}$; $Br_2/OH^{\ominus}$	

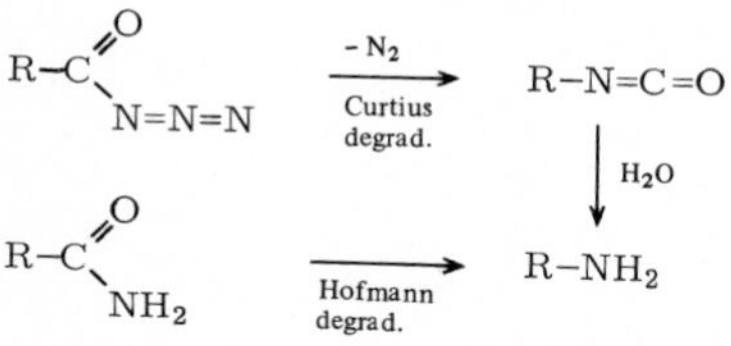

If $X = Y = H$, the reaction is called dehydrogenation. Synthetically useful are dehydrogenations of tertiary amines by $Hg(OAc)_2$ via immonium ions (see p. 111; A. Friedrich, 1975) and of carbonyl-, vinyl-, or aryl-activated C—C single bonds by SeO_2 (see p. 112f.; N. Rabjohn, 1949, 1976), or better by phenylselenium bromide (H.J. Reich, 1975, 1978, 1979), or by quinones (see p. 299; D. Walker, 1967; H.H. Stechl, 1975). Cross-conjugated cyclohexadienones from dehydrogenations of cyclohexanones are always easily isomerized into phenols by acids (J.N. Marx, 1974). A more special example is the exhaustive dehydrogenation of partially unsaturated polycyclic skeletons to give aromatic systems in the presence of Pd, Pt, or Se catalysts (W.E. Barth, 1971).

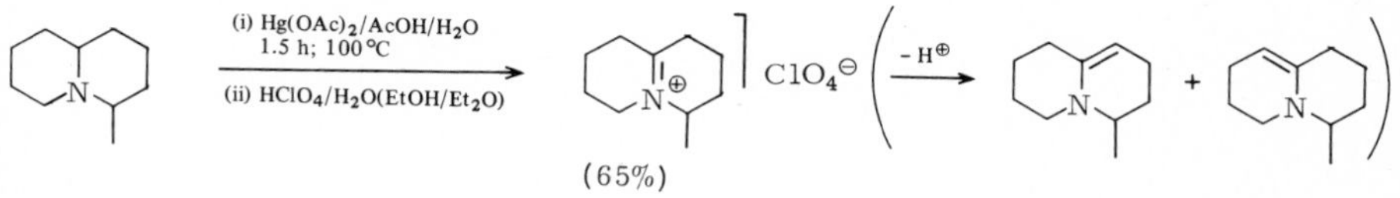

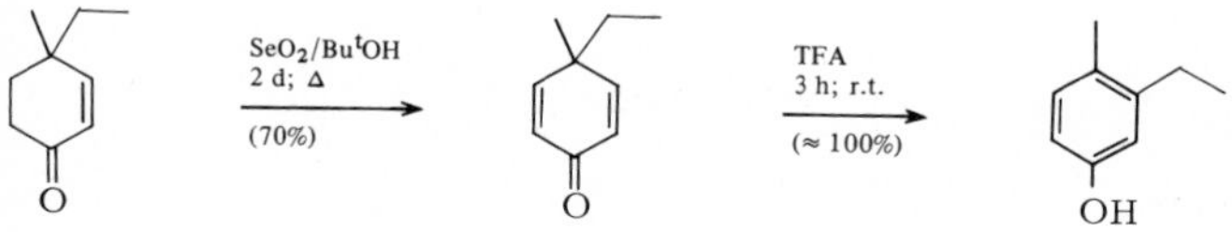

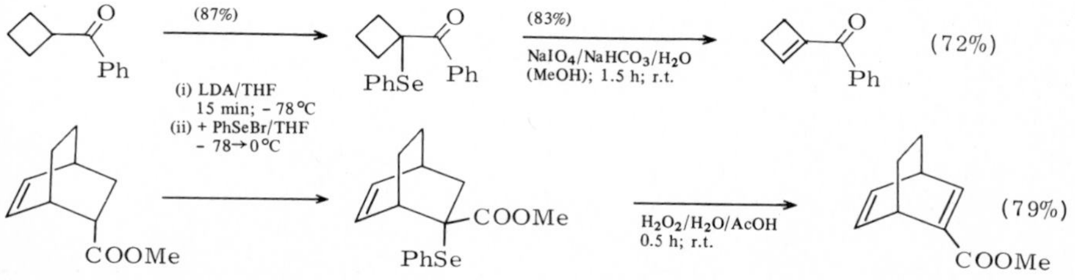

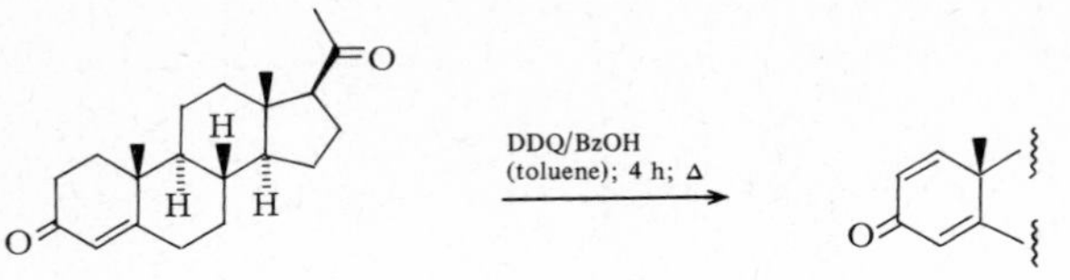

(88%)

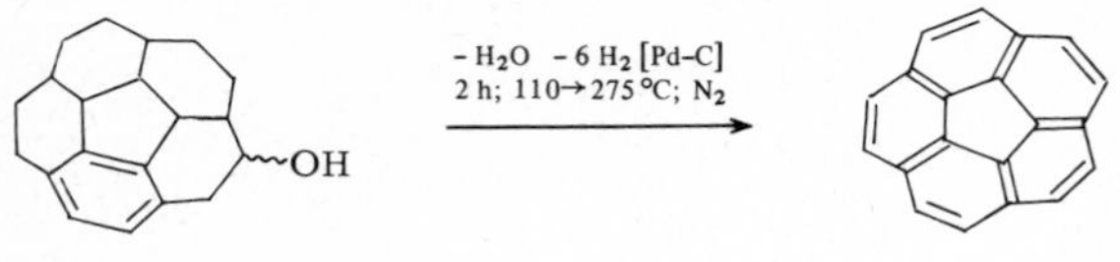

(50%)

More general procedures for the introduction of C=C double bonds involve removal of a leaving group X and a vicinal hydrogen atom, e.g. dehydrohalogenations (X = Cl/Br/I; with Li⁺: E.J. Corey, 1965; G. Jones, 1970), dehydrations (X = OH/OH_2^+, $MesOSO_2$: G.G. Hazen, 1964; in DMSO: V.J. Traynelis, 1962, 1964), base-induced Hofmann type eliminations (X = SMe_2^+: see p. 37; NMe_3^+: R.B. Woodward, 1961; $(N^+O^-)Me_2$: A.C. Cope, 1960 B), and ester pyrolyses (X = RCOO: W. Hückel, 1963; ROCSO: L.A. Paquette, 1977; RSCSO: H.R. Nace, 1962). Although some regioselectivity is found in these eliminations (Saytsev and Hofmann rules), they are generally only acceptable as synthetic reactions if only one hydrogen atom vicinal to the leaving group X is present or particularly activated.

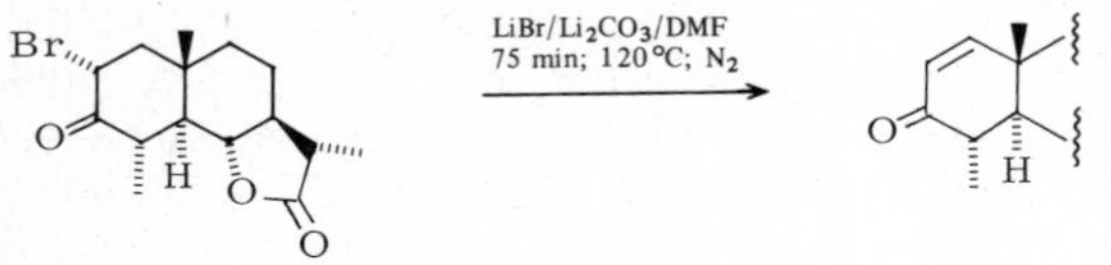

(80-90%)

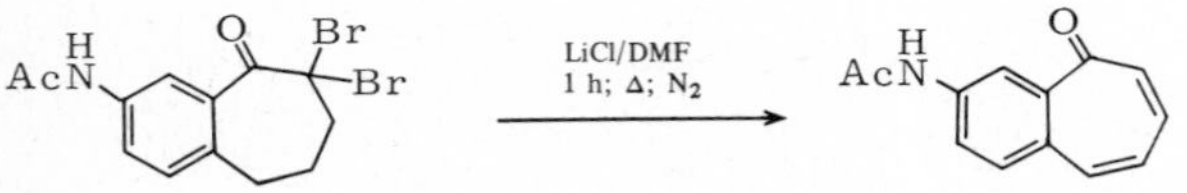

(95%)

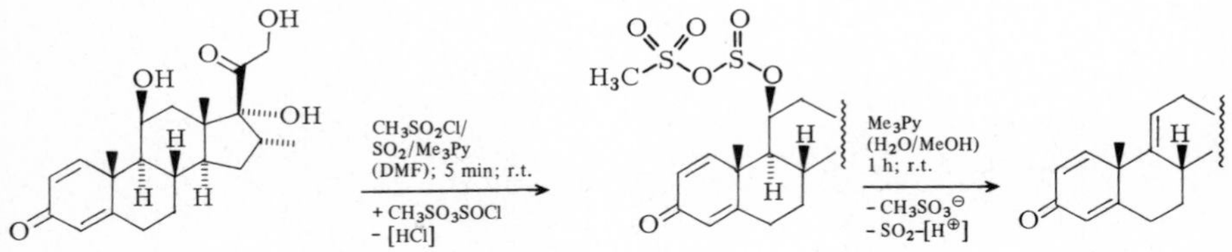

(96%)

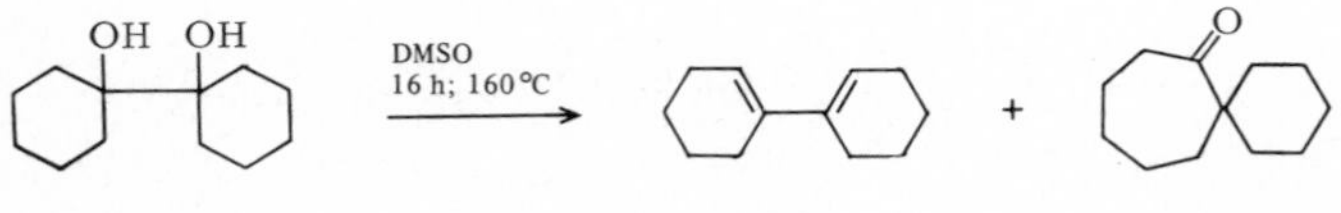

(85%)   +   only 4%
                    pinacolone

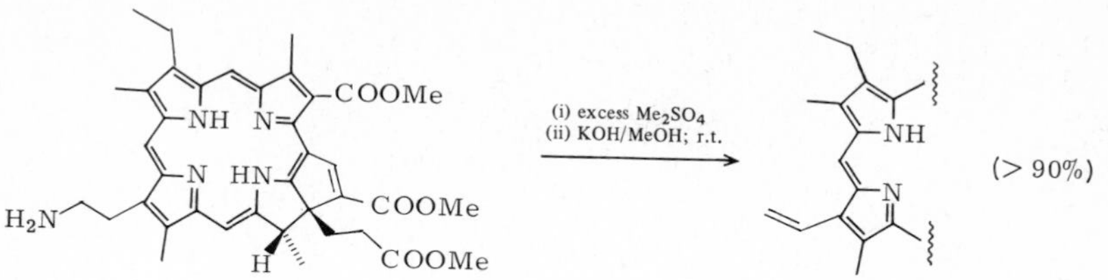

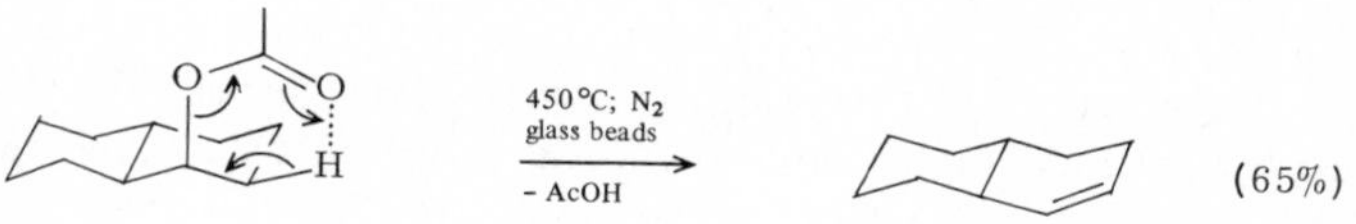

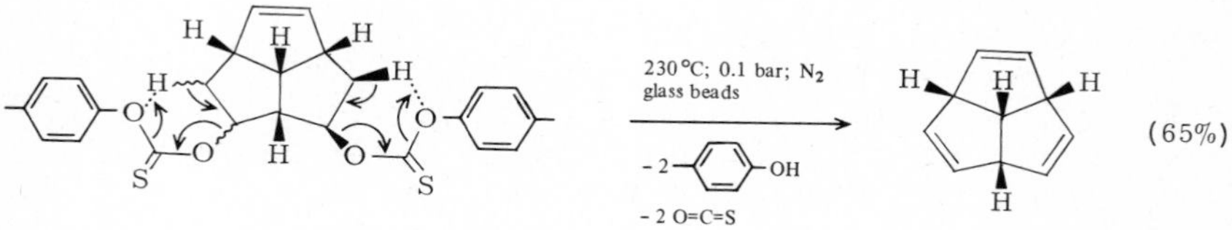

From ketones with an α-hydrogen atom alkenes are also obtained via the corresponding tosylhydrazones. Two equivalents of a strong base, e.g. MeLi, deprotonate the α-carbon atom as well as the hydrazone nitrogen, and upon loss of lithium p-toluenesulfinate and N_2 a lithium alkenide is formed. Olefins, which are difficult to obtain because of their tendency to undergo rearrangements, have been prepared in excellent yields by this method (R.H. Shapiro, 1976). Upon tosylhydrazone decomposition with sodium methoxide at elevated temperatures, however, carbenes are formed, which may undergo intramolecular insertion reactions, rearrangements, or fragmentations (e.g. α,β-epoxy toyslhydrazones, see pp. 81).

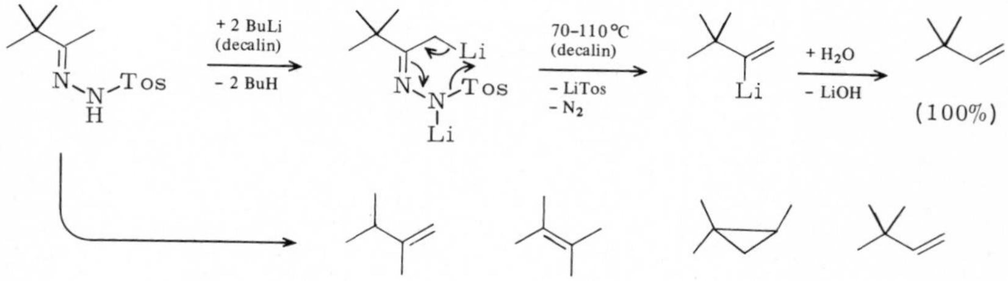

NaOMe/Et$_2$CO$_3$; Δ:	—	—	47%	52%
NaOMe/DEG; 140-180 °C:	56%	16%	11%	16%

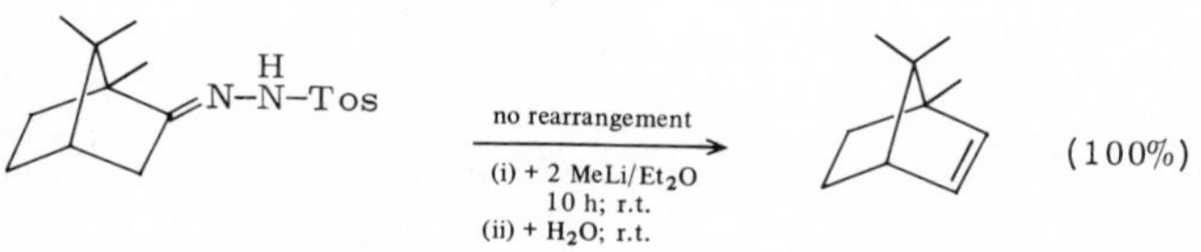

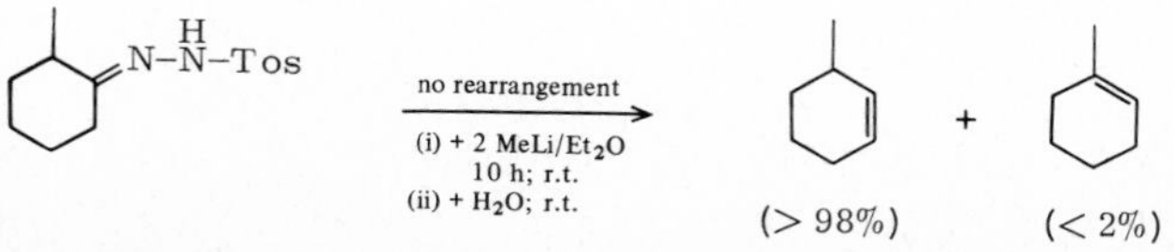

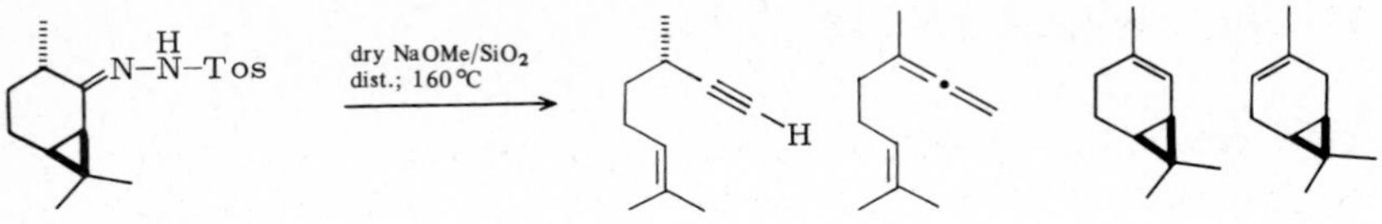

dry NaOMe/SiO$_2$
dist.; 160 °C

38%   1%   6%   5%

Regioselectivity of C=C double bond formation can also be achieved in the reductive or oxidative elimination of two functional groups from adjacent carbon atoms. Well established methods in synthesis include the reductive cleavage of cyclic thionocarbonates derived from glycols (E.J. Corey, 1968 C; W. Hartmann, 1972), the reduction of epoxides with Zn/NaI or of dihalides with metals, organometallic compounds, or NaI/acetone (see p. 143), and the oxidative decarboxylation of 1,2-dicarboxylic acids (C.A. Grob, 1958; S. Masamune, 1966; R.A. Sheldon, 1972) or their *t*-butyl peresters (E.N. Cain, 1969).

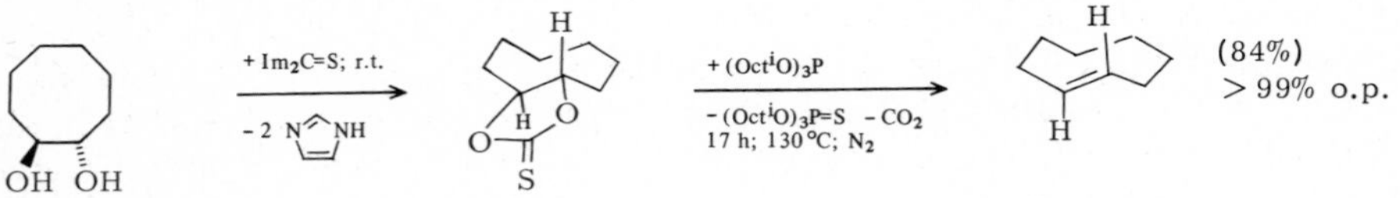

+ Im$_2$C=S; r.t.

− 2 N⁀NH

+ (Oct^{i}O)$_3$P

− (Oct^{i}O)$_3$P=S − CO$_2$
17 h; 130 °C; N$_2$

(84%)
> 99% o.p.

OH OH

(S,S)-trans

(−)-trans

(MeO)$_3$P

1 d; 110 °C; N$_2$

(91%)

Pb(OAc)$_4$; N$_2$
(C$_6$H$_6$/Py); 2 h; Δ

COOH

COOH

COOEt

COOEt

(71%)

COOH

COOH

Pb(OAc)$_4$/Py
2 h; 65 °C

(30%)

h·ν; r.t.
(C$_6$H$_6$); r.t.

(40%)

OOBu^t

OOBu^t

basketene

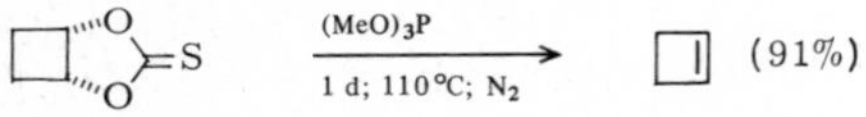

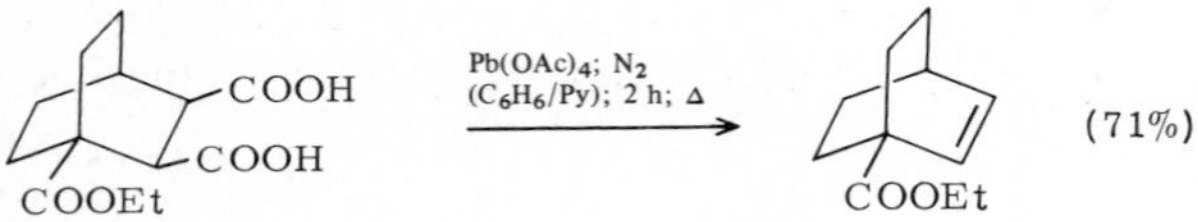

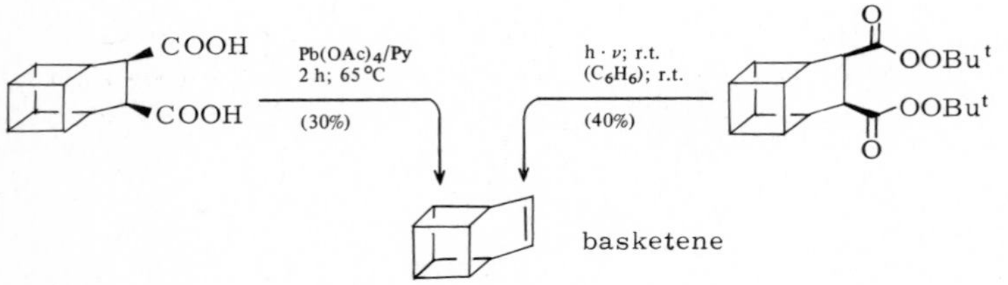

2.4 Synthesis of Carboxylic Acid Derivatives

In synthetic target molecules esters, lactones, amides, and lactams are the most common carboxylic acid derivatives. In order to synthesize them from carboxylic acids one has generally to produce an activated acid derivative, and an enormous variety of activating reagents is known, mostly developed for peptide syntheses (M. Bodanszky, 1976). In actual syntheses of complex esters and amides, however, only a small selection of these remedies is used, and we shall mention only generally applicable methods. The classic means of activating carboxyl groups are the acyl azide method of Curtius and the acyl chloride method of Emil Fischer.

The most practical route to acyl azides starts with the hydrazinolysis of esters. The usually rather poorly soluble and poorly stable hydrazides are dissolved in mixtures of organic solvents (e.g. THF, DMF, AcOH) and strong acids (e.g. HCl, TFA) and then mixed with an equimolar amount of sodium nitrite or amyl nitrite at -10 °C to yield the azide almost instantaneously. The coupling step with amines at room temperature may require several days. A great advantage of the acyl azide method is the lack of α-racemization (see p. 210f.). The acyl chloride method is quicker and may also be applied for the preparation of esters and amides. Here the free acid is used as starting material, and aprotic solvents (CHCl$_3$, DMF, pyridine) must be applied in the chlorination. Thionyl chloride and oxalyl chloride are the most common agents, and low temperatures are again advantageous. Nitrosation or chlorination of activated CH groups by nitrites or SOCl$_2$, resp., are sometimes troublesome side reactions.

Acyl azides may loose N$_2$ on heating and rearrange to isocyanates (Curtius rearrangement), which may be solvolyzed. Some of the possibilities of classical carboxyl conversions are exemplified in the schemes below, which are taken from a triquinacene synthesis (R. Russo, 1971; C. Mercier, 1973) and the ergotamine synthesis of A. Hofmann (1963).

In these cases the acyl azides formed have been used to prepare amines via Curtius rearrangement. The acyl chloride or azide intermediates can, however, also be reacted with amines or alcohols to form amides or esters.

As a catalyst for ester and amide formation from acyl chlorides or anhydrides, 4-(di-methylamino)pyridine has been recommended (DMAP; G. Höfle, 1978). In the presence of this agent highly hindered hydroxyl groups, e.g. of steroids and carbohydrates, are acylated under mild conditions, which is difficult to achieve with other catalysts.

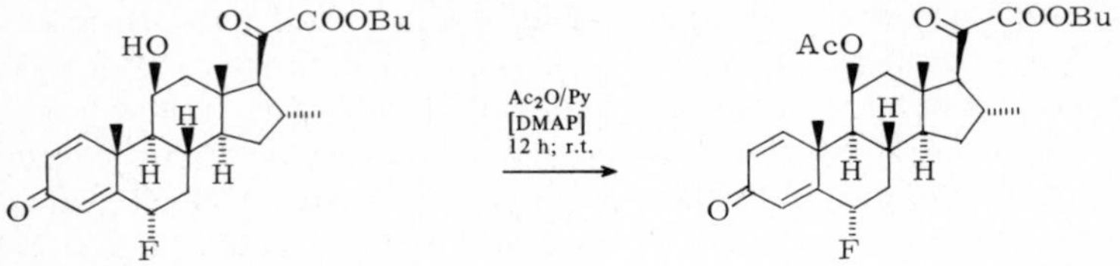

The most popular routes to esters and amides nowadays employ N,N'-dicyclo-hexylcarbodiimide (DCC) as coupling reagent. This reagent functions as a mild dehydrating agent and converts acids into active O-acylisoureas or alcohols into similarly active O-alkylisoureas, if Cu(I) is added as a catalyst. The pathway *via* O-alkylisoureas is more reliable than the one *via* O-acylisoureas because the latter easily rearrange to form inactive N-acylureas (L. Mathias, 1979). Condensations using DCC are also subject to catalysis by DMAP (G. Höfle, 1978). The fundamental synthetic advantage of DCC and related compounds as dehydrating agents is the possibility to work in aprotic, non-acidic and non-basic solutions. This fact allowed, for example, syntheses of acid sensitive β-lactam antibiotics, which are difficult to obtain by other methods (J.C. Sheehan, 1959). The sheme below also gives examples of some unusual dehydrations using DCC.

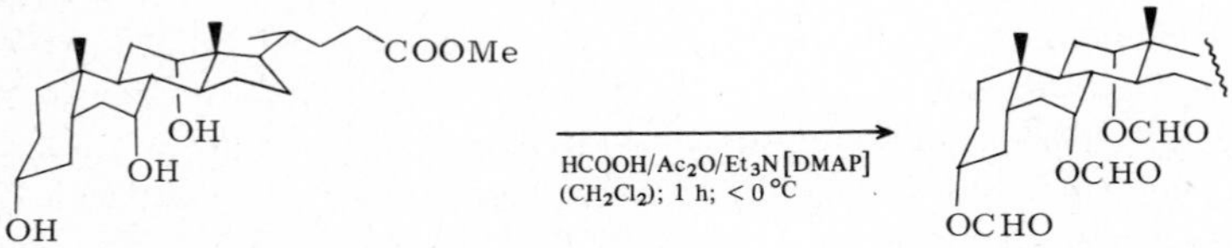

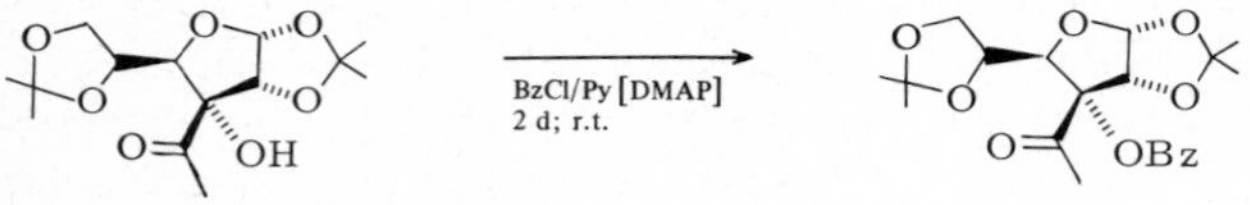

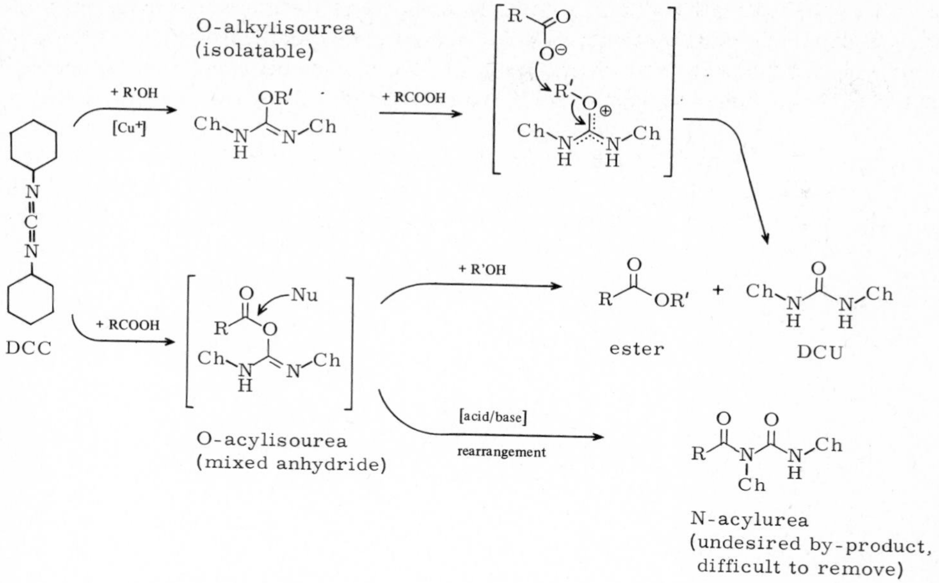

O-alkylisourea
(isolatable)

+ R'OH
[Cu⁺]

+ RCOOH

DCC

+ RCOOH

O-acylisourea
(mixed anhydride)

+ R'OH

$$R \overset{O}{\underset{}{\parallel}} OR' \quad + \quad Ch\overset{O}{\underset{}{\parallel}}Ch$$

ester

DCU

[acid/base]

rearrangement

N-acylurea
(undesired by-product,
difficult to remove)

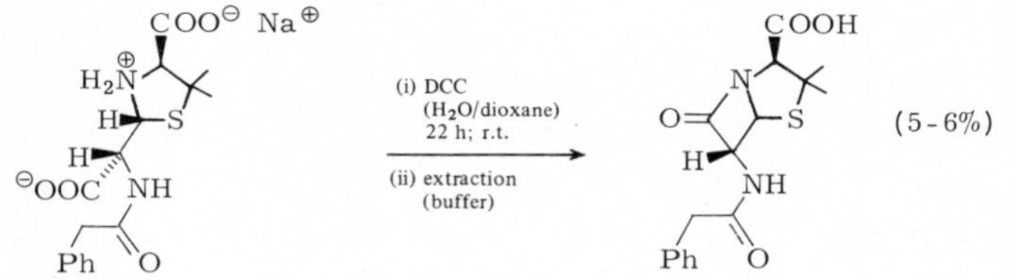

(i) DCC
(H₂O/dioxane)
22 h; r.t.

(ii) extraction
(buffer)

(5-6%)

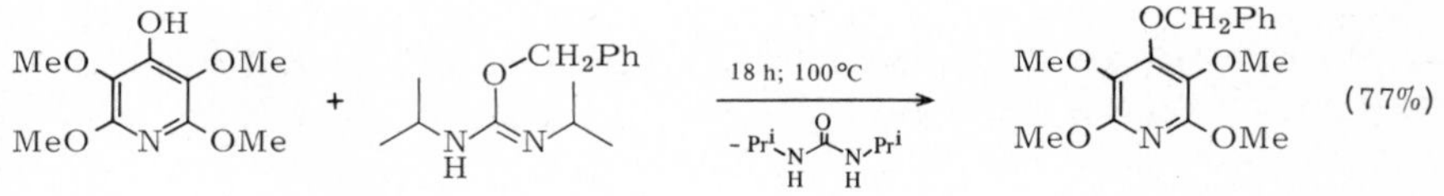

18 h; 100°C

(77%)

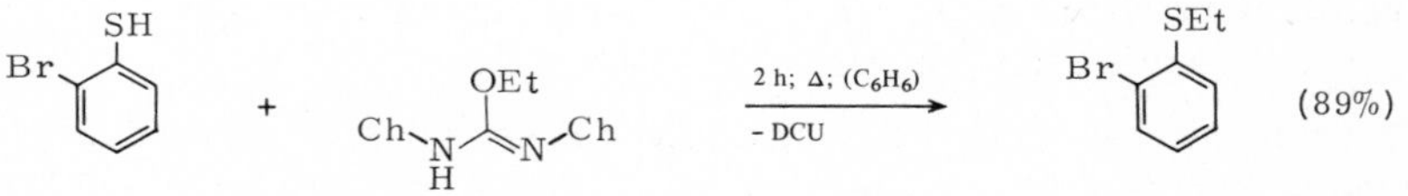

2 h; Δ; (C₆H₆)

– DCU

(89%)

The synthesis of five-, six-, and seven-membered cyclic esters or amides uses intramolecular condensations under the same reaction conditions as described for intermolecular reactions. Yields are generally excellent. An example from the colchicine synthesis of E.E. van Tamelen (1961) is given below. The synthesis of macrocyclic lactones (macrolides) and lactams ($n > 8$), however, which are of considerable biochemical and pharmacological interest, poses additional problems because of competing intermolecular polymerization reactions (see p. 221ff.). Inconveniently high dilution, which would be necessary to circumvent this side-reaction, can be avoided if both hydroxyl and the carboxyl groups are activated for mutual interaction before the actual condensation occurs. One way to achieve this is the use of a basic substituent bound to the active carboxyl derivative, which seizes and activates the remote hydroxyl group by hydrogen bonding. Successful derivatives of this kind are S-(2-pyridyl) and S-(2-imidazolyl) thioesters. If silver ions ($AgClO_4$, $AgBF_4$) are used to activate further the 2-pyridinethiol esters by complexation, cyclization occurs at room temperature in benzene solution. The best reagents to form the thioesters of 2-mercapto-pyridine and -imidazole with carboxylic acids are the corresponding disulfides together with triphenylphosphine (K.C. Nicolaou, 1977).

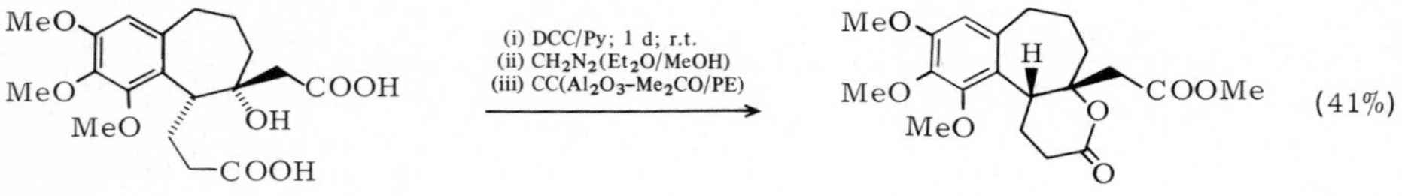

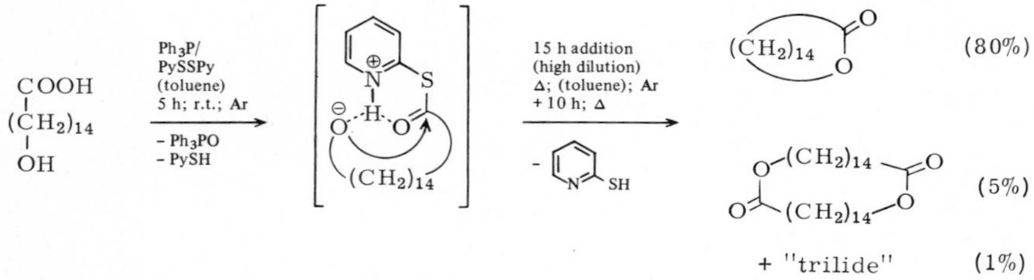

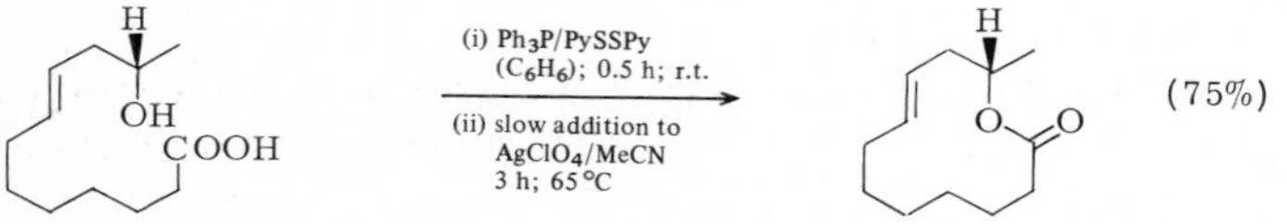

Syntheses of macrocyclic lactams and heterocycles are discussed in sections 4.1.2.5, 4.1.3, and 4.2.

2.5 Nitrogen Heterocycles

"Organic chemistry is largely comprised of heterocyclic chemistry. As a consequence of the size of the subject, any single volume is bound to omit more systems than it includes." This was said in the preface of a book on heterocycles of 1200 pages containing about 4000 references (P.G. Sammes, 1979). When we describe the synthesis of heterocycles within a few pages here, the picture given has to be vague and very probably trivial. It is nevertheless done because a general outline may put some system into the soul of the uninitiated student of heterocyclic chemistry. For further, more specific study the reader is referred to the recent excellent books by R.M. Acheson (1976 B) and P.G. Sammes (1979). We restrict ourselves to the synthesis of some common nitrogen heterocycles of more general interest.

Most syntheses of nitrogen heterocycles involve substitution and/or condensation reactions of nitrogen nucleophiles with difunctional halides or carbonyl compounds. Common nitrogen reagents are:

N_1: NH_3, $R-NH_2$, NH_2OH, $R-SO_2NH_2$

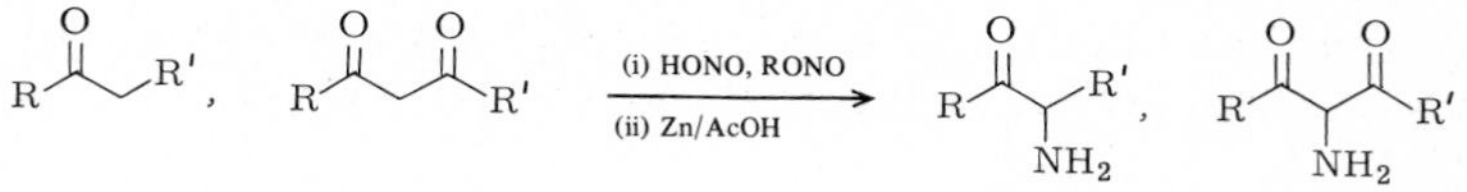

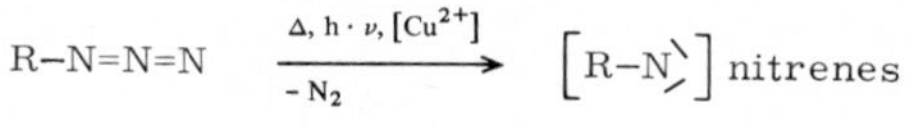

N_2: H_2N-NH_2, $R-NH-NH_2$, $Tos-NH-NH_2$

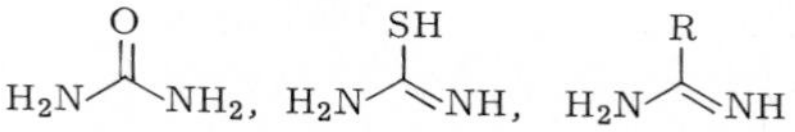

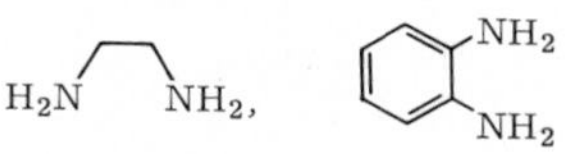

N_3: $R-N=N=N$ azides

The longest carbon chain within a heterocycle indicates possible open-chain precursors. We use this chain as a basis to classify heterocycles as 1,2- to 1,6-difunctional systems.

1,1

(diaziridines)   (sym-triazines)

1,2

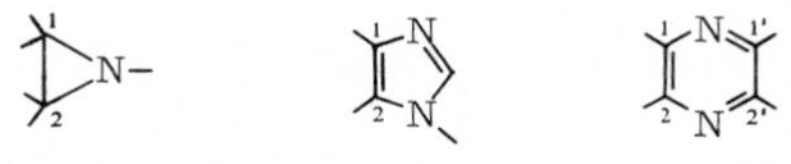

(aziridines)   (imidazoles) (pyrazines)

1,3

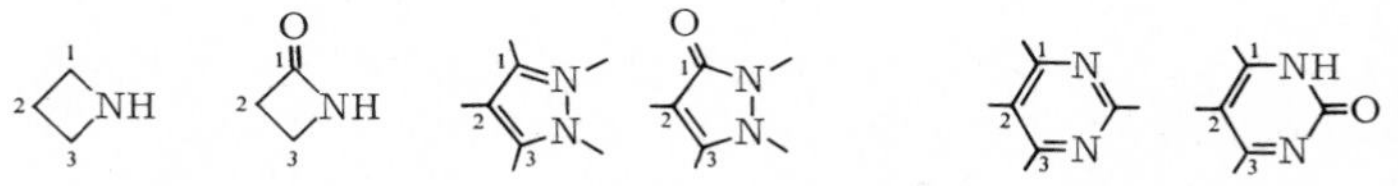

(azetidines, -ones)   (pyrazoles, -olones)   (pyrimidines, -ones)

1,4

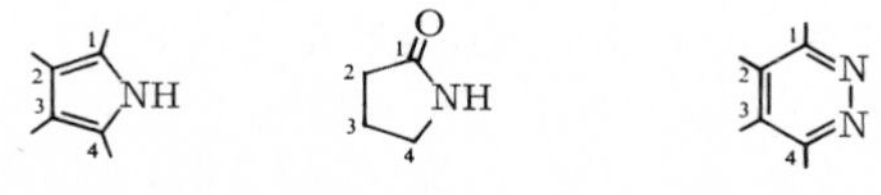

(pyrroles)   (γ-lactams)   (pyridazines)

1,5

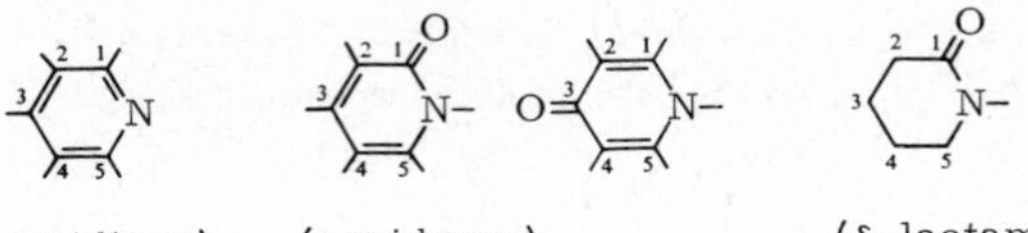

(pyridines)   (pyridones)   (δ-lactams)

1,6

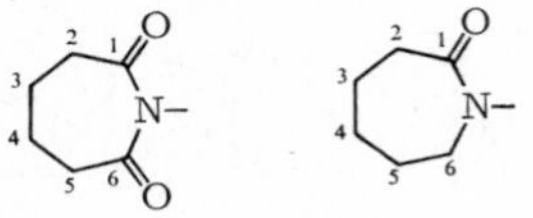

(adipimides)   (ε-lactams)

Since 1,2- to 1,6-difunctional open-chain compounds can be synthesized by general procedures (see chapter 1), it is useful to consider them as possible starting materials for syntheses of three- to seven-membered heterocycles: 1,2-heterocycles can be made from 1,2-difunctional compounds, e.g. olefins or dibromides; 1,3-difunctional compounds, e.g. 1,3-dibromides or 1,3-dioxo compounds, can be converted into 1,3-heterocycles etc.

For many heterocycles straightforward syntheses can be proposed employing only simple nitrogen containing reagents. Regioselectivity is mostly no problem in these synthetic reactions, since the nitrogen reagents react symmetrically. The oxidation states of the ring atoms in the heterocycle to be synthesized are determined by the educts. Another important fact is, that formation of heteroaromatic systems (e.g. imidazoles, pyrroles, pyridines) is thermodynamically favored and thus often achieved under acidic or oxidative reaction conditions. Oxygen as a substituent on nitrogen heterocycles, however, tends to be double-bonded, and the N-protonated tautomers generally prevail in equilibrium.

1,2

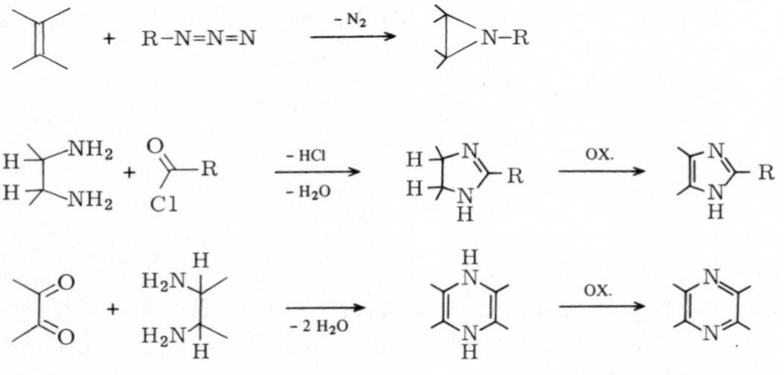

1,3

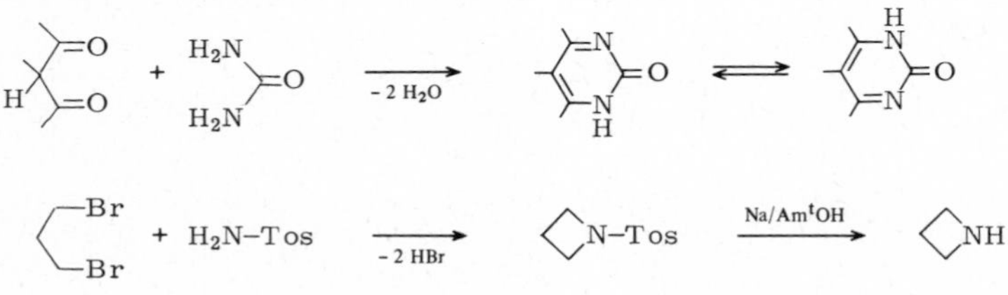

1,4

$$\text{(1,3-diketone)} + H_2N-R \xrightarrow{-H_2O} \left[\text{(hydroxy-pyrrolidine)} \right] \xrightarrow{-H_2O} \text{(N-R pyrrole)}$$

$$\text{(unsaturated 1,4-diketone)} + \begin{array}{c} NH_2 \\ NH_2 \end{array} \xrightarrow{-2 H_2O} \text{(pyridazine)}$$

1,5

$$\begin{array}{c} COOMe \\ R \underset{COOMe}{\overset{O}{\diagup}} \end{array} + NH_2OH \xrightarrow{-2 H_2O} \left[\begin{array}{c} MeOOC \\ R \\ MeOOC \end{array} N-OH \right] \xrightarrow{-H_2O} \begin{array}{c} MeOOC \\ R \\ MeOOC \end{array} N$$

1,6

$$\bigcirc \xrightarrow[\substack{\text{(ii) oxidation} \\ \text{(iii) SOCl}_2/\text{Py}}]{\text{(i) O}_3/\text{Me}_2\text{S}} \begin{array}{c} COCl \\ COCl \end{array} \xrightarrow[\text{(ii) LiAlH}_4]{\text{(i) + NH}_3 \ -2\,\text{HCl}} \text{(azepane, NH)}$$

Many saturated nitrogen heterocycles are commercially available from industrial processes, which involve, for example, nucleophilic substitution of hydroxyl groups by amino groups under conditions far from laboratory use, e.g.

$$NH_3 + \triangle O \xrightarrow{\Delta} \begin{array}{c} NH_2 \\ \\ OH \end{array} \xrightarrow[\substack{100-250 \text{ bar}}]{NH_3;\ 150-220\,^\circ C} \text{(piperazine)}$$

$$NH_3 + 2\,\triangle O \xrightarrow{\Delta} \underset{HO \qquad OH}{\overset{H}{N}} \xrightarrow[\substack{0.5 \text{ h};\ 220\,^\circ C;\ 200 \text{ bar}}]{NH_3/H_2\ [ZnO]} \text{(piperazine)} \quad (\approx 90\%)$$

Regioselectivity becomes important, if unsymmetric difunctional nitrogen components are used. In such cases two different reactions of the nitrogen nucleophile with the open-chain educt may be possible, one of which must be faster than the other. Hydrazone formation, for example, occurs more readily than hydrazinolysis of an ester. In the second example, on the other hand, the amide is formed very rapidly from the acyl chloride, and only one cyclization product is observed.

$$\begin{array}{c} R^1 \\ \underset{COOEt}{\overset{O}{\diagup}} \end{array} + \begin{array}{c} NH_2 \\ NH \\ R^2 \end{array} \xrightarrow{-H_2O} \left[\begin{array}{c} R^1 \\ \underset{COOEt}{\diagup} =N \diagdown NH-R^2 \end{array} \right] \xrightarrow{-EtOH} \begin{array}{c} R^1 \\ \text{(pyrazolone)} \\ N-R^2 \\ O \end{array}$$

$$\begin{array}{c} OEt \\ \overset{O}{\diagup} \\ NH_2 \end{array} + \begin{array}{c} CN \\ Cl \overset{O}{\diagup} \end{array} \xrightarrow{-HCl} \left[\begin{array}{c} OEt \\ \overset{O}{\diagup} CN \\ \underset{H}{N} \ O \end{array} \right] \xrightarrow{-H_2O} \begin{array}{c} OEt \\ CN \\ \underset{H}{N} \ O \end{array}$$

Many successful regioselective syntheses of heterocycles, however, are more complex than the examples given so far. They employ condensation of two different carbonyl or halide compounds with one nitrogen base or the condensation of an amino ketone with a second difunctional compound. Such reactions cannot be rationalized in a simple way, and the literature must be consulted.

The problems involved are exemplified here by Knorr's pyrrole synthesis (A. Gossauer, 1974). It has been known for almost a century that α-aminoketones (C_2N components) react with 1,3-dioxo compounds (C_2 components) to form pyrroles (C_4N-heterocycles). A side-reaction is the cyclodimerization of the α-aminoketones to yield dihydropyrazines (C_4N_2), but this can be minimized by keeping the concentration of the α-aminoketone low relative to the 1,3-dioxo compound. The first step in Knorr's pyrrole synthesis is the formation of an imine. This depends critically on the pH of the solution. The nucleophilicity of the amine is lost on protonation, whereas the carbonyl groups are activated by protons. An optimum is found around pH 5, where yields of about 60% can be reached. At pH 4 or 6 the yield of the pyrrole may approach zero. The ester groups of β-keto esters do not react with the amine under these conditions. If a more reactive 1,3-diketone is used, it has to be symmetrical, otherwise mixtures of two different imines are obtained. The imine formed rearranges to an enamine, which cyclizes and dehydrates to yield a 3-acylpyrrole as the "normal" Knorr product (A. Gossauer, 1974; G.W. Kenner, 1973 B).

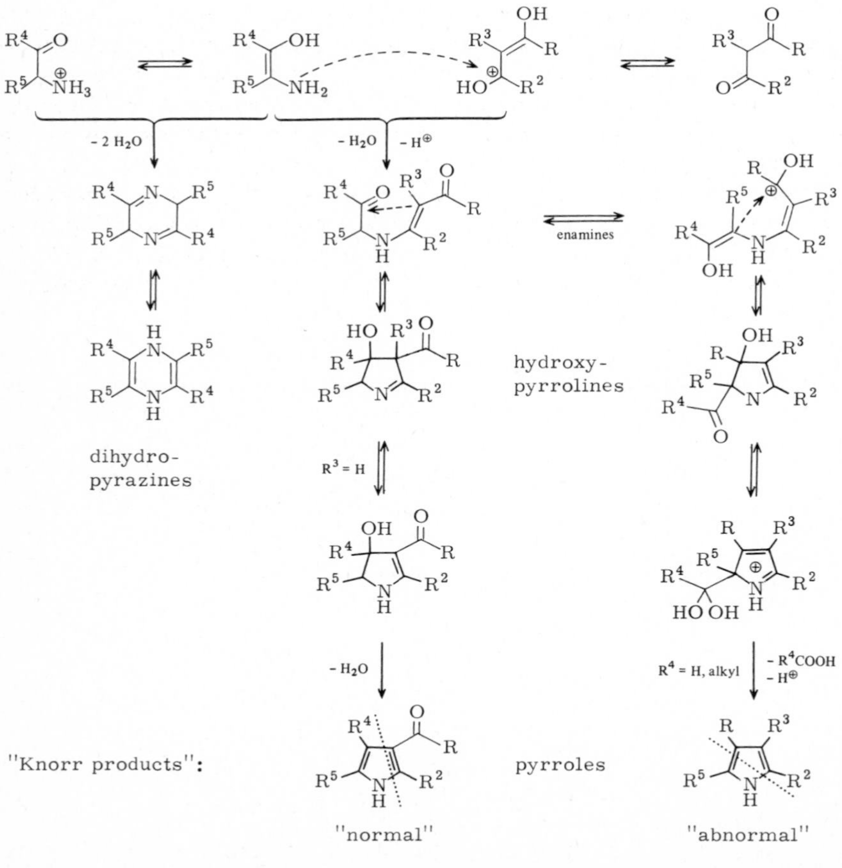

This simple reaction scheme accounts for three experimental observations:

(i) The amino component may be a mono- or a dicarbonyl compound, but the other component must be a 1,3-dioxo compound because the enamine would not form at pH 5 from an imine with a nonactivated neighboring methylene group. Therefore only 3-carboxy- or 3-acylpyrroles can be made efficiently (for a rearrangement see p. 121f.).

(ii) "Abnormal" products can be formed from 1,3-diketones or β-ketoaldehydes because an acyl or formyl group, resp., may be lost by *retro*-aldol type cleavage when the final dehydration step takes place.

(iii) Yields seldom exceed 60%. Several side products are observed in these complicated reactions because several intermediates may react in different ways.

The usefulness of the Knorr synthesis arises from the fact that 1,3-dioxo compounds and α-aminoketones are much more easily accessible in large quantities than "rational" 1,4-difunctional precursors. Such "practical" syntheses are known for several important heterocycles. They are usually limited to certain substitution patterns of the target molecules.

Similar enamine cyclization processes occur in several other successful heterocycle syntheses, e.g. in the Fischer indole synthesis. In this case, however, a labile N—N bond of a 1-aryl-2-vinylhydrazine is cleaved in a [3,3]-sigmatropic rearrangement, followed by cyclization and elimination of ammonia to yield the indole (B. Robinson, 1963, 1969; R.J. Sundberg, 1970). Regioselectivity is only observed if R^2 contains no enolizable hydrogen, otherwise two structurally isomeric indoles are obtained. Other related cyclization reactions are found in the Pechmann synthesis of triazoles (T.L. Gilchrist, 1974) and in G. Bredereck's (1959) imidazole synthesis (M.R. Grimmett, 1970).

Fischer
indole synthesis

Pechmann
triazole synthesis

Bredereck's
imidazole synthesis

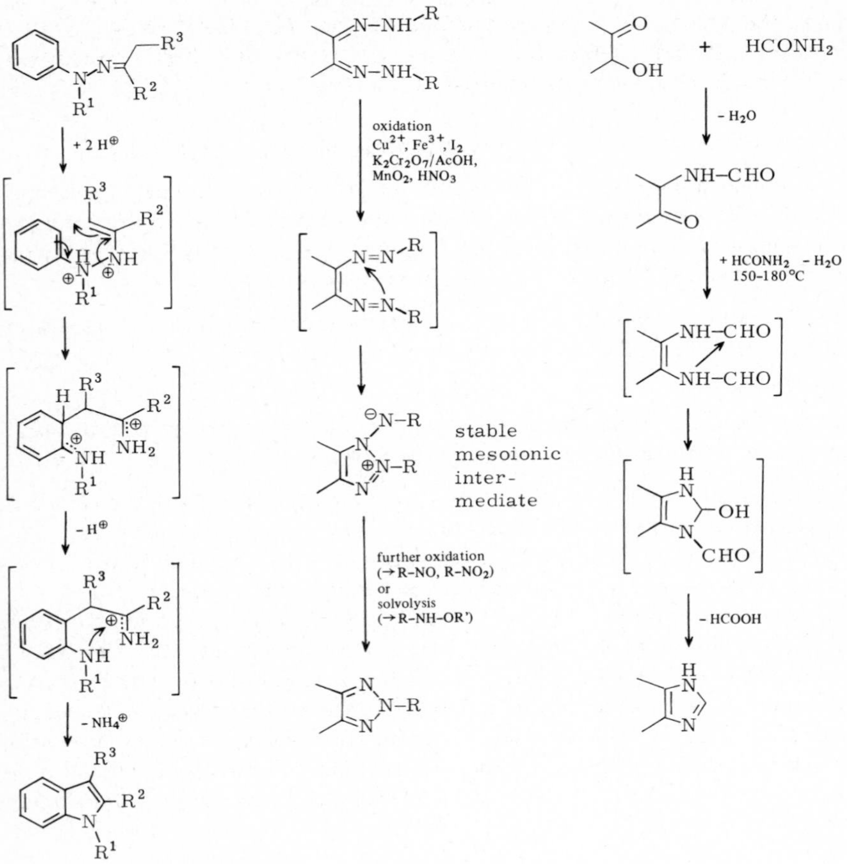

stable
mesoionic
inter-
mediate

A completely different, important type of synthesis, which was developed more recent-ly, takes advantage of the electrophilicity of nitrogen-containing 1,3-dipolar compounds rather than the nucleophilicity of amines or enamines. Such compounds add to multiple bonds, e.g. $C=C$, $C\equiv C$, $C=O$, in a $[2+3]$-cycloaddition to form five-membered hetero-cycles.

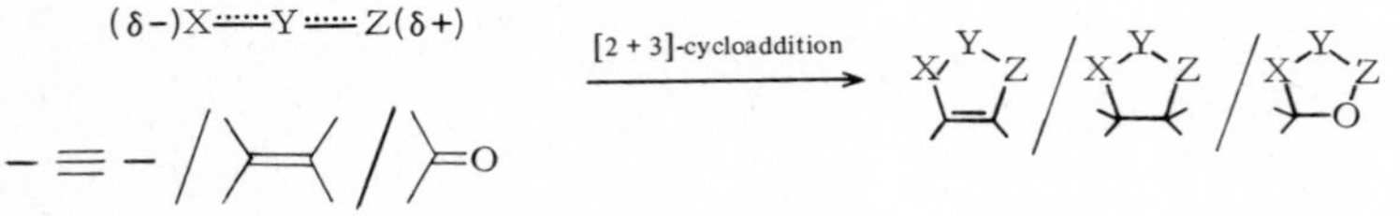

A few typical examples indicate the large variety of five-membered heterocycles, which can be synthesized efficiently by [2 + 3]-cycloadditions. [2 + 2]-Cycloadditions are useful in the synthesis of certain four-membered heterocycles (H. Ulrich, 1967), e.g. of β-lactams (J.R. Malpass, 1977). Diels-Alder type [2 + 4]-cycloadditions are possible with certain hetero-"ene" components (J.R. Malpass, 1977; S.F. Martin, 1980) or with highly reactive *o*-quinodimethanes as diene components (W. Oppolzer, 1978A).

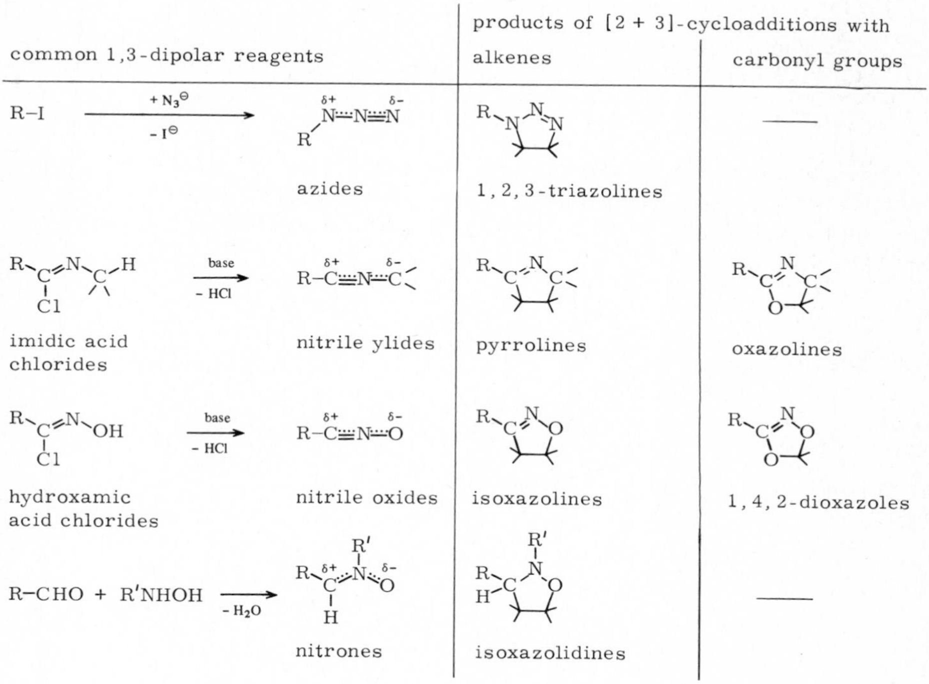

common 1,3-dipolar reagents		products of [2 + 3]-cycloadditions with	
		alkenes	carbonyl groups
R–I azides	nitrile ylides	1,2,3-triazolines	—
imidic acid chlorides	nitrile ylides	pyrrolines	oxazolines
hydroxamic acid chlorides	nitrile oxides	isoxazolines	1,4,2-dioxazoles
R–CHO + R′NHOH nitrones		isoxazolidines	—

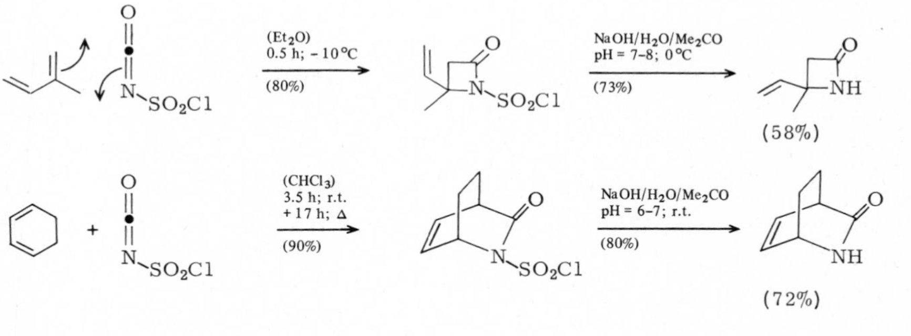

(Et₂O): (Et$_2$O) 0.5 h; –10 °C (80%) → NaOH/H$_2$O/Me$_2$CO pH = 7–8; 0 °C (73%) → (58%)

(CHCl$_3$) 3.5 h; r.t. +17 h; Δ (90%) → NaOH/H$_2$O/Me$_2$CO pH = 6–7; r.t. (80%) → (72%)

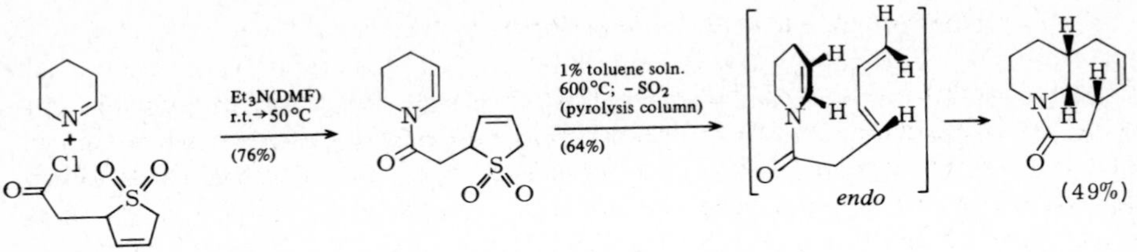

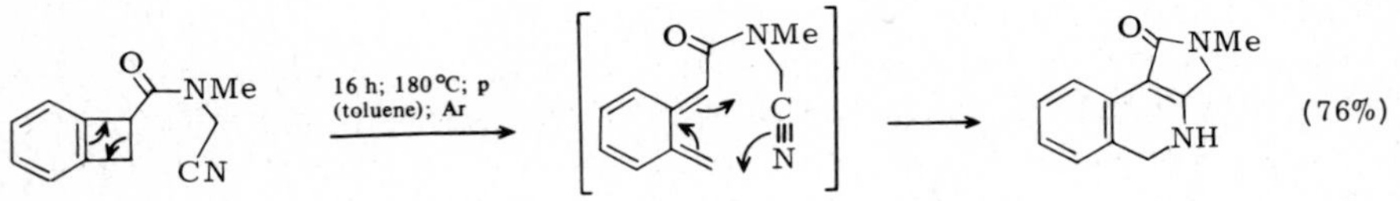

Sulfonium ylides may be added to C=N double bonds to yield aziridines in a formal [1 + 2]-cycloaddition. Alkyl azides are decomposed upon heating or irradiating to yield nitrenes, which may also undergo [1 + 2]-cycloaddition reactions to yield highly strained heterocycles (A.G. Hortmann, 1972).

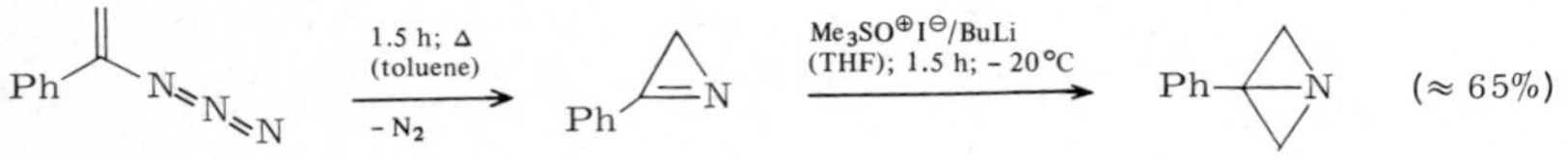

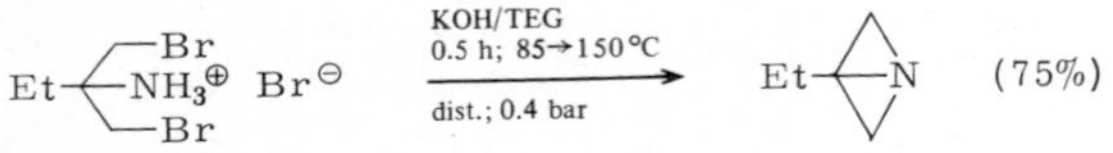

2.6 Protection of Functional Groups

The synthesis of complex organic molecules demands the availability of a variety of protective groups to ensure the survival of reactive functional groups during synthetic operations. An ideal protective group combines stability under a wide range of conditions with susceptibility to facile removal ("deblocking") by a specific, mild reagent. It is also desirable that the introduction of the "blocking" group is easy and that its reactions are complementary to other protecting groups. Within the last 20 years a large variety of new protective groups capable of removal under exceptionally mild and/or highly specific conditions has been developed mainly in syntheses of natural oligomers, e.g. peptides (see p. 209), nucleotides (see p. 198), and saccharides (see p. 241ff.), which has been summarized in a book by J.W.F. McOmie (1973). We shall discuss only a few protecting agents of common functional groups, which have proven to be useful in synthesis, and indicate some possible sequences for working with these blocking agents.

2.6.1 Reactive Carbon-Hydrogen and Carbon-Carbon Bonds

Only highly reactive CH groups, e.g. of terminal alkynes or of carbonyl compounds, have sometimes to be efficiently protected. Terminal alkynes are mostly used in organometallic syntheses (see p. 35) and in oxidative couplings (see p. 39). Protection of the terminal C—H bond is required, if another group is to be metallated, or if the oxidative coupling is to be performed with diterminal diynes. Without protection, rearranged organometallic compounds or polymers would be formed in uncontrollable side-reactions.

The most useful protecting group in these reactions is the trimethylsilyl group. It may be introduced by reaction of the acetylide anion with chlorotrimethylsilane and be removed by treatment with silver nitrate in quantitative yield. On the other hand, the C—Si bond remains intact during reactions with organometallic reagents (E.J. Corey, 1968 D) as well as during oxidative coupling of alkynes under controlled conditions (F. Sondheimer, 1970).

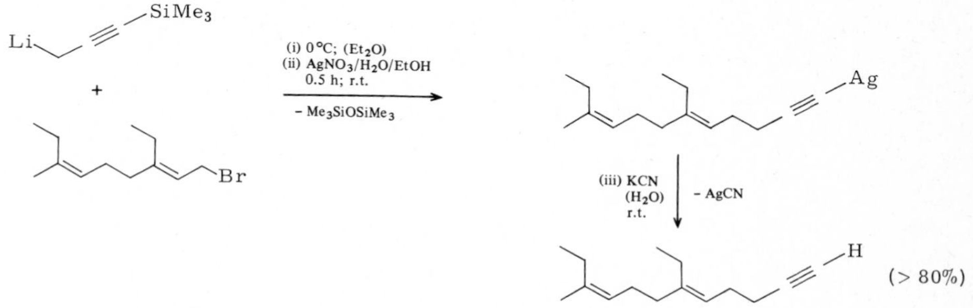

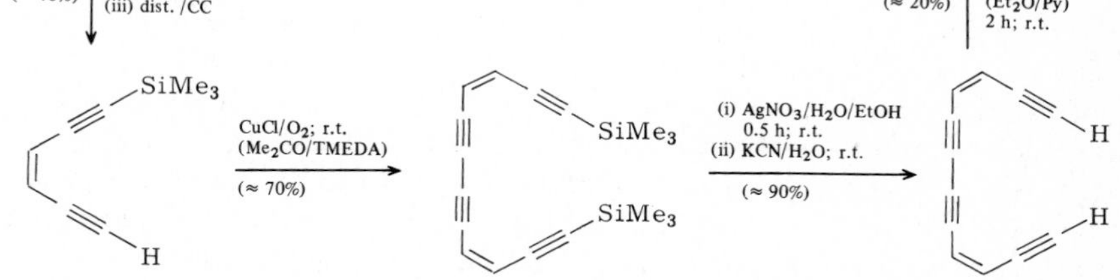

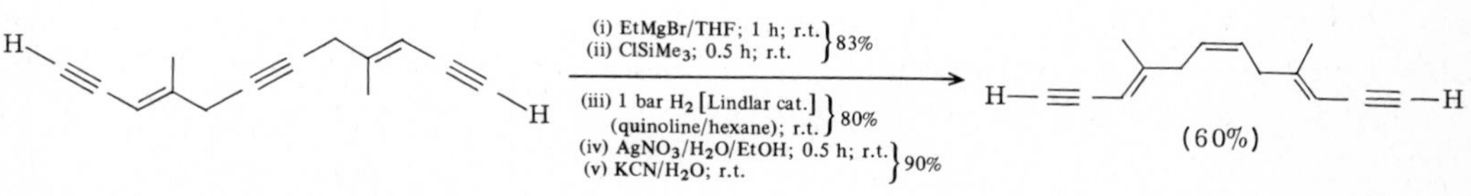

α'-Methylene groups of ketones containing a tertiary α-carbon atom, whose selective alkylation is desired, are protected by successive treatment with ethyl formate and 1-butanethiol. The butylthiomethylene derivatives formed are then deprotonated and alkylated at the tertiary α-carbon atom. The protecting group is finally removed as butanethiolate and formate by treatment with KOH in boiling diethylene gylcol. A useful application is the angular alkylation of 1-decalones (R.E. Ireland, 1962). For the alkylation of α,β-unsaturated keto-steroids at C^α of the double bond R.B. Woodward (1957, 1971) devised another methylene-protecting group. The α'-methylene group is again first condensed with ethyl formate, and the hydroxymethylene group is then replaced by a trimethylene dithioketal unit using 1,3-bis(tosylthio)propane. The usual alkylation of the dienolate (see p. 25) is then carried out, and finally the dithioketal is removed by reductive desulfurization with Raney nickel. J.A. Marshall (1969) used this dithioketal for a formal shift of a keto group to the neighboring carbon atom.

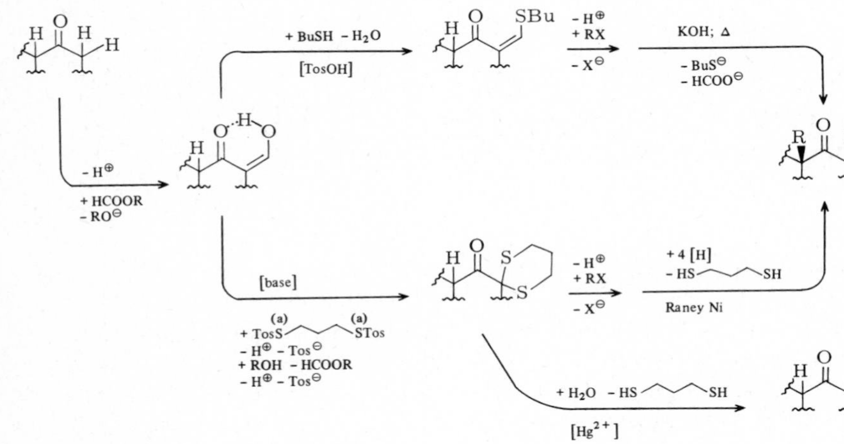

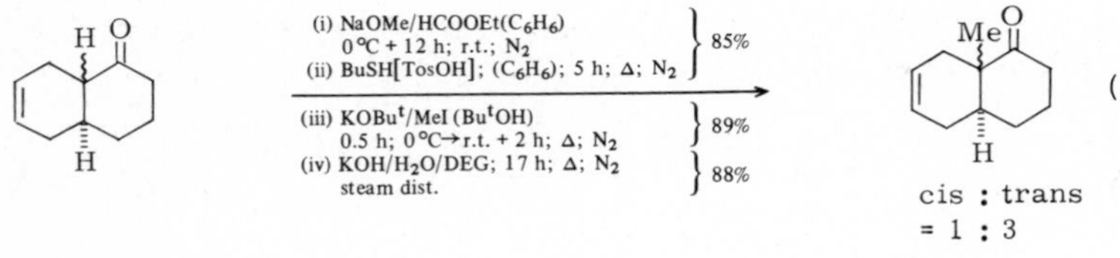

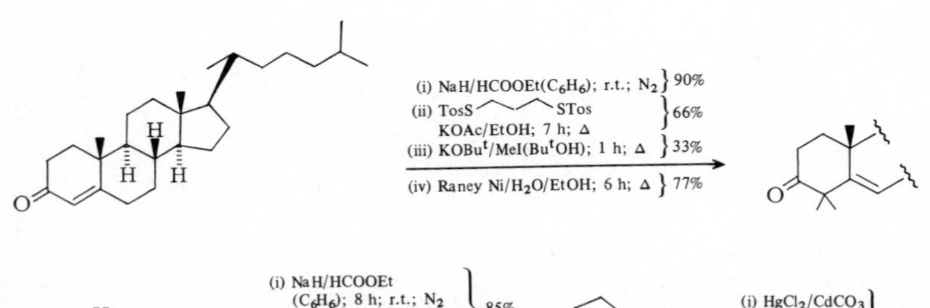

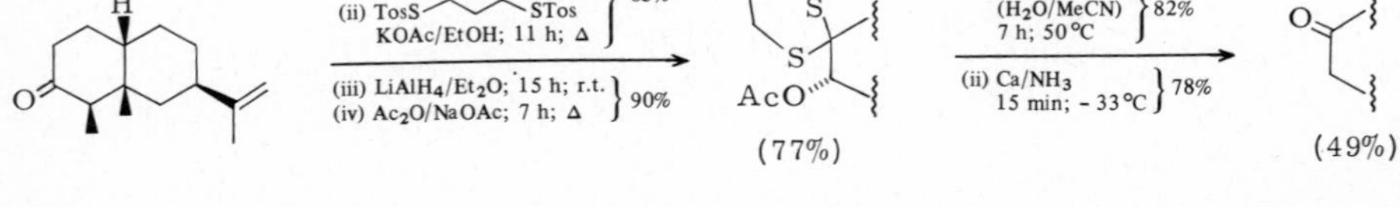

C=C double bonds may be protected against electrophiles by epoxidation and subsequent removal of the oxygen atom by treatment with zinc and sodium iodide in acetic acid (J.A. Edwards, 1972; W. Knöll, 1975). Halogenation has often been used for protection, too. The C=C double bond is here also easily regenerated with zinc (see p. 124, D.H.R. Barton, 1976).

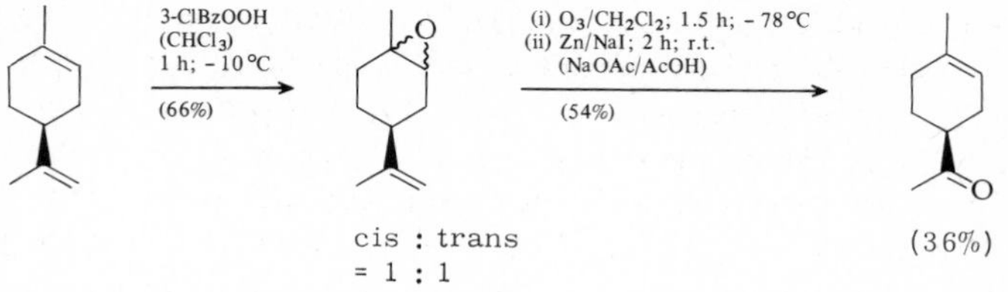

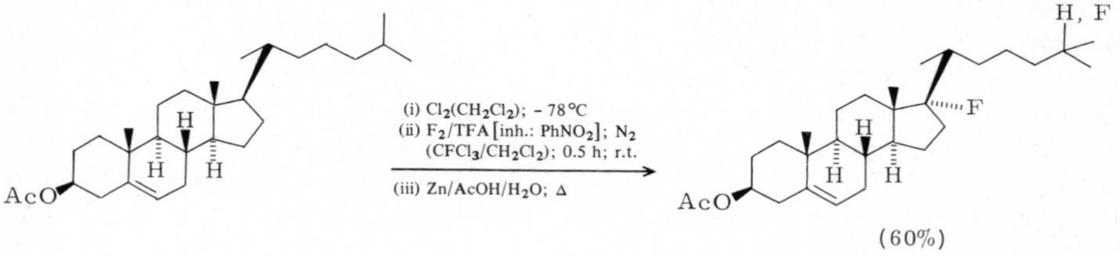

cis : trans
= 1 : 1

(36%)

(60%)

2.6.2 Alcoholic Hydroxyl Groups

Six protective groups for alcohols, which may be removed successively and selectively, have been listed by E.J. Corey (1972B). A hypothetical hexahydroxy compound with hydroxy groups 1 to 6 protected as (1) acetate, (2) 2,2,2-trichloroethyl carbonate, (3) benzyl ether, (4) dimethyl-t-butylsilyl ether, (5) 2-tetrahydropyranyl ether, and (6) methyl ether may be unmasked in that order by the reagents (1) K_2CO_3 or NH_3 in CH_3OH, (2) Zn in CH_3OH or AcOH, (3) H_2 over Pd, (4) F^-, (5) wet acetic acid, and (6) BBr_3. The groups may also be exposed to the same reagents in the order 4, 5, 2, 1, 3, 6.

protecting groups for R—OH		reagents	deblocking
Ac	H_3C—$\overset{O}{\underset{}{C}}$—O–R	$Ac_2O/Py\,[DMAP]$	NaOMe/MeOH K_2CO_3/MeOH NH_3/MeOH
Tceoc	Cl_3C—$\frown$—O—$\overset{O}{\underset{}{C}}$—O–R	TceocCl/Py	
Tbeoc	Br_3C—$\frown$—O—$\overset{O}{\underset{}{C}}$—O–R	TbeocCl/Py	Zn–Cu/AcOH
Bzl	—O–R	$PhCH_2Cl/KOH$ $PhCH_2Cl/Ag_2O$	$H_2\,[Pd]$ Na/NH_3

protecting groups for R–OH (continued)		reagents	deblocking
Me$_2$Bu^{t}Si	[structure: Si with O–R]	Me$_2$Bu^{t}SiCl [imidazole]	Bu$_4$N$^{\oplus}$F$^{\ominus}$ H$_2$O/AcOH
Thp	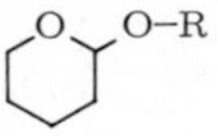 [structure: tetrahydropyranyl O–R]	 [structure] = DHP [TosOH]	AcOH/H$_2$O 0.1 M aq. HCl
Me	CH$_3$–O–R	Me$_2$SO$_4$/NaOH Me$_2$SO$_4$/Ba(OH)$_2$	BBr$_3$ BCl$_3$

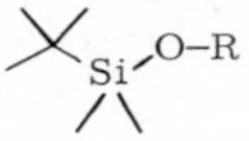

We shall describe a specific synthetic example for each protective group given above. Regioselective protection is generally only possible if there are hydroxyl groups of different sterical hindrance (prim < sec < tert; equatorial < axial). Acetylation has usually been effected with acetic anhydride. The acetylation of less reactive hydroxyl groups is catalyzed by DMAP (see p. 130). Acetates are stable toward oxidation with chromium trioxide in pyridine and have been used, for example, for protection of steroids (H.J.E. Loewenthal, 1959), carbohydrates (M.L. Wolfrom, 1963; J.M. Williams, 1967), and nucleosides (A.M. Michelson, 1963). The most common deacetylation procedures are ammonolysis with NH$_3$ in CH$_3$OH and methanolysis with K$_2$CO$_3$ or sodium methoxide.

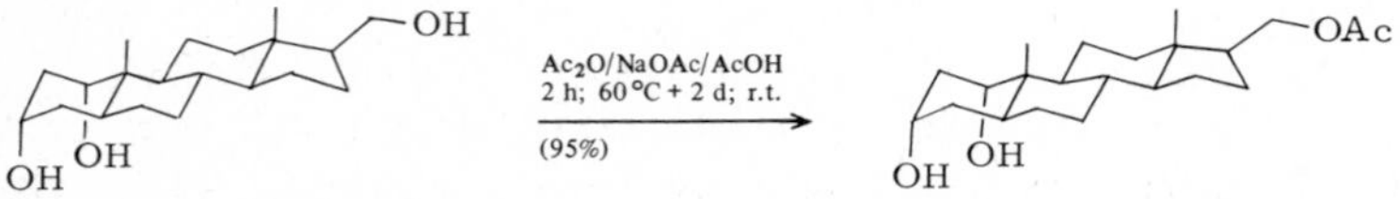

2,2,2-Trichloro- and 2,2,2-tribromoethoxycarbonyl (Tceoc and Tbeoc) protecting groups are introduced with the commercially available 2,2,2-trihaloethyl chloroformates. These derivatives are stable towards CrO$_3$ and acids, but can smoothly be cleaved by reduction with zinc in acetic acid at 20 °C to yield 1,1-dihaloethene and CO$_2$. Several examples in lipid (F.R. Pfeiffer, 1968, 1970) and nucleotide syntheses (A.F. Cook, 1968; see p. 201ff.) have been described.

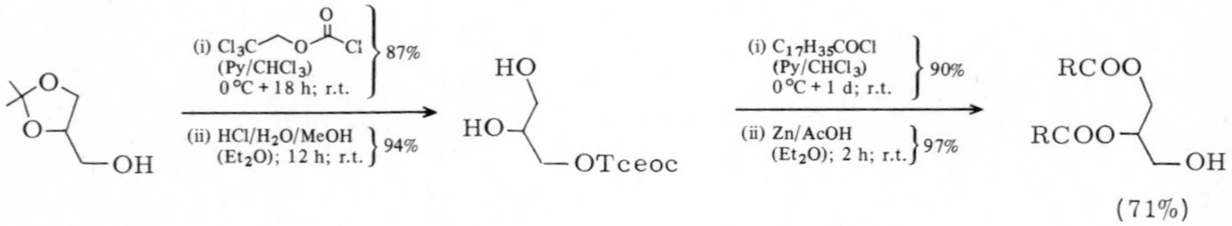

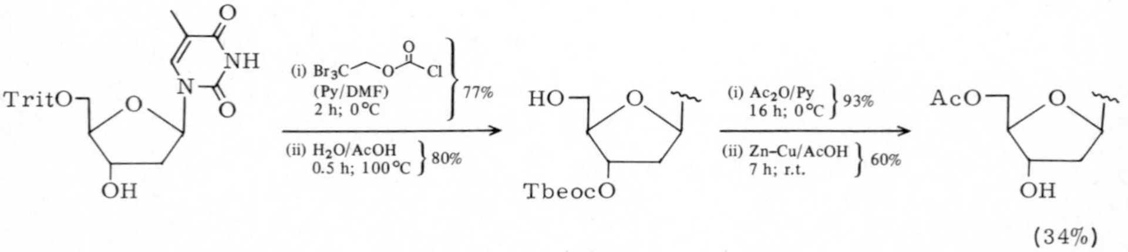

The benzyl group has been widely used for the protection of hydroxyl functions in carbohydrate and nucleotide chemistry (C.M. McCloskey, 1957; C.B. Reese, 1965; B.E. Griffin, 1966). A common benzylation procedure involves heating with neat benzyl chloride and strong bases. A milder procedure is the reaction in DMF solution at room temperature with the aid of silver oxide (E. Reinefeld, 1971). Benzyl ethers are not affected by hydroxides and are stable towards oxidants (e.g. periodate, lead tetraacetate), $LiAlH_4$, and weak acids. They are, however, readily cleaved in neutral solution at room temperature by palladium-catalyzed hydrogenolysis (S. Tejima, 1963) or by sodium in liquid ammonia or alcohols (E.J. Reist, 1964).

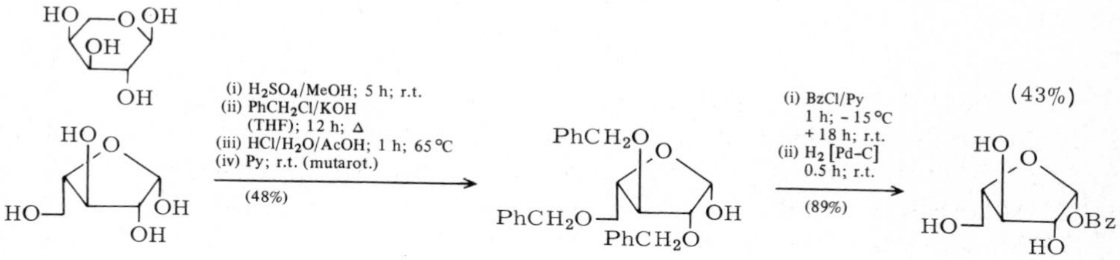

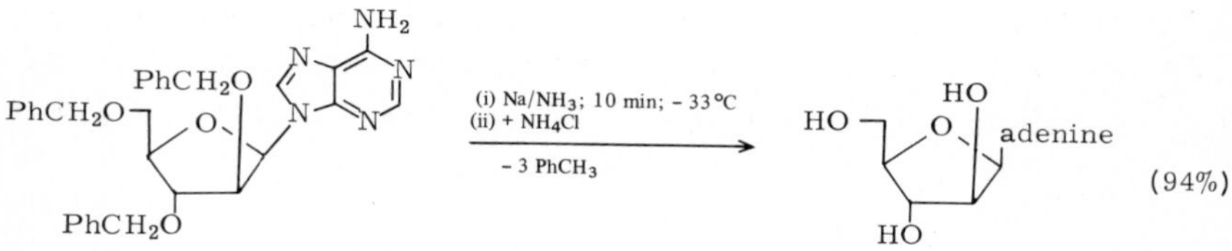

Trimethylsilyl ethers are too susceptible to solvolysis in protic media to be useful in syntheses, although they are widely used as volatile derivatives of alcohols in gas chromatography. The bulky dimethyl-*t*-butylsilyl ethers are about 10^4 times more stable. Primary and secondary alcohols are readily blocked with chlorodimethyl-*t*-butylsilane in DMF solution and imidazole as catalyst. These silyl ethers are stable toward water, CrO_3, hydrogenolysis, and mild reduction procedures. Upon treatment with tetrabutylammonium fluoride in THF at room temperature, they are rapidly cleaved into alcohols and the fluorosilane. E.J. Corey (1972B) demonstrated the highly selective cleavage of either of the benzyloxy and dimethyl-*t*-butylsiloxy groups at a prostaglandin precursor.

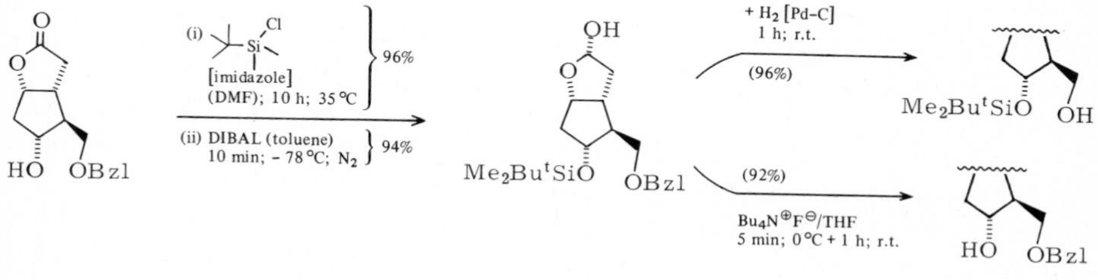

3,4-Dihydro-2H-pyran undergoes smooth acid-catalyzed addition to alcohols to give 2-tetrahydropyranyl ethers (Thp ethers; W.E. Parham, 1948). CHCl$_3$, ethers, and DMF may be used as solvents, and traces of HCl or *p*-toluenesulfonic acid as catalysts. This acetal system is stable toward bases, organometallic compounds, LiAlH$_4$, and alkylating agents. But it undergoes acid-catalyzed hydrolysis under very mild conditions. Aqueous acetic acid or 0.1 M HCl are sufficient. The example below is taken from a nucleotide synthesis by H.G. Khorana (M. Smith, 1959). It also involves selective tritylation of a primary alcohol group, which is often applied in carbohydrate and nucleoside chemistry (see p. 245f. and p. 201ff.).

The most stable protected alcohol derivatives are the methyl ethers. These are often employed in carbohydrate chemistry and can be made with dimethyl sulfate in the presence of aqueous sodium or barium hydroxides in DMF or DMSO. Simple ethers may be cleaved by treatment with BCl$_3$ or BBr$_3$, but generally methyl ethers are too stable to be used for routine protection of alcohols. They are more useful as volatile derivatives in gas-chromatographic and mass-spectrometric analyses. So the most labile (trimethylsilyl ether) and the most stable (methyl ether) alcohol derivatives are useful in analysis, but in synthesis they can be used only in exceptional cases. In synthesis, easily accessible intermediates of medium stability are most helpful.

2.6.3 Amino Groups

Primary and secondary amines are susceptible to oxidation and replacement reactions involving the N—H bonds. Within the development of peptide synthesis numerous protective groups for N—H bonds have been found (M. Bodanszky, 1976; L.A. Carpino, 1973), and we shall discuss five of the more general methods used involving the reversible formation of amides. A hypothetical pentaamino compound could have been blocked with the following protective reagents at all nitrogen atoms 1 to 5: (1) trifluoroacetic anhydride, (2) *t*-butyl azidoformate, (3) 2,2,2-trichloroethyl chloroformate, (4) benzyl chloroformate (= "carbobenzoxy chloride"), and (5) N-ethoxycarbonylphthalimide.

In all of these reagents a leaving group (CF$_3$COO$^-$, N$_3^-$, Cl$^-$, H$_2$NCOOEt) is smoothly substituted by the nucleophilic amino group under mild conditions, and an amide is formed in high yield. The amino groups may be unmasked in the given order by treatment with (1) aqueous Ba(OH)$_2$, (2) 25% CF$_3$COOH in CHCl$_3$, (3) Zn in acetic acid, (4) H$_2$ over Pd, and (5) HBr in acetic acid or aqueous N_2H_4. The five protecting agents will be discussed successively in the given order.

protecting groups for R–NH₂		reagents	deblocking
Tfac	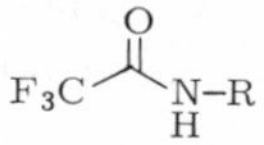 F₃C–C(=O)–N(H)–R	Tfac₂O/Py	Ba(OH)₂, NaHCO₃, NH₃/H₂O HCl/H₂O NaBH₄/MeOH
Boc(t-Box)	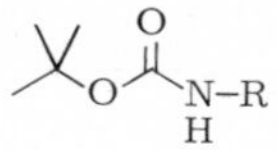 (CH₃)₃C–O–C(=O)–N(H)–R	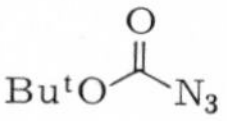 BuᵗO–C(=O)–N₃	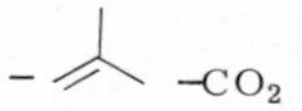 TFA/CHCl₃ } HF/H₂O } → (CH₃)₂C=CH₂ –CO₂
Tceoc	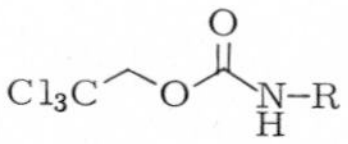 Cl₃C–CH₂–O–C(=O)–N(H)–R	TceocCl	Zn/AcOH cathodic red. 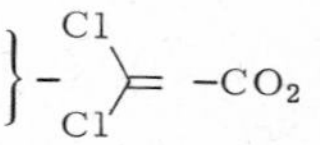 } → Cl₂C=CCl₂ –CO₂
Cbz("Z")	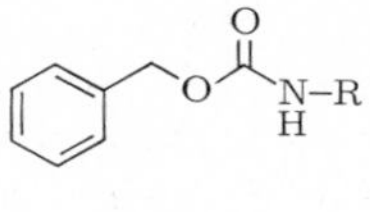 PhCH₂–O–C(=O)–N(H)–R	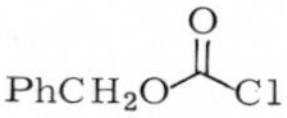 PhCH₂O–C(=O)–Cl	HBr/AcOH H₂[Pd]
Phth	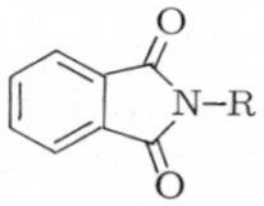 (phthalimido)N–R	PhthN–COOEt	HBr/AcOH N₂H₄/H₂O

Trifluoroacetamides are more stable toward nucleophiles than the corresponding esters and are easily formed from trifluoroacetic anhydride and the amine. The trifluoroacetyl group (Tfac) is slowly cleaved by aqueous or methanolic HCl, NH_3, or $Ba(OH)_2$ solutions as well as by $NaBH_4$ in methanol (M.L. Wolfrom, 1967).

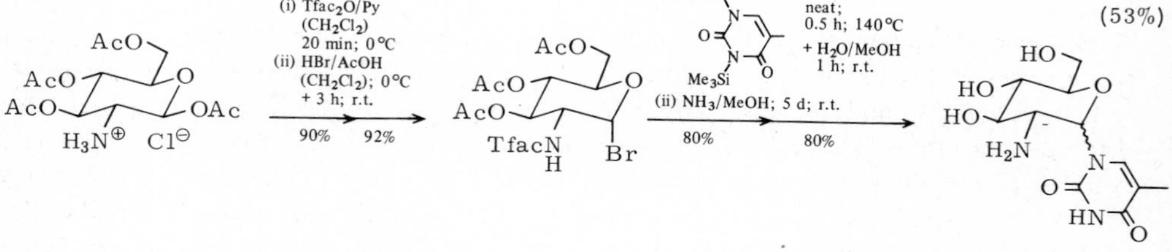

The *t*-butoxycarbonyl group (Boc, "*t*-box") has been extensively used in peptide synthesis, and Boc derivatives of many amino acids are commercially available. The customary reagent for the preparation from the amine is *t*-butyl azidoformate in water, dioxane/water, DMSO, or DMF. The cleavage by acids of medium strength proceeds with concomitant loss of isobutene and carbon dioxide (L.A. Carpino, 1957, 1973; see section 4.1.2.2).

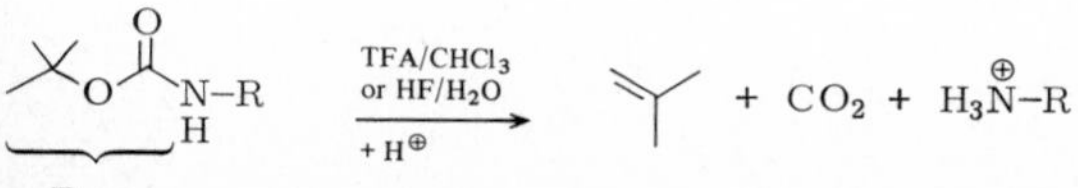

2,2,2-Trihaloethoxycarbonyl (Tbeoc, Tceoc) derivatives of amines are prepared with the corresponding chloroformates and have been used in the synthesis of β-lactam antibiotics (see section 4.8). They are cleaved by zinc in acetic acid or by electrolytic reduction (M.F. Semmelhack, 1972). Benzoxycarbonyl (Cbz, "Z") derivatives of amines are formed with benzyl chloroformate in alkaline aqueous solution or other polar solvents. They are more stable towards reducing agents than 2,2,2-trihaloethoxycarbonyl derivatives. Many Cbz protected amino acids and peptides are commercially available. The most widely used deblocking procedures are solvolysis with HBr (G. Losse, 1968; G.A. Olah, 1970) and hydrogenolysis over palladium catalyst. The ester bond of the Cbz group is remarkably stable towards alkaline hydrolysis.

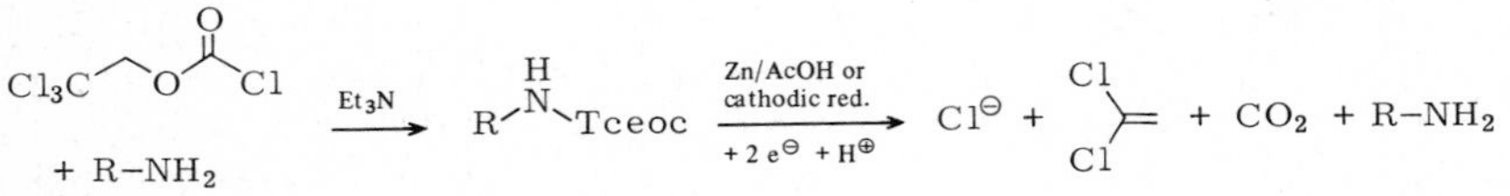

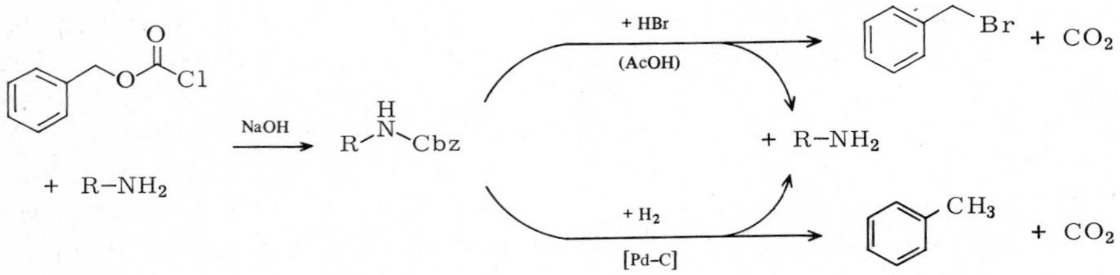

The phthaloyl (Phth) derivatives of amines, formed from amines and N-ethoxy-carbonylphthalimide (G.H.L. Nefkens, 1960), are acid-resistant imides, which can be easily deblocked by nucleophilic reagents, most conveniently by hydrazine.

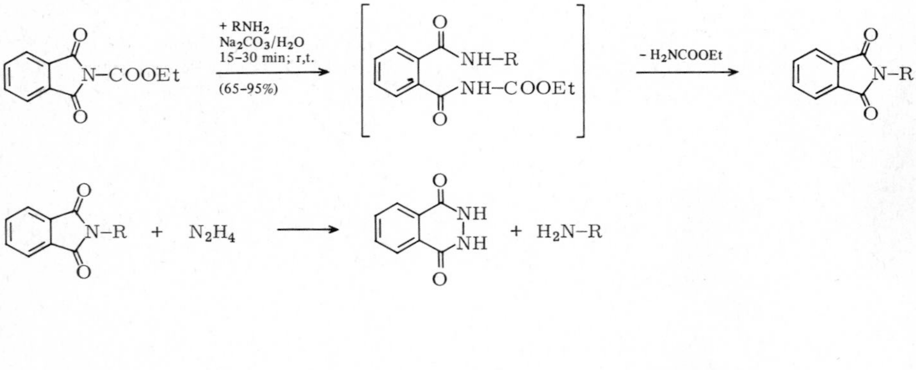

2.6.4 Carboxyl Groups

The blocking and deblocking of carboxyl groups occurs by reactions similar to those described for hydroxyl and amino groups. The most important protected derivatives are *t*-butyl, benzyl, and methyl esters. These may be cleaved in this order by trifluoroacetic acid, hydrogenolysis, and strong acid or base (J.F.W. McOmie, 1973). 2,2,2-Trihaloethyl esters are cleaved electrolytically (M.F. Semmelhack, 1972) or by zinc in acetic acid like the Tbeoc- and Tceoc-protected hydroxyl and amino groups.

2.6.5 Aldehyde and Keto Groups

The most commonly used protected derivatives of aldehydes and ketones are 1,3-dioxolanes and 1,3-oxathiolanes. They are obtained from the carbonyl compounds and 1,2-ethanediol or 2-mercaptoethanol, respectively, in aprotic solvents and in the presence of catalysts, e.g. BF$_3$ (L.F. Fieser, 1954; G.E. Wilson, Jr., 1968), and water scavengers, e.g. orthoesters (P. Doyle, 1965). Acid-catalyzed "exchange dioxolanation" with dioxolanes of low boiling ketones, e.g. acetone, which are distilled during the reaction, can also be applied (H.J. Dauben, Jr., 1954). Selective monoketalization of diketones is often used with good success (C. Mercier, 1973). Even from diketones with two keto groups of very similar reactivity monoketals may be obtained by repeated acid-catalyzed equilibration (W.S. Johnson, 1962; A.G. Hortmann, 1969). Most aldehydes are easily converted into acetals. The ketalization of ketones is more difficult for sterical reasons and often requires long reaction times at elevated temperatures. α,β-Unsaturated ketones react more slowly than saturated ketones. 2-Mercaptoethanol is more reactive than 1,2-ethanediol (J. Romo, 1951; C. Djerassi, 1952; G.E. Wilson, Jr., 1968).

1,3-Dioxolanes and 1,3-oxathiolanes are stable under most alkaline and neutral reaction conditions. They may, however, be cleaved by organometallic compounds. Deblocking is achieved with strong acids, preferably in the presence of acetone. The ease of acidic cleavage parallels the ease of formation. Ketals of highly hindered ketones may need boiling in mineral acids. The 1,3-oxathiolanes are again more reactive and may be cleaved even in neutral or slightly basic acetone or alcohol solutions, or most efficiently using Raney nickel (C. Djerassi, 1952, 1958; G.R. Pettit, 1962 A). 1,3-Dithianes, which are used in the "umpolung" (see p. 17) and removal (see p. 101) of carbonyl groups, may also be used as protected derivatives (D. Seebach, 1969).

(70%)   (47%)

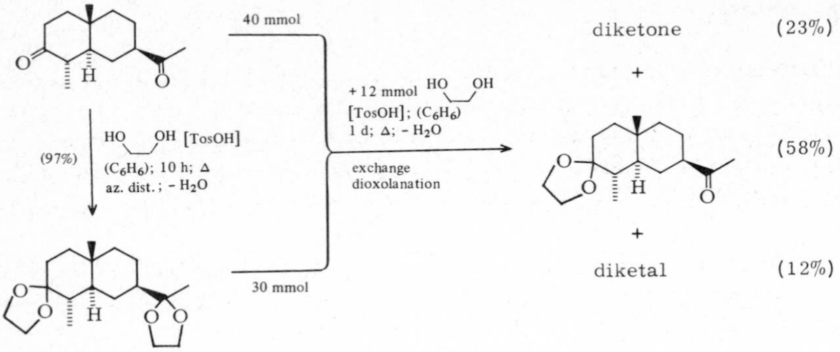

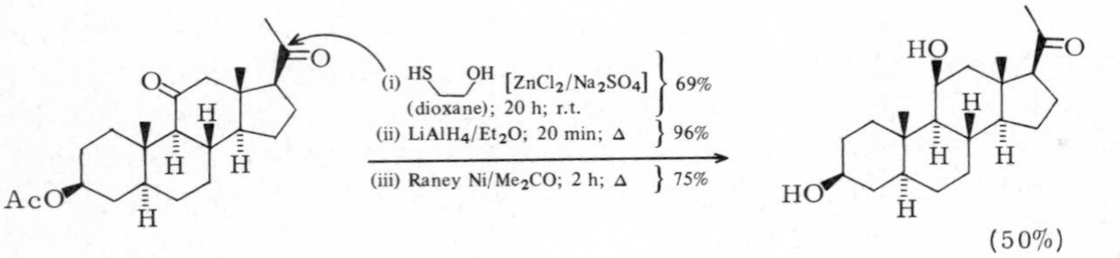

The only acid-resistant protective group for carbonyl functions is the dicyanomethylene group formed by Knoevenagel condensation with malononitrile. Friedel-Crafts acylation conditions, treatment with hot mineral acids, and chlorination with sulfuryl chloride do not affect this group. They have, however, to be cleaved by rather drastic treatment with concentrated alkaline solutions (J.B. Bastús, 1963; H. Fischer, 1932; R.B. Woodward, 1960, 1961).

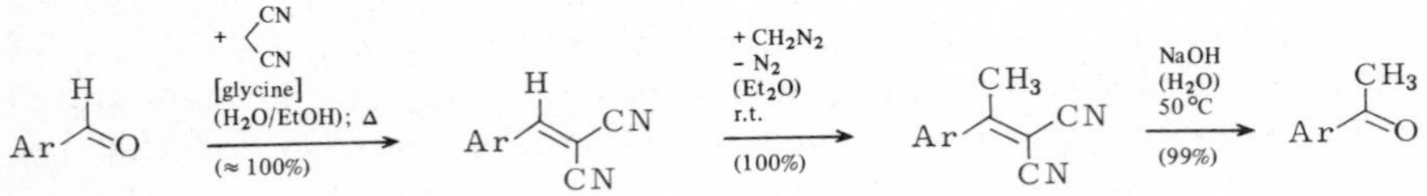

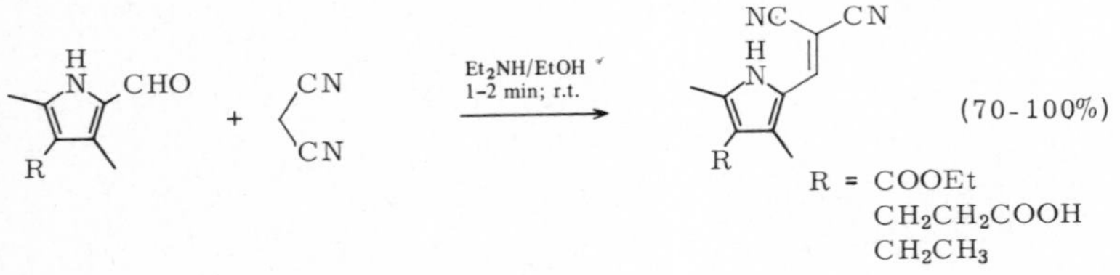

2.6.6 Phosphate Groups

If the three esterifiable OH groups of phosphoric acid have to be esterified successively with different alcohols, they have to be protected.

A widely used protecting agent is 2-cyanoethanol (= 3-hydroxypropanonitrile, hydracrylonitrile), which is condensed with phosphates with the aid of DCC. The 2-cyanoethyl (Ce) group is quantitatively removed as acrylonitrile by treatment with weak bases (H.G. Khorana, 1965).

2,2,2-Trichloroethanol may be used analogously. The 2,2,2-trichloroethyl (Tce) group is best removed by reduction with copper-zinc alloy in DMF at 50 °C (F. Eckstein, 1967). For specific protection in the strategy of nucleic acid synthesis see section 4.1.1.

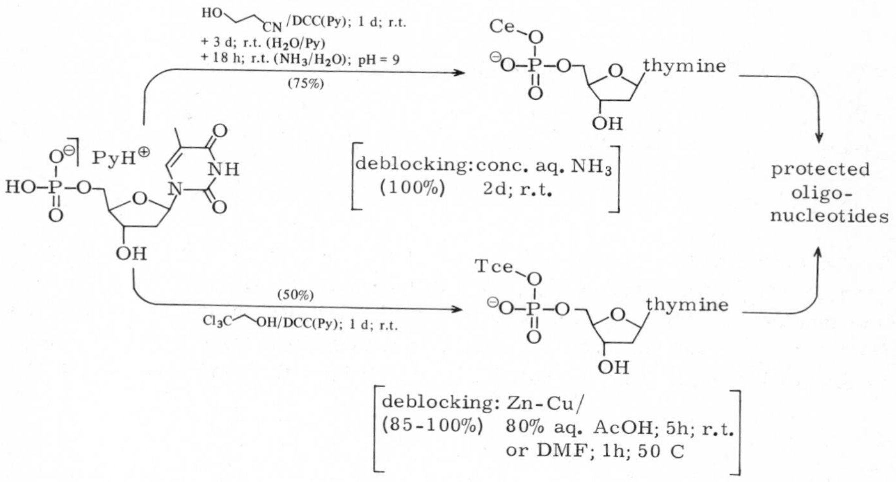

3 *Retro*-Synthetic Analysis of Simple Organic Compounds

Introduction and Summary

This chapter does not introduce new chemical reactions. On the contrary, mainly elementary reactions are employed. The attempt is made here to provide an introduction into the planning of syntheses of simple "target molecules" based upon the synthon approach of E.J. Corey (1967A, 1971) and the knowledge of the market of "fine chemicals".

- *A semi-systematic survey of inexpensive, commercial starting materials for organic syntheses is given. About seven hundred di- and oligofunctional reagents are listed.*

- *Retro-synthetic analyses (= "antitheses") of relatively simple, but non-commercial target molecules are described. Stepwise disconnections and functional group interconversions are used to transform the desired molecule into inexpensive commercial compounds.*

- *In antithetical analyses of carbon skeletons the synthon approach described in chapter 1 is used in the reverse order, e.g. 1,3-difunctional target molecules are "transformed" by imaginary retro-aldol type reactions, cyclohexene derivatives by imaginary retro-Diels-Alder reactions.*

- *Functional group interconversions (FGI) on the target molecule are used in such a way as to facilitate disconnections into common synthons.*

"Simple" compounds are defined here in an unusual but practical way: *a simple molecule is one, that may be obtained by four or less synthetic reactions from inexpensive commercial compounds.* We call a commercial compound "inexpensive" if it costs less or not much more than one German mark per gram. This also implies, that only those compounds that cannot be purchased inexpensively are considered as synthetic "target molecules" in this book.

3.1 Starting Materials

The first presupposition for the planning of a simple synthesis is the knowledge of the articles of commerce. We shall try to mediate this in a practical, semi-systematic way. As main sources we use the Aldrich-Europe, Merck, and Fluka catalogues. In our selected lists we indicate approximate prices of the chemicals in German marks per 10 g. Although deviations from actual prices are always possible and prices may be very different from other manufacturers, we do consider these numbers an important guide-line. The argument is that as divergent and widespread use as possible should be made of inexpensive reagents. This would clearly enhance the industrial and therefore social impact of academic research on synthesis.

We divide the approximately 10,000 compounds, which cost less than one German mark per gram, into five groups:

(1)   Open-chain carbon skeletons (including cyclic acetals, lactones, lactams, cyclic anhydrides, etc.)
        (1.1) Monofunctional
        (1.2) Difunctional
        (1.3) Trifunctional
        (1.4) Polyfunctional (for sugars see p. 239f.)
(2)   Silicon and phosphorus compounds
(3)   Cyclic carbon skeletons (for alkaloids and steroids see chapter 4)
        (3.1) Monocyclic
        (3.2) Bi-, tri-, and tetracyclic
(4)   Aromatic compounds
(5)   Heteroaromatic compounds

About 700 chemicals of synthetic potential which belong to subgroups 1.2, 1.3, 1.4, to group 2, and to subgroups 3.1 and 3.2 will be listed. Open-chain monofunctional (subgroup 1.1), aromatic, and heteroaromatic compounds (groups 4 and 5) will be treated only in short surveys, because the listing would become too long.

3.1.1 Monofunctional Open-Chain Reagents

Most compounds with six or less carbon atoms and one usual functional group (e.g. $C=C$, $C\equiv C$, OH, Br, Cl, NH_2, CHO, $C=O$, COOR, CN) can be purchased inexpensively. Ethers and nitroalkanes are less available. Unbranched carbon chains with up to sixteen carbon atoms and one terminal functional group (CH_2OH, CH_2NH_2, COOH, CN) are frequently available at low cost. All symmetrical straight-chain ketones up to 9-heptadecanone, all 2-alkanones up to 2-dodecanone, 3-heptanone, 3-octanone, and 4-decanone are also good value. Compounds with less common functional groups are seldom available and generally quite expensive, e.g. sulfones, sulfonic acids, organometallic and organoboron compounds. A list of low-cost commercial phosphorus and silicon compounds is given on p. 169.

3.1.2 Difunctional, Trifunctional, and Oligofunctional Open-Chain Reagents

Most synthetically useful reagents belong to this group. Many complicated reagents are surprisingly inexpensive, and a general correlation between price and molecular structure cannot be given. *All symmetric α, ω -difunctional straight-chain compounds are omitted. Many low-cost reagents of this type from C_2 up to C_{12} are commercially available.*
The lists are arranged according to the occurrence of functional groups.

(1) $C=C$ double and $C\equiv C$ triple bonds.
(2)   C—O single bonds (alcohols, epoxides, ethers, acetates, etc.).
(3) $C=O$ double bonds and derivatives, e.g. acetals.
(4)   Carboxylic acids and derivatives (including amides and nitriles).

(5) Nitrogen-containing functional groups.
(6) Halogen-containing functional groups.
(7) Sulfur-containing functional groups.

The first table lists compounds that contain no elements other than C, H, and O, in the above order (1) to (4). Since all molecules are difunctional, we use the second functional group for further subdivision again in the order (1) to (4). The next table lists compounds with nitrogen-containing functional groups in the order (1) to (5) according to their second functional group. The same procedure is repeated for compounds with halogen- and sulfur-containing functional groups. Within one subgroup, e.g. haloketones, the compounds are arranged in order of increasing chain length and increasing distance between the two functional groups. Exactly the same procedure is used for ordering trifunctional and oligofunctional compounds.

A summary of the tables is given below, while the tables with structural formulae, names, and approximate prices* follow thereafter.

Open-chain difunctional compounds
containing no other elements than
- C, H, O   table 4
- C, H, O, N   table 5
- C, H, O, N, halogen   table 6
- C, H, O, N, halogen, S   table 7

Open-chain trifunctional compounds
containing no other elements than
- C, H, O   table 8
- C, H, O, N, halogen, S   table 9

Open-chain oligofunctional compounds
containing no other elements than
- C, H, O   table 10
- C, H, O, N. halogen, S   table 11

---

* The numbers in brackets after the names of the compounds indicate the approximate prices in German marks per 10g: [0] means less than DM 0.50, [1] means DM 1.00 ± 0.50, and so on. For prices in US dollars divide by two.

Table 4. Open-chain difunctional C,H and C,H,O compounds.

● C=C/C≡C + C=C/C≡C

1,3-butadiene [0-2]

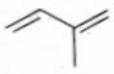

isoprene [0-1]

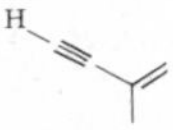

2-methyl-1-buten-3-yne [3]

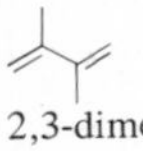

2,3-dimethyl-1,3-butadiene [9]

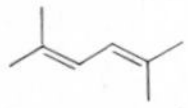

2,5-dimethyl-2,4-hexadiene [6-7]

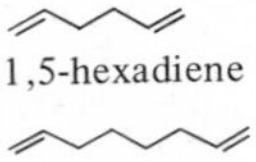

1,5-hexadiene [5]

1,7-octadiene [2-6]

● C=C/C≡C + C—O

alkyl vinyl ethers: R=Me [1],
Et, Bu^i [0], Bu [0-1]; vinyl acetate [0]

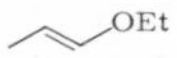

1-ethoxypropene [1]

2-methoxypropene [1-2];
isopropenyl acetate [0]

R=H: allyl alcohol [0];
R=Ac: allyl acetate [1]

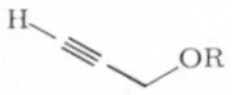

R=H: 2-propyn-1-ol [0-1];
R=Me: methyl propargyl ether [6]

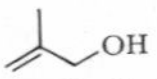

methallyl alcohol [1-2]

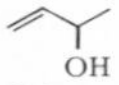

3-buten-2-ol [3]

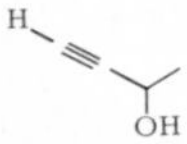

3-butyn-2-ol [1-2]

2-methyl-3-buten-2-ol [1]

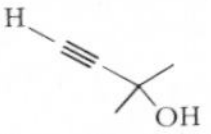

2-methyl-3-butyn-2-ol [0]

crotyl alcohol [2]

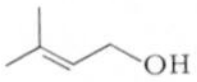

3-methyl-2-buten-1-ol [4]

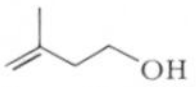

3-methyl-3-buten-1-ol [2-3]

3-methyl-1-penten-3-ol [1]

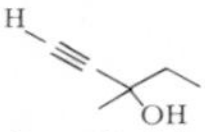

3-methyl-1-pentyn-3-ol [1-2]

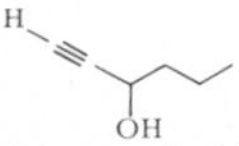

1-hexyn-3-ol [2]

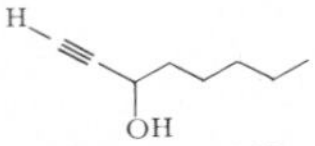

1-octyn-3-ol [6]

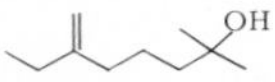

dihydromyrcenol [4]

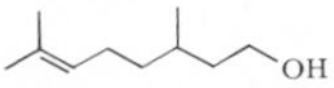

(±)-citronellol [2]

10-undecen-1-ol [1]

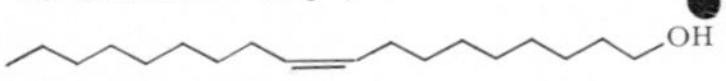

oleyl alcohol [0-1]

● C=C/C≡C + C=O/C(OR)₂/C=C—OR

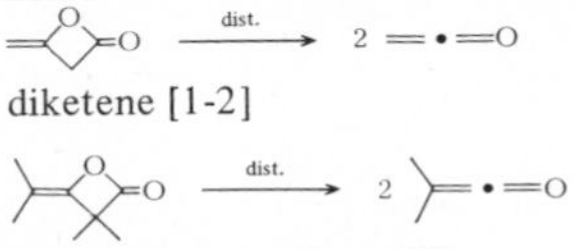

diketene [1-2]   2 =•=O

dimethylketene dimer [3]   2

acrolein [0]; acetals: R=Me [9],
Et [8], pentaerythrityl [1]

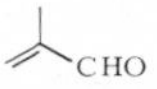

methacrolein [1-2]

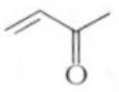

methyl vinyl ketone [1-2]

crotonaldehyde [0]
1,1-diacetoxy-2-butene [4]

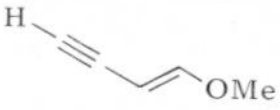

1-methoxy-1-buten-3-yne [1]

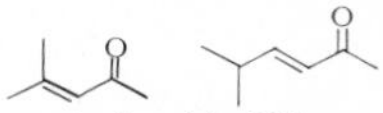

mesityl oxide [0];
5-methyl-3-hexen-2-one [0]

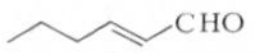

trans-2-hexenal [8]

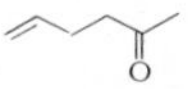

allylacetone [6-8]

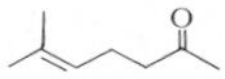

6-methyl-5-hepten-2-one [1]

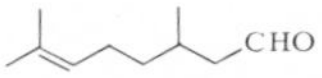

(±)-citronellal [1-2]

10-undecenal [13]

● C=C/C≡C + C—X/C≡N

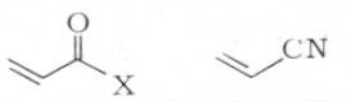

acrylic acid: X=OH [0],
OAlkyl [0-1], NH₂ [0], Cl [2];
acrylonitrile [0-1]

H—≡—COOH
propiolic acid [21]

able 4. (Continued)

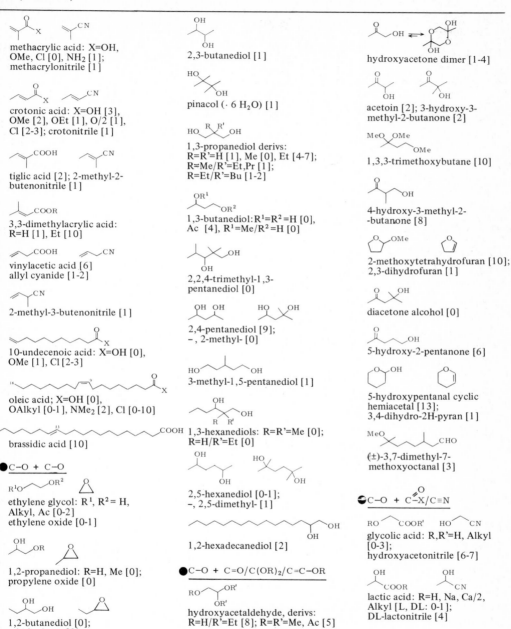

methacrylic acid: X=OH,
OMe, Cl [0], NH₂ [1];
methacrylonitrile [1]

crotonic acid: X=OH [3],
OMe [2], OEt [1], O/2 [1],
Cl [2-3]; crotonitrile [1]

tiglic acid [2]; 2-methyl-2-
butenonitrile [1]

3,3-dimethylacrylic acid:
R=H [1], Et [10]

vinylacetic acid [6]
allyl cyanide [1-2]

2-methyl-3-butenonitrile [1]

10-undecenoic acid: X=OH [0],
OMe [1], Cl [2-3]

oleic acid; X=OH [0],
OAlkyl [0-1], NMe₂ [2], Cl [0-10]

brassidic acid [10]

● C−O + C−O

ethylene glycol: R¹, R² = H,
Alkyl, Ac [0-2]
ethylene oxide [0-1]

1,2-propanediol: R=H, Me [0];
propylene oxide [0]

1,2-butanediol [0];
ethyloxirane [0]

2,3-butanediol [1]

pinacol (· 6 H₂O) [1]

1,3-propanediol derivs:
R=R'=H [1], Me [0], Et [4-7];
R=Me/R'=Et,Pr [1];
R=Et/R'=Bu [1-2]

1,3-butanediol: R¹=R²=H [0],
Ac [4], R¹=Me/R²=H [0]

2,2,4-trimethyl-1,3-
pentanediol [0]

2,4-pentanediol [9];
−, 2-methyl- [0]

3-methyl-1,5-pentanediol [1]

1,3-hexanediols: R=R'=Me [0];
R=H/R'=Et [0]

2,5-hexanediol [0-1];
−, 2,5-dimethyl- [1]

1,2-hexadecanediol [2]

● C−O + C=O/C(OR)₂/C=C−OR

hydroxyacetaldehyde, derivs:
R=H/R'=Et [8]; R=R'=Me, Ac [5]

hydroxyacetone dimer [1-4]

acetoin [2]; 3-hydroxy-3-
methyl-2-butanone [2]

1,3,3-trimethoxybutane [10]

4-hydroxy-3-methyl-2-
-butanone [8]

2-methoxytetrahydrofuran [10];
2,3-dihydrofuran [1]

diacetone alcohol [0]

5-hydroxy-2-pentanone [6]

5-hydroxypentanal cyclic
hemiacetal [13];
3,4-dihydro-2H-pyran [1]

(±)-3,7-dimethyl-7-
methoxyoctanal [3]

● C−O + C−X/C≡N

glycolic acid: R,R'=H, Alkyl
[0-3];
hydroxyacetonitrile [6-7]

lactic acid: R=H, Na, Ca/2,
Alkyl [L, DL: 0-1];
DL-lactonitrile [4]

Table 4. (Continued)

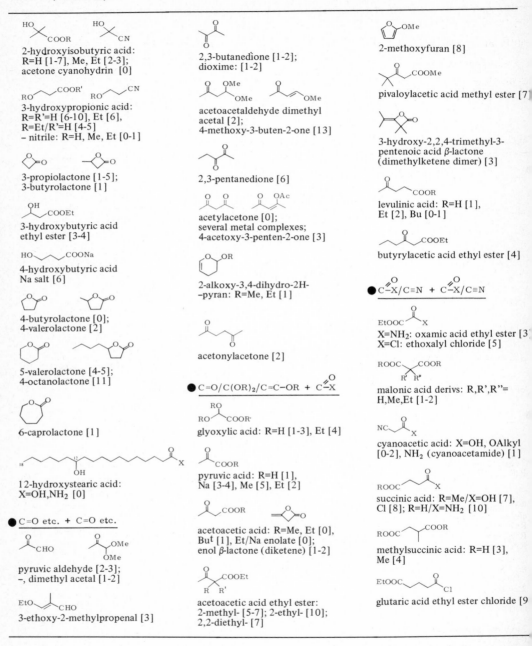

2-hydroxyisobutyric acid:
R=H [1-7], Me, Et [2-3];
acetone cyanohydrin [0]

3-hydroxypropionic acid:
R=R'=H [6-10], Et [6],
R=Et/R'=H [4-5]
– nitrile: R=H, Me, Et [0-1]

3-propiolactone [1-5];
3-butyrolactone [1]

3-hydroxybutyric acid
ethyl ester [3-4]

4-hydroxybutyric acid
Na salt [6]

4-butyrolactone [0];
4-valerolactone [2]

5-valerolactone [4-5];
4-octanolactone [11]

6-caprolactone [1]

12-hydroxystearic acid:
X=OH,NH₂ [0]

● C=O etc. + C=O etc.

pyruvic aldehyde [2-3];
–, dimethyl acetal [1-2]

3-ethoxy-2-methylpropenal [3]

2,3-butanedione [1-2];
dioxime: [1-2]

acetoacetaldehyde dimethyl
acetal [2];
4-methoxy-3-buten-2-one [13]

2,3-pentanedione [6]

acetylacetone [0];
several metal complexes;
4-acetoxy-3-penten-2-one [3]

2-alkoxy-3,4-dihydro-2H-
–pyran: R=Me, Et [1]

acetonylacetone [2]

● C=O/C(OR)₂/C=C–OR + C=O
                              ⁄

glyoxylic acid: R=H [1-3], Et [4]

pyruvic acid: R=H [1],
Na [3-4], Me [5], Et [2]

acetoacetic acid: R=Me, Et [0],
Buᵗ [1], Et/Na enolate [0];
enol β-lactone (diketene) [1-2]

acetoacetic acid ethyl ester:
2-methyl- [5-7]; 2-ethyl- [10];
2,2-diethyl- [7]

2-methoxyfuran [8]

pivaloylacetic acid methyl ester [7]

3-hydroxy-2,2,4-trimethyl-3-
pentenoic acid β-lactone
(dimethylketene dimer) [3]

levulinic acid: R=H [1],
Et [2], Bu [0-1]

butyrylacetic acid ethyl ester [4]

● C=X/C≡N + C=X/C≡N

X=NH₂: oxamic acid ethyl ester [3
X=Cl: ethoxalyl chloride [5]

malonic acid derivs: R,R',R''=
H,Me,Et [1-2]

cyanoacetic acid: X=OH, OAlkyl
[0-2], NH₂ (cyanoacetamide) [1]

succinic acid: R=Me/X=OH [7],
Cl [8]; R=H/X=NH₂ [10]

methylsuccinic acid: R=H [3],
Me [4]

glutaric acid ethyl ester chloride [9

Table 4. (Continued)

CN

methylglutaronitrile [0-1]

OOC‚‚‚‚COOH

2-dimethylglutaric acid [2]

HOOC‚‚‚‚COOH

3,3-dimethylglutaric acid [8]

ROOC‚‚‚‚COOH

adipic acid monoesters:
R=Me [6], Et [9]

HOOC‚‚‚‚COOH

3-methyladipic acid [7]

Table 5. Open-chain difunctional C,H,N and C,H,O,N compounds.

C=C/C≡C + C–N

‚‚‚NH₂

allylamine [1]

H‚‚‚NR₂

propargylamines: R=Me [2], Et [5]

H‚‚‚NH₂

1,1-dimethylpropargylamine [3-5]

H‚‚‚NH₂

3-amino-3-ethyl-1-pentyne [4-6]

‚‚‚NH₂

oleylamine [1]

C–O + C–N

R'
|
RO‚‚‚N‚R"

2-aminoethanol derivs:
R,R',R"=H, Alkyl [0-1]

⊕NMe₃] X⊖
|
RO

choline derivatives:
R=H/X=OH [1], Cl [0], Br [7], I [3];
R=Ac/X=Cl [2], Br [3], I [6];
R=C₃H₇CO/X=Cl [10], Br [9], I [6]

HO
O₂N NO₂

2,2-dinitropropanol [3]

HO
NR₂

2-amino-2-methyl-1-propanol:
R=H [0], Me [1]

OH
‚‚‚NR₂

DL-1-amino-2-propanol:
R=H [0], Me [0-1], Et [2]

HO‚‚‚NR₂

3-amino-1-propanol: R=H [1],
Me [1-2], Et [8]

HO
H NH₂

2-amino-1-butanol [(–)-L, DL: 1]

OH

NO₂

3-nitro-2-pentanol [8]

HO‚‚‚NH₂

5-amino-1-pentanol [7]

HO
⊕NH₃] Cl⊖

6-amino-2-methyl-2-heptanol [3]

C=O/C(OR)₂/C=C–OR + C–N/C=N

R' OR"
| |
R‚‚‚N‚‚‚OR"

aminoacetaldehyde derivs:
R,R'=H,Me,Et/R"=Me,Et [6-11]

O‚H
|
N

butanedione monoxime [3]

O
‖
Et₂N‚‚‚

4-diethylamino-2-butanone [8]

O
‖
Me₂N‚‚‚

4-dimethylamino-3-methyl-2-
butanone [2]

H₃N⊕ O COO⊖
] |
COOH

diacetonamine hydrogen
oxalate [9]

O
‖
Et₂N‚‚‚

5-diethylamino-2-pentanone [5]

C–X/C≡N + C–N

R'
|
R‚N‚COOR"

glycine derivatives:
R=R'=H/R"=H [0],
Me, Et [1];
R=R"=H/R'=Me, Ac [1];
R"=Et/R=R'=Me [5], Et [8]

R'
|
R‚N‚CN

aminoacetonitrile: R,R'= H, Me
[2-6]

Me₃N⊕‚‚‚COO⊖

betaine [1], · H₂O [1], · HCl [0],
ethyl ester chloride [6]

⊖N‚N⊕‚‚‚COOEt

diazoacetic acid ethyl ester [8]

H₃N⊕ H
‚‚‚
COO⊖

alanine [L: 3-4, DL: 1]

H₃N⊕
‚‚‚COO⊖

α-aminoisobutyric acid [5]

COOR'
R₂N‚‚‚

β-alanine: R=R'=H [0-1];
R'=Et/R=Me [6], Et [2-3]

Table 5. (Continued)

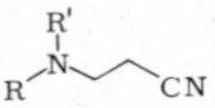

3-aminopropionitrile:
R=Me/R'=H [2], Me [1]

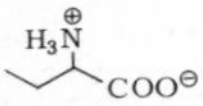

DL-2-aminobutyric acid [2]

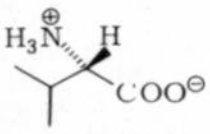

valine [L: 3-4, DL: 1]

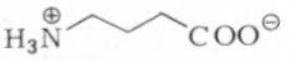

γ-aminobutyric acid [2]

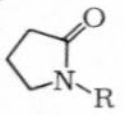

2-pyrrolidinone:
R=H, Me [0], Et [6-9]

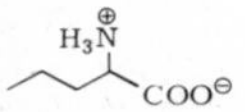

DL-2-aminovaleric acid [2]

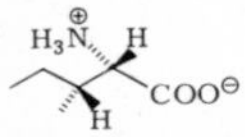

isoleucine [L: 8, DL: 7]

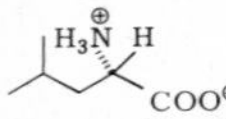

leucine [L, DL: 2];
–, N-acetyl-DL- [3]

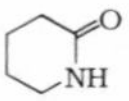

5-valerolactam [9]

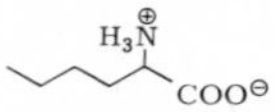

DL-2-aminocaproic acid [3]

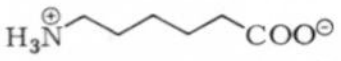

6-aminocaproic acid [3]

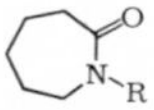

6-caprolactam: R=H [0],
Me, Et [1]

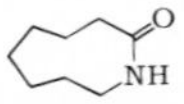

8-caprylolactam [11]

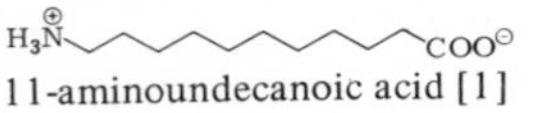

11-aminoundecanoic acid [1]

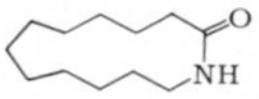

12-laurolactam [1]

●C–N + C–N

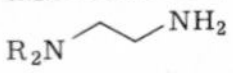

N,N-dialkylethylenediamine: R=Me
[3-6], Et [2]

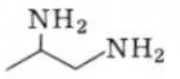

1,2-diaminopropane [0-1];
propylene imine [3]

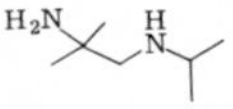

2-amino-1-isopropylamino-2-
methylpropane |1-2|

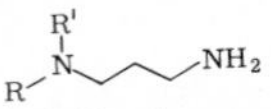

1,3-diaminopropane derivs:
R=H/R'=Me [1-2], Et [3];
R=R'=Me, Et [0], Bu [1-2]

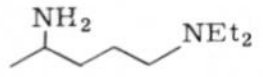

4-amino-1-diethylamino-
pentane [3]

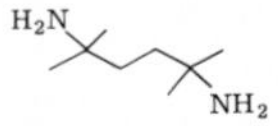

2,5-diamino-2,5-dimethyl-hexane [2-4]

Table 6. Open-chain difunctional halides.

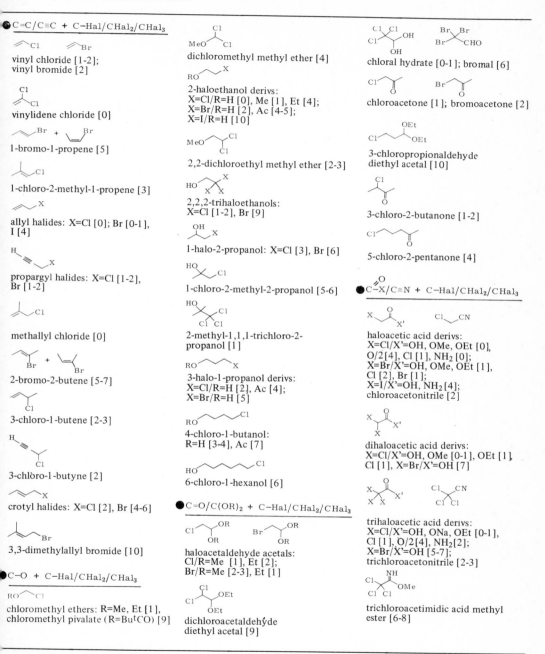

● C=C/C≡C + C−Hal/CHal₂/CHal₃

vinyl chloride [1-2];
vinyl bromide [2]

vinylidene chloride [0]

1-bromo-1-propene [5]

1-chloro-2-methyl-1-propene [3]

allyl halides: X=Cl [0]; Br [0-1],
I [4]

propargyl halides: X=Cl [1-2],
Br [1-2]

methallyl chloride [0]

2-bromo-2-butene [5-7]

3-chloro-1-butene [2-3]

3-chloro-1-butyne [2]

crotyl halides: X=Cl [2], Br [4-6]

3,3-dimethylallyl bromide [10]

● C−O + C−Hal/CHal₂/CHal₃

chloromethyl ethers: R=Me, Et [1],
chloromethyl pivalate (R=Bu^{t}CO) [9]

dichloromethyl methyl ether [4]

2-haloethanol derivs:
X=Cl/R=H [0], Me [1], Et [4];
X=Br/R=H [2], Ac [4-5];
X=I/R=H [10]

2,2-dichloroethyl methyl ether [2-3]

2,2,2-trihaloethanols:
X=Cl [1-2], Br [9]

1-halo-2-propanol: X=Cl [3], Br [6]

1-chloro-2-methyl-2-propanol [5-6]

2-methyl-1,1,1-trichloro-2-
propanol [1]

3-halo-1-propanol derivs:
X=Cl/R=H [2], Ac [4];
X=Br/R=H [5]

4-chloro-1-butanol:
R=H [3-4], Ac [7]

6-chloro-1-hexanol [6]

● C=O/C(OR)₂ + C−Hal/CHal₂/CHal₃

haloacetaldehyde acetals:
Cl/R=Me [1], Et [2];
Br/R=Me [2-3], Et [1]

dichloroacetaldehyde
diethyl acetal [9]

chloral hydrate [0-1]; bromal [6]

chloroacetone [1]; bromoacetone [2]

3-chloropropionaldehyde
diethyl acetal [10]

3-chloro-2-butanone [1-2]

5-chloro-2-pentanone [4]

● C−X/C≡N + C−Hal/CHal₂/CHal₃

haloacetic acid derivs:
X=Cl/X'=OH, OMe, OEt [0],
O/2[4], Cl [1], NH₂ [0];
X=Br/X'=OH, OMe, OEt [1],
Cl [2], Br [1];
X=I/X'=OH, NH₂ [4];
chloroacetonitrile [2]

dihaloacetic acid derivs:
X=Cl/X'=OH, OMe [0-1], OEt [1],
Cl [1], X=Br/X'=OH [7]

trihaloacetic acid derivs:
X=Cl/X'=OH, ONa, OEt [0-1],
Cl [1], O/2[4], NH₂[2];
X=Br/X'=OH [5-7];
trichloroacetonitrile [2-3]

trichloroacetimidic acid methyl
ester [6-8]

Table 6. (Continued)

2-halopropionic acid derivs:
Cl/X'=OH, OMe, OEt [0-3], Cl [3];
Br/X'=OH, OMe, OEt [1],
Cl [11], Br [1]

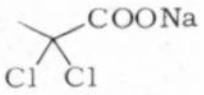

2,2-dichloropropionic acid Na
salt [0-1]

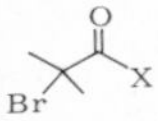

2-bromoisobutyric acid
X=OH, OEt [1], Br [3]

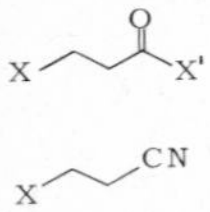

3-halopropionic acid derivs:
X=Cl/X'=OH, Cl [0-1], OEt [2-3];
X=Br/X'=OH, OEt [1], Cl [10];
nitriles: X=Cl [1], Br [2-3]

β-chloropivalic acid [8]

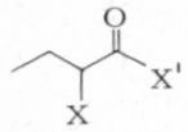

2-halobutyric acid derivs:
X=Cl/X'=OH [2]
X=Br/X=OH, OEt, Br [1]

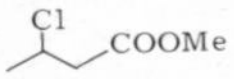

3-chlorobutyric acid methyl
ester [4]

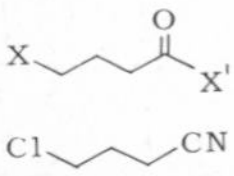

4-halobutyric acid derivs:
X=Cl/X'=OH [5], OMe, OEt [2],
Cl [0-1]; nitrile [1-3];
X=Br/X'=OEt [1-2]

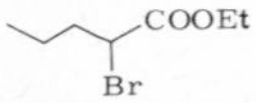

2-bromovaleric acid ethyl ester [8]

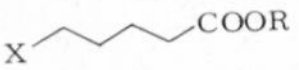

5-halovaleric acid derivs:
X=Cl/R=H [9];
X=Br/R=Et [10]

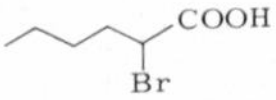

2-bromohexanoic acid [2]

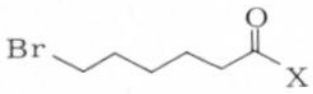

6-bromohexanoic acid: X=OH [12],
Cl [8]

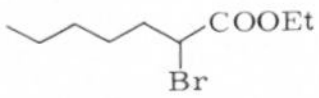

2-bromoheptanoic acid ethyl ester [1]

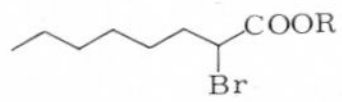

2-bromooctanoic acid: R=H [6], Et [3]

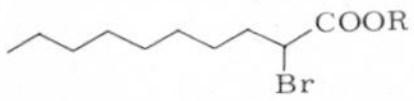

2-bromodecanoic acid:
R=H [9], Et [10]

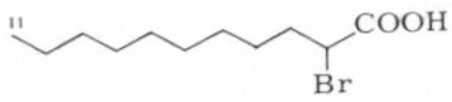

2-bromoundecanoic acid [6]

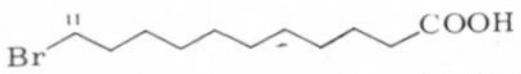

11-bromoundecanoic acid [1-5]

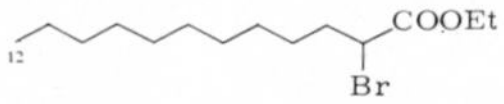

2-bromododecanoic acid ethyl
ester [7]

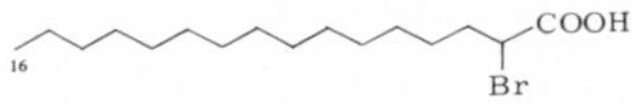

2-bromohexadecanoic acid [5-6]

● C–N + C–Hal

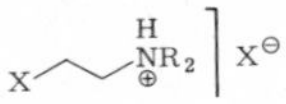

2-haloethylamines:
X=Cl/R=H, Me, Et, Pr^i [1-2];
X=Br/R=H [2]

(2-chloroethyl)trimethylammonium
chloride [9]

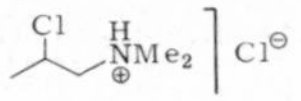

2-chloro-1-dimethylamino-
propane · HCl [1-2]

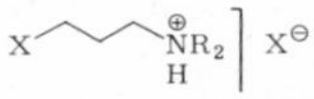

3-halopropylamines:
X=Cl/R=H [3], Me [1], Et [6];
X=Br/R=H [6]

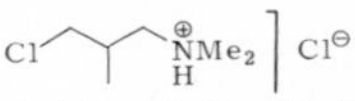

1-chloro-3-dimethylamino-
-2-methylpropane · HCl [3]

● C–Hal/C Hal$_2$/C Hal$_3$

+ C–Hal/C Hal$_2$/C Hal$_3$

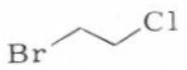

1-bromo-2-chloroethane [1-2]

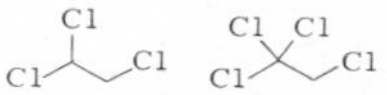

1,1,2-trichloroethane [0];
1,1,1,2-tetrachloroethane [10]

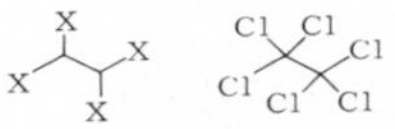

1,1,2,2-tetrahaloethanes:
X=Cl [0], Br [0-1];
pentachloroethane [0-2]

hexachloroethane [0-2];
1,2-dibromotetrachloroethane [7]

1,2-dichloropropane [0];
1,2-dibromopropane [0]

1-bromo-3-chloropropane
[0-1]

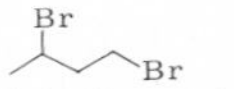

1,3-dibromobutane [1]

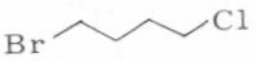

1-bromo-4-chlorobutane [9]

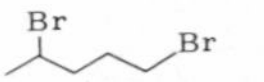

1,4-dibromopentane [5]

Table 7. Open-chain difunctional sulfur compounds.

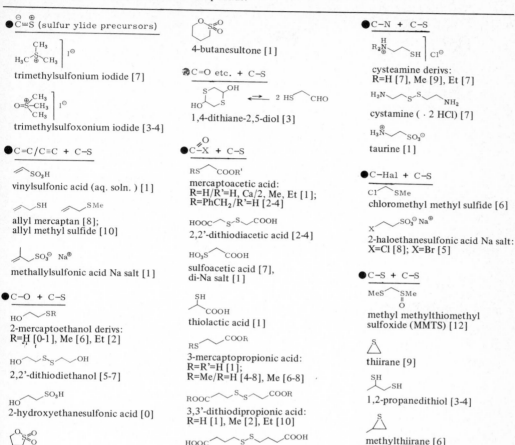

● C≏S (sulfur ylide precursors)

trimethylsulfonium iodide [7]

trimethylsulfoxonium iodide [3-4]

● C=C/C≡C + C–S

vinylsulfonic acid (aq. soln.) [1]

allyl mercaptan [8];
allyl methyl sulfide [10]

methallylsulfonic acid Na salt [1]

● C–O + C–S

2-mercaptoethanol derivs:
R=H [0-1], Me [6], Et [2]

2,2'-dithiodiethanol [5-7]

2-hydroxyethanesulfonic acid [0]

3-propanesultone [2]

4-butanesultone [1]

C=O etc. + C–S

1,4-dithiane-2,5-diol [3]

2 HS⌒CHO

● C–X + C–S

mercaptoacetic acid:
R=H/R'=H, Ca/2, Me, Et [1];
R=PhCH₂/R'=H [2-4]

2,2'-dithiodiacetic acid [2-4]

sulfoacetic acid [7],
di-Na salt [1]

thiolactic acid [1]

3-mercaptopropionic acid:
R=R'=H [1];
R=Me/R'=H [4-8], Me [6-8]

3,3'-dithiodipropionic acid:
R=H [1], Me [2], Et [10]

4,4'-dithiodibutyric acid [8]

● C–N + C–S

cysteamine derivs:
R=H [7], Me [9], Et [7]

cystamine (· 2 HCl) [7]

taurine [1]

● C–Hal + C–S

chloromethyl methyl sulfide [6]

2-haloethanesulfonic acid Na salt:
X=Cl [8]; X=Br [5]

● C–S + C–S

methyl methylthiomethyl
sulfoxide (MMTS) [12]

thiirane [9]

1,2-propanedithiol [3-4]

methylthiirane [6]

Table 8. Open-chain trifunctional C,H and C,H,O compounds.

● C=C/C≡C + two FGs

$$\begin{array}{c}\quad\quad O \\ (C=C,\ C{-}O,\ C{=}O,\ C{-}X,\ C{\equiv}N)\end{array}$$

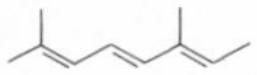

alloocimene [0]

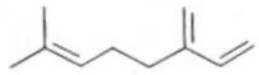

myrcene [0]

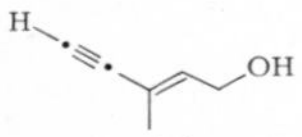

3-methyl-2-penten-4-yn-
-1-ol [(E): 2, (Z): 2-3]

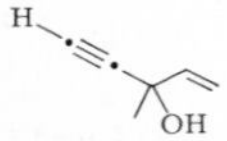

3-methyl-1-penten-4-yn-
-3-ol [3]

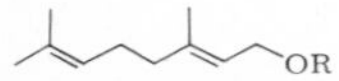

geraniol: R=H [2-5];
geranyl acetate R=Ac [2-3]

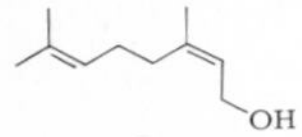

nerol [8]

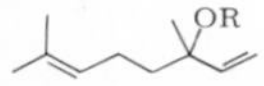

(±)-linalool: R=H, Ac [1-2]

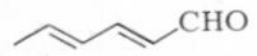

sorbic aldehyde [5-6]

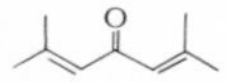

phorone [1-2]

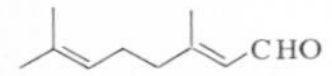

citral (E+Z) [2]

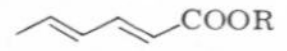

sorbic acid: R=H [0-1], Et [6]

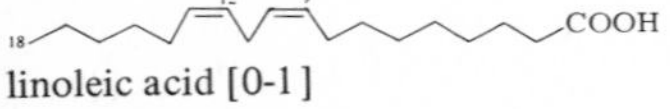

linoleic acid [0-1]

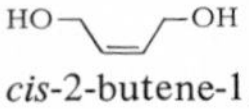

---

cis-2-butene-1,4-diol [0]

2-butyne-1,4-diol [0-3]

3-hexyne-2,5-diol [1]

-, 2,5-dimethyl- [1-3]

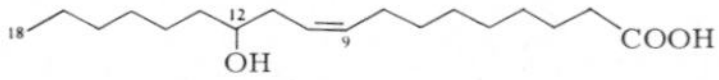

ricinoleic acid [1]

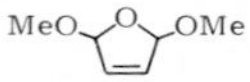

maleic dialdehyde dimethyl
acetal [2-3]

maleic acid derivs: R=H, Me, Et,
Bu [0]; anhydride [0]

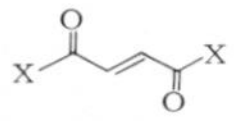

fumaric acid: X=OH [0],
OEt [1-2], Cl [2-3]

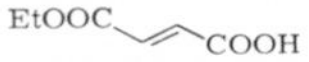

–monoethyl ester [3];

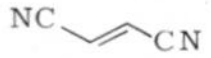

– dinitrile [9]

acetylenedicarboxylic acid:
R=H [2-4], Me [8]

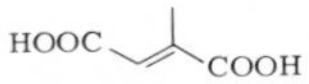

mesaconic acid [10]

citraconic acid [2];
– anhydride [1-2]

---

itaconic acid: R=H [0], Me [1];
– anhydride [1-2]

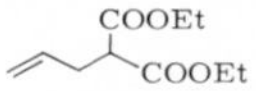

allylmalonic acid diethyl ester [1-3]

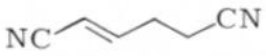

2-hexenedioic acid
dinitrile [(E)+(Z): 1]

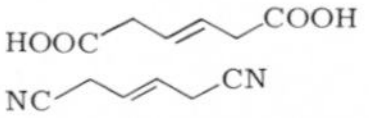

(E)-3-hexenedioic acid [4];
– dinitrile [2]

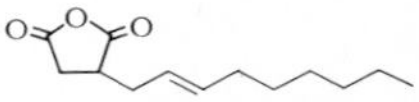

2-nonen-1-ylsuccinic acid
anhydride [1]

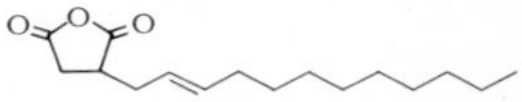

2-dodecen-1-ylsuccinic acid
anhydride [1-4]

● C–O/C=O/C–X/C≡N (3 FGs)

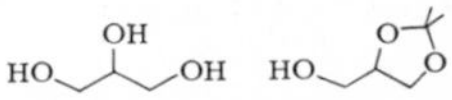

glycerol, several triesters [0-1];
1-acetate [3]; acetonide [0-1]

glycidol [1]

1,1,1-tris(hydroxymethyl)- ethane,
-propane [0]

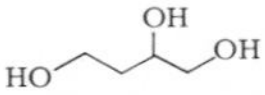

1,2,4-butanetriol [6]

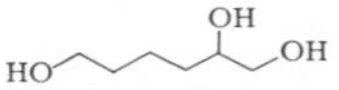

1,2,6-hexanetriol [1]

ble 8. (Continued)

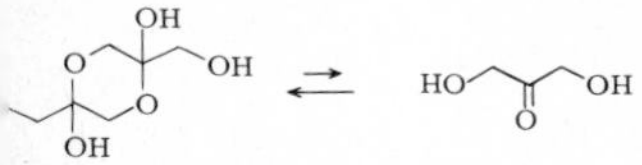

-dihydroxyacetone dimer [2]

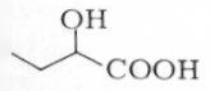

-glyceric acid [9]

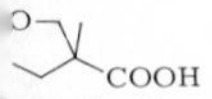

?-bis(hydroxymethyl)-
opionic acid [0-1]

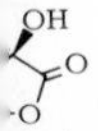

ntoic acid γ-lactone
)-D: 12, DL: 1]

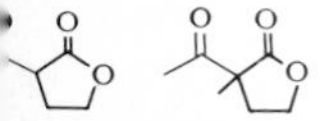

cetyl-4-butyrolactone [1];
cetyl-2-methyl-4-butyrolactone [10]

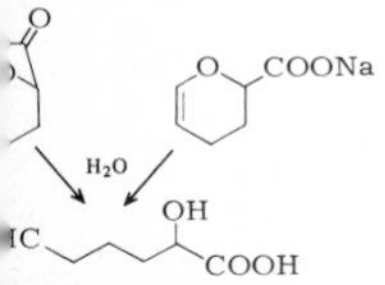

ydroxytetrahydropyran-2- carboxylic
d lactone [8];
-dihydro-2H-pyran-2-carboxylic
d Na salt [2-3]

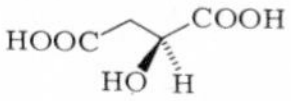

malic acid [L: 4, DL: 0]

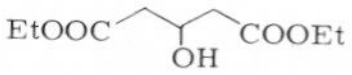

3-hydroxyglutaric acid diethyl ester [9]

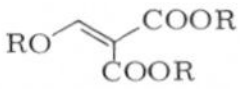

alkoxymethylenemalonic esters:
R=Me [4] , Et [1-2]

(ethoxymethylene)-cyanoacetic acid
ethyl ester [2], -malonitrile [8]

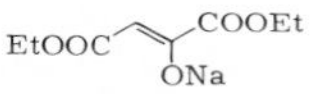

oxaloacetic acid diethyl ester Na
enolate [1]

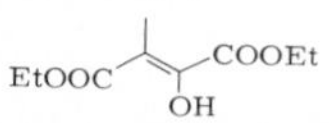

oxalopropionic acid diethyl ester [3-4] •

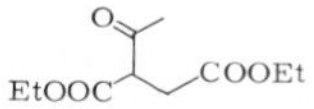

acetylsuccinic acid diethyl ester [8]

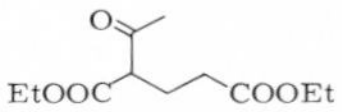

2-acetylglutaric acid diethyl ester [5]

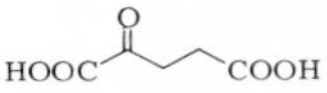

2-oxoglutaric acid [2-3]

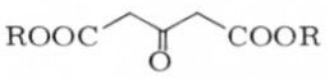

1,3-acetonedicarboxylic acid: R=H
[3-4], Me, Et [2-3]

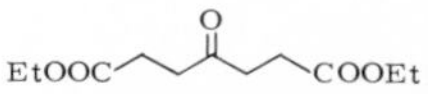

4-oxopimelic acid diethyl ester [10]

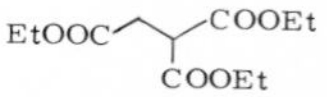

1,1,2-ethanetricarboxylic acid triethyl
ester [14]

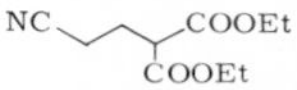

(2-cyanoethyl)malonic acid
diethyl ester [10]

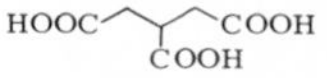

tricarballic acid [14]

Table 9. Open-chain trifunctional nitrogen, halogen, and sulfur compounds.

● C–N/C=N

3-aminocrotonic acid: X=OMe [1-2], OEt [2], NH₂ [5-7]; – nitrile [2]

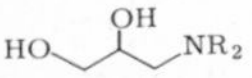

3-dialkylamino-1,2-propanediols: R=Me, Et [4]

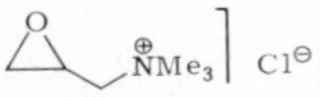

glycidyltrimethylammonium chloride [1]

2-amino-2-alkyl-1,3-propanediols: R=Me [2], Et [0-1]

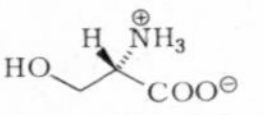

serine [L: 6-7, DL: 3]

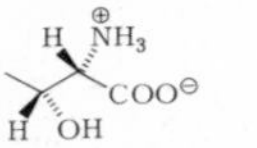

threonine [D: 8, L: 5-7, DL:5]

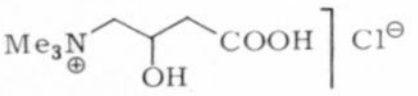

DL-carnitine · HCl [2]

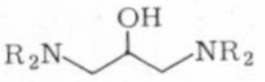

1,3-diamino-2-propanol: R=H [3-4], Me [6-9]

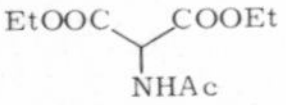

acetamidomalonic acid diethyl ester [2]

acetamido- [5]; oximino-cyanoacetic acid ethyl ester [7]

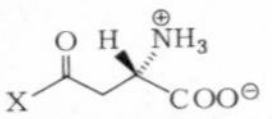

X=OH: aspartic acid [L, DL: 1]
X=NH₂: asparagine [D: 3, L: 1-4, DL: 2]

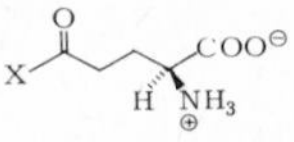

X=OH: glutamic acid [L: 0-1, DL: 2, N-acetyl-L: 6];
X=OMe [9], OEt [8]: 5-esters;
X=NH₂: glutamine [L: 2-3]

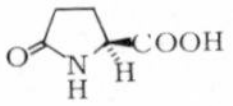

pyroglutamic acid [L: 4, DL: 2]

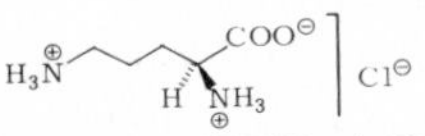

ornithine · HCl [L: 4, DL: 9]

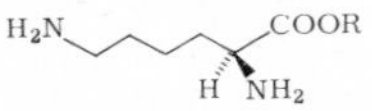

R=H: lysine [L: 4-10], · HCl [L: 1, DL: 3-4]
R=Me: methyl ester, · HCl [L:10]

● C–Hal/CHal₂/CHal₃

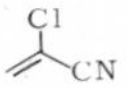

2-chloroacrylonitrile [1]

1,2-dihaloethenes: X=Cl
[*E*: 1, *Z*: 6, *E*+*Z*: 0];
X=Br [*E*+*Z*: 4]

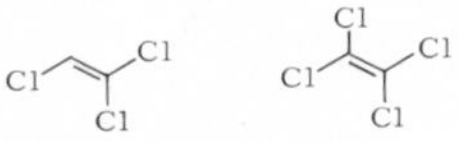

trichloroethene [0];
tetrachloroethene[0]

2,3-dihalopropenes: Cl [1]; Br [9]

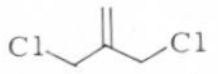

3-chloro-2-chloromethyl-1-propene [1]

1,4-dihalo-2-butenes:
X=Cl [*Z*: 10, *E*: 1];
X=Br [*E*: 9, *E*+*Z*: 8]

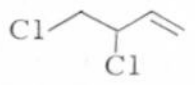

3,4-dichloro-1-butene [1-2]

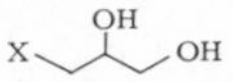

3-halo-1,2-propanediols:
X=Cl [0]; X=Br [9]

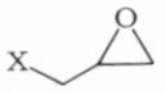

epihalohydrins: X=Cl [0], Br [1

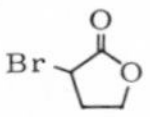

2-bromo-4-butyrolactone [7]

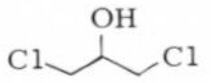

1,3-dichloro-2-propanol [0-1]

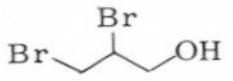

2,3-dibromo-1-propanol [1]

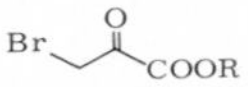

bromopyruvic acid: R=H [9-14
R=Et [6]

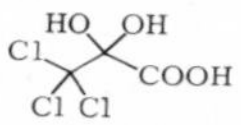

2,2-dihydroxy-3,3,3-trichloro-propionic acid [5]

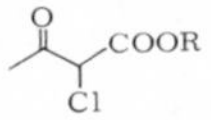

2-chloroacetoacetic esters:
R=Me [0-1], Et [1]

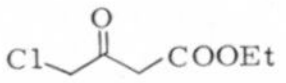

4-chloroacetoacetic acid ethyl ester [2-3]

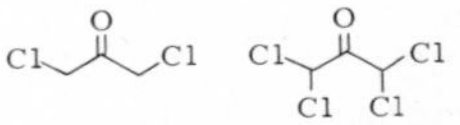

1,3-dichloroacetone [2-4];
1,1,3,3-tetrachloroacetone [9]

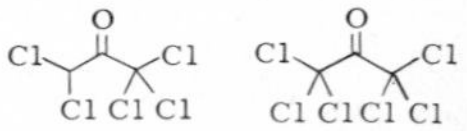

pentachloroacetone [7];
hexachloroacetone [0-1]

able 9. (Continued)

OOC COOR
X

alomalonic acid esters:
=Cl/R=Me [2], Et [5];
=Br/R=Me [8], Et [3]

OOC COOEt
Br

-bromo-2-methylmalonic
cid diethyl ester [11]

OOC COOH
X

=Cl: chlorosuccinic acid [2];
=Br: bromosuccinic acid [13]

COOR
X

3-dihalopropionic acid:
=Cl/R=Me [6];
=Br/R=H [4-5], Me [7],Et [1-2]

Cl, Cl, Cl
1,2,3-trichloropropane [0-1];
1,2,3-tribromopropane [1]

Br, Br, Br

● C–S

OH
HO SH
3-mercapto-1,2-propanediol [2]

OH
Cl SH
1-chloro-3-mercapto-2-propanol [4]

COOH
HOOC SH
mercaptosuccinic acid [1]

COO⊖
HS H NH₃
⊕
L-cysteine [2-3];
–, S-benzyl-[7]; N-acetyl-[5];
– ethyl ester [7]

$H_3\overset{\oplus}{N}$ H S–S COO⊖
⊖OOC H $\overset{\oplus}{N}H_3$
L-cystine [2]

⊖O₃S COOH
H NH₃
⊕
(+)-L-cysteic acid [10]

RS COO⊖
H NH₃
⊕
R=Me: methionine [L: 2, DL: 0];
S-methylsulfonium halides:
· MeCl [DL: 3], · MeBr [DL: 2-3];
R=Et: ethionine [DL: 10];
disulfide=homocystine [DL: 8]

NH₂ NHAc
DL-homocysteine thiolactone
(· HCl) [4]; –, N-acetyl- [3]

---

ble 10. Open-chain oligofunctional C,H and C,H,O compounds.

squalene [3]

farnesol [9]

nerolidol [11]

linolenic acid [1]

2,7-dimethyl-3,5-octadiyne
-2,7-diol [6]

EtOOC COOEt
diallylmalonic acid diethyl
ester [2]

COOH
coumalic acid [9],
(malonaldehydic acid dimer]

COOH
isodehydracetic acid:
R=H [9], Me [5]
(acetoacetic acid dimer)

COOH
HOOC COOH
trans-aconitic acid [8]

HO OH
HO OH
pentaerythritol [0];
– monoformal [3-5]

HO OH
aleuritic acid [3-4]

HO OH
EtOOC COOEt
bis(hydroxymethyl)malonic
acid diethyl ester [10]

OR
R'OOC COOR'
OR
tartaric acid derivs:
R=R'=H [D: 10-12, L: 0, DL: 1-2,
meso: 10];

R=H/R'=Me [L: 3-4], Et, Bu [L: 1-2];
R=PhCO/R'=H [D: 5, L: 2];

AcO
AcO
O,O'-diacetyl-L-tartaric acid anhydride
[2]

OH
3-hydroxy-2-methyl-4-pyrone
(maltol) [10]

Table 10. (Continued)

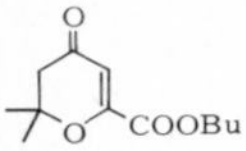

butopyronoxyl (indalone) [2]

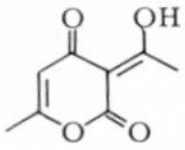

dehydracetic acid [1]

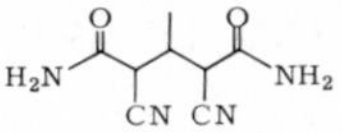

2,4-dicyano-3-methyl-
glutaramide [4]

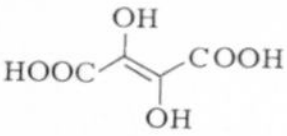

HOOC—COOH

dihydroxyfumaric acid
(dihydroxymaleic acid) [11]

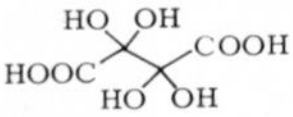

dihydroxytartaric acid di-Na
salt [3]

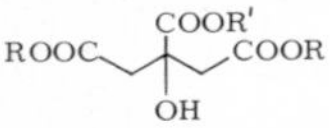

citric acid: R=R'=H, Et, Bu
[0]; R=H/R'=Pr^i [1]

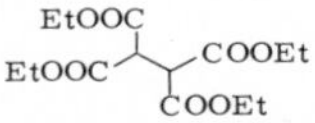

1,1,2,2-ethanetetracarboxylic acid
tetraethyl ester [12]

Table 11. Open-chain oligofunctional nitrogen and halogen compounds.

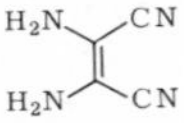

diaminomaleonitrile [5]

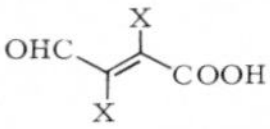

X=Cl: mucochloric acid [1]
X=Br: mucobromic acid [5]

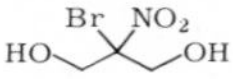

bronopol [9]

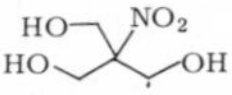

tris(hydroxymethyl)methylamine
(TRIS) [0-1, · HCl: 1-2]

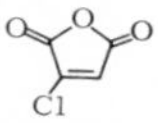

chloromaleic acid anhydride [2]

2,2-bis(chloromethyl) -
-1,3-propanediol [9]

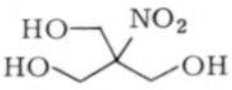

tris(hydroxymethyl)nitromethane [0]

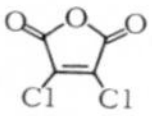

dichloromaleic acid anhydride [2-3]

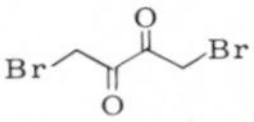

1,4-dibromo-2,3-butanedione [11

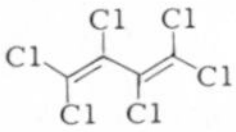

hexachloro-1,3-butadiene [0]

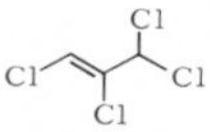

1,2,3,3-tetrachloropropene [9]

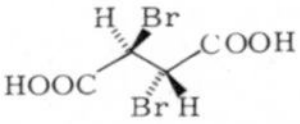

meso-2,3-dibromosuccinic acid [2

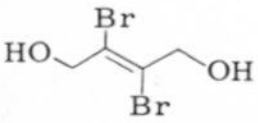

trans-2,3-dibromo-2-butene-1,4-
diol [2]

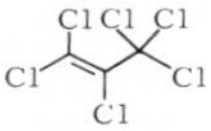

hexachloropropene [3-4]

pentaerythrityl tetrabromide [14]

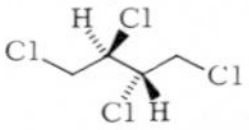

meso-1,2,3,4-tetrachlorobutane [

1.3 Silicon and Phosphorus Reagents

nly a few of these compounds are commercially available at reasonable prices.

able 12. Silicon compounds.

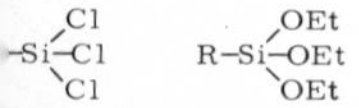

kyltrichlorosilanes: R=Me, Et [0-1],
, $C_{18}H_{37}$ [4]

kyltriethoxysilanes: R=Me [1], Et [2]

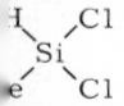

chloromethylsilane [1]

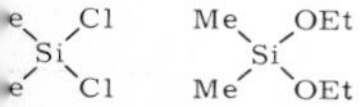

chlorodimethylsilane [0-1]
ethoxydimethylsilane [1-2]

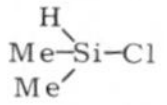

chlorodimethylsilane [0-1]

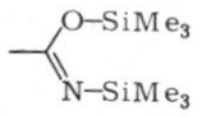

chlorotrimethylsilane [1];
N, O-bis(trimethylsilyl)-
acetamide [8]

trichlorovinylsilane [1];
triethoxyvinylsilane [1]

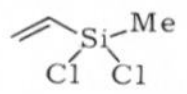

dichloromethylvinylsilane [1-2]

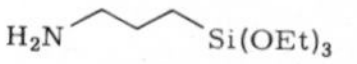

H_2N $Si(OEt)_3$
(3-aminopropyl)triethoxysilane [1]

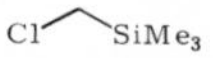

Cl $SiMe_3$
(chloromethyl)trimethylsilane [5]

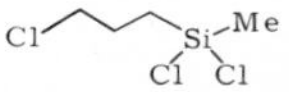

(3-chloropropyl) dichloromethyl-
silane [2]

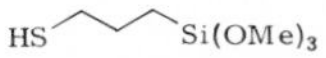

HS $Si(OMe)_3$
(3-mercaptopropyl)-
trimethoxysilane [4]

able 13. Phosphorus compounds.

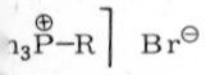

kyltriphenylphosphonium bromides
=Me [3], Et [4], Pr, Bu [5]

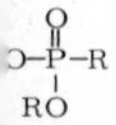

kanephosphonic acid dialkyl esters:
=Me [1-2], Et [1]

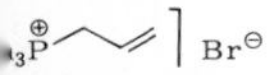

lyltriphenylphosphonium
omide [5]

iethoxymethyl) triphenyl-
iosphonium bromide [7]

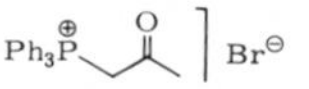

acetonyltriphenylphosphonium
bromide [6]

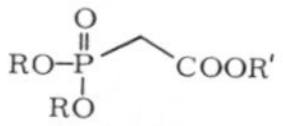

phosphonoacetic acid triesters:
R=R'=Me [9], Et [1-2];
R=Et/R'=Me [9]

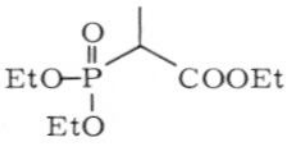

2-phosphonopropionic acid
triethyl ester [15]

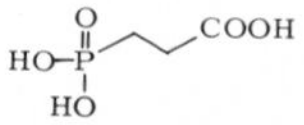

3-phosphonopropionic acid [1]

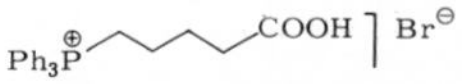

(4-carboxybutyl) triphenyl-
phosphonium bromide [8]

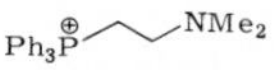

(2-dimethylaminoethyl) triphenyl-
phosphonium bromide [14]

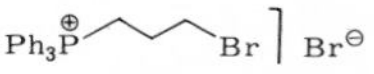

(3-bromopropyl) triphenyl-
phosphonium bromide [9]

3.1.4 Nonaromatic Carbocyclic Reagents

Mono- and oligocyclic carbon compounds are listed in order of increasing ring size. The lists are further subdivided in the same order of functional groups as described in section 3.1.2.

Table 14. Monocyclic hydrocarbon derivatives.

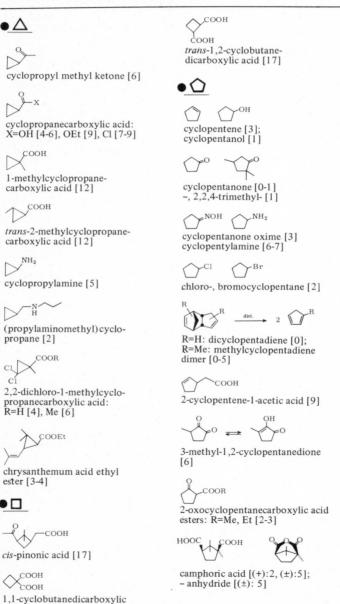

● △

cyclopropyl methyl ketone [6]

cyclopropanecarboxylic acid:
X=OH [4-6], OEt [9], Cl [7-9]

1-methylcyclopropane-
carboxylic acid [12]

trans-2-methylcyclopropane-
carboxylic acid [12]

cyclopropylamine [5]

(propylaminomethyl)cyclo-
propane [2]

2,2-dichloro-1-methylcyclo-
propanecarboxylic acid:
R=H [4], Me [6]

chrysanthemum acid ethyl
ester [3-4]

● ☐

cis-pinonic acid [17]

1,1-cyclobutanedicarboxylic
acid [16]

trans-1,2-cyclobutane-
dicarboxylic acid [17]

● ⬠

cyclopentene [3];
cyclopentanol [1]

cyclopentanone [0-1]
–, 2,2,4-trimethyl- [1]

cyclopentanone oxime [3]
cyclopentylamine [6-7]

chloro-, bromocyclopentane [2]

R=H: dicyclopentadiene [0];
R=Me: methylcyclopentadiene
dimer [0-5]

2-cyclopentene-1-acetic acid [9]

3-methyl-1,2-cyclopentanedione
[6]

2-oxocyclopentanecarboxylic acid
esters: R=Me, Et [2-3]

camphoric acid [(+):2, (±):5];
– anhydride [(±): 5]

2-(aminomethyl)-3,3,5-trimethyl
cyclopentylamine [1]

(*all-cis*)-1,2,3,4-cyclopentanetetra
carboxylic acid [3];
– 1,2:3,4-dianhydride [2]

hexachlorocyclopentadiene [0]

● ⬡

monofunctional

cyclohexene [0-1]; 1-methyl- [8]
4-methyl- [4-7]

cis-trans mixtures

cyclohexanol (R=H) [0-1];
alkylcylclohexanols: R=1-Me [7]
2-Me, 3-Me, 4-Me [1]; 3-Et, 4-Et
[5]; 4-But [1];
dimethylcyclohexanols:
2,3-[5]; 2,4- [7]; 2,5- [5];
2,6-[3]; 3,4- [10]; 3,5- [6-7]

menthol [(–)-(1R.2S,5R), (±):2]
3,3,5-trimethylcyclohexanol
[*cis*+*trans*: 0]

cyclohexylmethanol [4];
2-cyclohexylethanol [5]

able 14. (Continued)

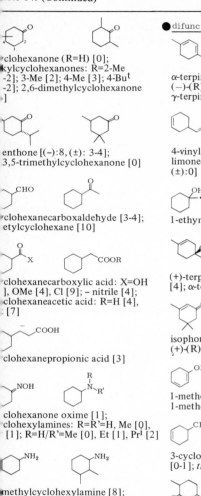

clohexanone (R=H) [0];
kylcyclohexanones: R=2-Me
-2]; 3-Me [2]; 4-Me [3]; 4-Bu^t
-2]; 2,6-dimethylcyclohexanone
»]

enthone [(−):8, (±): 3-4];
3,5-trimethylcyclohexanone [0]

clohexanecarboxaldehyde [3-4];
etylcyclohexane [10]

clohexanecarboxylic acid: X=OH
], OMe [4], Cl [9]; – nitrile [4];
clohexaneacetic acid: R=H [4],
 [7]

clohexanepropionic acid [3]

clohexanone oxime [1];
clohexylamines: R=R'=H, Me [0],
[1]; R=H/R'=Me [0], Et [1], Pr^i [2]

methylcyclohexylamine [8];
3-dimethylcyclohexylamine [3]

lorocyclohexane [0-1];
omocyclohexane [0];
docyclohexane [3]

clohexanethiol [1-2]

● difunctional

α-terpinene [1-2];
(−)-(R)-α-phellandrene [2];
γ-terpinene [2]

4-vinylcyclohexene [1];
limonene [(+)-(R),(−)-(S): 1;
(±):0]

1-ethynylcyclohexanol [0]

(+)-terpinen-4-ol [6];(−)-isopulegol
[4]; α-terpineol [1]

isophorone [0];
(+)-(R)-pulegone [4-8]

1-methoxy-1,3-cyclohexadiene [13],
1-methoxy-1,4-cyclohexadiene [3-5]

3-cyclohexene-1-carboxaldehyde
[0-1]; *trans*-6-methyl- [2]

(+)-dihydrocarvone [11]

3-cyclohexene-1-carboxylic acid
methyl ester [6];
3-cyclohexene-1-carbonitrile [10]

1,2-cyclohexanediol [E+Z:2];
cyclohexene oxide [2]

1,3-cyclohexanediol [E+Z:6];
1,4-cyclohexanediol [E+Z:2-3]

1,4-cyclohexanedimethanol
[E+Z: 0-2]

cyclohexanone cyanohydrin [10]

1,3-cyclohexanedione [4];
−, 5,5-dimethyl- (dimedone) [4]

2-oxocyclohexanecarboxylic acid
ethyl ester [6]

cis-1,2-cyclohexanedicarboxylic
acid anhydride [0]; −, 4-methyl- [1]

1,4-cyclohexanedicarboxylic acid
dimethyl ester [E+Z:1]

1-ethynylcyclohexylamine [3-5]

4-aminocyclohexanol [5];
3-aminomethyl-3,5,5-trimethyl-
cyclohexanol [1]

2-(dimethylaminomethyl)-
cyclohexanone [4]

Table 14. (Continued)

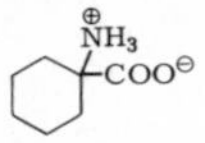

1-aminocyclohexanecarboxylic
acid [7]

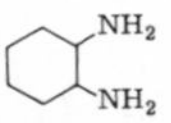

1,2-diaminocyclohexane [0-3]

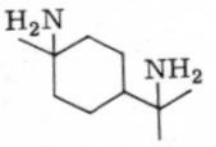

1,8-diamino-*para*-menthane [1]

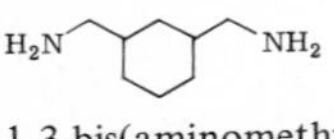

1,3-bis(aminomethyl)-
cyclohexane [1-2]

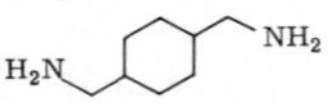

1,4-bis(aminomethyl)-
cyclohexane [2]

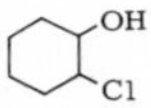

2-chlorocyclohexanol [10]

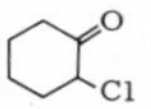

2-chlorocyclohexanone [4]

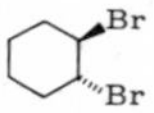

(±)-*trans*-1,2-dibromo-
cyclohexane [1-2]

●trifunctional

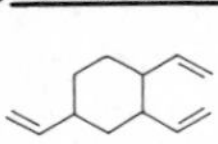

1,2,4-trivinylcyclohexane [2]

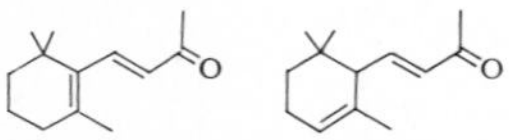

β-ionone [4]; α-ionone [2]

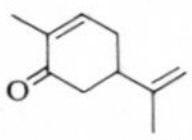

carvone [(+): 7, (−): 3]

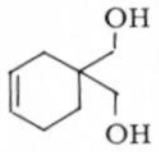

3-cyclohexene-1,1-dimethanol [4]

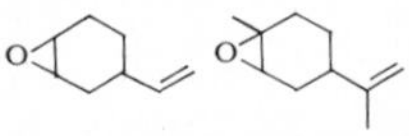

3-vinyl-7-oxabicyclo [4.1.0]-
heptane [1]; limonene oxide
(*E+Z*) [(+):4, (−):6]

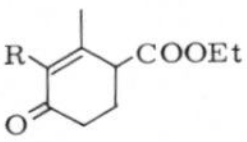

R=H: Hagemann's ester [16];
R=Et: 3-ethyl- [12]

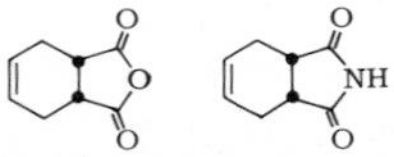

cis-4-cyclohexene-1,2-
dicarboxylic acid anhydride [0],
imide [1-5]

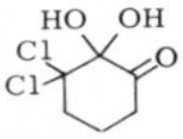

3,3-dichloro-2,2-dihydroxy-
cyclohexanone [10]

●oligofunctional

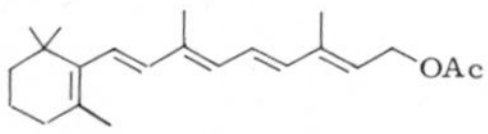

retinyl acetate [4]

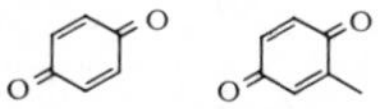

1,4-benzoquinone [1];
toluquinone [8]

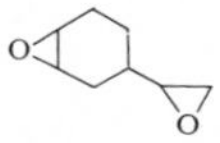

4-vinylcyclohexene dioxide [1-2]

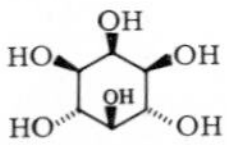

meso-inositol [2]; hexaphosphate=
phytic acid [hexa-Na salt: 5]

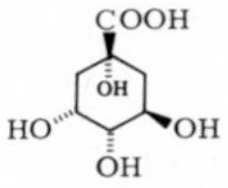

(−)-D-quinic acid [5]

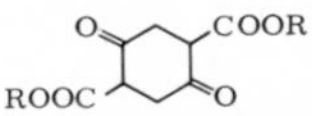

succinylosuccinic acid diesters:
R=Me [3], Et [2]

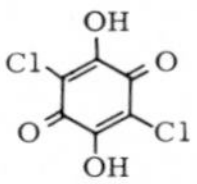

2,5-dichloro-3,6-dihydroxy-
1,4-benzoquinone [4]

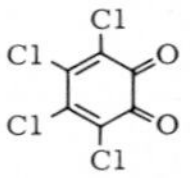

ortho-chloranil [11]

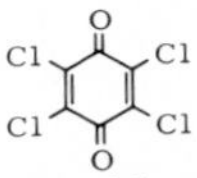

para-chloranil [1-2]

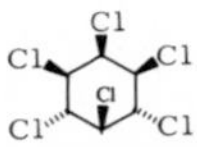

γ-hexachlorocyclohexane
(gammexane, lindane) [1]

able 14. (Continued)

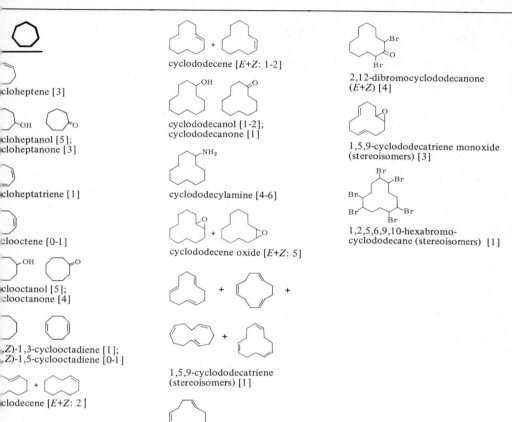

cycloheptene [3]

cycloheptanol [5];
cycloheptanone [3]

cycloheptatriene [1]

cyclooctene [0-1]

cyclooctanol [5];
cyclooctanone [4]

,Z)-1,3-cyclooctadiene [1];
,Z)-1,5-cyclooctadiene [0-1]

cyclodecene [E+Z: 2]

cyclododecene [E+Z: 1-2]

cyclododecanol [1-2];
cyclododecanone [1]

cyclododecylamine [4-6]

cyclododecene oxide [E+Z: 5]

1,5,9-cyclododecatriene
(stereoisomers) [1]

(E,E,Z)-1,5,9-cyclododecatriene [0-1]

2,12-dibromocyclododecanone
(E+Z) [4]

1,5,9-cyclododecatriene monoxide
(stereoisomers) [3]

1,2,5,6,9,10-hexabromo-
cyclododecane (stereoisomers) [1]

---

ble 15. Oligocyclic hydrocarbon derivatives.

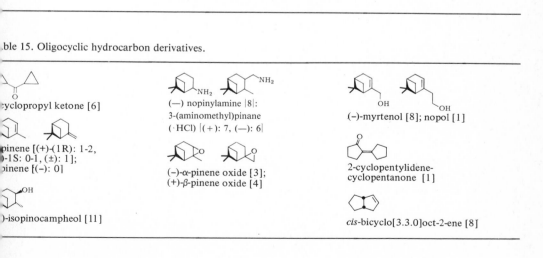

cyclopropyl ketone [6]

pinene [(+)-(1R): 1-2,
-1S: 0-1, (±): 1];
pinene [(−): 0]

)-isopinocampheol [11]

(−) nopinylamine |8|:
3-(aminomethyl)pinane
(·HCl) |(+): 7, (−): 6|

(−)-α-pinene oxide [3];
(+)-β-pinene oxide [4]

(−)-myrtenol [8]; nopol [1]

2-cyclopentylidene-
cyclopentanone [1]

cis-bicyclo[3.3.0]oct-2-ene [8]

Table 15. (Continued)

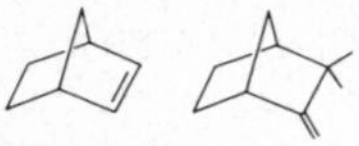

norbornene [0-1];
camphene [(+): 0]

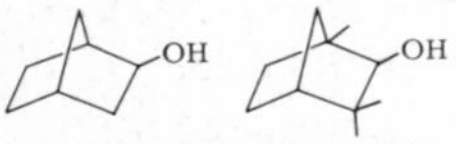

exo-norborneol [(±): 1];
fenchol [(+): 1-2]

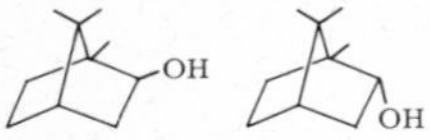

isoborneol [(+): 1];
borneol [(-): 2, (±): 1]

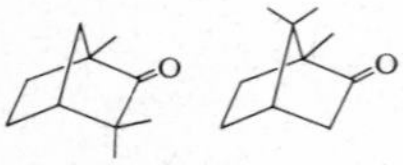

fenchone [(+): 1-4, (-): 2];
camphor [(+): 1, (±): 0]

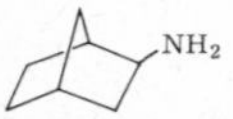

exo-2-norbornylamine [4]

norbornadiene [1]

5-methylene-2-norbornene [4];
5-ethylidene-2-norbornene [0-1]

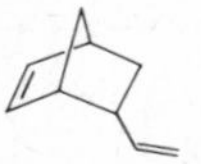

endo-5-vinyl-2-norbornene [1]

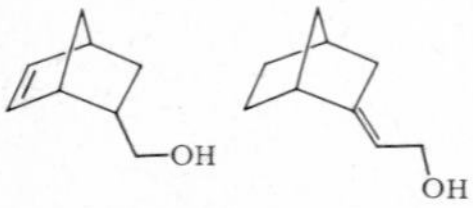

endo-5-norbornene-2-methanol [1];
patchenol [±]: 2]

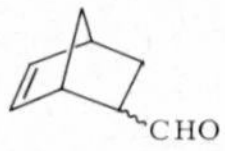

5-norbornene-2-carboxaldehyde
(*endo+exo*) [8]

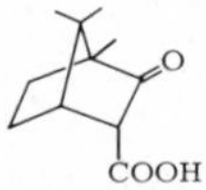

(±)-camphor-3-carboxylic acid [1-2]

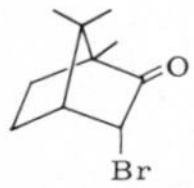

(+)-*endo*-3-bromocamphor [2-3]

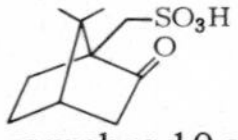

camphor-10-sulfonic acid
[(+): 2-3, (±): 1-4]

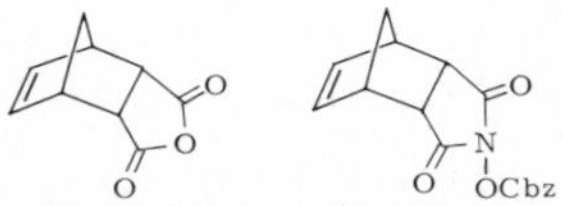

endo-5-norbornene-2,3-dicarboxylic
acid anhydridre [1];
endo-N-(benzyloxy-
carbonyloxy)-5-norbornene-2,3-
dicarboximide [4]

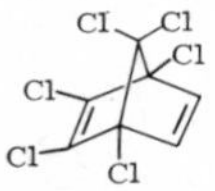

1,2,3,4,7,7-hexachloro-2,5-
norbornadiene [8]

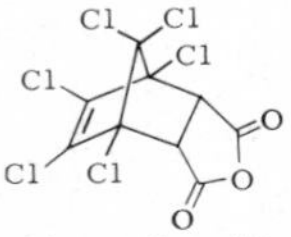

chlorendic acid anhydride [0-1]

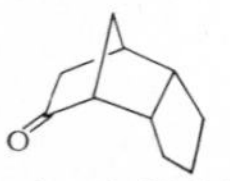

tricyclo[5.2.1.0^{2,6}]decan-8-one [1-2]

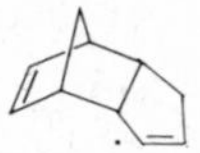

dicyclopentadiene [0]

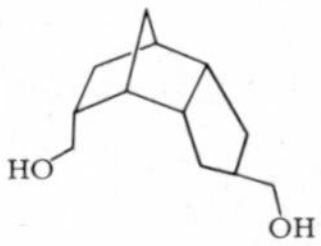

4,8-bis(hydroxymethyl)tricyclo-
[5.2.1.0^{2,6}]decane [1]

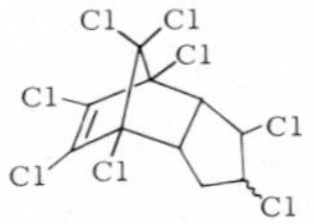

chlordan [0]

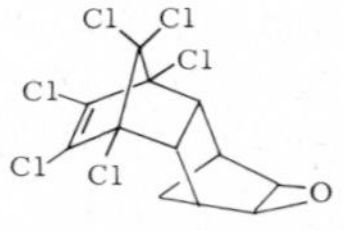

dieldrin [4]

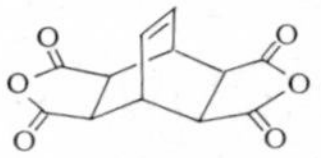

bicyclo[2.2.2]oct-7-ene-
-2,3,3,6-tetracarboxylid acid
2,3:5,6-dianhydride [1]

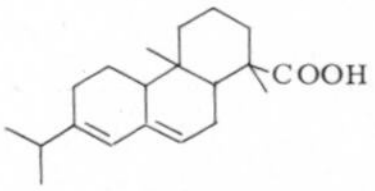

abietic acid [2]

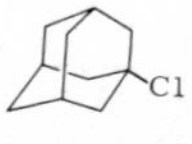

1-chloroadamantane [9]

3.1.5 Aromatic and Heterocyclic Compounds

If target molecules contain benzene rings and/or heterocyclic rings, it is always advisable to check in manufacturers' catalogues for inexpensively available precursors. Very often it is not advisable to synthesize such parts of the target molecule from open-chain precursors, but rather to use functional group interconversions, alkylation, or acylation procedures on commercial products. Since the lists of these compounds are very long and would contain a large number of exotic compounds, they are not given here. No generalizations are attempted either. One may only state that different manufacturers of fine chemicals offer a large variety of different products.

3.2 *Retro*-Synthetic Analysis (= "Antithesis")

Section 3.2.1 covers a similar area of synthesis planning as S. Warren's (1978) book "Designing Organic Synthesis". This book contains many examples of antitheses of achiral carbo-and heterocyclic compounds that are largely ignored in this chapter (but see sections 2.5, 4.6, and 4.7 for heterocycles). Warren's much more elaborate introduction into the "synthon approach" to *retro*-synthetic analysis is based on E.J. Corey's work (1967A), 1971) and is highly recommended for further study.

3.2.1 Antithesis of Mono- and Difunctional Achiral Open-Chain Target Molecules

The organization of a synthetic plan starts, of course, with the structural formula of a target molecule. The target molecule is methodically broken apart in such a way that reassembling the pieces can be done by known or conceivable reactions. This analytical reverse-synthetic procedure is called *retro*-synthesis or antithesis (E.J. Corey, 1967A, 1971). The structural changes in the antithetic direction are named *transforms*, whereas the same operations in the synthetic direction are reactions. A double-lined arrow (⇒) will be used to indicate the direction associated with a transform in contrast to a single arrow in the direction of a synthetic reaction. Since the structural units within a molecule, which are combined by a synthetic operation, are called *synthons* (see p. 4), one can also say that antithesis is the analytical process, by which a molecule is converted into synthons or into their *equivalent reagents*. The flow of events (a computer language term) within antithesis may be regarded as:

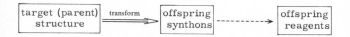

Each transform should lead to reagents, which are more easily accessible than the target molecule. In the subsequent steps of antithesis the reagents are defined as new target molecules, and the transform procedure is repeated until the reagents needed are identical with commercially available starting materials.

The following terms are used to specify the type of transform in the subsequent discussions of antitheses:

- Antithetical *disconnections* are thought to divide the target molecule at a C—C bond into an electron acceptor synthon (a) and an electron donor synthon (d), into two electroneutral radical synthons (r), or into two electroneutral non-radical synthons (e). The latter may undergo electrocyclic reactions and may be stable reagents, whereas other synthons are unstable carboions or radicals. The disconnection site in the structural formula of the target molecule may be indicated by a wavy line and the symbols (a, d, r, e) for the respective synthons. If the target molecule contains only one functional group, it is generally split by an antithetical *"one-group" disconnection* of the α- or β-carbon atom. This corresponds to a synthetic alkylation reaction. If the target molecule contains two proximate functional groups (1,2- to 1,6-difunctional), a *"two-group" disconnection* between the functional groups may be applied, which usually produces two carbonyl-derived synthons. The synthons (mostly a carbanion and a carbocation) may be written down before the structures of the proposed equivalent reagents or may be left out. Disconnection operations are indicated by a double-lined arrow. Between "formal" synthons and their equivalent reagents dashed arrows may be introduced and reaction conditions proposed.

transform type	target molecule	synthons	reagents and reaction conditions
one-group disconnection	(structure with OH) *retro-Grignard transform*	(a) ⊕ OH / H + C₂H₅ ⊖ (d)	CH_3CHO + C_2H_5MgBr (i) 0°C (THF) (ii) NH_4Cl/H_2O
two-group disconnection (heterolytic)	(structure =O, OH) *retro-aldol type transform*	(d) ⊖ =O + ⊕ H / OH (a)	(structure) OLi + CH_3CHO (i) -78°C → r.t. (THF) (ii) NH_4Cl/H_2O
two-group disconnection (homolytic)	(ring structure =O, OH) *retro-acyloin transform*	(r) •=O + • OH (r)	(ring) COOEt / COOEt (i) Na/Me₃SiCl (toluene); Δ (ii) H_2O
electrocyclic disconnection	(structure COOMe, COOMe) *retro-Diels-Alder transform*	(e) + (e) COOMe ‖‖ COOMe	(synthons ≡ reagents) (C₆H₆); Δ [hydroquinone]

- Antithetical *connections* (the reversal of synthetic cleavages) and *rearrangements* are indicated by a "con" or "rearr" on the double-lined arrow. Here it is always practical to draw right away the reagents instead of synthons. A plausible reaction mechanism may, of course, always be indicated.

transform type	target molecules	reagents and reaction conditions
connection	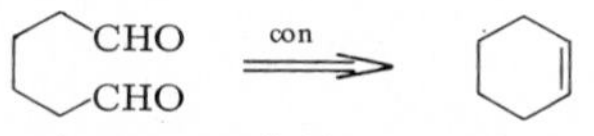 retro-ozonolytic transform	O_3/Me_2S (CH_2Cl_2); $-78°C$
rearrangement	 retro-Beckmann transform	H_2SO_4; Δ

- Antithetical *functional group interconversions* (FGI), *additions* (FGA), or *removals* (FGR) are symbolized by the given abbreviations on the double-lined arrow.

transform type	target molecules	reagents and reaction conditions
functional group interconversion (FGI)		CrO_3/H_2SO_4 (acetone)
		$HgCl_2(CH_3CN)$
		$HgCl_2$ (aq. H_2SO_4)
functional group addition (FGA)		$PhNH_2$; Δ
		$H_2[Pd-C]$ (EtOH)
functional group removal (FGR)	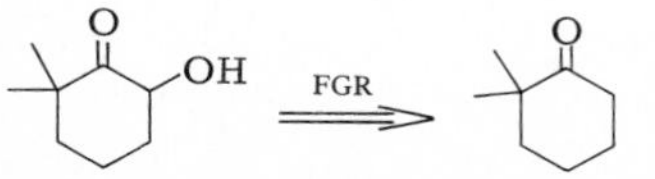	(i) LDA(THF) $-25°C$ (ii) O_2; $-25°C$ (iii) $I^{\ominus}/H_2O$

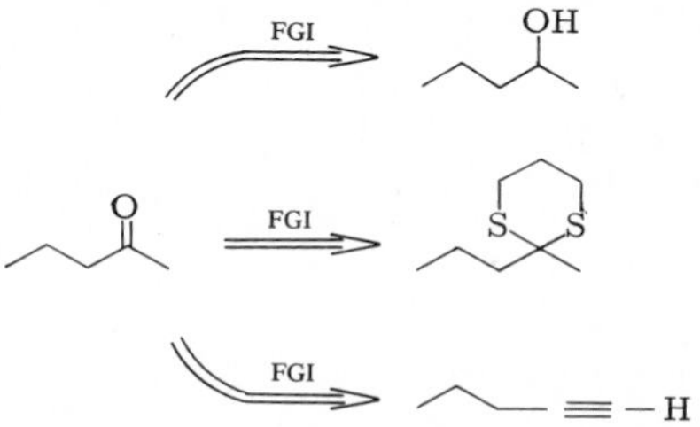

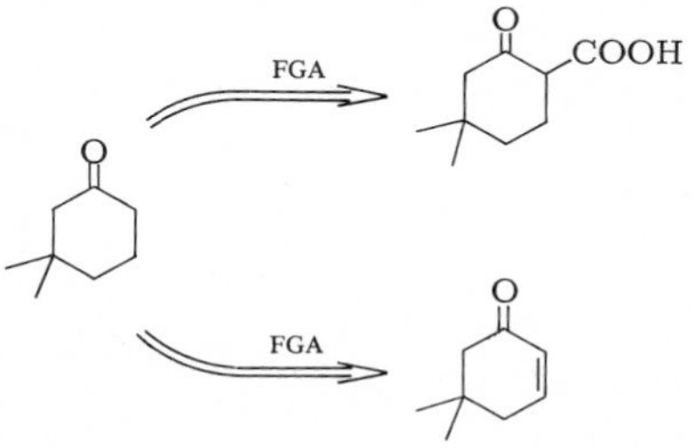

Table 16. Terms and symbols in antithetical schemes.

Terms	Symbols
disconnection (of C—C bonds)	(a) ⌇ (d) C—╫—C ⟹ (r) ⌇ (r) C—╫—C ⟹ (e) ⌇ (e) C—╫—C ⟹
connection (of C—C bonds)	con ⟹
rearrangement	rearr ⟹
functional group interconversion	FGI ⟹
functional group addition	FGA ⟹
functional group removal	FGR ⟹
transform	all of the operations above taken collectively
synthon	structural formula of molecular fragment with (a), (d), (r), or (e)
(equivalent) reagent	structural formula of a reactive molecule, which produces the synthon under the reaction conditions. It is often, although incorrectly, called a synthon. Only in electrocyclic transforms both terms may denote the same.

In later sections of this chapter we shall concentrate on the stereochemistry of target molecules, although the functional group chemistry will, of course, remain the basis of all synthetic operations. In this section we shall analyze synthetic functional group chemistry in two ways:

(i) Systematic variation of functional groups (FGI, FGA, FGR) in the target molecule to produce "alternative target molecules", which may be easier to synthesize and which can be converted into the target molecule by a known or conceivable reaction.

(ii) Systematic generation of synthons from the target molecule and all "alternative target molecules" by disconnections, connections, and rearrangements and formulation of the corresponding reagents (= educts, intermediates). The reagents should be more easily available from simple starting materials than the target molecule.

Both items are repeated until the intermediates are reasonably priced commercial starting materials. The resulting antithetic schemes are then evaluated and the most promising ones can be converted into synthetic plans and be investigated in the laboratory.

The systematic application of both antithetic steps will now be exemplified with the admittedly trivial synthesis of 3-methylbutanal (isovaleraldehyde). Functional group operations would yield the following alternative target molecules:

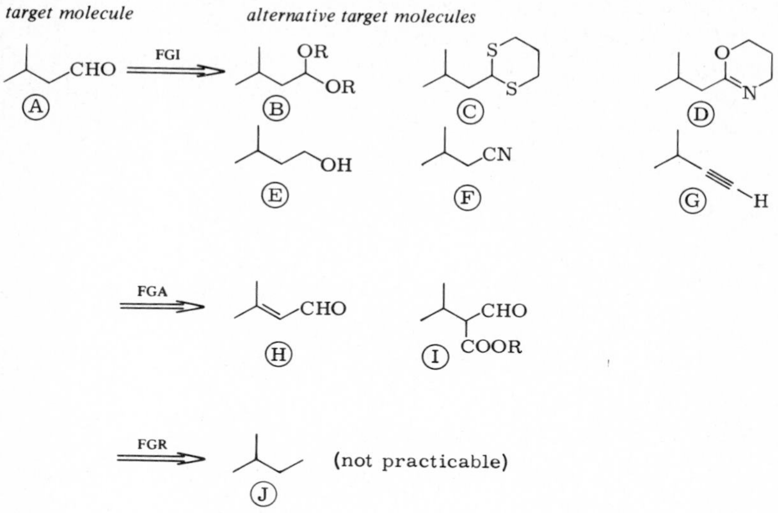

The following reactions could be used to convert compounds (B)—(I) into the target molecule (A). Doubtful or difficult reactions are indicated by a question mark.

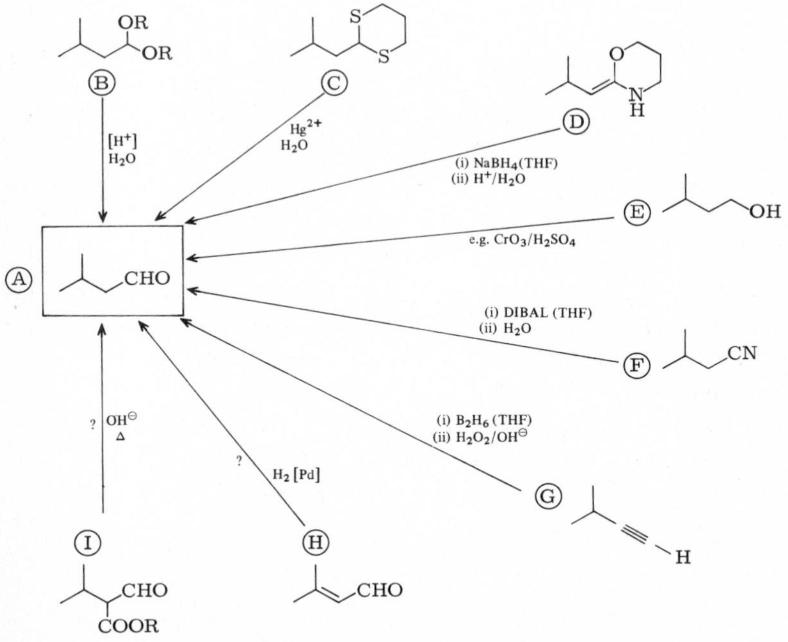

Since (A) does not contain any other functional group in addition to the formyl group, one may predict that suitable reaction conditions could be found for all conversions into (A). Many other "alternative" target molecules can, of course, be formulated. The reduction of (H), for example, may require introduction of a protecting group, e.g. acetal formation. The industrial synthesis of (A) is based upon the oxidation of (E) since 3-methylbutanol (isoamyl alcohol) is a cheap distillation product from alcoholic fermentation ("fusel oils"). The second step of our simple antithetic analysis — systematic disconnection — will now be exemplified with all target molecules of the scheme above. For the sake of brevity we shall omit the synthons and indicate only the reagents and reaction conditions.

(A) $\bigvee$CHO ⟹ $\bigvee$Br + Fe(CO)$_4^{2-}$

(i) CO/Ph$_3$P(THF); low T
(ii) AcOH

$\bigvee$CHO ⟹ $\bigvee$MgBr + HCONR$_2$

(i) (THF); low T
(ii) H$^+$/H$_2$O

$\bigvee$CHO ⟹̸ $\bigvee$Br + $^{\ominus}$CH$_2$CHO

$\bigvee$CHO ⟹̸ $\left(\bigvee\right)_2$CuLi + Br$\diagdown$CHO

protection of the
formyl group would
be necessary.

$\bigvee$CHO ⟹ Me$_2$CuLi + $\diagdown\diagup$CHO

(i) (THF); low T
(ii) NH$_4$Cl/H$_2$O

(B) $\bigvee\begin{smallmatrix}OR\\OR\end{smallmatrix}$ ⟹ $\bigvee$Li + HC(OR)$_3$

(THF); −LiOR

(C) $\bigvee\begin{smallmatrix}S\\S\end{smallmatrix}$ ⟹ $\bigvee$Br + Li$\begin{smallmatrix}S\\S\end{smallmatrix}$

(THF); −LiBr

(D) $\bigvee\overset{O}{\diagdown}$N H ⟹ $\bigvee$Br + $\overset{O}{\diagup}$N Li

(i) (THF); low T
(ii) H$_2$O

(E) $\bigvee$OH ⟹ $\bigvee$MgBr + CH$_2$O

(Et$_2$O); low T

$\bigvee$OH ⟹ $\bigvee$MgBr + $\triangle$O

(Et$_2$O); low T

(F) ![structure] CN (a)⌇(d) $\Longrightarrow$ Br + CN$^\ominus$ KCN(MeCN)

(G) ![structure] H (a)⌇(d) $\Longrightarrow$ Br + Li−≡−H   (TMEDA/THF)

(H) ![structure] CHO (a)⌇(d) $\Longrightarrow$ O + ![structure] N−R, Li

(i) (THF); low T
(II) H$^+$/H$_2$O; Δ

(I) (d)⌇(a) CHO, COOR $\xLeftarrow{//}$ RO⌇OLi + HCOOR'

(a)⌇(d) CHO, COOR $\xLeftarrow{//}$ Br + RO⌇O⌇Li, H

⎫
⎬ unprotected
  formyl groups
  would undergo
  side-reactions
⎭

It is clearly evident from the extremely simple example of the synthesis of isovaleralde-hyde that a fully systematic approach to antithesis is not very useful. Chemists are not inter-ested in encyclopedic catalogues of synthetic routes. We shall now discuss a few simple ex-amples, where availability and price of starting materials are considered. This restriction gen-erally reduces long lists of "alternative target molecules" and "precursors" to a few propo-sals.

2-Ethylpentanoic acid is our next target molecule. No commercial compound of accept-able price with all seven carbon atoms and a functional group in the right order could be found. The usual α- and β-disconnections of monofunctional target molecules yield either sec-ondary halides (disconnections 1) or primary halides (disconnections 2 and 3) as precursors. Usually the reagents with terminal functional groups are much more abundant and inexpens-ive than chemicals with branched substituents. Butanoic and pentanoic acid esters and nitriles, ethyl and propyl bromides are all rather low-cost reagents. Alkylations with these reagents, however, all need highly purified, aprotic solvents and very strong bases as auxiliary reagents. The pathway via dialkylmalonic ester is less demanding. It could be carried out in ethanol and would probably be the reaction of choice for a low-cost preparation in large quantity.

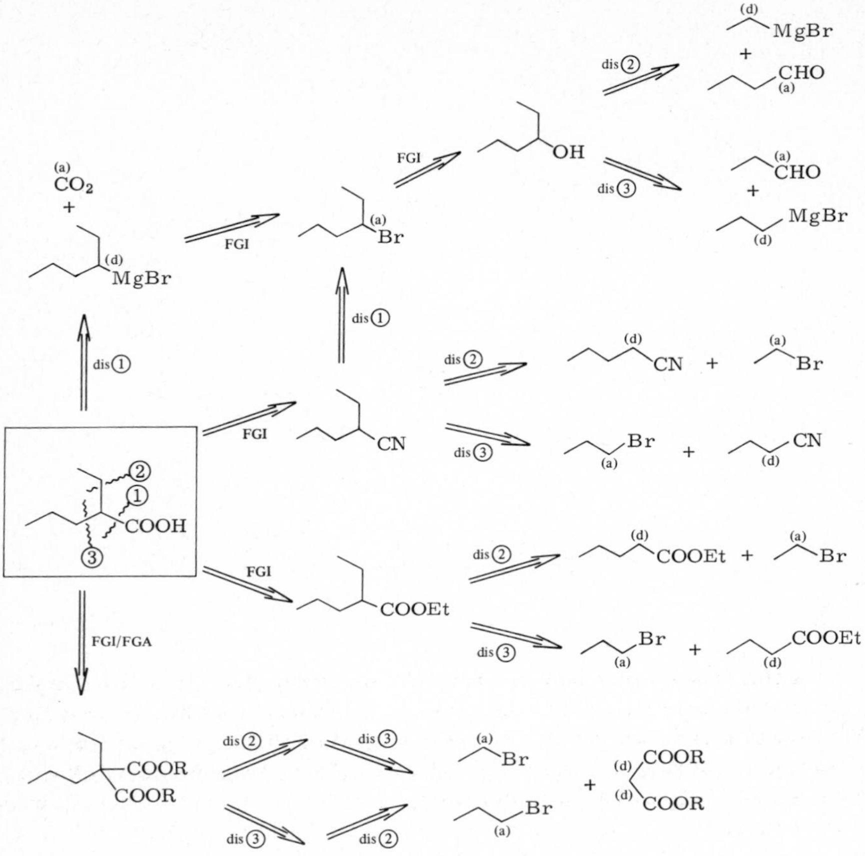

Now we turn to syntheses of simple, noncommercial or very expensive difunctional open-chain compounds. We have chosen one or two examples each of 1,2- up to 1,6-difunctional target molecules.

Our first example is 2-hydroxy-2-methyl-3-octanone. 3-Octanone can be purchased, but it would be difficult to differentiate the two activated methylene groups in alkylation and oxidation reactions. Usual syntheses of acyloins are based upon addition of terminal alkynes to ketones (disconnection 1; see p. 49). For syntheses of unsymmetrical 1,2-difunctional compounds it is often advisable to look also for reactive starting materials, which do already contain the right substitution pattern. In the present case it turns out that 3-hydroxy-3-methyl-2-butanone is an inexpensive commercial product. This molecule dictates disconnection 3. Another practical synthesis starts with acetone cyanohydrin and pentylmagnesium bromide (disconnection 2). Many 1,2-difunctional compounds are accessible via oxidation of C—C multiple bonds. In this case the target molecule may be obtained by simple permanganate oxidation of 2-methyl-2-octene, which may be synthesized by Wittig reaction (disconnection 1).

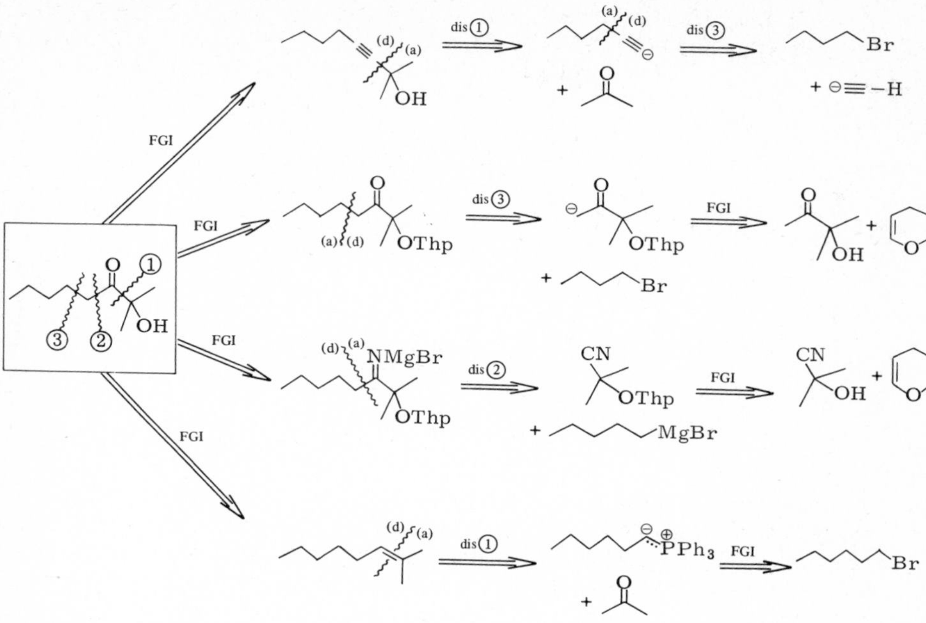

As another 1,2-difunctional target molecule we examine butyl glyoxylate. This simple compound is quite expensive in pure state because it is rather unstable. It is best prepared freshly in solution and used immediately for further reactions (F.J. Wolf, 1963). Obvious FGI-precursors are dibutyl oxalate and butyl glycolate. Aldehydes, however, can also be prepared very efficiently by cleavage of olefins or glycols. Our list of starting materials contains three symmetrical diacids, which would be ideally suited for this purpose, namely fumaric, maleic, and tartaric acids. Cleavages of olefins and glycols are such common synthetic procedures, that they would probably be preferable to partial reduction or oxidation procedures, which often are difficult to control.

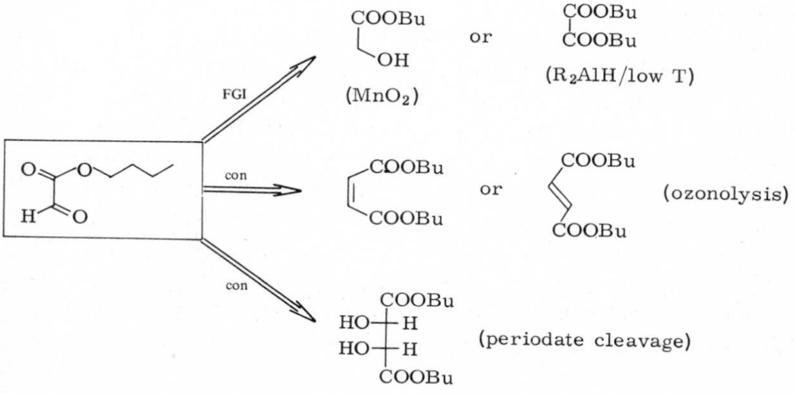

1,3-Difunctional products often come from aldol type reactions. The following example, 2-(hydroxymethyl)-2-methylbutanal, needs no further comment (F. Nerdel, 1968).

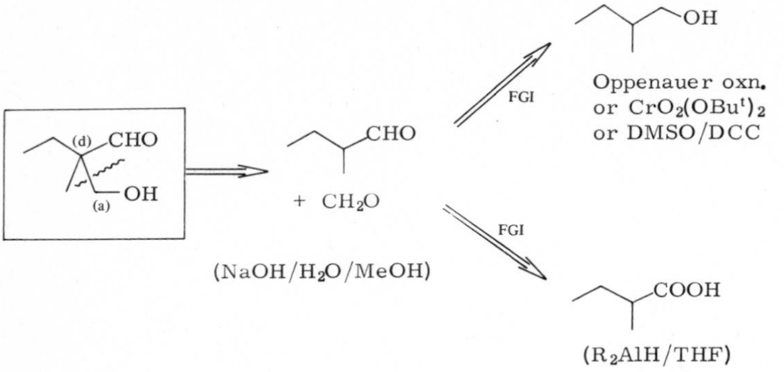

2,4-Nonanedione, another 1,3-difunctional target molecule, may be obtained from the reaction of hexanoyl chloride with acetonide anion (disconnection 1). The 2,4-dioxo substitution pattern, however, is already present in inexpensive, symmetrical acetylacetone (2,4-pentanedione). Disconnection 2 would therefore offer a tempting alternative. A problem arises because of the acidity of protons at C-3 of acetylacetone. This, however, would probably not be a serious obstacle if one produces the dianion with strong base, since the strongly basic terminal carbanion would be a much more reactive nucleophile than the central one (K.G. Hampton, 1973; see p. 9f.).

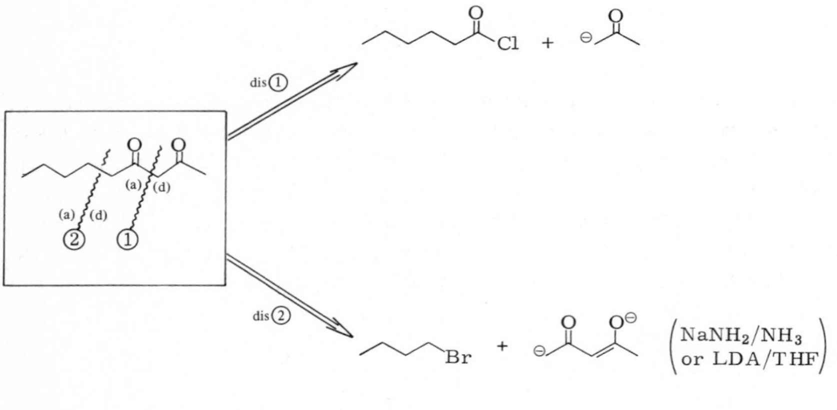

Our example of a 1,4-difunctional target molecule is 6,6-dimethyl-2,5-heptanedione (B. Hankinson, 1972). It should be synthesized by conventional combination (section 1.11) of a d^2- and an a^2-synthon (disconnection 1) or of a d^1- and an a^3-synthon (disconnection 2). Disconnection 3 produces more problems. The keto group of 4-oxopentanoyl chloride (levulinic acid chloride) would have to be protected, before reaction with *t*-butylzinc or -copper reagents or with $Na_2Fe(CO)_4$ and *t*-butyl halide could be tried.

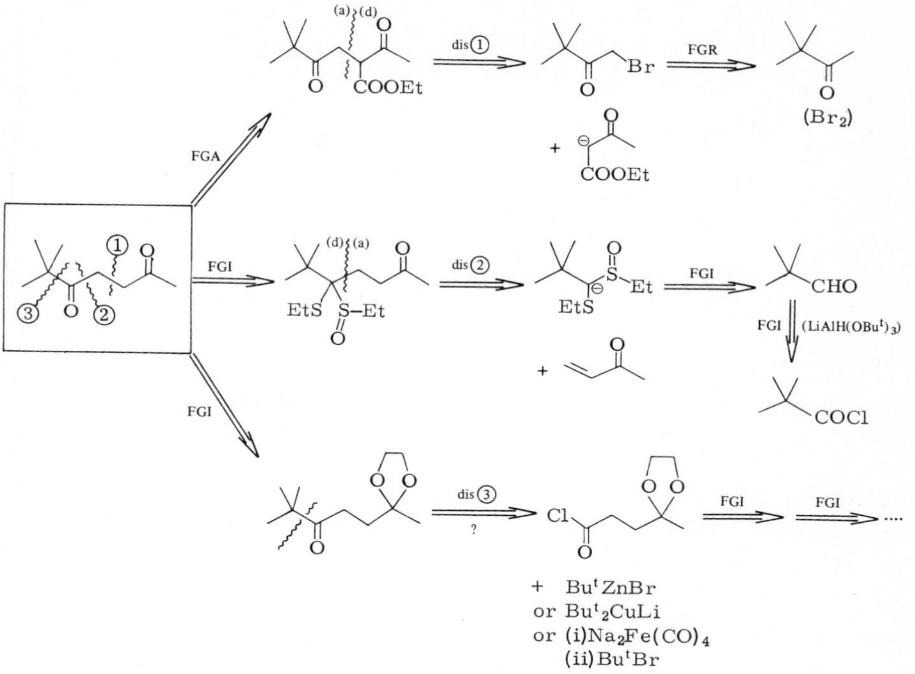

1,5-Difunctional target molecules are generally easily disconnected in a *retro*-Michael type transform. As an example we have chosen a simple symmetrical molecule, namely 4-(4-methoxyphenyl)-2,6-heptanedione. Only *p*-anisaldehyde and two acetone equivalents are needed as starting materials. The antithesis scheme given below is self-explanatory. The aldol condensation product must be synthesized first and then be reacted under controlled conditions with a second enolate (e.g. a silyl enolate plus $TiCl_4$ or a lithium enolate), enamine (M. Pfau, 1979), or best with acetoacetic ester anion as acetone equivalents.

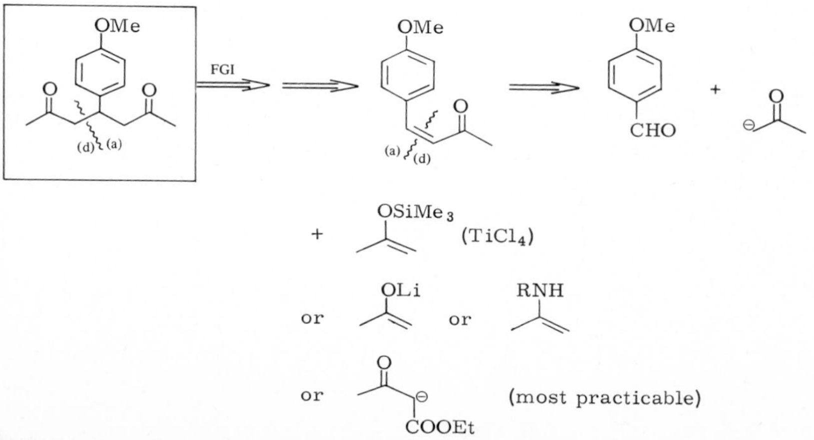

Methyl 6-hydroxy-3-methylhexanoate is our 1,6-difunctional target molecule. Obvious precursors are cyclohexene and cyclohexadiene derivatives (section 1.14). Another possible starting material, namely citronellal, originates from the "magic box" of readily available natural products (C.G. Overberger, 1967, 1968; E.J. Corey, 1968D; R.D. Clark, 1976).

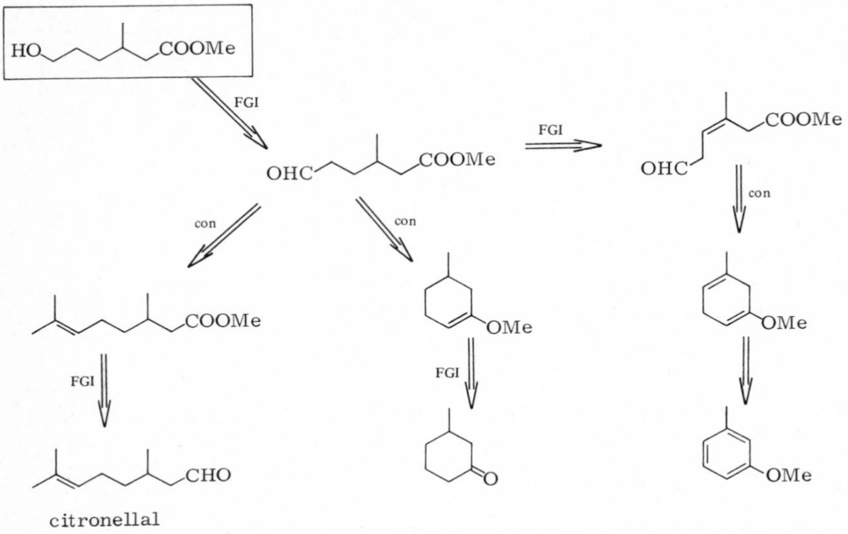

citronellal

We close the section on open-chain molecules with an example of a trifunctional target molecule. This does not include any fundamentally new problem. In antithetic analysis one simply chooses an appropriate difunctional starting material, which may be further disconnected into monofunctional starting materials.

Diethyl 3-oxoheptanedioate, for example, is clearly derived from glutaryl and acetic acid synthons (e.g. acetoacetic ester; M. Guha, 1973; disconnection 1). Disconnection 2 leads to acrylic and acetoacetic esters as reagents. The dianion of acetoacetic ester could, in principle, be used as described for acetylacetone (p. 9f.), but the reaction with acrylic ester would inevitably yield by-products from aldol-type side-reactions.

EtOOC COOEt

② ①

FGA (NH₃) ⟹ EtOOC (a) COOEt (d)

dis ②

EtOOC

+

⊖ O O⊖ OEt

dis ①

EtOOC + ⊖ COOEt O

Cl

FGI

EtOOC COOH

FGI

O O O + EtOH

From the above discussion it should be obvious that antithetical analysis of difunctional molecules simply is a reversal of the synthesis scheme already described in chapter 1.

3.2.2 Mono- and Bicyclic Target Molecules. The Problem of Regio- and Stereoselective Antithesis

The conformational stability of substituted cyclic compounds can be used in synthesis to take advantage of neighboring group effects. The directing effect of ring substituents is the basis of most regio- and stereoselective syntheses. In this section we shall again only analyze target molecules that can be obtained in four or less steps from commercial, inexpensive starting materials. The examples have been chosen mostly from the recent literature, and references are given for the actual syntheses. We shall restrict ourselves to a *retro*-synthetic analysis.

We begin with two examples of combined electronic and sterical effects in benzene-derived target molecules. These are generally obtained from starting materials already containing a benzene ring. An obvious precursor for our next target molecule, methyl 2-formyl-4,6-dimethoxybenzoate, would be 3,5-dimethoxyphthalic acid (A) (H. Brockmann, 1957). One might predict, that the less hindered carboxyl group could be reduced selectively, e.g. by a bulky hydride such as DIBAL. However, neither diacid (A) nor alternative derivatives are commercially available. Monoacids (B) and (C) can be purchased. Acid (B), however, is no suitable substrate for the introduction of a carbon substituent at C-6 since all three substituents would direct an incoming electrophile towards C-3 and C-5. Substrate (C) is more promising since the electron-donating methoxy groups activate all unsubstituted positions (C-2, C-4, C-6) and since its symmetry equalizes positions C-2 and C-6. If the monoformylation of (C) occurred statistically at positions 2, 4, and 6, about 67 % of the product would be the desired isomer.

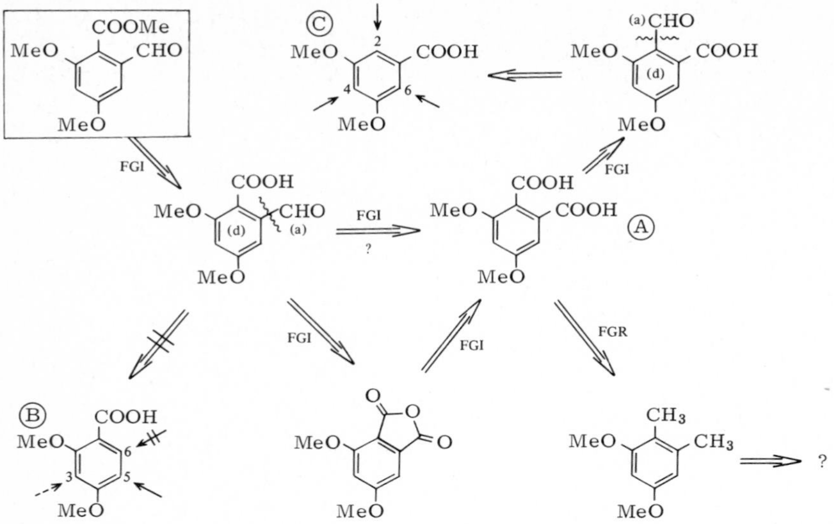

Our second aromatic target molecule is a trisubstituted tetralin derivative with two vicinal *trans*-oriented substituents at the reduced ring (platyphyllide; F. Bohlmann, 1979). Its antithetic analysis is dictated by the requirement to transform it into the only reasonable starting material 1-naphthoic acid (A). Naphthalenes can be reduced to give tetralins, which in turn can be oxidized at the benzylic methylene groups. If a metal salt is used as oxidant, one might hope that a carboxylate or amide group may bind the metal ion, which then would preferably attack the neighboring, sterically hindered *peri*-position. Thus molecule (B) should be available. The alternative target molecule (C) could now be obtained from (B) by aldol addition with acetone. We need, however, the enolate of the tetralone compound and a reaction that leads stereoselectively to the *trans* product. The actual synthesis has been carried out by a Lewis acid catalyzed aldol addition with the trimethylsilyl enolate of the tetralone. Reduction of the ketol with LiAlH$_4$ produces the *trans*-diol because the tertiary alcohol group binds the hydride reagent below the ring plane. The final steps are re-oxidation of the primary alcohol group and dehydration.

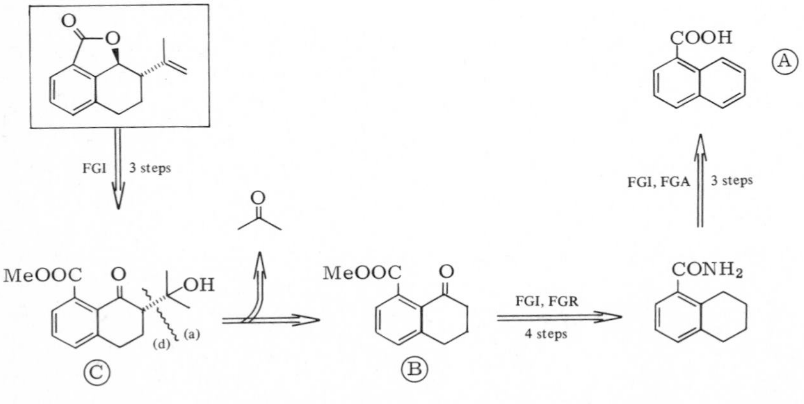

Our first nonaromatic carbocyclic target molecule contains two condensed five-membered rings, an angular methyl group and an *α,β*-unsaturated ketone moiety (S.C. Welch, 1979). The first disconnection is clearly of the *retro*-aldol type. The resulting diketone poses no problem since 2-methylcyclopentanone can be purchased and selective formation of the thermodynamically more stable, more substituted enolate is possible (see p. 11). The a²-synthon could be bromoacetone, but its keto group must be protected, e.g. as a ketal. Nucleophilic substitution of this bulky bromide by the highly hindered enolate, however, could be difficult. Inexpensive 2,3-dibromopropene is a good alternative. The allylic bromine should be substituted much faster than the vinylic one. Conversion of the vinylic bromide into the ketone is catalyzed by acids and mercury(II) ions.

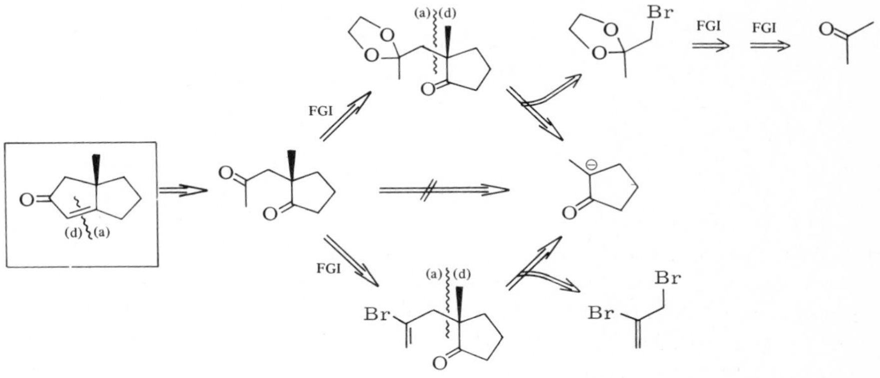

The most common stereoselective syntheses involve the formation and cleavage of cyclopentane and cyclohexane derivatives or their unsaturated analogues. The target molecule (*all-cis*)-2-methyl-1,4-cyclohexanediol has all of its substituents on the same side of the ring. Such a compound can be obtained by catalytic hydrogenation of a planar cyclic precursor. Methyl-1,4-benzoquinone is an ideal choice (*p*-toluquinone; M. Nakazaki, 1966).

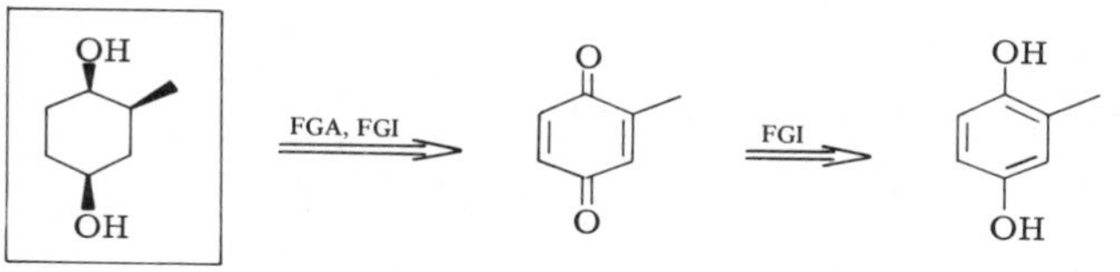

The following example of an antithesis of a complicated polycyclic and chiral (although racemic) compound, which can be transformed into commercially available starting materials in a few steps, provides convincing evidence for the power of the retro-Diels-Alder transform. Its application can result simultaneously in a decrease of the number of rings and of chiral centres, in disconnection of a molecule into two stable fragments, and finally in simplification of functionality patterns. The target molecule (S. Danishefsky, 1979) is first transformed into the more symmetrical dicarboxylic acid (A) by FGA. Synthetic decarboxylation of this vinylogous β-keto acid should occur readily in mildly basic media. The cyclohexenone system must then be transformed into a substituted cyclohexene derivative to prepare for the anticipated *retro*-Diels-Alder disconnection. The choice of (B) and the corresponding disubstituted butadiene is dictated by the (commercial) availability of 4-methoxy-3-buten-2-one (see p. 158), which can be converted into the silyl enol ether. The next transforms are FGI of the cyclic anhydride, removal of the isopropylidene group, and FGI of the *cis*-glycol. As a synthetic reaction osmium tetroxide catalyzed *cis*-dihydroxylation of the C=C double bond is advisable since oxidation should occur at the less hindered *exo*-side. The glycol would be ketalized with acetone and acidic catalysts. A final *retro*-Diels-Alder transform yields the readily available acetylenedicarboxylic ester and cyclopentadiene.

Antithesis of the cyclopentene derivative given below with its three chiral centres is another example of the utility of the *retro*-Diels-Alder transform, which is used here for simultaneous stereospecific disconnection of two ring substituents. The acetic acid and hydroxyl groups are *cis* to each other and can therefore be transformed into a lactone ring system, which can be produced from a bicyclic ketone precursor by Baeyer-Villiger oxidation. The cyclopentadiene derivative from the following *retro*-Diels-Alder-type disconnection can be synthesized easily from cyclopentadienyl anion and chloromethyl methyl ether. One should, however, choose the mildest possible reaction conditions for the alkylation of cyclopentadiene (low temperature) and for the Diels-Alder reaction (acid catalysis), since substituted cyclopentadienes isomerize quickly. The "ene" reagent could not be ketene itself but "protected" equivalents like 1,1-dichloroethene or 2-chloropropenonitrile. The latter would probably be the reagent of choice because of its activating, electron-withdrawing cyano group. The third chiral centre could be expected to be formed predominantly with the correct relative stereochemistry since addition of the "ene" should occur preferably on the less hindered side of the cyclopentadiene ring (E.J. Corey, 1969).

Antithetic analysis of another chiral cyclopentane derivative (G.W.K. Cavill, 1967) is simple, if one remembers that cyclopentanecarboxylic acids are easily obtained by ring contraction from cyclohexanones in a Favorskii rearrangement (see p. 77f.) and if table 14 is consulted. Otherwise it would be quite complex. The obvious precursor is low-cost pulegone. It has the right number and arrangement of carbon atoms and contains only one enolizable carbon suitable as electron donor in the Favorskii rearrangement. The electron acceptor for this ring contraction must be C-4, which is, however, an electron-rich olefinic carbon atom. An "umpolung" is obtained if the double bond is epoxidized. This epoxide is useful since its opening yields the desired hydroxyl group at C-8. The only remaining problem is the stereoselectivity of the epoxidation of pulegone. The directing sterical influence of the remote methyl group at the chiral center C-1 upon peroxide attack can be expected to be quite small. Thus a nearly equimolar mixture of *cis-* and *trans-*epoxidized products must be separated, e.g. by distillation or chromatography.

3.2.3 Stereoselective Antithesis of Open-Chain Target Molecules

In stereoselective antitheses of chiral open-chain molecules transformations into cyclic precursors should usually be tried. The *erythro*-configurated acetylenic alcohol given below, for example, is disconnected into an acetylene monoanion and a symmetrical oxirane (M.A. Adams, 1979). Since nucleophilic substitution occurs with inversion of configuration this oxirane must be *trans*-configurated; its precursor is commercially available *trans*-2-butene.

(erythro)

The 1,6-difunctional hydroxyketone given below contains an octyl chain at the keto group and two chiral centers at C-2 and C-3 (G. Magnusson, 1977). In the first step of the antithesis of this molecule it is best to disconnect the octyl chain and to transform the chiral residue into a cyclic synthon simultaneously. Since we know that ketones can be produced from acid derivatives by alkylation (see p. 42f.), an obvious precursor would be a seven-membered lactone ring, which is opened in synthesis by octyl anion at low temperature. The lactone in turn can be transformed into *cis*-2,3-dimethylcyclohexanone, which is available by FGI from (2,3-*cis*)-2,3-dimethylcyclohexanol. The latter can be separated from the commercial *cis-trans* mixture, e.g. by distillation or chromatography.

All procedures based upon achiral reagents would, of course, yield racemates. Only the application of chiral reagents, e.g. chiral Wilkinson catalysts (see p. 95) or enzymes (see p. 250f.), would make enantioselective syntheses possible (D. Valentine, 1978; P.A. Bartlett, 1980). Another approach to direct synthesis of optically active compounds involves the use of low-cost chiral starting materials, e.g. α-amino acids, hydroxy acids, sugar derivatives, or terpenes.

A simple example of an optically active target molecule is the triol given below (U. Ravid, 1977; D. Valentine, 1978). The amino acid resembling this molecule is glutamic acid, of which the expensive (R)- and the low-cost (S)-enantiomers both can be purchased. The primary alcohol groups are transformed to the carboxyl groups and the chiral secondary alcohol group into the amino group by FGI. An obvious reaction for the synthetic conversion of the amine into the alcohol is treatment with nitrous acid. The stereoselectivity of this reaction varies, but retention of configuration can often be achieved to a high degree.

The last example is a chiral olefinic alcohol, which is disconnected at the double bond by a *retro*-Wittig transform. In the resulting 4-hydroxypentanal we recognize again glutamic acid, if methods are available to convert regio- and stereoselectively

(1)   the γ-carboxyl group into the aldehyde group,
(2)   the α-carboxyl group into a methyl group, and
(3)   the amino group into the hydroxyl group with retention of configuration.

Conversion (3) has already been substantiated in the example above. For conversions (1) and (2) we need a synthetically useful differentiation of the α- and γ-carboxyl groups. This is quite simple, because only the γ-carboxyl group can react with the α-hydroxyl group to form a lactone. The remaining free α-carboxyl group could be converted into the chloride, then selectively reduced to a hydroxymethyl group, converted via the tosylate into an iodomethyl group, which is finally reduced to yield the desired methyl group. The lactone can be converted into the cyclic hemiacetal by DIBAL (K. Mori, 1975).

3.2.4 Bridged Polycyclic Molecules

Topological analysis is the first task in antitheses of complex molecular structures having sets of interconnecting bridges. Both bridged and fused oligocyclic precursors have to be defined, from which the desired skeleton can be produced by formation of one or two connecting bonds. We shall exemplify the analytical procedure with a simple bridged carbocyclic and two heterocyclic target molecules.

6-Methyltricyclo[4.4.0.0^{2,7}]decan-3-one is our first target molecule. We designate all carbon atoms, which belong to more than one ring, by a dot. First, disconnections of bonds between these *"common atoms"* are examined, which lead to the decalones (A) and (B) in our example. Synthesis of the target molecule from (A) would be relatively simple, since C-2 is already activated by the keto group and therefore liable to act as a donor group. The enolate would readily substitute an appropriate leaving group, e.g. tosylate, at C-7. (B) would give many more problems, e.g. synthesis of a suitable trifunctional *cis*-decalone derivative, prevention of *cis-trans* isomerization, and protection of the keto group.

Next we consider disconnections of bonds ending in only one "common atom". These yield bicyclo[3.1.1]heptane derivatives with a carbon substituent at a methylene bridge. Compounds of type (C) could be made from cyclohexanone derivatives, but again this synthetic route would be much more difficult than via (A). For precursors of type (D) and (E) even lengthier syntheses would certainly be required. Therefore we settled on precursor (A) and called the 2,7-bond (or the equivalent 1,2-bond) our *"strategic"* bond since in the development of a synthetic strategy we would first have to consider how to establish this bond. Since precursors of type (B) and (C) should also be synthesizable, we designate the 1,6-(= 6,7-) and 1,10-(= 7,8-)bonds as our *"reserve" strategic bonds* if we meet problems with our first choice (A). The *retro*-synthetic analysis of (A) given below, however, does not show any serious difficulties (C.H. Heathcock, 1966, 1967; S. Ramachandran, 1961).

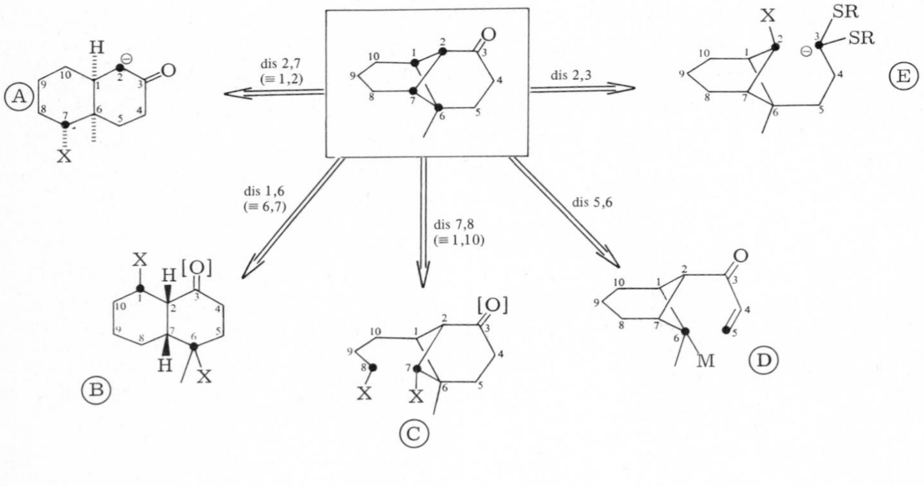

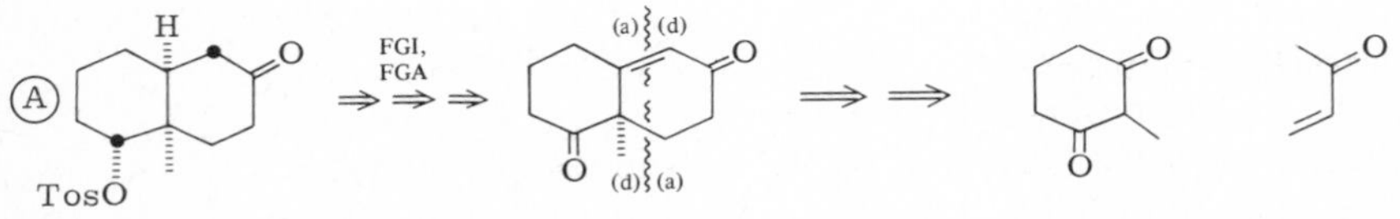

In the preceding example we did not consider cycloaddition reactions since these would not offer any suitable alternative synthetic pathway. The bicyclic isoquinuclidine derivative given below (G. Büchi, 1965, 1966A) contains only unstrained six-membered rings, and the *retro*-Diels-Alder transform is obviously the furthest-reaching simplification and the fastest antithetical route to commercial starting materials. Both bridgehead atoms can be introduced in one step.

The two-bond disconnection (*retro*-cycloaddition) approach also often works very well if the target molecule contains three-, four-, or five-membered rings (see section 1.13 and 2.5). The following tricyclic aziridine can be transformed by one step into a monocyclic amine (W. Nagata, 1968). In synthesis one would have to convert the amine into a nitrene, which would add spontaneously to a C=C double bond in the vicinity.

3.2.5 Summary of Antithetical Analysis of Simple Molecules

We have considered the following structural features affecting the choice of antithetical steps starting from simple target molecules:

(i) Arrangement of functionality and interconversion of functionality ("alternative target molecules").

(ii) Presence of special substructures close to commercial starting materials.

(iii) Stereochemical configuration.

(iv) Symmetry in (alternative) target molecules.

(v) Presence of "strategic" bonds in polycyclic structures.

The same considerations are fundamental in the analysis of more complex molecules, which we shall discuss in the following chapter, with the exception that in item (ii) the "special substructures" are generally related to more complex, non-commercial precursors, which are known from the literature.

3.2.6 Learning from Research Papers

The best textbooks for the advanced student of chemistry are the scientific journals. One may, for example, take any recent issue of the Journal of the American Chemical Society and look for structural formulas. On my desk I have Vol. 103, No. 25 (1981). The following seven compounds are discussed in this particular issue (page numbers in brackets).

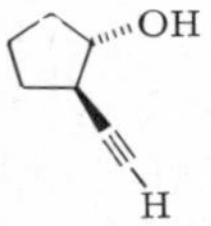

(p. 7523) 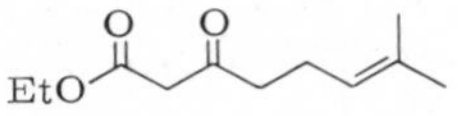   (p. 7552) 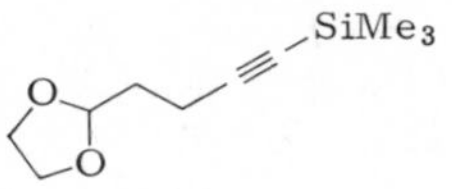   (p. 7560)

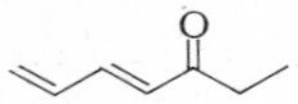

(p. 7573) 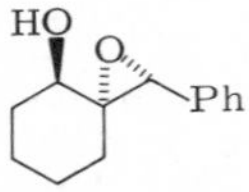   (p. 7663) 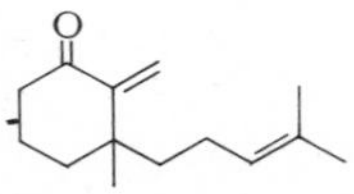   (p. 7668)   (from

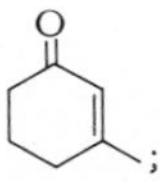

                                                         enolate quenching
                                                         with CH₂O)

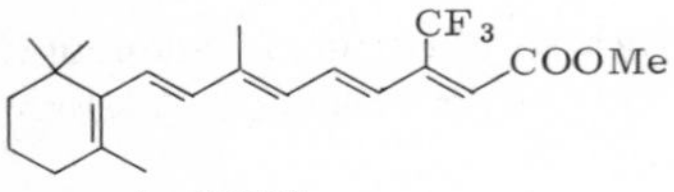

(p. 7642)   (from β-ionone)

These compounds can be made from the starting materials listed in this chapter in a few steps. Try to find out starting materials and propose synthetic procedures of your own! Then compare with the procedures given in the journal. If you regularly make up your own problems from scientific journals, work them through seriously, and slowly get to more complicated target molecules, you cannot fail to learn a lot about solving synthetic problems in a realistic manner!

4 Methods in the Construction of Complex Molecules

The large majority of chemists involved in synthesis are specialized in particular classes of compounds. One works as an alkaloid chemist, a porphyrin chemist, a steroid chemist, and so on. This intellectually unfortunate situation arises because complex molecules can hardly ever be constructed by general methods. The most powerful "logical" method in synthesis is still conclusion from analogy. The intelligent and creative application of such reasoning in a given field, however, is only possible, if one knows intimately at least a few hundred publications in the field. The second presupposition is, of course, a good knowledge of classical and modern synthetic reactions, their mechanisms, and their limitations. Thirdly specialized knowledge of separation techniques and spectroscopic analysis is needed. Therefore productive chemists mainly read and write reviews on their special fields, and general textbooks on organic chemistry nowadays do not go far beyond the description of structures of complex molecules.

A major trend in organic synthesis, however, is the move towards complex systems. It may happen that one needs to combine a steroid and a sugar molecule, a porphyrin and a carotenoid, a penicillin and a peptide. Also the specialists in a field have developed reactions and concepts that may, with or without modifications, be applied in other fields. If one needs to protect an amino group in a steroid, it is advisable not only to search the steroid literature but also to look into publications on peptide synthesis. In the synthesis of corrin chromophores with chiral centres, special knowledge of steroid, porphyrin, and alkaloid chemistry has been very helpful (R.B. Woodward, 1967; A. Eschenmoser, 1970).

In this chapter some important synthetic reactions specific to each class of compounds are described. Only small parts of certain total syntheses will be discussed. With the given references, however, the interested reader will easily locate the complete descriptions of the syntheses. I. Fleming's (1973) book is recommended as a guide through some ingenious classic total syntheses.

4.1 Syntheses by Functional Group Interconversions (Condensation Reactions)

Conceptually the most simple syntheses of complex molecules involve the joining of structural units in which all functional groups and all asymmetric centres are preformed. This technique can usually only be applied to compounds in which these units are connected by —C—X— bonds rather than C—C. It is illustrated here by the standard syntheses of oligonucleotides, peptides, and polydentate macrocyclic ligands.

4.1.1 Oligonucleotides

The fundamental problem of oligodeoxyribonucleotide synthesis is the efficient formation of the internucleotidic phosphodiester bond specifically between C-3' and C-5' positions of two adjacent nucleosides. Any functional group (NH_2 of nucleic base; "the other" OH of deoxyribose; other phosphate groups) must be protected. In oligoribonucleotide synthesis the additional protection of the 2'-hydroxyl function is necessary. This group must be deblocked under conditions mild enough to avoid isomerization of the 3' → 5' phosphodiester linkage to the unnatural 2' → 5' position. The following table summarizes protecting groups commonly used nowadays in nucleic acid synthesis (S.A. Narang, 1973; H. Köster, 1979).

Table 17. Protecting groups used for nucleotides (see also section 2.6.).

Functional group	Protecting group	Abbreviation	Reaction conditions for deprotection	
5'-OH (primary)	Triphenylmethyl (= trityl)	Trit	ac.	80% AcOH; 100 °C
	Di-p-methoxytrityl	Dmtr	ac.	80% AcOH; r.t.
	Dimethylpropanoyl (= pivaloyl)	Piv	bas.	$Et_4N^+OH^-$/MeOH; r.t.
3'-OH (secondary)	Acetyl	Ac	bas.	2M NaOH; 0 °C
	β-Benzoylpropionyl	Bzpr	ac.	N_2H_4/Py/AcOH
2'-OH (secondary)	Tetrahydropyranyl	Thp	ac.	0.01M HCl; r.t.
NH_2 (nucleic bases A,C,G)	Acetyl	Ac	bas.	1M NaOH; r.t.
	Benzoyl	Bz		
	p-Methoxybenzoyl (= anisoyl)	MeOBz		
Phosphodiester	β-Cyanoethyl	Ce	bas.	K_2CO_3
	β,β,β-Trichloroethyl	Tce	red.	Zn-Cu/AcOH; r.t.
5'-Phosphate	2-(p-Tritylphenyl-thio)ethyl (p. 200)	Tpte	ox. + bas.	(i) NCS (→ sulfone) (ii) NaOH; r.t.

Homopolymeric deoxyribonucleotides containing a single nucleotide were prepared by the polymerization of mononucleotides with a free 3'-OH group, a 5'-phosphoric ester, and a protected nucleic base. With DCC (see p. 130f.) or mesitylenesulfonyl chloride as condensing agents and pyridine as solvent two homologous series of oligonucleotides are observed. The first are the desired linear oligonucleotides (A), while the second series of compounds (B) contains cyclic oligonucleotides resulting from intramolecular phosphodiester bond formation between the terminal 5'-phosphomonoester and the 3'-OH. The cyclization can be suppressed by adding some 25% of 3'-O-acetyl derivative at the start of polymerization. The 3'-O-acetyl-nucleotides form terminating units, thus blocking cyclization, and the acetyl group can subsequently be removed by mild alkali treatment. Other commonly encountered by-products (C) are oligonucleotides linked to each other by pyrophosphate bonds. They are removed by treating the entire polymerization reaction mixture with excess acetic anhydride in pyridine, which cleaves the pyrophosphate bonds. This produces another series of minor contaminants (D), namely terminal pyridinium cations. The yield of linear oligonucleotides longer than tetranucleotides is generally less than 50%. Separation and purification of components are generally the major problems with non-selective polymerizations of this kind. A typical reaction mixture contains about twenty different compounds. The largest molecules have, depending on condensation agent and reaction time, between twelve and thirty units. Chromatography at cellulose derivatives is used for the separation of polynucleotides, and only milligram quantities can usually be isolated in pure form (S.A. Narang, 1973).

linear oligonucleotides

cyclic oligonucleotides

pyrophosphate-linked
oligonucleotides

5'-pyridinium-
substituted
oligonucleotides

The description above indicates that uncontrolled chemical polymerization is, in general, not a good route to defined products. Cross-linking of growing chains, cyclization reactions, and reactions with other nucleophiles can often not be prevented, and inhomogeneous chain lengths are invariably the result of such reactions. Therefore the modern approach to the synthesis of homo- or blockpolymeric*) nucleotides involves the enzymatic polymerization of the requisite nucleoside or oligonucleotide 5'-tri- or diphosphates by use of phosphorylase enzymes. The chain lengths (approx. 10 units) are more homogeneous, and side-reactions do not occur.

For the synthesis of oligodeoxynucleotides with defined sequences, topical in connection with "genetic engineering", a different strategy of synthesis has to be applied. It is often necessary that the phosphorylated nucleic acid component has all its hydroxyl groups and amino groups protected to avoid self-condensation, and that in the alcohol component all functional groups are protected, except the hydroxyl group which one wants to esterify with a

---

* polynucleotides containing repeating di- or trinucleotides.

phosphomonoester. The protecting groups are also used to solubilize synthetic intermediates in organic solvents, e.g. methylene chloride. Chromatography is then possible on a larger scale, since silica gel can be used as adsorbent. Six synthetic strategies are currently used (H. Köster, 1979):

(i) the diester method.
(ii) the triester method.
(iii) the phosphite method.
(iv) the 1,3,2-dioxaphosphole method.
(v) the solid-phase method.
(vi) the combined chemical-enzymatic method.

4.1.1.1 Diester and Triester Methods

In the diester method a deoxynucleoside-5′-monophosphate is condensed with the 3′-OH group of a deoxynucleotide to produce a 3′,5′-phosphodiester. This is illustrated by a general method for dinucleotide synthesis developed by H.G. Khorana (K.L. Agarwal, 1976). One N-protected mononucleotide is condensed with an excess of 2-(*p*-tritylphenylthio)ethanol (Tpte-OH), and the 3′-OH group of the other N-protected mononucleotide is acetylated. Both components are then condensed with triisopropylbenzenesulfonyl chloride (TpsCl) in pyridine. Because of the hydrophobic trityl group the protected dinucleotide product can be chromatographed on silica gel. The O-acetyl and N-protecting groups are removed with ammonia at 50 °C, and the Tpte group is first oxidized to the sulfone by N-chlorosuccinimide and then hydrolyzed with NaOH. Yields of isolated dinucleotides are around 70%.

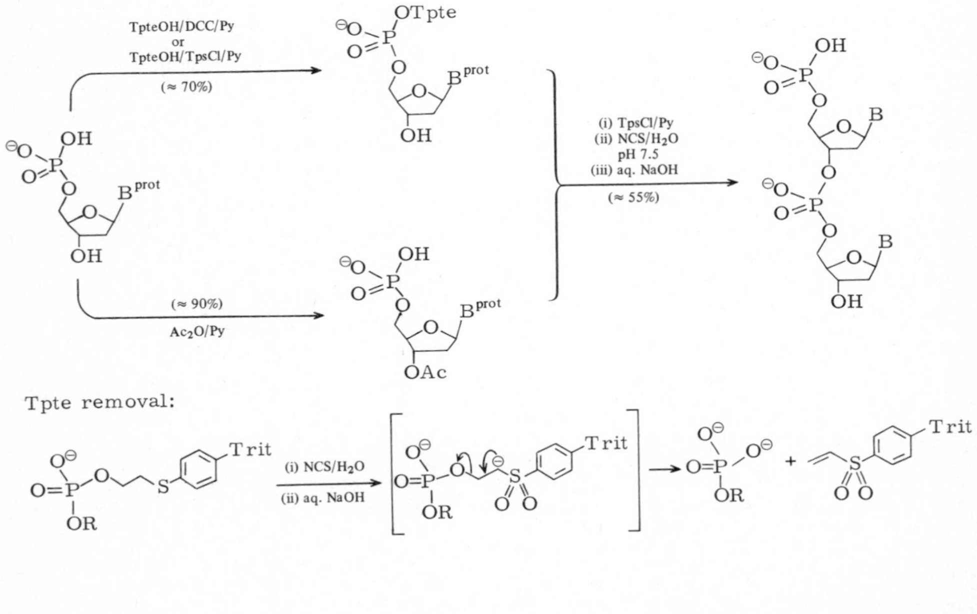

Tpte removal:

A major problem with the diester approach is the fact that one P—O⁻ group of the phosphate always remains free. Since this group is also activated by condensing agents branched pyrophosphates or triesters may also be formed. Cyclization may also occur when the protecting group of the 3′-OH group is removed before a third nucleotide is condensed to it. These side-reactions drastically reduce yields, especially if the condensing reagent is used repeatedly on a growing oligonucleotide chain with several phosphodiester groups.

The internucleotidic phosphodiester may be converted into a triester and thus be protected. In the usual ''triester-method'' a 3′-phosphodiester of a protected nucleoside is condensed with a nucleoside-3′-phosphate protected on all functional groups except for the primary 5′-OH group. The quantitative removal of several protecting groups (e.g. β,β,β-trichloroethyl, 4-chlorophenyl) from an oligomer, however, is still a problem. It may be incomplete or be accompanied by isomerizations of the terminal phosphodiester groups, e.g. by formation of 3′,3′- or 5′,5′-phosphodiesters. Furthermore, long-chain oligomers of phosphotriesters are more difficult to separate by chromatography than are the corresponding diesters since diesters of different chain lengths are differentiated by different numbers of negative charges, whereas triesters are all electroneutral. The triester method therefore generally gives better yields but is more time consuming; it is also more difficult to obtain pure products.

A major problem in the development of phosphotriester syntheses has been the lack of appropriate condensing agents. DCC cannot be used, because it will not activate phosphodiester functions. Triisopropylbenzenesulfonyl chloride (TpsCl) has been extensively applied, but gave low yields (10-20%) when condensations of products containing purine bases, especially guanine were attempted. These low yields have been attributed to the liberation of hydrogen chloride, but trials with better or less innocuous leaving groups, e.g. azide, have also been unsuccessful. Only when triazoles were introduced by S.A. Narang (N. Katagiri, 1975) the preparation of phosphotriesters became a general and high-yield procedure. Benzenesulfonyl chlorides are condensed with $1H$-1,2,4-triazole in the presence of triethylamine in chloroform solution to form sulfonyltriazoles, e.g. MsT or pNbsT. These condensing agents will yield dinucleotides from protected mononucleotides, e.g. from mono(p-chlorophenyl) 5′-O-(di-p-methoxytrityl)deoxythymidine-3′-phosphate (A) (1 mol equiv) and thymidine-3′-phosphotriester (B) (1.2 mol equiv) in $\geq$ 80% yield. Hexanucleotides have been made in 100 mg quantities by the triester method.

MsT: $R^2 = R^4 = R^6 = CH_3$
pNbsT: $R^2 = R^6 = H; R^4 = NO_2$

(i)   coupling: MsT or pNbsT; (Py); 1-2d; r.t.
(ii) 5'-O-(di-p-methoxytrityl) removal: 80% aq. AcOH; 20 min.; r. t.
(iii) 2,2,2-trichloroethyl removal: Zn/AcOH/Py
(iv) 2-cyanoethyl and 4-chlorophenyl removal:
        0. 1 M NaOH/H$_2$O/dioxane; 3-6 h; r. t.
(v)   N-deblocking of nucleic bases B^{prot}: conc. aq. NH$_3$; 3 h; 50°C

4.1.1.2 Phosphite and 1,3,2-Dioxaphosphole Methods

The chemical basis for the phosphite triester approach is the observation, that dialkyl phosphorochloridites such as (C$_2$H$_5$O)$_2$PCl react very rapidly at the 3'-OH of nucleosides in pyridine even at low temperatures. In contrast, the reactions of analogous chloridates, e.g. (C$_2$H$_5$O)$_2$POCl, require several hours at room temperature. It was later found that phosphite esters can be oxidized quantitatively to the phosphates by using iodine in water and that clean condensation of phosphorochloridites with nucleosides can be achieved in THF at -78 °C. To develop this chemistry into a useful synthetic procedure it was necessary to establish which protecting groups are compatible with the highly reactive phosphorochloridites. It was found that O-trityl, methoxytrityl, acetyl, phenoxyacetyl, and benzoyl are stable.

In the condensation of 5'-O-(phenoxyacetyl)thymidine with 3'-O-(mono-p-methoxytrityl)thymidine, outlined below, β,β,β-trichloroethyl phosphorodichloridite (Cl$_3$CCH$_2$OPCl$_2$) was used. The reaction time at -78 °C was only 5 min. Oxidation with iodine in water and chromatography on silica gel gave the desired dinucleoside phosphate triester. Cleavage of the methoxytrityl ether with aqueous acetic acid afforded the 3'-OH compound; similarly treatment with ammonium hydroxide yielded the 5'-OH product. The synthetic sequence was repeated at the 5'-OH compound with more 5'-O-(phenoxyacetyl)thymidine and phosphorodichloridite. This cycle yielded the trinucleotide derivative (69%), the next cycle gave the tetranucleotide derivative (75%). The removal of the trichloroethyl groups from the phosphate esters and of the mono-p-methoxytrityl group from 3'-oxygen to yield the trinucleotidyl-nucleoside dTpTpTpT was achieved in 70% yield by reduction with sodium-naphthalene in hexamethylphosphoric triamide (R.L. Letsinger, 1976).

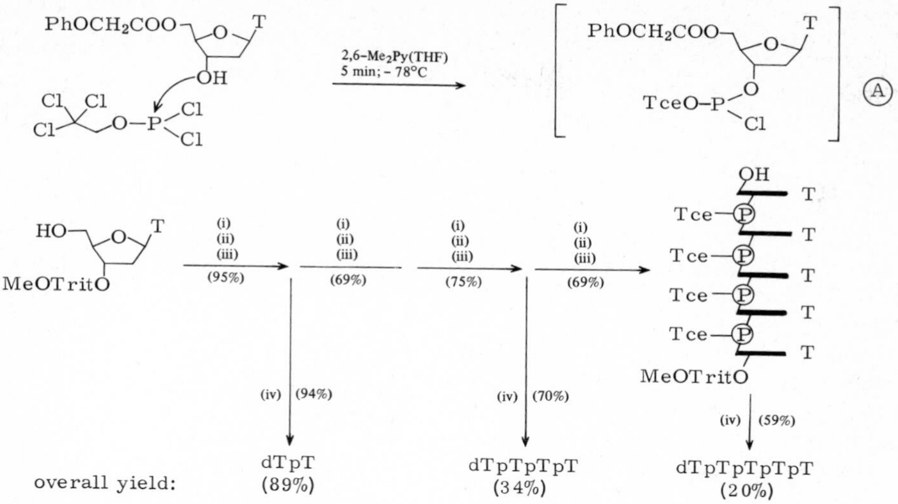

(i)   coupling: Ⓐ /2,6-Me$_2$Py(THF); 15min; -78°C
(ii)   phosphite triester oxidation: + I$_2$/H$_2$O; 5min; 0°C
(iii) 5'-O-phenoxyacetyl removal: conc. NH$_3$/H$_2$O/dioxane; 10min; r.t.
(iv)   Tce and 3'-O-(p-methoxytrityl) removal: Na/ ⬡⬡ /HMPTA; 5min; r.t.

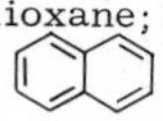

The ideal phosphorylating reagents for phosphodiester syntheses should meet the following criteria:

(i)   They should differentiate between primary and secondary alcohols.
(ii)   After the phosphorylation of a nucleoside the phosphorylating power of the reagent should be preserved or easily be restored to allow a second esterification.
(iii) The reagent should carry an auxiliary group which prevents side-reactions of synthetic esters and can be easily removed when it is no longer needed.

In the 1,3,2-dioxaphosphole method a bis(2-buten-2,3-diyl) pyrophosphate (A) is used as the condensing agent. It allows two successive esterifications of one phosphate group to be performed without additional activation. First a 5'-O-protected nucleoside is added in methylene chloride; in the second reaction an unprotected nucleoside can be used, since only the 5'-OH group is able to attack the cyclic enediol 3'-nucleosidyl phosphotriester. Protected dinucleoside triesters are obtained in 80% yield. Removals of protective groups, methoxytrityl by means of trifluoroacetic acid in methylene chloride and 1-methylacetonyl by aqueous triethylamine, also give about 80% yield (F. Ramirez, 1975, 1977).

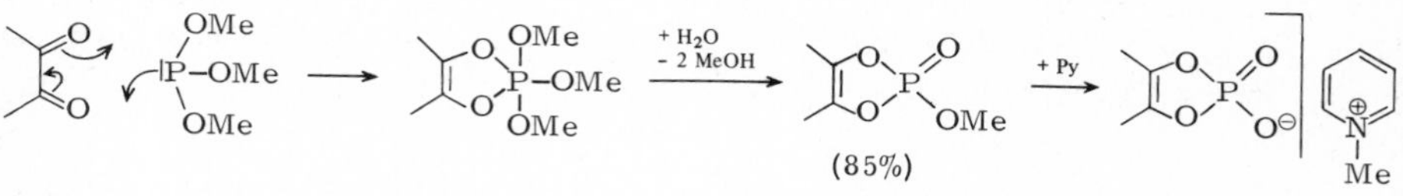

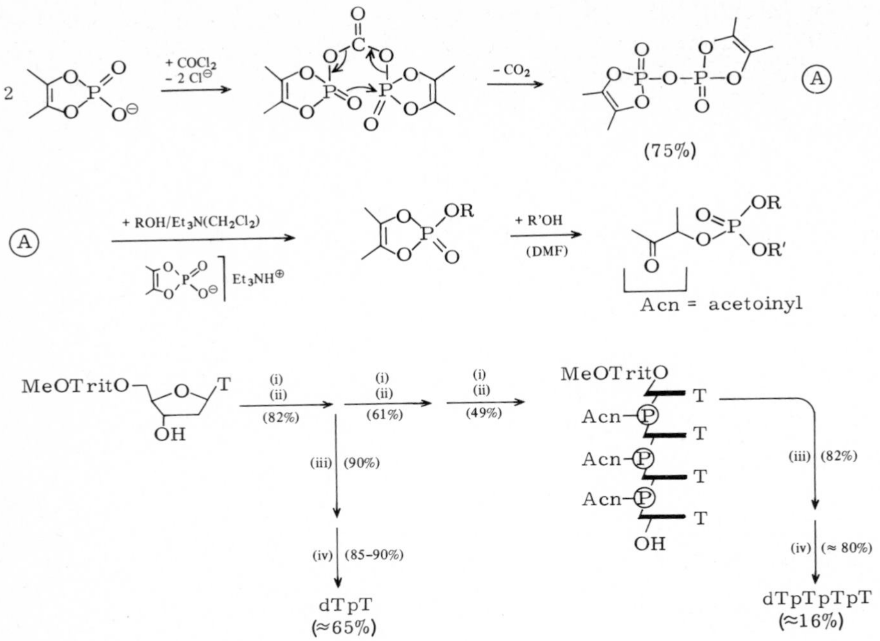

(75%)

Acn = acetoinyl

dTpT
(≈65%)

dTpTpTpT
(≈16%)

(i)   phosphorylation: (A)/Et$_3$N(CH$_2$Cl$_2$); 5h; 0°C
(ii) coupling with *unprotected* deoxythymidine:
      Et$_3$N/DMF; 14h; 0°C + 2h; r.t.
(iii) 5'-O-(p-methoxytrityl) removal: TFA(CH$_2$Cl$_2$); 20 min; 0°C
(iv) acetoinyl removal: Et$_3$N/H$_2$O/MeCN; 2.5h; 0°C

4.1.1.3 Solid-Phase Synthesis

Stepwise oligonucleotide synthesis may also be carried out on polymeric supports. A cross-linked poly(N,N-dimethylacrylamide) resin has been used, which swells to about ten times its dry bed volume in polar solvents (E. Atherton, 1975). It also contains long spacer groups with terminal (2-hydroxyethylthio)phenyl groups for the reversible binding of nucleotides (M.J. Gait, 1976). In oligodeoxyribonucleotide synthesis a "preactivated" 5'-phosphate, e.g. by TpsCl (p. 200), of a protected nucleotide is bound to the OH-group of the β-hydroxyethylthio side-chain. The usual reaction sequence of the diester approach (p. 200) is then followed: the 3'-O-protecting group is removed, activated nucleotide phosphomonoester with protected functional groups is added and its 3'-OH group is again liberated. Since the condensation product is insoluble, it can be separated from non-reacted phosphomonoester and condensing reagents by simple filtration. Yields are comparable with those obtained in solution synthesis, e.g. 87, 79, 68, 56, and 43% overall yields in five steps of the synthesis of the hexanucleotide d(pT)$_6$. The heptanucleotide d(pC—A—G—T—G—A—T) has been obtained from polymer bound dpC in 20% overall yield (M.J. Gait, 1976).

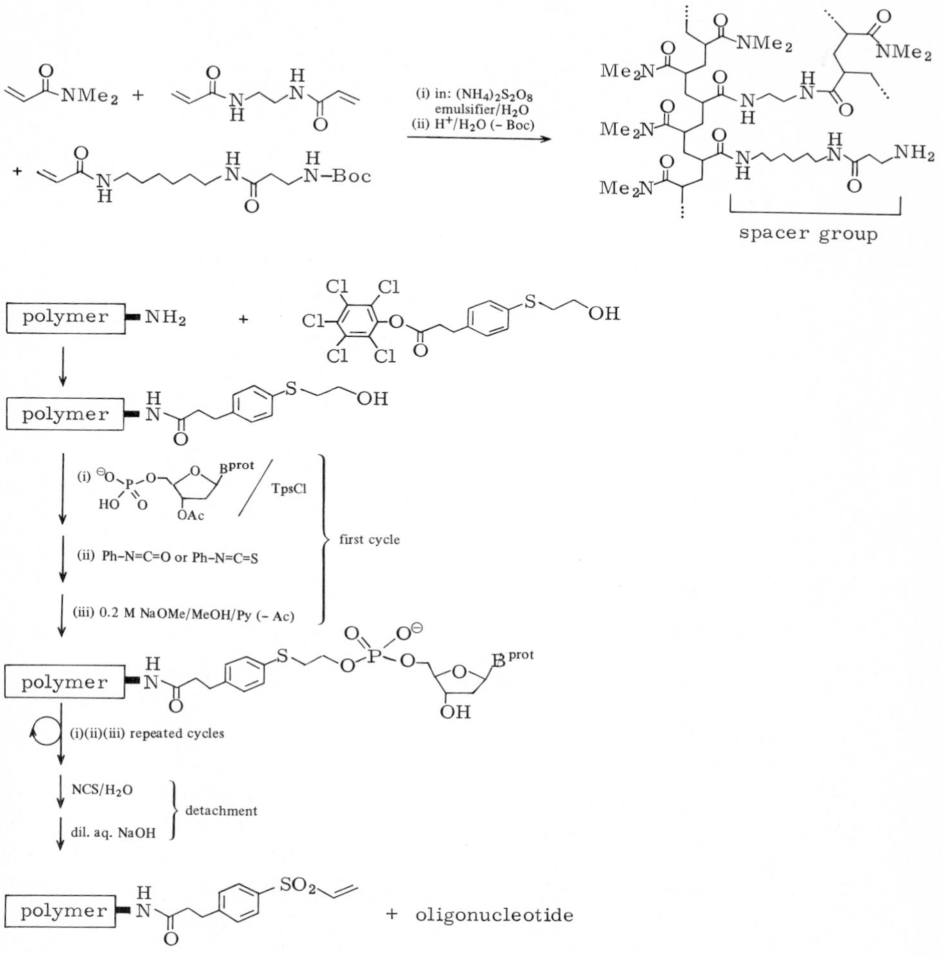

The whole mixture of oligomers may, after completion of the synthesis, be detached from the polymer support and separated by chromatography. Removal of various oligonucleotides with incomplete sequences ($\approx$ 80% of products) is, however, very tedious. The procedure may be simplified in two ways:

(i) After each condensation step the non-phosphorylated portion of the polymer-bound oligonucleotide is blocked, e.g. by irreversible carbamate formation of unreacted 3′-OH groups (see scheme above). This procedure drastically reduces the percentage of oligonucleotides with wrong sequences.

(ii) Di- and trinucleotides are used as units instead of the monomers. This "convergent" synthetic strategy simplifies the purification of products, since they are differentiated by a much higher jump in molecular mass and functionality from the educts than in monomer additions, and it raises the yield. We can illustrate the latter effect with an imaginary sequence of seven synthetic steps, e.g. nucleotide condensations, where the yield is 80% in each step. In a converging seven-step synthesis an octanucleotide would be obtained in $0.8^3 \times 100 = 51\%$ yield, compared with a $0.8^7 \times 100 = 21\%$ yield in a linear synthesis.

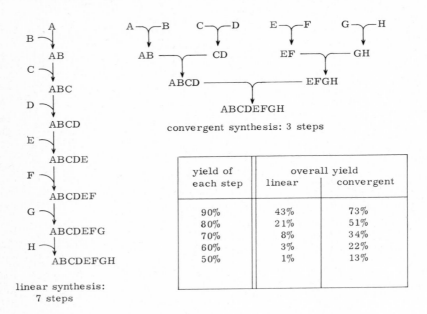

convergent synthesis: 3 steps

linear synthesis:
   7 steps

yield of each step	overall yield	
	linear	convergent
90%	43%	73%
80%	21%	51%
70%	8%	34%
60%	3%	22%
50%	1%	13%

This "trick" can, of course, be applied in all multiple-step syntheses.

4.1.1.4 Combined Chemical-Enzymatic Syntheses

Synthetic oligonucleotides may be used as "primers" and be elongated stepwise with the aid of polynucleotide phosphorylase (PNPase) and nucleoside diphosphates.

$$d(TACG) + pp\text{-}dA \xrightarrow[{[Mn^{2+}]}]{[PNPase]^{*)}} d(TACGA) + P_i$$

   * polynucleotide phosphorylase

More important is the enzymatic condensation of synthetic oligonucleotides by the "sticky end" approach. The term "sticky ends" indicates that oligonucleotides with at least four complementary bases ($A=T$ and $G\equiv C$)* at their ends aggregate spontaneously and spe-

---

* The bonds indicate the number of hydrogen bonds.

cifically in aqueous solution. One oligonucleotide molecule can bind two other oligonucleotides and bring them in close and well defined contact with each other. An enzyme can then be used to condense the preformed aggregates to produce long, double-stranded DNA segments. Structural genes have already been synthesized from several oligonucleotides by simple mixing of all components and addition of DNA-ligase, which accepts all paired oligonucleotides as substrates (K. Itakura, 1977).

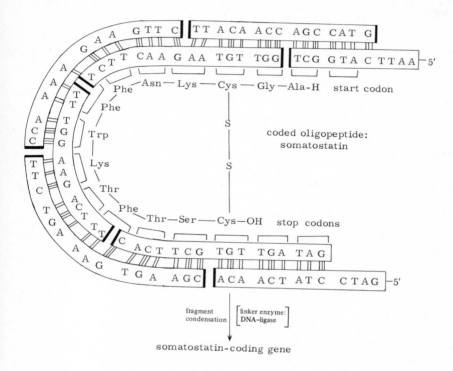

4.1.2 Peptides

In 1969 a series of short communications appeared in the Journal of the American Chemical Society, which announced the total syntheses of a few milligrams of complete or almost complete enzymes containing about 120 amino acid units (B. Gutte, 1969; R. Hirschmann, 1969). Their biological activities ranged from 1% to 13% of the natural enzyme. The papers were received by the editor between November 25th and December 19th, 1968 and were printed in the copy of the journal that appeared on January 15th, 1969. This is, to our knowledge, a record. The community of chemists obviously liked the idea "that a protein molecule with true enzymatic activity toward its natural substrate can be totally synthesized".

Seven years later, in a critical review on the synthesis of peptides, the following statement was made: *"Chemists in particular should respect the classical criteria of what constitutes synthesis of a natural product, i.e., that synthesis of a natural product has been achieved, when the physical, chemical and biological properties of the synthetic compound match those of the natural prototype. Unfortunately not a single one of the "synthetic proteins" satisfies these criteria. It is frequently argued that these criteria are not applicable to more complex situations, but lowering standards of purity is not likely to advance the field. Presently available analytical methods cannot adequately detect inhomogeneity in a high molecular peptide that is produced by stepwise synthesis. Consequently, the synthetic method must be chosen so that the product can be purified and critically evaluated by the available analytical techniques .... " F.M. Finn, 1976).*

The following short descriptions of the steps involved in the synthesis of a tripeptide will demonstrate the complexity of the problem as well as the fragility of the (protected) amino acid units. In the later parts of this section we shall describe actual syntheses of well defined oligopeptides by linear elongation reactions and of less well defined polypeptides by fragment condensation.

4.1.2.1 Stages of Peptide Synthesis

One starts with individual amino acids or with peptides and tries to achieve the regioselective formation of a new amide bond. In its most general form such syntheses of peptides involve the following stages:

(i) Preparation of a "carboxyl component" by blocking the amino group of an amino acid or a peptide with a group Y (see p. 146ff).

$$H_3\overset{\oplus}{N}-\underset{\underset{R^2}{|}}{\overset{\overset{H}{|}}{C}}-COO^{\ominus} \quad \xrightarrow{N^\alpha\text{-protection}} \quad \boxed{Y}-\underset{\underset{R^2}{|}}{\overset{\overset{H}{|}}{\underset{}{N}}}\overset{\overset{H}{|}}{\underset{}{C}}-COOH$$

(ii) Synthesis of an "amino component" by protecting the carboxyl group of another amino acid or peptide by a group Z (see p. 149).

$$H_3\overset{\oplus}{N}-\underset{\underset{R^1}{|}}{\overset{\overset{H}{|}}{C}}-COO^{\ominus} \quad \xrightarrow[\text{protection}]{\text{carboxyl}} \quad H_2N-\underset{\underset{R^1}{|}}{\overset{\overset{H}{|}}{C}}-\overset{\overset{O}{||}}{C}-O-\boxed{Z}$$

(iii) Protection (= prot.) of all other functional groups that are reactive in condensations. The "best" reagents for side-chain protections are summarized in table 18.

$$\boxed{Y}-\underset{\underset{R^2}{|}}{\overset{\overset{H}{|}}{N}}\overset{\overset{H}{|}}{\underset{|}{C}}-COOH \quad \xrightarrow[\text{protection}]{\text{side-chain}} \quad \boxed{Y}-\underset{\underset{R^2 - \boxed{prot}}{|}}{\overset{\overset{H}{|}}{N}}\overset{\overset{H}{|}}{\underset{}{C}}-COOH$$

$$H_2N-\underset{\underset{R^1 - \boxed{prot}}{|}}{\overset{\overset{H}{|}}{C}}-\overset{\overset{O}{||}}{C}-O-\boxed{Z} \quad \xrightarrow[\text{protection}]{\text{side-chain}} \quad H_2N-\underset{\underset{R^1 - \boxed{prot}}{|}}{\overset{\overset{H}{|}}{C}}-\overset{\overset{O}{||}}{C}-O-\boxed{Z}$$

Table 18. Protection of amino acid side-chains.

functional group	amino acid	protected derivative	reagent	conditions for removal
–NH$_2$	lysine	N$^\omega$ -(diisopropyl-methoxycarbonyl)		HF/anisole
		N$^\omega$ -benzyloxycarbonyl (= Cbz,Z) etc. X = H,Cl,NO$_2$		HBr/AcOH or H$_2$/Pd
$\overset{H}{-N}\overset{}{\underset{\underset{NH}{\|}}{C}}NH_2$	arginine	guanidinium cation	pH < 10	
		N,N'- bis (adamantyl-oxycarbonyl) = Adoc		H$^+$
		N-nitro	HNO$_3$/H$_2$SO$_4$	H$_2$/Pd
–SH	cysteine	triphenylmethyl (= trityl, Trit)	Ph$_3$CCl	H$^+$ /Hg^{2+}/I$_2$
		acetamidomethyl (= Acm)	AcN$\overset{H}{}\diagup$OH	Hg^{2+}/I$_2$
		ethylcarbamoyl (= Ec)	EtNCO	Hg^{2+}
–S–CH$_3$	methionine	sulfoxide	H$_2$O$_2$	HSCH$_2$COOH
–COOH	aspartic acid glutamic acid	benzyl and t-butyl esters or ethers, resp.	PhCH$_2$Br; $\diagdown\diagup$/BF$_3$·Et$_2$O	HBr/TFA
⟨⟩–OH –OH	tyrosine serine theonine			

The phenolic hydroxyl group of tyrosine, the imidazole moiety of histidine, and the amide groups of asparagine and glutamine are often not protected in peptide synthesis, since it is usually unnecessary. The protection of the hydroxyl group in serine and threonine (O-acetylation or O-benzylation) is not needed in the azide condensation procedure but may become important when other activation methods are used.

(iv) Activation of the carboxyl component by a group X (see p. 129 ff).

$$\boxed{Y}-\overset{H}{\underset{\underset{\boxed{prot}}{\underset{|}{R^2}}}{\underset{|}{N}}-\overset{H}{\underset{|}{C}}-COOH \xrightarrow{\text{carboxyl activation}} \boxed{Y}-\overset{H}{\underset{\underset{\boxed{prot}}{\underset{|}{R^2}}}{\underset{|}{N}}-\overset{H}{\underset{|}{C}}-\overset{O}{\overset{\|}{C}}\sim\circ\!\!\!\!X$$

(v) Coupling of the activated carboxylic acid with the amino component to give a protected peptide. This is usually performed in an organic solvent in the presence of a slight excess of a base.

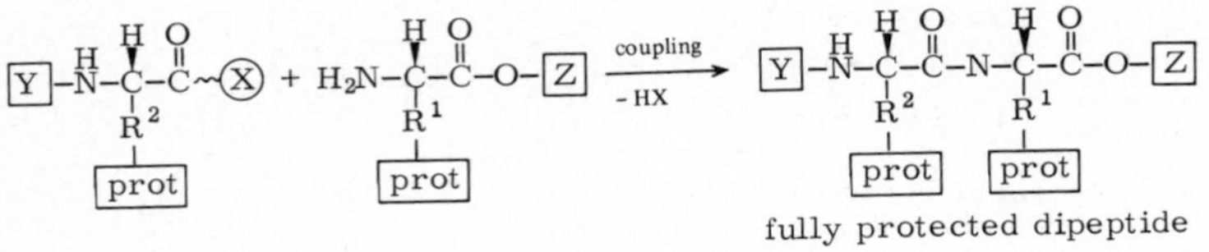

$$Y-N-C-C\sim X + H_2N-C-C-O-Z \xrightarrow[-HX]{coupling} Y-N-C-C-N-C-C-O-Z$$

fully protected dipeptide

(vi) Removal of the blocking group Y (p. 146ff.).

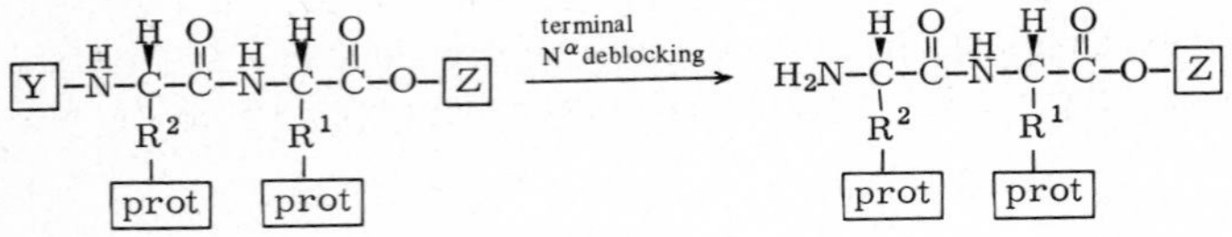

$$Y-N-C-C-N-C-C-O-Z \xrightarrow{terminal\ N^\alpha deblocking} H_2N-C-C-N-C-C-O-Z$$

(vii) Repeated coupling with an activated acid.

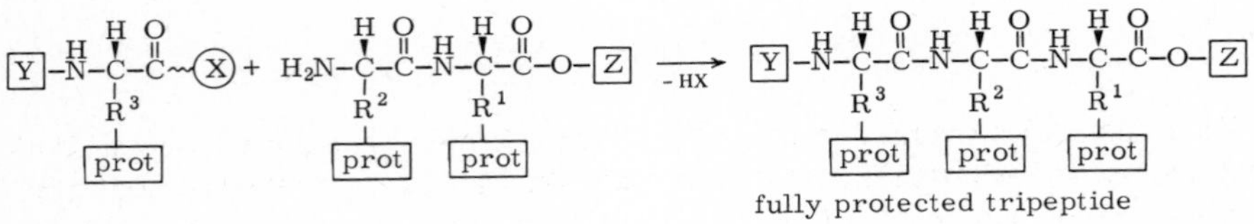

$$Y-N-C-C\sim X + H_2N-C-C-N-C-C-O-Z \xrightarrow{-HX} Y-N-C-C-N-C-C-N-C-C-O-Z$$

fully protected tripeptide

(viii) Removal of protecting groups from side-chains R (p. 209), amino (p. 146ff.), and carboxyl groups (p. 149) to give the free peptide.

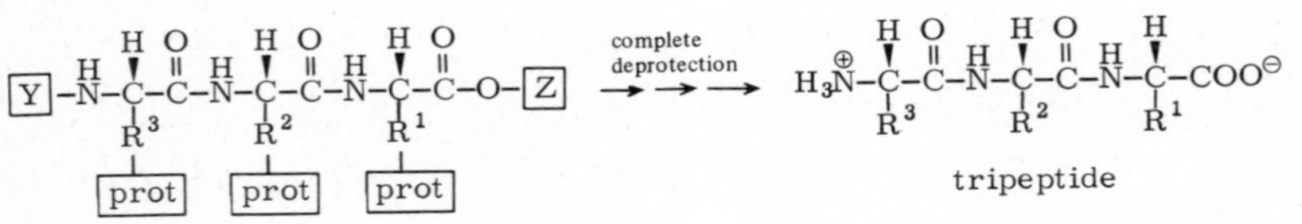

$$Y-N-C-C-N-C-C-N-C-C-O-Z \xrightarrow{complete\ deprotection} H_3\overset{\oplus}{N}-C-C-N-C-C-N-C-COO^{\ominus}$$

tripeptide

Protected amino acids with either a free amino or carboxyl function can usually be prepared by proven methods or are even commercially available. Therefore stages (i) - (iii) may be considered as simple routine nowadays, although great care must be taken that the protected starting materials are pure enantiomers. The reactions that cause most trouble are in stages (iv), (v) and (vii). In these stages an activated carboxyl group is involved and the chiral centre adjacent to it is at peril from racemization. A typical reaction which causes epimerization is azlactone formation. With acids or bases these cyclization products may reversibly enolize and racemize. Direct racemization of amino acids has also been observed.

Groups R, Y, and X affect the kinetics of these processes as do reaction conditions. The best X is azide, which gives little or no racemization. This has been rationalized with the assumption that the relatively electropositive azide group attracts the electronegative amide oxygen. This effect could stabilize the cyclic conformation given below, which does prohibit azlactone formation. Another explanation is that the azides are only moderately activated and therefore differentiate between the weakly nucleophilic oxygen and the strong amine-nucleophile.

stable conformation of
N^α-acylated amino acid azides

With the dicyclohexylcarbodiimide (DCC) reagent racemization is more pronounced in polar solvents such as DMF than in CH_2Cl_2, for example. An efficient method for reduction of racemization in coupling with DCC is to use additives such as N-hydroxysuccinimide or 1-hydroxybenzotriazole. A possible explanation for this effect of nucleophilic additives is that they compete with the amino component for the acyl group to form active esters, which in turn react without racemization. There are some other condensation agents (e.g. 2-ethyl-7-hydroxybenz[d]isoxazolium and 1-ethoxycarbonyl-2-ethoxy-1,2-dihydroquinoline) that have been found not to lead to significant racemization. They have, however, not been widely tested in peptide synthesis.

The protecting group Y of the amine is generally an alkoxycarbonyl derivative since their nucleophilicity is low. Benzyloxy- or *tert*-butoxycarbonyl derivatives usually do not undergo azlactone formation.

The best preventive measure against racemization in critical synthetic steps (e.g. fragment condensation, see p. 219) is to use glycine (which is achiral) or proline (no azlactone) as the activated carboxylic acid component. The next best choice is an aliphatic monoamino monocarboxylic acid, especially with large alkyl substituents (valine, leucine). Aromatic amino acids (phenylalanine, tyrosine, tryptophan) and those having electronegative substituents in the β-position (serine, threonine, cysteine) are, on the other hand, most prone to racemization. Reaction conditions that inhibit azlactone formation and racemization are non-polar solvents, a minimum amount of base, and low temperature. If all precautions are taken, one still has to reckon with an average inversion of 1% per condensation reaction. This means, for example, that a synthetic hectapeptide contains only 0.99^{100} x 100% = 37% of the fully correct diastereomer (see p. 213f.).

We now turn from the general problems of peptide synthesis to specific problems connected with three currently important procedures, namely:

(i) Solid-phase peptide synthesis.
(ii) Solution methods for peptide synthesis.
(iii) Condensation of peptide fragments.

4.1.2.2 Solid-Phase Peptide Synthesis

To illustrate the specific operations involved, the scheme below shows the first steps and the final detachment reaction of a peptide synthesis starting from the carboxyl terminal. N-Boc-glycine is attached to chloromethylated styrene-divinylbenzene copolymer resin. This polymer swells in organic solvents but is completely insoluble.*) Treatment with HCl in acetic acid removes the *tert*-butoxycarbonyl (Boc) group as isobutene and carbon dioxide. The resulting amine hydrochloride is neutralized with triethylamine in DMF.

Then N-Boc-O-benzylserine is coupled to the free amino group with DCC. This concludes one cycle (N^α-deprotection, neutralization, coupling) in solid-phase synthesis. All three steps can be driven to very high total yields ($\approx$ 99.5%) since excesses of Boc-amino acids and DCC (about fourfold) in CH_2Cl_2 can be used and since side-reactions which lead to soluble products do not lower the yield of condensation product. One side-reaction in DCC-promoted condensations leads to N-acylated ureas. These products will remain in solution and not react with the polymer-bound amine. At the end of the reaction time, the polymer is filtered off and washed. The times consumed for 99% completion of condensation vary from 5 min for small amino acids to several hours for a bulky amino acid, e.g. Boc-Ile, with other bulky amino acids on a resin. A new cycle can begin without any workup problems (R.B. Merrifield, 1969; B.W. Erickson, 1976; M. Bodanszky, 1976).

---

* Among alternative supports for solid-phase synthesis are non-swelling porous glass-beads. They exhibit higher rates of mass transfer and higher reaction rates. Silanol groups on the surface are used to attach organic substituents.

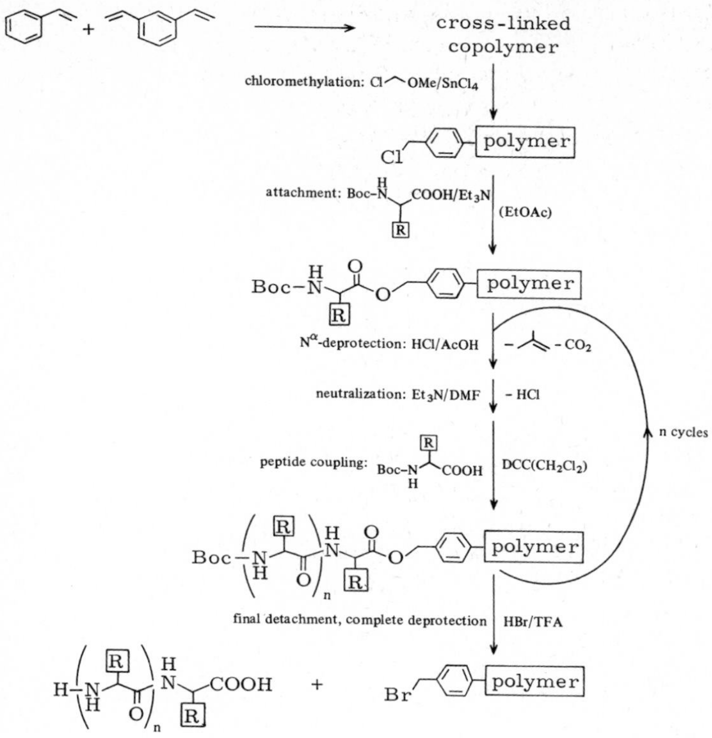

The purity of the peptide finally obtained depends critically on the yield of each cycle. It must be extraordinarily good to produce even moderately pure products (K. Lübke, 1975). If the average yield of amide formation in the synthesis of an undecapeptide ($n = 10$) is, for example, 98%, the product will contain already about 20% of different impurities which may be difficult to remove.

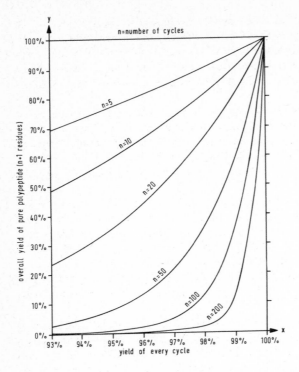

To keep reaction times low, one generally uses DCC as condensing agent. With easily racemizing amino acids, however, this procedure is not safe. The safe azide method, on the other hand, is too slow. Commercial 4-nitrophenyl esters are usually selected as simple alternatives (M. Bodanszky, 1976). 1,2,4-Triazole can serve as a bifunctional catalyst which accelerates the coupling rate of 4-nitrophenyl esters. A 50% condensation needs, nevertheless, in the order of hours, and completion can rarely be expected in less than 24 hours, and often 4-nitrophenyl esters do not couple quantitatively at all. On the other hand these esters offer two more advantages over the DCC method: protection of side-chain hydroxyl groups is not required, and troublesome side-reactions such as the formation of nitriles by dehydration of asparagine or glutamine are not observed.

The N-to-C assembly of the peptide chain is unfavorable for the chemical synthesis of peptides on solid supports. This strategy can be dismissed already for the single reason that repeated activation of the carboxyl ends on the growing peptide chain would lead to a much higher percentage of racemization. Several other more practical disadvantages also tend to disfavor this approach, and acid activation on the polymer support is usually only used in one-step fragment condensations (p. 221).

In each step of the usual C-to-N peptide synthesis the N-protecting group of the newly coupled amino acid must be selectively removed under conditions that leave all side-chain protecting groups of the peptide intact. The most common protecting groups of side-chains (p. 209) are all stable towards 50% trifluoroacetic acid in dichloromethane, and this reagent is most commonly used for N^α-deprotection. Only *tert*-butyl esters and carbamates (= Boc) are solvolyzed in this mixture.

Peptide chains that fail to couple with the N-protected amino acid still bear the free amino group, while chains that do couple have a blocked α-amino group. If the resin is reacted irreversibly with a great excess of an acylating agent, these amino groups will be permanently terminal because chain growth is inhibited. Acetylation with acetic anhydride and triethylamine in DMF after each coupling step reduces the concentration of deletion peptides by a factor of at least ten. If 3-nitrophthalic anhydride is used, the "terminated" peptides bear an additional negative charge and are colored. They are easily separated by chromatography on an anion-exchange resin.

Final detachment of the peptide bound by a benzyl ester link from the solid support usually occurs by acidolysis. If a side-chain protected peptide is needed, trifluoroacetic acid is used. The usual reaction time at 25 °C is 60 min, although after 5 min a high percentage may already be set free. If the unprotected peptide is to be liberated, liquid HF for 1 hr at 0 °C removes all side-chain protecting groups and cleaves the anchoring bond in one step. For obvious reasons this method cannot be used with glass-bead supports. Anisole or ethyl methyl sulfide may be added as nucleophilic scavengers to protect tyrosine, tryptophan, histidine, and methionine residues from alkylation by carbonium ions (e.g. from Boc) produced during cleavage. It is usually advisable to go first for protected peptides, since they frequently crystallize whereas the free peptides rarely do so.

We shall now exemplify the solid-phase peptide synthesis approach by *cyclo*-[-L-Val-D-Pro-D-Val-L-Pro-]$_3$, which was prepared by Merrifield himself, the inventor of the method (B.F. Gisin, 1972).

The linear peptide was synthesized first starting with L-proline at the C-terminal. Commercial polystyrene-co-1% divinylbenzene was treated with several solvents at 90-100 °C to transfer it to a swollen state and was then chloromethylated. The chloromethyl groups were converted into acetoxymethyl. Any remaining traces of chloromethyl groups were aminolyzed with boiling diethylamine, thus eliminating the chance of subsequent reactions of amines on the C—Cl bond later on. The resin was mixed at low temperature with *tert*-butoxycarbonyl-L-proline (= Boc-L-Pro) in methylene chloride and with carbonyl-diimidazole as condensing agent. Reaction time was three days at room temperature. Hydrolysis of the resin indicated a substitution of approx. 130 mh (= 0.6 mmol) of Boc-LPro per gram of resin. The starting material was now 4.5 g of Boc-L-prolyl resin.

The peptide chain was built up using the following cycle: deprotection with 50% trifluoroacetic acid in CH$_2$Cl$_2$, neutralization with diisopropylamine in CH$_2$Cl$_2$, coupling with a twofold excess of both DCC and Boc-D-(or L)-Val-OH or Boc-D-(or L)-Pro-OH and again with 1 equiv of each for 2 hours. In order to block unreacted amino groups and also hydroxyl groups that may have formed by detachment of peptide from the resin, acetylation steps (Ac$_2$O/Py) were added after the tri-, tetra-, penta-, and nona-peptide stages. Afterwards the bound Boc-peptide was deblocked and the free amino groups were determined as picrates with picric acid. The total amount of amine was as follows: H-L-Pro-resin = 100%; tri-peptidyl resin 72%; pentapeptidyl resin 62%; heptapeptidyl resin 62%; nonapeptidyl resin 59%; dodeca-peptidyl resin 60%. 30% of educt was lost in the first two condensation steps (see below). The later con-densation steps gave essentially quantitative yields.

The amino acid analysis of all peptide chains on the resins indicated a ratio of Pro:Val = 6.6:6.0 (calcd. 6:6). The peptides were then cleaved from the resin with 30% HBr in acetic acid and chromatogra-phed on sephadex LH-20 in 0.001 M HCl. 335 mg dodecapeptide was isolated. Hydrolysis followed by quantitative amino acid analysis gave a ratio of Pro:Val = 6.0:5.6 (calcd. 6:6). Cyclization in DMF with Woodward's reagent K (see scheme below) yielded after purification 138 mg of needles of the desired cyc-lododecapeptide. The compound yielded a yellow adduct with potassium picrate, and here an analytically more acceptable ratio Pro: Val of 1.03: 1.00 (calcd. 1:1) was found. The mass spectrum contained a molecular ion peak. No other spectral measurements (lack of ORD, NMR) have been reported. For a thirty-six step synthesis in which each step may cause side-reactions the charac-terization of the final product should, of course, be more elaborate.

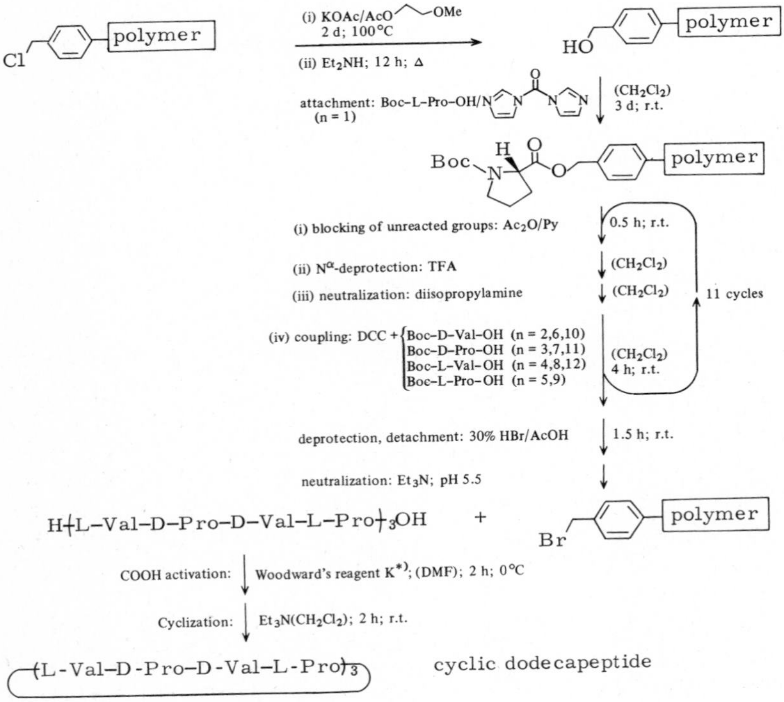

H-[L-Val–D-Pro–D-Val–L-Pro]₃OH   +   Br-⬡-[polymer]

cyclic dodecapeptide

$$\{L\text{-Val}-D\text{-Pro}-D\text{-Val}-L\text{-Pro}\}_3$$

* Woodward's reagent K:

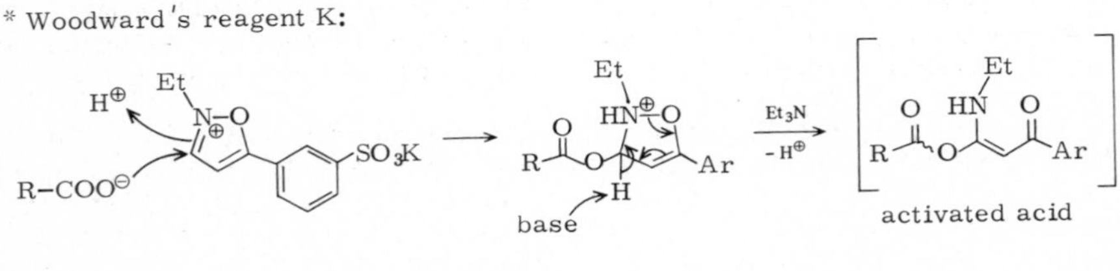

The synthesis described met some difficulties. D-Valyl-L-prolyl resin was found to undergo intramolecular aminolysis during the coupling step with DCC. 70% of the dipeptide was cleaved from the polymer, and the diketopiperazine of D-valyl-L-proline was excreted into solution. The reaction was catalyzed by small amounts of acetic acid and inhibited by a higher concentration (protonation of amine). This side-reaction can be suppressed by adding the DCC prior to the carboxyl component. In this way, the carboxyl component is "consumed" immediately to form the DCC adduct and cannot catalyze the cyclization.

cyclic dipeptide
(diketopiperazine)

4.1.2.3 Solution (= Liquid-Phase) Methods for Peptide Synthesis

The major disadvantage of solid-phase peptide synthesis is the fact that all the by-products attached to the resin can only be removed at the final stages of synthesis. Another problem is the relatively low local concentration of peptide which can be obtained on the polymer, and this limits the turnover of all other educts. Preparation of large quantities ($>$ 1 g) is therefore difficult. Thirdly, the racemization-safe methods for acid activation, e.g. with azides, are too mild (= slow) for solid-phase synthesis. For these reasons the convenient Merrifield procedures are quite generally used for syntheses of small peptides, whereas for larger polypeptides many research groups adhere to "classic" solution methods and purification after each condensation step (F.M. Finn, 1976).

We shall describe here one step in the total synthesis of a protected heptatetracontapeptide (= 47 amino acids) by K. Hofmann (H.T. Storey, 1972), and compare the techniques and the results with those of solid-phase synthesis.

A protected serine hydrazide was condensed by the azide method to an S-protected tripeptide H-Asn-Cys-Tyr-NHNH-Cbz to form a protected tetrapeptide Boc-Ser-Asn-Cys-Tyr-NHNH-Cbz: 1.02 g of N-*tert*-butoxycarbonylserine hydrazide (Boc-Ser-NH-NH$_2$) in DMF containing HCl in dioxane was mixed at -20 °C with *tert*-butyl nitrite. This mixture containing the azide Boc-Ser-N$_3$ was neutralized with triethylamine, and a solution of 3.4 g asparaginyl-S-(ethylcarbamoyl)cysteinyl-tyrosinyl 2-(benzyloxy-carbonyl)hydrazide trifluoroacetate was added. After 72 hours at 4 °C a simple work-up procedure and precipitation from methanol-petroleum ether yielded 3 g of impure protected tetrapeptide hydrazide. It was deblocked at N^{α} with 90% trifluoroacetic acid and again precipitated. Chromatography on an ion-exchange resin yielded 1.1 g of S-protected Ser-Asn-Cys-Tyr-NHNH-Cbz acetate. The material was dried in the cold and the only analytical data given are optical rotation, elemental analysis, and amino acid ratios in acid hydrolysate: $Ser_{1.0}$ $Asn_{1.1}$ $Cys_{1.0}Tyr_{1.0}$. Thin layer chromatography in two systems showed only a single spot.

The cyclopeptide described above was tailored to form stable potassium complexes. It is one of the very few examples of complex peptide syntheses which do not lead to a natural compound.

The example clearly demonstrates the main advantages of solution chemistry over the solid-phase approach:

(i) Most important, thin-layer chromatography can be used immediately to demonstrate homogeneity.
(ii) Less protecting groups (Ser, Tyr unprotected) are sometimes needed.
(iii) The azide method which gives only little racemization (p. 129) is complete within the reasonable time of 72 hours at 4 °C.

On the other hand, a lot of material is lost in the chromatography and work-up procedures ($\approx 40\%$), and the analytical data provided are not very convincing, either (except for TLC).

In more recent publications the analytical standards have been raised considerably (HPLC, ¹H-NMR, CD; R.F. Nutt, 1980), and one may predict that in the near future it will be possible to characterize fully synthetic oligopeptides of moderate size.

4.1.2.4 Peptide Fragment Condensation

The modern stepwise methods provide excellent routes to peptides up to the pentadecapeptide range. Homogeneous peptides containing a hundred or more amino acid residues can only be made from rigorously purified smaller protected peptides. In the fragment assembly technique peptides in the range of decapeptides are condensed. The separation of unreacted small peptide educts from the peptide product of much higher molecular mass is without problems and rigorous proof of homogeneity can often be accomplished.

Problems in the condensation of oligopeptide blocks are the frequently observed low coupling yields of large fragments and the sparing solubility of large protected peptides. The latter difficulty can be overcome by the use of DMSO or HMPTA as solvents. These highly polar solvents, however, favor racemization, and only peptides with carboxyl-terminal glycine or proline should be selected for construction of complex peptides when DCC or similar condensing agents are used. Other N-protected peptides can only be coupled by time consuming azide methods, and nonracemization has to be demonstrated after each fragment condensation.

An example of fragment condensation is taken from a convergent synthesis of bovine insulin (H. Zahn, 1970). The triacontapeptide B-chain of this hormone was built up from six protected oligopeptides. The following reaction sequence was used for every azide coupling cycle: (i) hydrazinolysis of the carboxyl-terminal methyl ester for 2 to 6 days, (ii) azide formation with isoamyl nitrite and HCl for 30 min at -20 °C, (iii) neutralization with triethylamine at -40 °C, and (iv) coupling with the amino component, first 12 hours warming up from -40 to 0 °C, then 3 days at 0 °C. The thiol groups of cysteine residues were protected by formation of disulfide-linked dimers or polymers which were finally cleaved by "oxidative sulfitolysis".

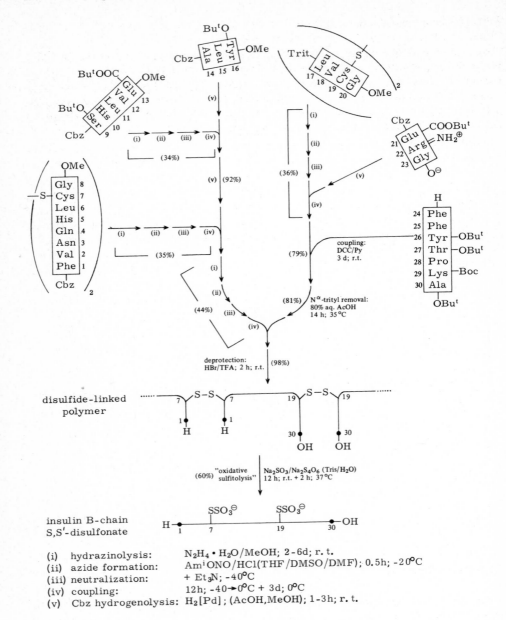

(i)   hydrazinolysis: $N_2H_4 \cdot H_2O/MeOH$; 2-6d; r. t.
(ii) azide formation: $Am^iONO/HCl(THF/DMSO/DMF)$; 0.5h; $-20°C$
(iii) neutralization: $+ Et_3N$; $-40°C$
(iv) coupling: 12h; $-40 \to 0°C$ + 3d; $0°C$
(v)   Cbz hydrogenolysis: $H_2[Pd]$; (AcOH,MeOH); 1-3h; r. t.

4.1.2.5 Macrocyclic Peptides and Depsipeptides

Macrocyclic peptides and depsipeptides (=macrocyclic peptides with amide *and* ester linkages) are important natural compounds. They have been synthesized in low yield from open-chain precursors by DCC treatment at high dilution (E. Schröder, 1963; M.M. Shemyakin,

1961). More successful are solid-phase methods in which the linear precursor is attached through a labile ester bond (e.g. o-nitrophenyl) to a polymer.

First the protected oligopeptide is coupled with polymer-bound nitrophenol by DCC. N^α-Deblocking leads then to simultaneous cyclization and detachment of the product from the polymer (M. Fridkin, 1965). Recent work indicates that high dilution in liquid-phase cyclization is only necessary, if the cyclization reaction is sterically hindered. Working at low temperatures and moderate dilution with moderately activated acid derivatives is the method of choice for the formation of macrocyclic lactams (R.F. Nutt, 1980).

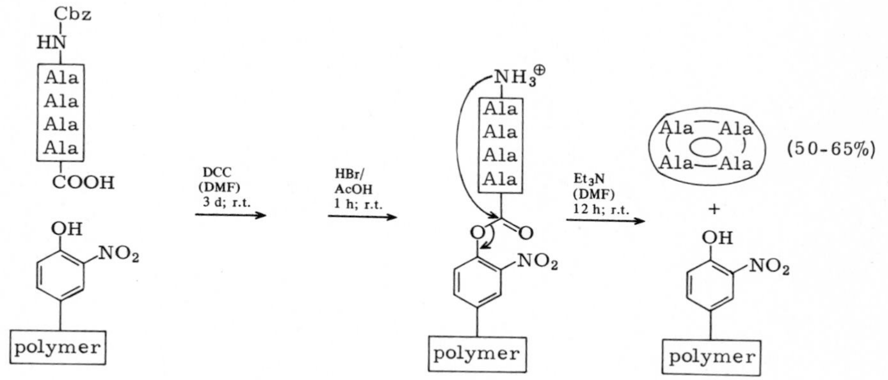

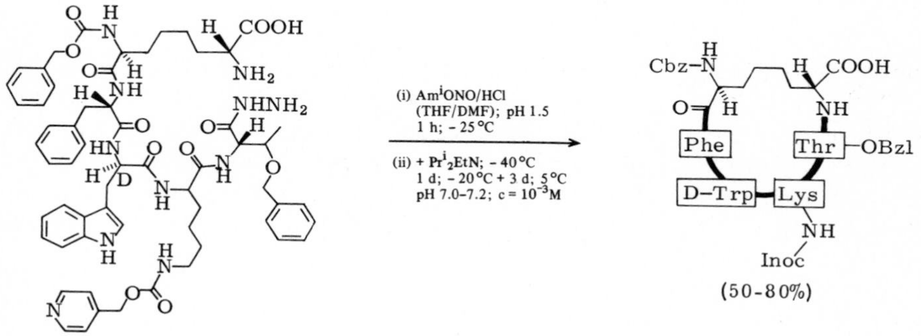

4.1.3 Macro-Heterocycles

Metal complexes of synthetic macrocycles are of topical interest, since there is some hope that useful reactions found in nature (e.g. photolysis of water with sunlight; catalysis of hydrocarbon oxygenation; ion transport through membranes) can be reproduced with these easily accessible compounds. A recent book (G.A. Melson, 1979) summarizes current knowledge of this subject. We shall discuss some general problems in the synthesis of macrocycles together with specific examples for porphyrin-like tetraaza ligands and alkali-ion binding polyethers.

The cyclization reactions discussed here either involve the intramolecular reaction of a donor group D with an acceptor group A or a cyclizing dimerization of two molecules with two terminal acceptors and two donors. A polymerization reaction will always compete with cyclization. For macrolides see p. 132 and p. 291.

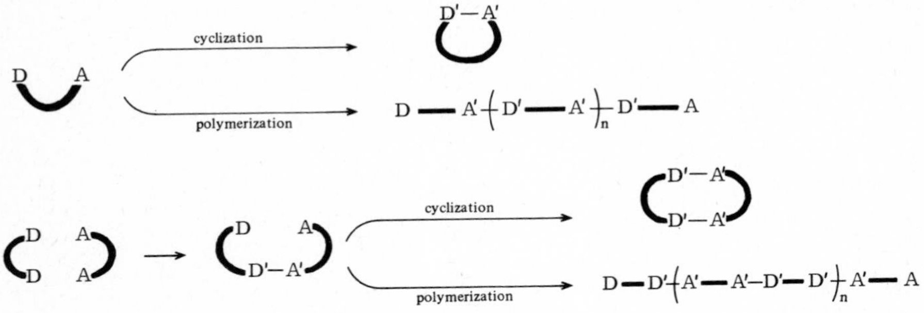

The intramolecular cyclization can be favored by (i) high dilution techniques, (ii) the action of a template which forces the reacting ends together, and (iii) stepwise condensations.

4.1.3.1 High-Dilution Methods

Synthesis by high-dilution techniques requires slow admixture of reagents ($\approx$ 8-24 hrs) or very large volumes of solvents ($\approx$ 100 l/mmol). Fast reactions can also be carried out in suitable flow cells (J.L. Dye, 1973). High dilution conditions have been used in the dilactam formation from 1,8-diamino-3,6-dioxaoctane and 3,6-dioxaoctanedioyl dichloride in benzene. The amide groups were reduced with lithium aluminum hydride, and a second cyclization with the same dichloride was then carried out. The new bicyclic compound was reduced with diborane. This ligand envelops metal ions completely and is therefore called a "cryptand" (B. Dietrich, 1969).

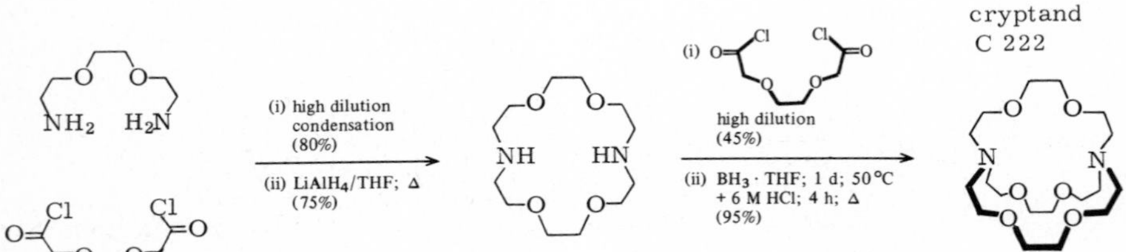

Another bicyclic compound with two condensed macrocycles was constructed as described above. Its secondary amino groups were deblocked by reductive detosylation and connected by a third cyclization with a diacid dichloride again under high-dilution conditions. Upon reduction a tricyclic cryptand was obtained with four nitrogen bridgeheads, which could form a tetrahedral ligand field, and six octahedrally arranged ether bridges (E. Graf, 1975).

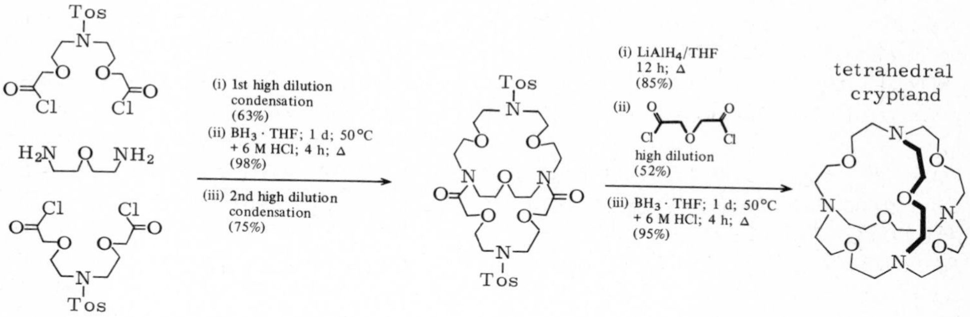

4.1.3.2 Template Reactions

The expression "template reaction" indicates mostly a reaction in which a complexed metal ion holds reactive groups in the correct orientation to allow selective multi-step reactions. The template effect of the metal is twofold: (i) polymerization reactions are suppressed, since the local concentration of reactants around the metal ion is very high; (ii) multi-step reactions are possible, since the metal holds the reactants together. In the following one-step synthesis eleven molecules (three ethylenediamine = "en", six formaldehyde, and two ammonia molecules) react with each other to form one single compound in a reported yield of 95%. It is obvious that such a reaction is dictated by the organizing power of the metal ion (I.I. Creaser, 1977).

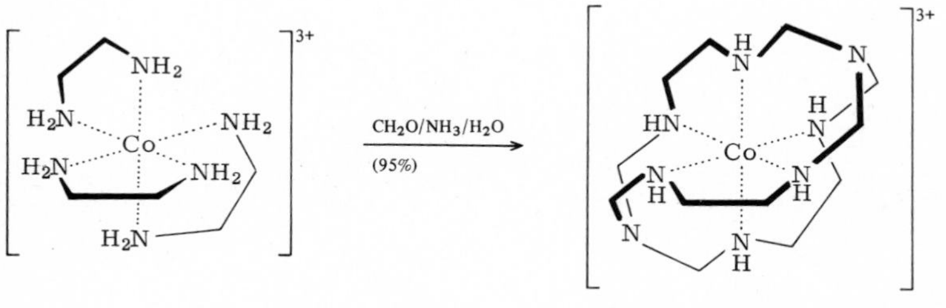

clathrochelate, cryptate

A common disadvantage of many template reactions is that it is often difficult to remove the metal ion. Such syntheses are therefore *in situ* syntheses of metal complexes and can only occasionally be used for the synthesis of the metal-free ligands.

4.1.3.3 Stepwise Condensations

Synthesis of large heterocycles usually involves condensation reactions of two difunctional molecules. Such molecules tend to polymerize. So far two special techniques have been described above to avoid this important "side-reaction", namely high dilution and use of templates. The general procedure to avoid polymerizations in reactions between difunctional molecules is, of course, the application of protecting groups as described in sections 4.1.2 and 2.6.

In 1,3-dicarbonyl compounds a special situation arises, because the prototropic ketoenols are formed. Both carbonyl groups are deactivated as acceptor synthons since neither the enol nor the α,β-conjugated ketone are very electrophilic. It is usually possible to condense one of the keto groups with amines. Acetylacetone and ethylenediamine will, for example, give bis(acetylacetone) ethylenediimine. The second keto group, however, will not react. If one wants to produce macrocyclic compounds by two nucleophilic reactions on the carbonyl groups, one has to activate them. In a successful synthesis of this kind acetylacetone was converted to its enamine and O-alkylated with Meerwein's reagent. Treatment with 1 equivalent of ethylenediamine and subsequently with 2 equivalents of sodium methoxide and a second equivalent of diamine gave the desired macrocycle in 35% yield. O-Alkylation made the carbonyl group more electrophilic, and the ketimino group is also more reactive toward nucleophiles than a keto group.

The resulting macrocyclic ligand was then metallated with nickel(II) acetate. Hydride abstraction by the strongly electrophilic trityl cation and proton elimination resulted in the formation of carbon-carbon double bonds (T.J. Truex, 1972).

4.1.3.4 The "Zip-Reaction"

A recently developed "zip-reaction" (U. Kramer, 1978, 1979) leads to giant macrocycles. Potassium 3-(aminopropyl)amide = "KAPA" ("superbase") in 1,3-diaminopropane is used to deprotonate amines. The amide anions are highly nucleophilic and may, for example, be used to transamidate carboxylic amides. If N-(39-amino-4,8,12,16,20,24,28,32,36-nonaaza-nonatriacontyl)dodecanolactam is treated with KAPA, the amino groups may be deprotonated and react with the macrocyclic lactam. The most probable reaction is the intramolecular formation of the six-membered ring intermediate indicated below. This intermediate opens spontaneously to produce the azalactam with seventeen atoms in the cycle. This reaction is repeated nine times in the presence of excess KAPA, and the 53-membered macrocycle is formed in reasonable yield.

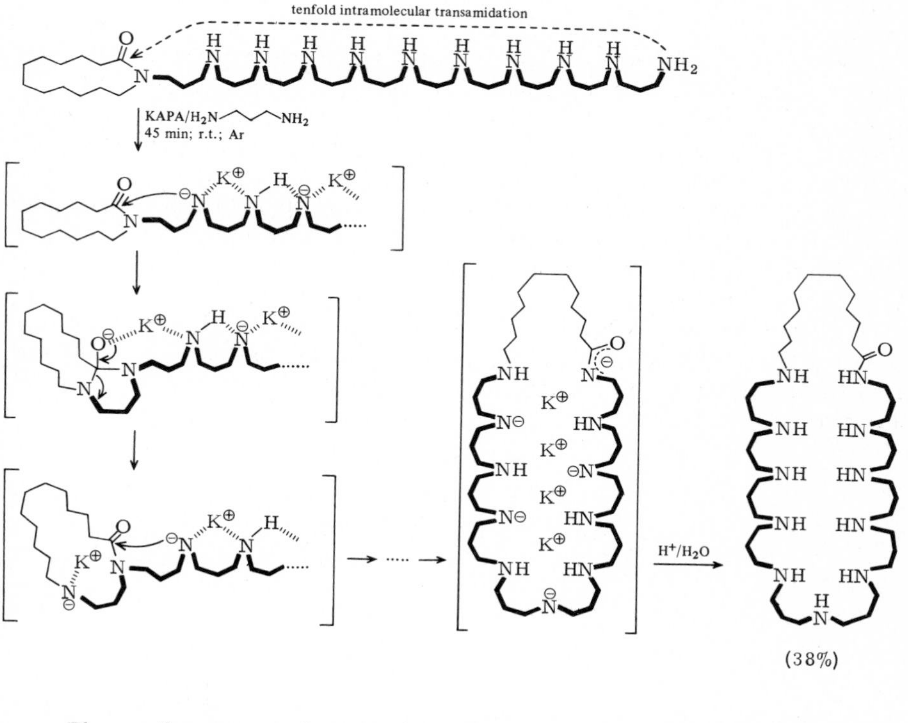

(38%)

The reaction sequence is successful because reverse, ring-contraction reactions are unlikely and because only the final product contains a secondary lactam group, which is deprotonated under the reaction conditions.

4.2 Porphyrins, Chlorophyll a, and Corrins

Porphyrins and chlorophylls are the most widespread natural pigments. They are associated with the energy-converting processes of respiration and photosynthesis in living organisms, and the synthesis of specific porphyrin derivatives is often motivated by the desire to perform similar processes in the test tube. The structurally and biosynthetically related corrins (e.g. vitamin B$_{12}$) catalyze alkylations and rearrangements of carbon skeletons via organocobalt intermediates. The biosyntheses of these chromophores are also of topical interest.

4.2.1 Porphyrins and Porphyrinogens

If one heats acetone and pyrrole in the presence of catalytic amounts of acid, so-called "acetone pyrrole" is formed in over 80% yield. This colorless, macrocyclic compound contains four pyrrole units which are connected by dimethylmethylene bridges. It is formed by electrophilic α-substitution of pyrrole by acetone, acid-catalyzed oligomerization, and spontaneous, non-template cyclization wherein four pyrrole units are combined. The reason for internal reaction instead of chain elongations with more acetone and pyrrole units is a purely statistical one: the intramolecular reaction is more probable (D. Mauzerall, 1960; J.-H. Fuhrhop, 1974). No dilution technique is needed in this synthesis presumably because the alkylated pyrrole unit at the chain end reacts faster than the unsubstituted pyrrole.

[HCl, MesOH, or TosOH]
acetone; 1-2 d; r. t.

"acetonepyrrole"
($\approx$ 90%)

Acetone pyrrole belongs to the class of tetrapyrrolic macrocycles in which the pyrrole units are connected by sp^3-hybridized carbon bridges. Such compounds are called porphyrinogens. Since the methyl groups cannot be removed, acetone pyrrole is stable. Similar porphyrinogens are formed, if the α-aminomethyl substituted pyrrole porphobilinogen is dissolved under anaerobic conditions in slightly acidic water. Since this pyrrole, which is the precursor of almost all tetrapyrrole pigments in nature, contains two different β-pyrrolic substituents a mixture of four porphyrinogen isomers is formed, which differ in the arrangement of the β-pyrrolic substituents. If this mixture is treated with an oxidant, e.g. oxygen, iodine, or a high-potential quinone, the methylene bridges are oxidized to form methine bridges, and red porphyrins are obtained in almost quantitative yield. Since porphyrinogens are oxidized in air and tend to isomerize in protic solutions, they are usually prepared from the corresponding porphyrins by metal reduction (Zn/AcOH) or catalytic hydrogenation immediately before use, e.g. in biosynthetic studies. Pure uroporphyrinogen I can be obtained anaerobically from porphobilinogen by the action of a deaminating enzyme (uroporphyrinogen I synthetase, K.M. Smith, 1975; A.R. Battersby, 1979).

porphobilinogen   uroporphyrinogens I-IV   uroporphyrins I-IV

.*pH 0:   1 M HCl; 0.5h; 98°C ⟶ statistical ratio = 1:1:4:2;   (78%)
 pH 7.6: phosphate buffer; 21h; 60°C ⟶ mainly I + III;   (55%)
 pH 10: 0.001 M NaOH; 2h; 98°C ⟶ selectively I;   (70%)

A = −CH₂−COOH
P = −CH₂−CH₂−COOH

A = $-CH_2-COOH$
P = $-CH_2-CH_2-COOH$

Because of the occurrence of isomers, the synthesis from pyrroles is only useful for porphyrins with eight identical β-pyrrolic and four identical methine bridge substituents. Famous examples are the syntheses of chloroform-soluble β-octaethyl-porphyrin* and *meso*-tetraphenylporphyrin* which have been used in innumerable studies on porphyrin reactivity (K.M. Smith, 1975). Porphyrins with four long *meso*-alkyl side-chains can be obtained by use of analogous reactions. These porphyrins have melting points below 100 °C and are readily soluble in petroleum ether. Sulfonation of olefinic double bonds leads to highly charged, water soluble porphyrins (J.-H. Fuhrhop, 1976).

---

* ''β'' means β-pyrrolic, *meso* indicates the methine bridge.

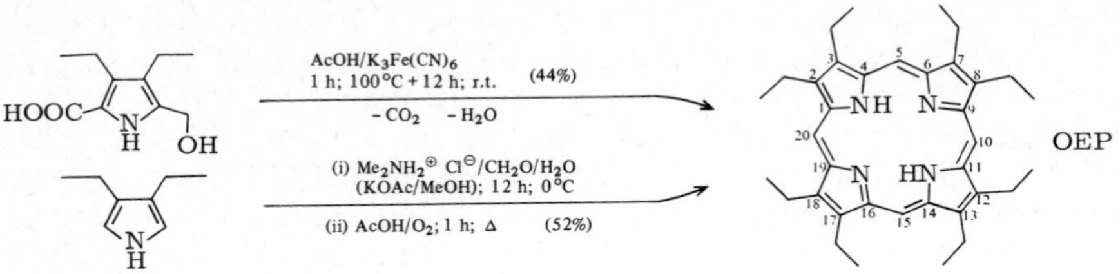

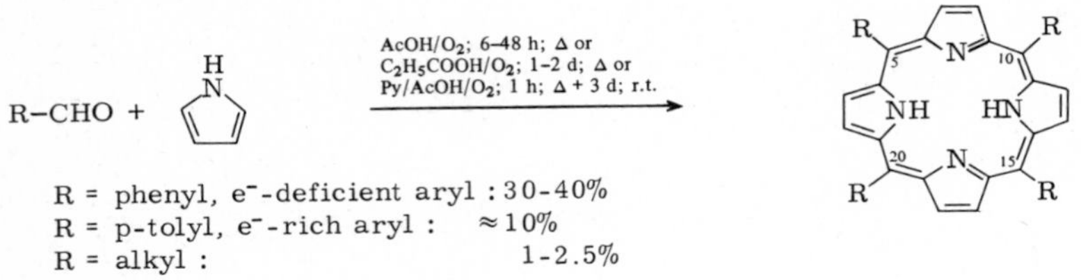

R = phenyl, e⁻-deficient aryl : 30-40%
R = p-tolyl, e⁻-rich aryl :   ≈10%
R = alkyl :   1-2.5%

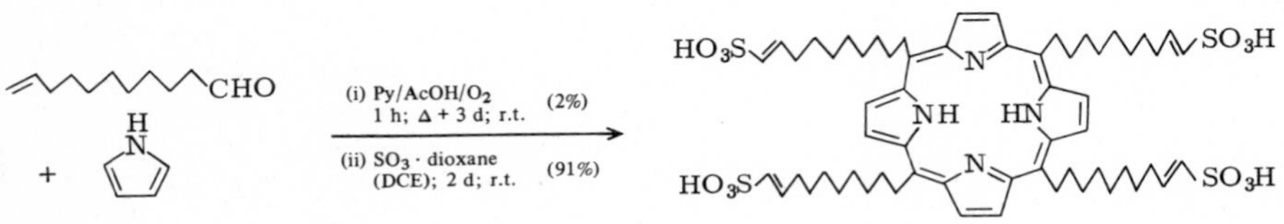

This apparently extremely simple synthesis of symmetrical porphyrins from aldehydes and pyrrole has also been used to produce porphyrins with interesting stereochemical properties. *ortho*-Nitrobenzaldehyde yielded the corresponding tetrakis(nitrophenyl) porphyrin which was reduced to the tetraamine and converted into a tetraamide using pivaloyl chloride. Since the porphyrin macrocycle tends to retain a planar conformation, and because of strong sterical interactions between β-pyrrolic and methine bridge substituents the planes of the phenyl groups are forced into a position approximately perpendicular to the porphyrin plane. This minimizes interactions between β-pyrrolic H-atoms and phenyl substituents. If the phenyl substituents are large, e.g. pivaloylamide in the *ortho*-position, rotation of the phenyl ring is strongly impeded and rotational isomers are stable. The *all-cis* isomer is the famous "picket fence" porphyrin (J.P. Collman, 1975B, 1977). Another, so-called "capped" porphyrin with substituents on one side of the porphyrin plane was obtained in 2% yield from a tetraaldehyde and pyrrole (J. Almog, 1975).

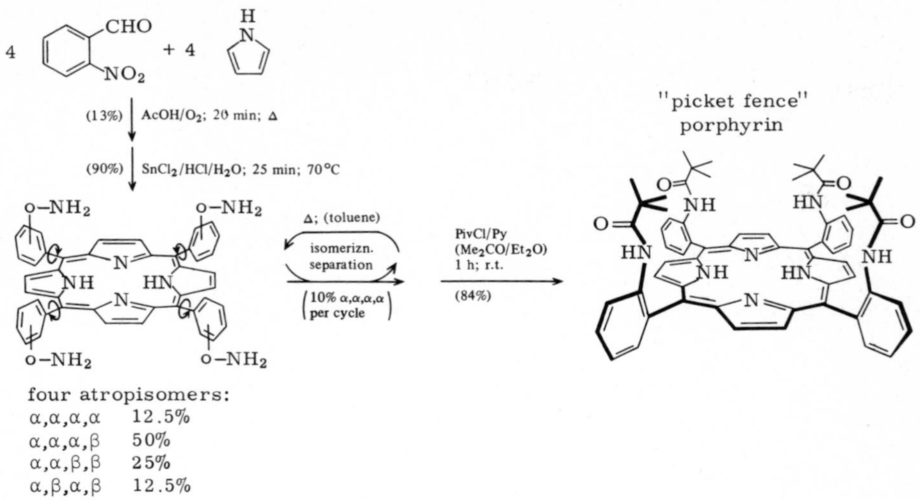

"picket fence" porphyrin

four atropisomers:
α,α,α,α   12.5%
α,α,α,β   50%
α,α,β,β   25%
α,β,α,β   12.5%

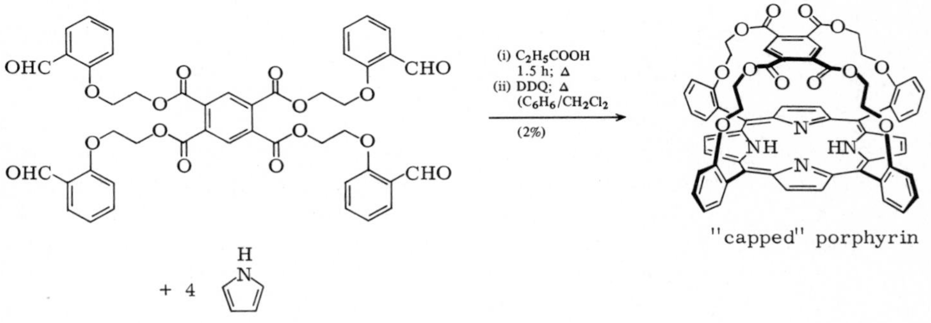

"capped" porphyrin

Naturally occurring porphyrins are usually symmetrically substituted about the 15-methine bridge. These porphyrins can be synthesized by the condensation of two dipyrrolic intermediates. Typical dipyrrolic intermediates in current use are the dipyrromethanes and the dipyrromethenes. Both methods will shortly be described. This again is a highly specialized topic, discussed here mainly because general problems in the synthesis of complex aromatic molecules and the influence of π-electron distribution in the educts on reactivity can be exemplified.

Unsymmetrically substituted dipyrromethanes are obtained from α-unsubstitued pyrroles and α-(bromomethyl)pyrroles in hot acetic acid within a few minutes. These reaction conditions are relatively mild and the α-unsubstituted pyrrole may even bear an electron withdrawing carboxylic ester function. It is still sufficiently nucleophilic to substitute bromine or acetoxy groups on an α-pyrrolic methyl group. Hetero atoms in this position are extremely reactive leaving groups since the α-pyrrolylmethenium(= "azafulvenium") cation formed as an intermediate is highly resonance-stabilized.

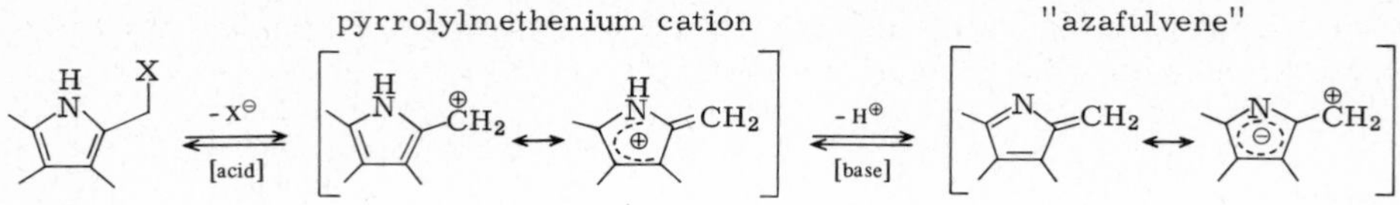

pyrrolylmethenium cation   "azafulvene"

X = NH₂, OH, OAc, OTos, halogen, etc.

$X = NH_2, OH, OAc, OTos, halogen, etc.$

(NaOAc/AcOH)
1.5 h; Δ; N₂

(72%)

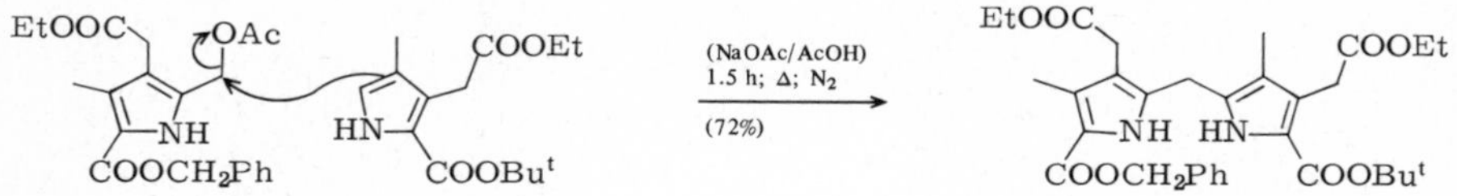

Dipyrromethanes are inherently unstable toward "jumbling" or "redistribution" in the presence of acidic reagents. The dipyrromethanes are easily protonated at α-pyrrolic positions, and rearrangements occur through the relatively long-lived pyrrolylmethenium ions. This process is not only found with dipyrromethanes but is equally pronounced in porphyrinogens (see p. 227) and open-chain tetrapyrrole pigments with methylene bridges (A.H. Jackson, 1973).

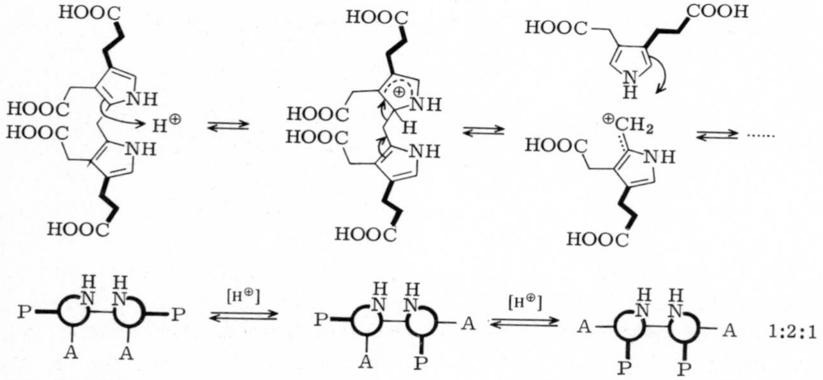

P = CH₂–CH₂–COOH
A = CH₂–COOH

1:2:1

A mild procedure which does not involve strong acids, has to be used in the synthesis of pure isomers of unsymmetrically substituted porphyrins from dipyrromethanes. The best procedure having been applied, e.g. in unequivocal syntheses of uroporphyrins II, III, and IV (see p. 227), is the condensation of 5,5′-diformyldipyrromethanes with 5,5′-unsubstituted dipyrromethanes in a very dilute solution of hydroiodic acid in acetic acid (A.H. Jackson, 1973). The electron-withdrawing formyl groups disfavor protonation of the pyrrole and therefore isomerization. The porphodimethene that is formed during short reaction times isomerizes only very slowly, since the pyrrole units are part of a dipyrromethene chromophore (see below). Furthermore, it can be oxidized immediately after its synthesis to give stable porphyrins.

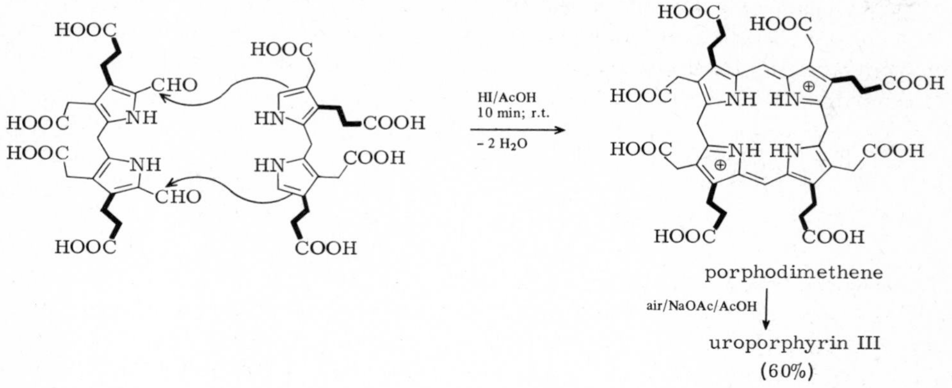

porphodimethene

uroporphyrin III
(60%)

Conjugated dipyrrolic pigments, the dipyrromethenes, are synthesized by acid-catalyzed condensation of an α-formyl pyrrole and an α-unsubstituted pyrrole. They are readily protonated and deprotonated and are difficult to purify by chromatography.

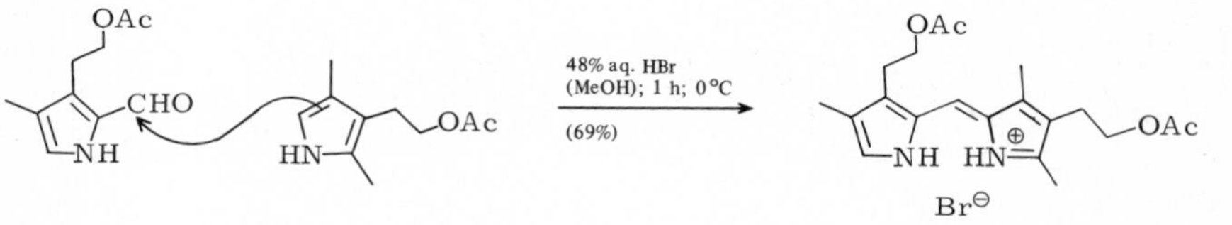

The pyridine-like nitrogen of the 2H-pyrrol-2-ylidene unit tends to withdraw electrons from the conjugated system and deactivates it in reactions with electrophiles. The acid-catalyzed condensations described above for pyrroles and dipyrromethanes therefore do not occur with dipyrromethenes. Vilsmeier formylation, for example, is only successful with pyrroles and dipyrromethanes but not with dipyrromethenes.

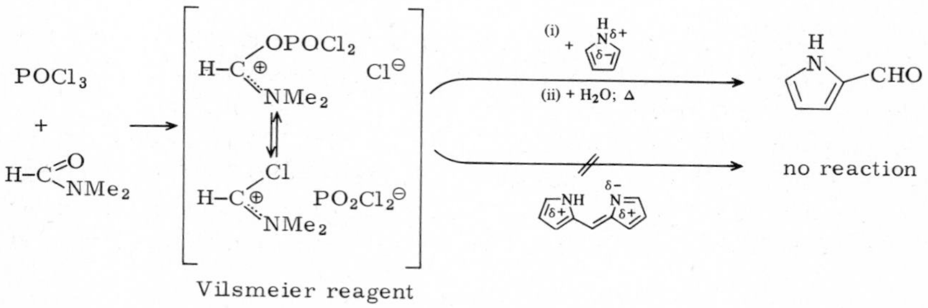

Vilsmeier reagent

α-Pyrrolic bromine substituents on dipyrromethenes, on the other hand, are reactive in nucleophilic substitutions. Under forcing conditions, i.e. fusion in a succinic acid melt, α-methyldipyrromethene cations can be deprotonated at the methyl group, and the resulting enamine develops some nucleophilic reactivity. This is the basis of H. Fischer's classic porphyrin synthesis. The formation of isomer mixtures can only be avoided, if one pyrromethene unit is symmetrically substituted about the methine carbon, or if the desired porphyrin is centrosymmetrical (H. Fischer, 1927, 1940; A. Treibs, 1971).

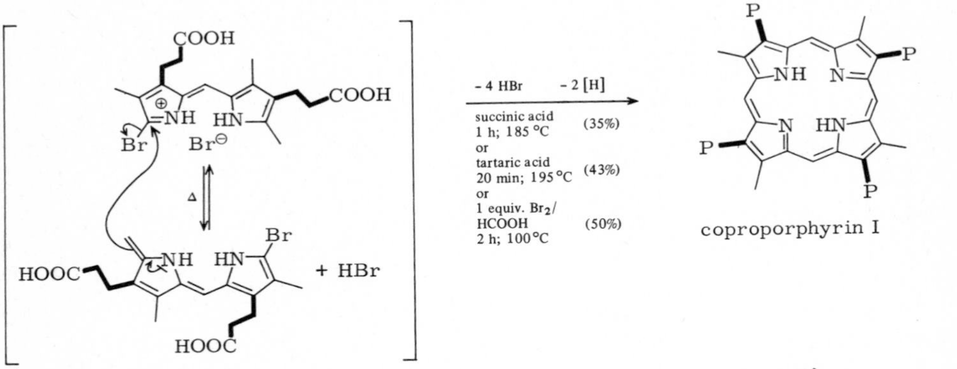

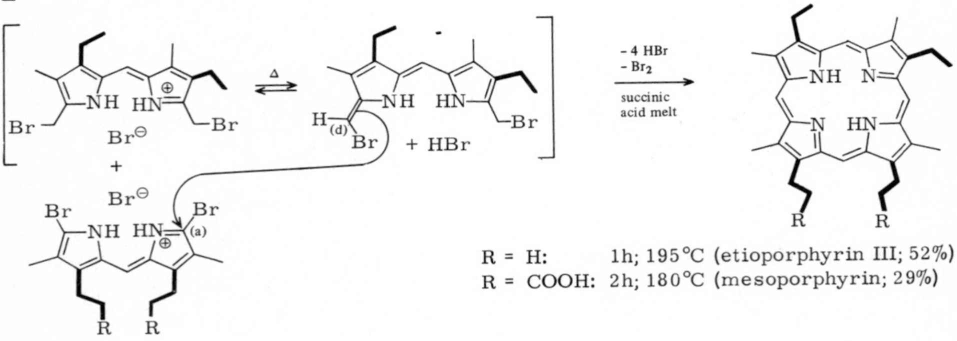

coproporphyrin I

R = H: 1h; 195 °C (etioporphyrin III; 52%)
R = COOH: 2h; 180 °C (mesoporphyrin; 29%)

This reaction sequence is much less prone to difficulties with isomerizations since the "pyridine-like" carbons of dipyrromethenes do not add protons. Yields are often low, however, since the intermediates do not survive the high temperatures. The more reactive, "faster" but "less reliable" system is certainly provided by the dipyrromethanes, in which the reactivity of the pyrrole units is comparable to activated benzene derivatives such as phenol or aniline. The situation is comparable with that found in peptide synthesis where the "slow" azide method gives cleaner products than the "fast" DCC-promoted condensations (see p. 214).

With the catalysis of strong Lewis acids, such as tin(IV) chloride, dipyrromethenes may also be alkylated. A very successful porphyrin synthesis involves 5-bromo-5'-bromomethyl and 5'-unsubstituted 5-methyl-dipyrromethenes. In the first alkylation step a tetrapyrrolic intermediate is formed which cyclizes to produce the porphyrin in DMSO in the presence of pyridine. This reaction sequence is useful for the synthesis of completely unsymmetrical porphyrins (K.M. Smith, 1975).

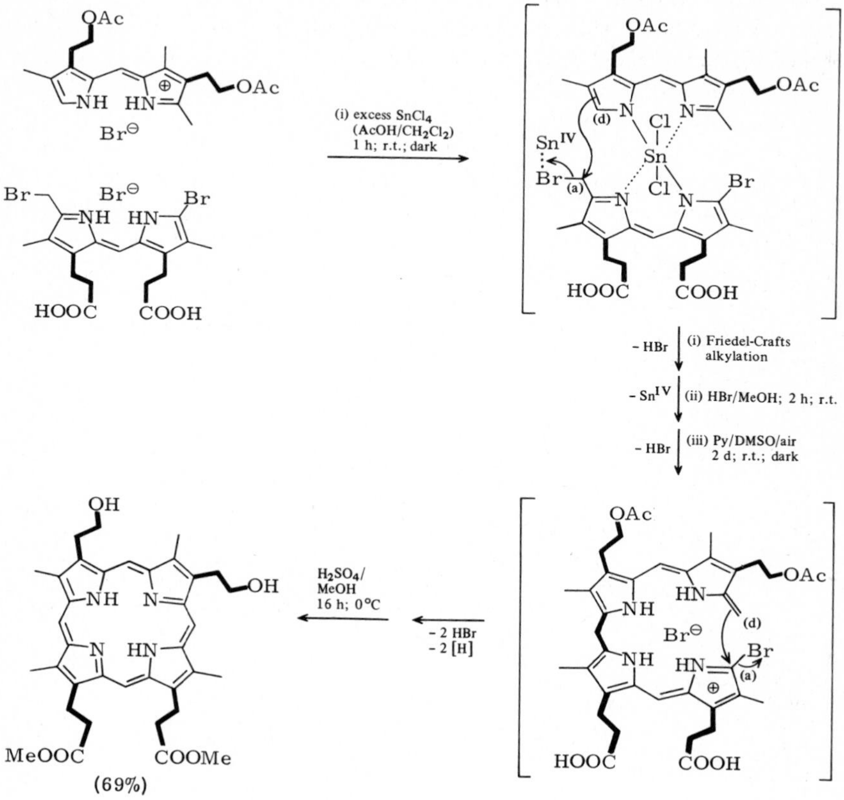

4.2.2 Chlorophyll *a* — Comprehension of Structural Features and Synthesis

Chlorophyll *a* (L.P. Vernon, 1966) contains an unsymmetrical porphyrin chromophore with two special features: the double bond between C-17 and C-18 is hydrogenated and carbon atoms 13 and 15 bear a carboxylated, "isocyclic" cyclopentanone ring E.

chlorophyll a : R = CH₃
chlorophyll b : R = CHO

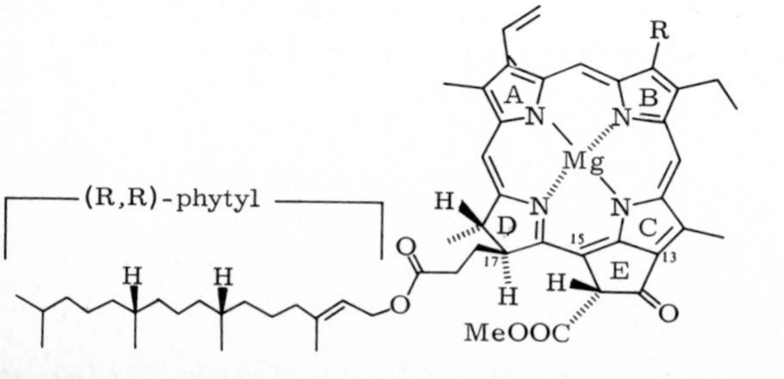

Since several high-yield porphyrin syntheses are known, chlorophyll *a* syntheses start with the preparation of an appropriately substituted porphyrin. The following porphyrin with an acetic acid side-chain on C-15 and a 13-methoxycarbonyl group has been used to solve the major problem, namely the regioselective hydrogenation of ring D at C-17/C-18. This reaction should be simplified by the effect of interactions between the 15-methine bridge and the β-pyrrolic 13,17-substituents, i.e. the distortion of the plane of the aromatic porphyrin ring. Pyrrole rings C and D have to be twisted out of the plane. This "sterical strain" could be released, if one of the three carbon atoms 13, 15, or 17 became sp³-hybridized. Its substituents would then be above and below the plane, and the chromophoric π-system could again be planar. This is exactly the situation found in chlorophyll, and R.B. Woodward (1960) founded his synthetic plan on this fact. It was hoped that the porphyrin below could be hydrogenated to the dihydroporphyrin (= chlorin) corresponding to chlorophyll *a*. Ring D should react much faster than ring C, because in the latter the double bond is deactivated by the carboxylic ester group. It turned out, however, and this is nowadays a very well established fact in porphyrin chemistry, that the methine bridges react much faster in all types of reactions than the aromatic pyrrole rings (J.-H. Fuhrhop, 1974; see also p. 227). On hydrogenation under various conditions one could only obtain the "phlorin" with a reduced carbon bridge but no "chlorin" with a reduced pyrrole ring. This failure led to a new concept on the same basis of "stereochemical strain release". Instead of the acetic acid, an acrylic acid side-chain was introduced at C-15. An acid-catalyzed intramolecular reaction between the olefinic double bond and the aromatic pyrrole unit D led to a methoxycarbonylated cyclopentene ring and a reduced ring D. The proton on C-18 is probably derived from an acid-catalyzed tautomerization of the acidic proton adjacent to the methoxycarbonyl group (R.B. Woodward, 1960).

a phlorin

AcOH; 30 h; 110°C; N₂ (repeated isomerizations)

(70%)
a chlorin

The C=C double bond in the cyclopentene ring can be cleaved by the osmium tetroxide-periodate procedure or by photooxygenation. The methoxalyl group on C-17 can, as a typical α-dicarbonyl system, be split off with strong base and is replaced by a proton. Since this elimination occurs with retention of the most stable configuration of the cyclization equilibrium, the substituents at C-17 and C-18 are located *trans* to one another. The critical introduction of both hydrogens was thus achieved regio- and stereoselectively.

In principle, the direct hydride addition or catalytic hydrogenation, which did not give chlorins, was replaced by an electrocyclic intramolecular addition which is much easier with the above system. Complete regioselectivity was also achieved since electrocyclization did not occur with the resonance-stabilized ring C.

This case history presents only a simple account of one of R.B. Woodward's adventures based on ingenious understanding of structural features and experimental findings described in the literature. The hydrogenation of porphyrins is still one of the most active subjects in heterocyclic natural products chemistry, and the interested reader may find some modern developments in the publications of A. Eschenmoser (C. Angst, 1980; J.E. Johansen, 1980).

4.2.3 Corrins

"Corrin" is the "porphyrinoid" chromophore of the vitamin B_{12} parent compound cobyrinic acid. Corrin itself has not yet been synthesized, but routes to cobyrinic acid and several other synthetic corrins have been described by A. Eschenmoser (1970, 1974) and R.B. Woodward (1967).

corrin

corrole
= decadehydro-
corrin

cobyrinic acid

It is conceivable that related ligands, e.g. dehydrocorrins, could be obtained from pyrrolic units using pathways similar to those used for porphyrins and could be hydrogenated to corrins. This has indeed been achieved (I.D. Dicker, 1971), but it is, of course, impossible to introduce the nine chiral centres of cobyrinic acid by such procedures.

The most straightforward synthesis of the corrin nucleus would be one in which the pyrroline rings are preformed acceptor and donor synthons similar to the α-substituted and α-unsubstituted pyrrole rings in porphyrin synthesis. The double-bonds of the corrin ring should be present in the condensation product since subsequent oxidation of a saturated system to the conjugated tetraazapolyene system of the corrins is difficult: corrins are easily destroyed by oxidants. The donor could be an enamine d^2-synthon in which the nitrogen atom belongs to the pyrrolidine ring and the double bond corresponds to an exocyclic methylene group, which would become the methine bridge. The choice of the a^1-acceptor synthon is more difficult. A good leaving group on a C=N-bond is provided, for example, by alkoxy groups. Since imidic esters can be synthesized from amides with Meerwein's trialkyloxonium reagents they were the first choice of A. Eschenmoser (1970). The condensation of imidic esters with enamines yields diimines connected via a methylene bridge. In order to facilitate isomerization to the desired α,β-unsaturated imines and the deprotonation of the enamine which raises its nucleophilicity a nitrile group was introduced.

Later it turned out that activation of enamine components could not only be achieved by deprotonation of the nitrogen atom but also by connecting it with certain metals, e.g. Ni(II), Pd(II), or Co(II), and subsequent treatment with base.

The scheme below shows how the "eastern" and "western" parts of a corrin chromophore can be combined regioselectively. The western part has a more acidic enamine than the eastern part, whereas the imidic ester of the eastern part is more electrophilic.

Corrins bearing additional alkyl substituents next to the critical methylene reaction centres could not be synthesized by the approach described above. Eschenmoser therefore applied his own principle (p. 34) and replaced the intermolecular coupling step by an intramolecular reaction. The lactam group of one condensation partner was converted into the thiolactim (via imidic ester; see second last reaction in the scheme below). The nucleophilic sulfur atom could be either condensed with a bromomethyl group adjacent to an imine or converted itself into an acceptor group: in the presence of hydrogen chloride it was dimerized oxidatively with dibenzoylperoxide to give the disulfide. This disulfide reacted spontaneously with nucleophilic enamides. The "sulfide bridge" was converted into the desired "methine bridge" by sulfide contraction (see p. 56f).

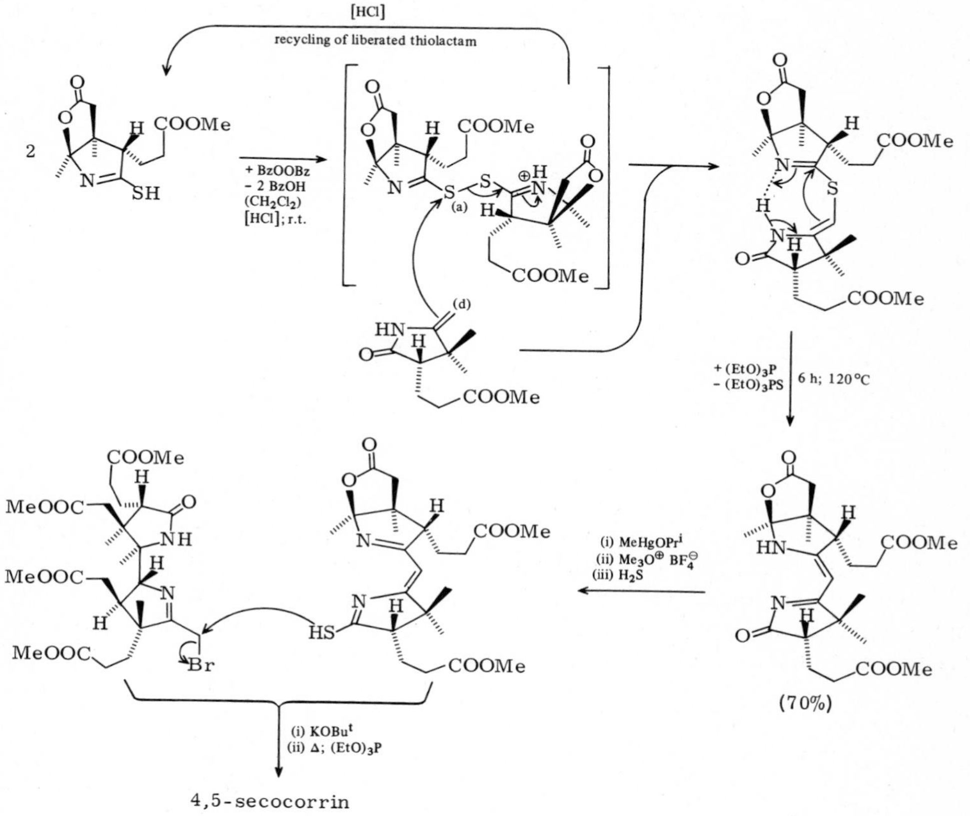

4,5-secocorrin

The direct connection of rings A and D at C-1 cannot be achieved by enamine or sulfide couplings. This reaction has been carried out in almost quantitative yield by electrocyclic reactions of A/D-secocorrinoid metal complexes and constitutes a magnificent application of the Woodward-Hoffmann rules. First an *antarafacial* hydrogen shift from C-19 to C-1' is induced by light (sigmatropic 18-electron rearrangement), and second, a *conrotatory* thermally allowed cyclization of the mesoionic 16 π-electron intermediate occurs. Only the 1,19-*trans*-isomer is formed (A. Eschenmoser, 1974; A. Pfaltz, 1977).

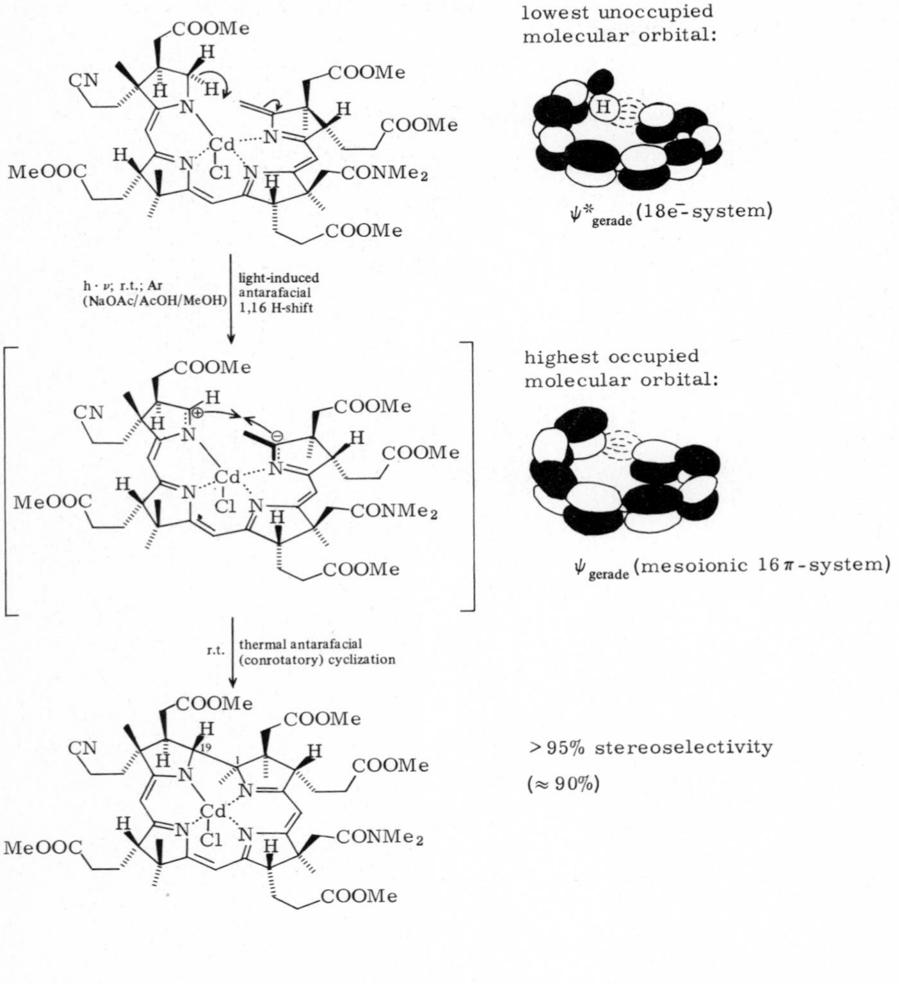

lowest unoccupied
molecular orbital:

ψ^*_{gerade} (18e⁻-system)

h·ν; r.t.; Ar
(NaOAc/AcOH/MeOH) | light-induced antarafacial 1,16 H-shift

highest occupied
molecular orbital:

ψ_{gerade} (mesoionic 16 π-system)

r.t. | thermal antarafacial (conrotatory) cyclization

> 95% stereoselectivity

(≈ 90%)

4.3 Natural Product Synthesis from Carbohydrates

Some carbohydrates are inexpensive and provide a great variety of functional and stereo-chemical features. Faithful to our philosophy that low-cost starting materials should be utilized, we shall discuss some reactions of glucose and recent syntheses with glucose as starting material (S. Hanessian, 1977, 1979). Structures of glucose derivatives and other commercially available* carbohydrates are given below.

Table 19. Inexpensive sugars and sugar derivatives.

Reduced sugar derivatives	Pentoses	Glucose derivatives	Other hexoses
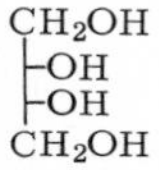 CH$_2$OH $\vert$OH $\vert$OH CH$_2$OH Erythritol; [5]	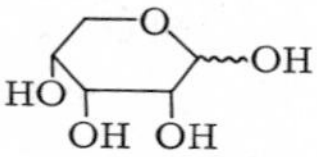 D-(–)-Ribose; [9]	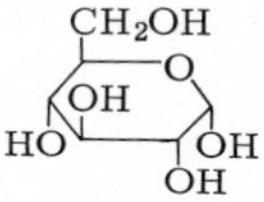 α-D-Glucose; [0] Penta-O-acetyl; [2] α-Methyl D-glucoside; [1]	Aldohexoses 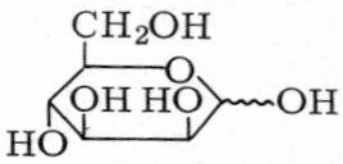 D-(+)-Mannose; [5] α-Metnyl D-mannoside; [5]
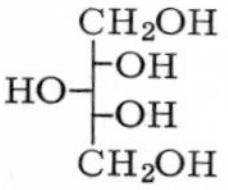 CH$_2$OH HO$\vert$OH $\vert$OH CH$_2$OH Xylitol; [1]	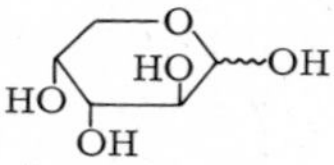 D-(–)-Arabinose; [5]	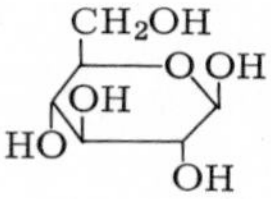 β-D-Glucose; [4] Penta-O-acetyl; [2]	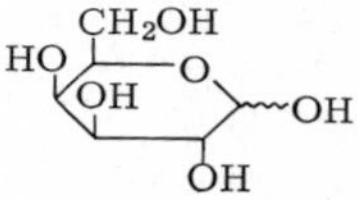 D-(+)-Galactose; [1] Penta-O-acetyl; [7]
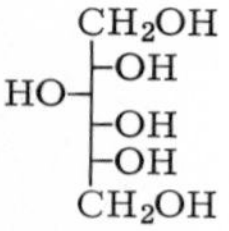 CH$_2$OH $\vert$OH HO$\vert$OH $\vert$OH CH$_2$OH D-Glucitol; [0] –, Hexa-O-acetyl; [6]	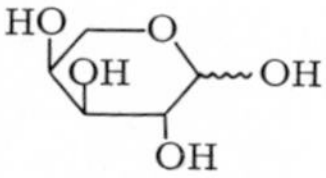 L-(+)-Arabinose; [3]	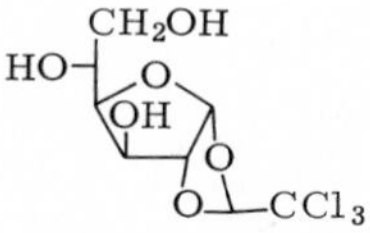 α-Chloralose; [2]	Deoxyaldohexose 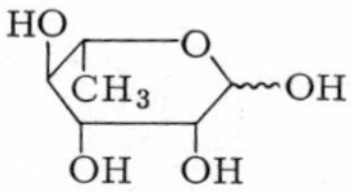 L-(+)-Rhamnose; [11]
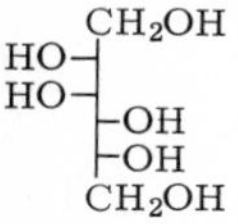 CH$_2$OH HO$\vert$ HO$\vert$ $\vert$OH $\vert$OH CH$_2$OH D-Mannitol; [0]	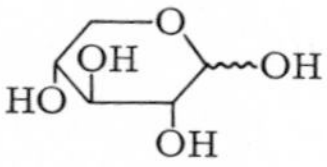 D-(+)-Xylose; [1]	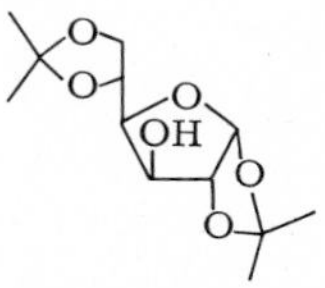 1,2:5,6-Di-O--isopropylidene--glucofuranose; [7]	Ketohexoses 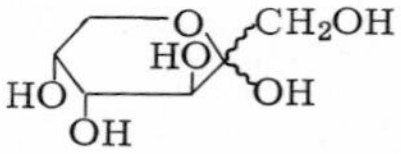 D-(–)-Fructose; [0]
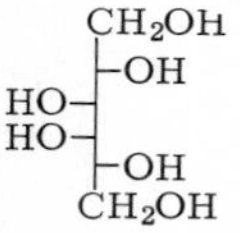 CH$_2$OH $\vert$OH HO$\vert$ HO$\vert$ $\vert$OH CH$_2$OH Galactitol; [2]		CH$_2$OH α-D-Glucosamine, hydrochloride; [2]	L-(–)-Sorbose; [1]
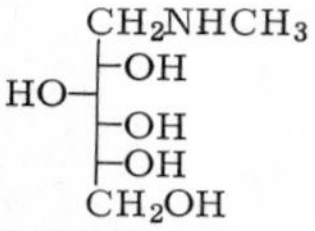 CH$_2$NHCH$_3$ HO$\vert$OH $\vert$OH $\vert$OH CH$_2$OH N-Methyl--D-glucamine; [1]		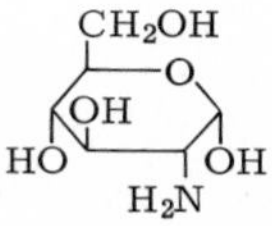	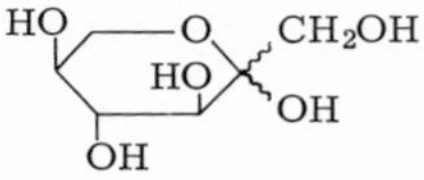

---

* The numbers in brackets indicate the approximate prices in German marks per 10 g.

Table 19. (Continued)

Monosaccharide aldonic acids	Other oxidized monosaccharides	Oligosaccharide derivatives
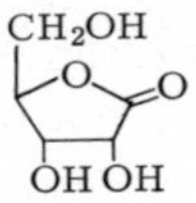 D-Ribonic acid γ-lactone [5]	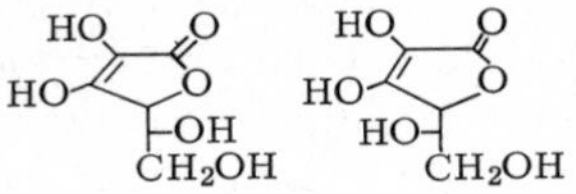 D-Araboascorbic acid = isoascorbic acid [1] L-Ascorbic acid [1]	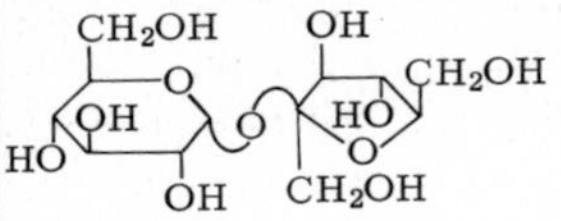 Sucrose [0] Octa-O-acetyl: [0]
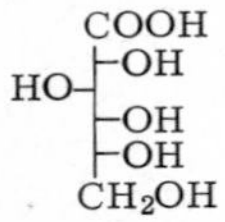 D-Gluconic acid Na salt: [0]	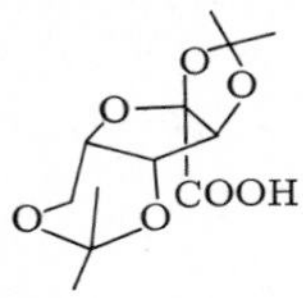 2-Keto-L-gulonic acid diacetonide [2]	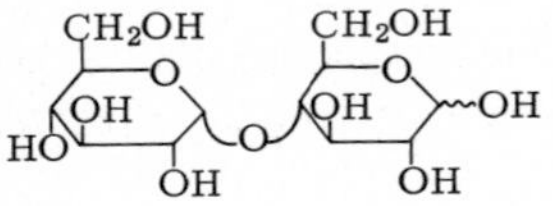 D-(+)-Maltose [1]
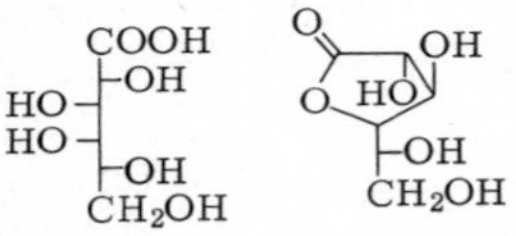 D-Galactonic acid Ca salt: [2] γ-Lactone: [9]	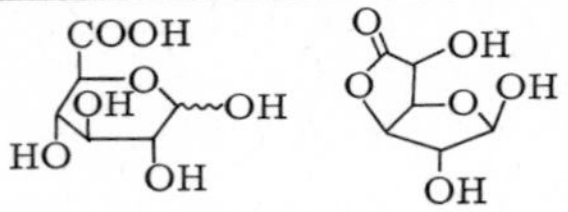 D-Glucuronic acid Na salt: [4] γ-Lactone: [2] D-Galacturonic acid [10]	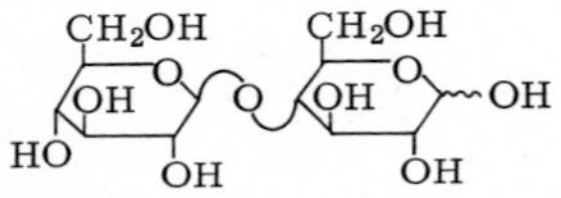 D-(+)-Cellobiose [10] Octa-O-acetyl: [8]
D-Gulonic acid γ-Lactone [7]	D-Glucaric acid = saccharic acid Hemi-K salt: [2]	Lactose [0] Lactobionic acid [7] Ca salt: [3]
D-Glucoheptonic acid Ca salt: [1] γ-Lactone: [2]	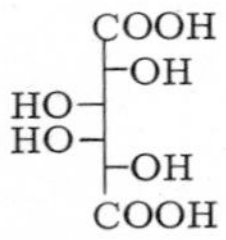 Galactaric acid = mucic acid [2]	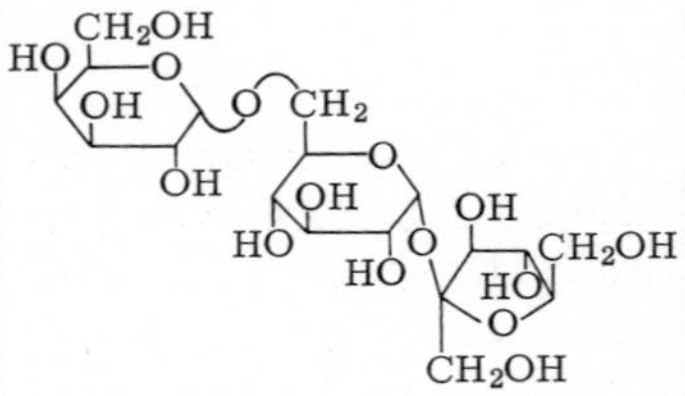 D-(+)-Raffinose [5] Undeca-O-acetyl: [2]

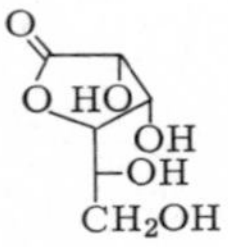

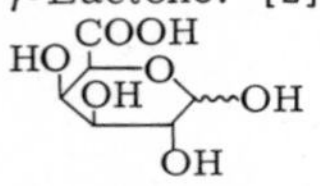

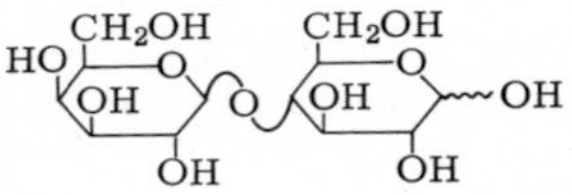

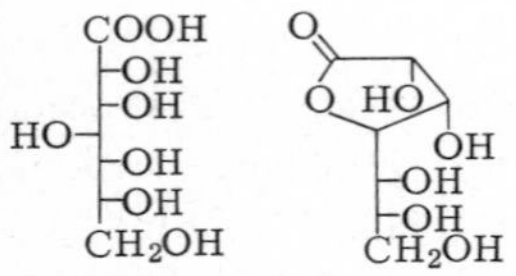

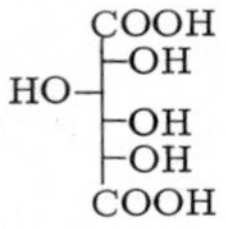

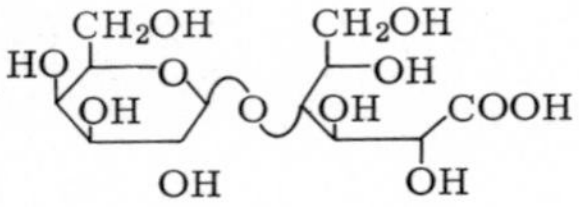

The hydroxyl groups of glucose (and, of course, other saccharides) must be regio- and stereoselectively attacked, if this most abundant natural carbon compound is to be used as starting material. We shall first show with a few selected examples, how this can be achieved (A.H. Haines, 1976; J. Lehmann, 1976; L. Hough, 1979).

The hemiacetal hydroxyl group on C-1 reacts selectively with alcohols in the presence of catalytic amounts of hydrochloric acid. The resulting acetal is called a glycoside or, in the case of glucose, a glucoside. These glycosides are relatively stable toward bases and nucleophiles. The same hydroxyl group, however, can also be activated selectively to nucleophiles. If glucose is peracetylated, only the C-1 acetate group is substituted by bromide anions. Usually only the more stable α-bromoglucose is formed. The corresponding β-chloro compound can be obtained by treatment of the β-acetate with hydrogen chloride and phosphorus trichloride. The primary alcohol group on C-6 has been converted to halides in high yields with carbon tetrahalides and triphenylphosphine. It is in general quite easy to protect or activate selectively C-1 or C-6 hydroxyl groups in aldohexoses.

In aqueous solution glucose occurs only ($> 99\%$) as six-membered ring hemiacetal (pyranose). With acetone and catalytic amounts of H_2SO_4 a five-membered ring derivative ("furanoside") is the major product. Septanose forms have also been isolated from the reaction mixture (J.D. Stevens, 1972). A general rule in carbohydrate chemistry says, that acetone tends to form five-membered 1,3-dioxolane rings ("isopropylidene derivatives") and reacts only with *cis*-1,2-diol groupings. Therefore, even though the most stable form of the hemiacetal of glucose itself is a pyranose, the reaction with acetone leads to a furanose derivative, because it provides two *cis*-diol groupings instead of one.

In 1,2:5,6-di-O-isopropylidene-α-D-glucofuranose five of the six functional groups of glucose are protected. The free 3-hydroxyl group is a popular starting point in synthesis. It can be oxidized, e.g. with RuO₄ (D.C. Baker, 1976), tosylated, substituted etc. Examples are the conversions of 3-keto groups of sugar derivatives to an α-hydroxy aldehyde (H. Paulsen, 1972, 1977) or to an acetic acid side-chain (G.J. Lourens, 1975).

The exocyclic 1,3-dioxolane ring is much more vulnerable to acid hydrolysis than the ring connected with the acetal group. Partial deprotection of the side-chain is easily achieved by treatment with sulfuric acid.

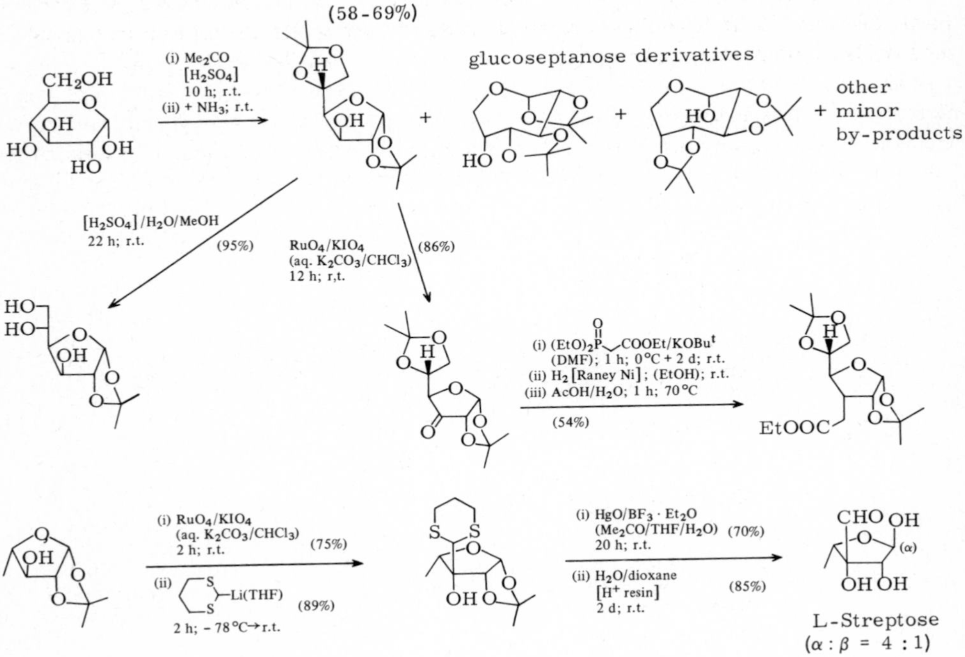

When the 3-hydroxyl group has been converted to the desired functional group, e.g. to an amino group, the diacetonide is hydrolyzed, and the pyranose is recovered. The example synthesis of a steroid glycoside below indicates a typical pattern of protection, deprotection, and activation steps. The acetylated glycosidic hydroxyl group can be quantitatively exchanged with HBr, whereas the other acetyl groups remain intact (W. Meyer zu Reckendorf, 1971).

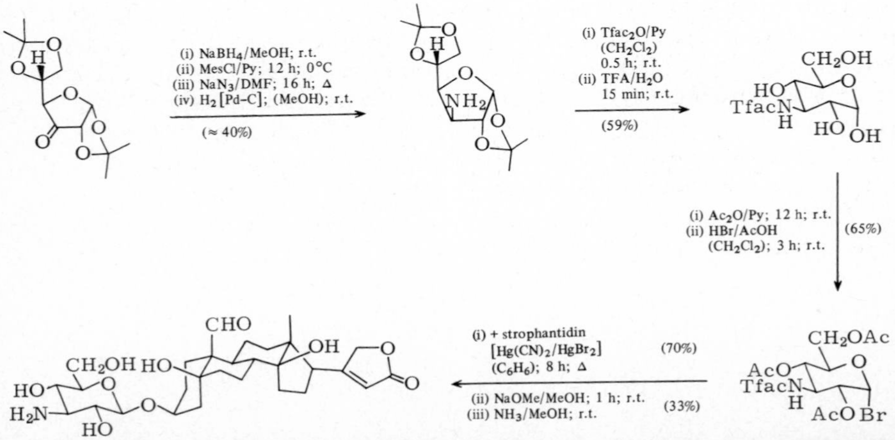

1,2-Dideoxy 1-enopyranoses (''glycals'') are readily prepared from the corresponding glycosyl bromides by reduction with zinc in acetic acid (E. Fischer, 1914; B. Helferich, 1952). 2-Hydroxyglucal tetraacetate was obtained from tetra-O-acetylglucosyl bromide by a base catalyzed elimination of HBr (R.U. Lemieux, 1965) or via the corresponding iodide (R.J. Ferrier, 1966). In general, deoxy sugars can be synthesized from the corresponding tosylates or nitrobenzene-*p*-sulfonates by an iodide exchange followed by a reduction step. The addition of nitrosyl chloride to glycals gives 2-nitroso pyranosyl chlorides with high selectivity (R.U. Lemieux, 1968). The nitroso compounds may dimerize. Reduction with zinc in acetic acid yields 2-amino-2-deoxypyranoses.

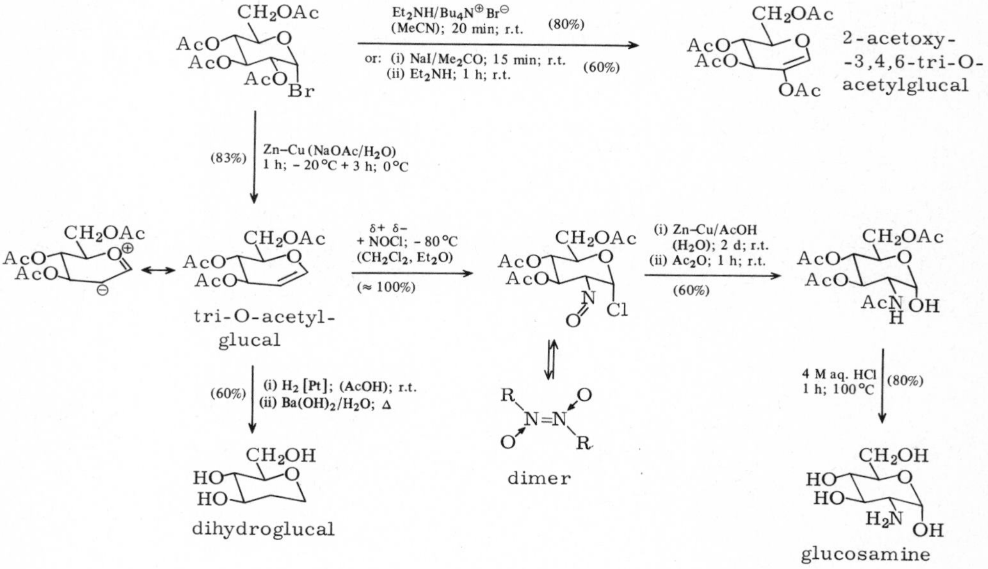

Another common carbonyl compound for the protection of two sugar hydroxyl groups is benzaldehyde. It gives six-membered acetal rings (1,3-dioxanes) but *no bridged* bicyclic compounds. With glucose or methyl glucoside the 4,6-O-benzylidene derivatives can be obtained (H.G. Fletcher, 1963, M.E. Evans, 1980). Benzylidene protecting groups can be removed by catalytic hydrogenation.

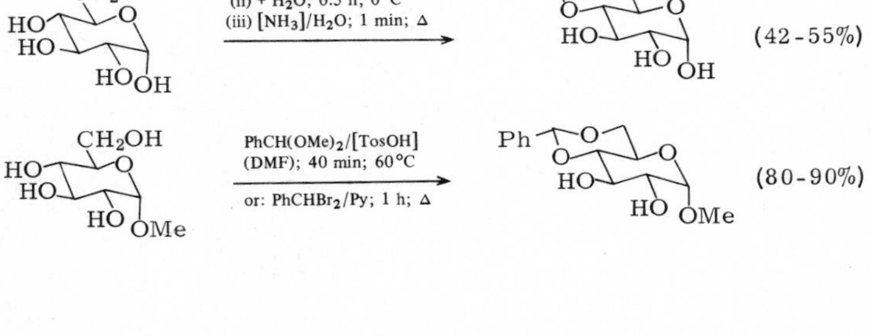

The benzylidene derivative above is used, if both hydroxyl groups on C-2 and C-3 are needed in synthesis. This *trans*-2,3-diol can be converted to the sterically more hindered α-epoxide by tosylation of both hydroxy groups and subsequent treatment with base (N.R. Williams, 1970; J.G. Buchanan, 1976). An oxide anion is formed and displaces the sulfonyloxy group by a rearside attack. The oxirane may then be re-opened with nucleophiles, e.g. methyl lithium, and the less hindered carbon atom will react selectively. In the following sequence starting with an α-glucoside only the 2-methyl-2-deoxyaltrose is obtained (S. Hanessian, 1977).

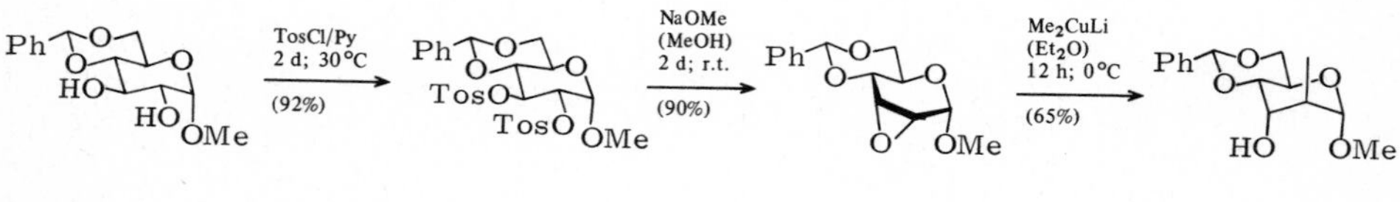

With the background of the selected glucose reactions we can now discuss syntheses starting with this compound.

The synthesis of 11-oxaprostaglandins from D-glucose uses the typical reactions of glucofuranose diacetonide outlined on p. 242. Reduction of the hemiacetal group is achieved via a thioacetal. The carbon chains are introduced by Wittig reactions on the aldehyde groups, which are liberated by periodate oxidation and lactone reduction (S. Hanessian, 1979; G.J. Lourens, 1975).

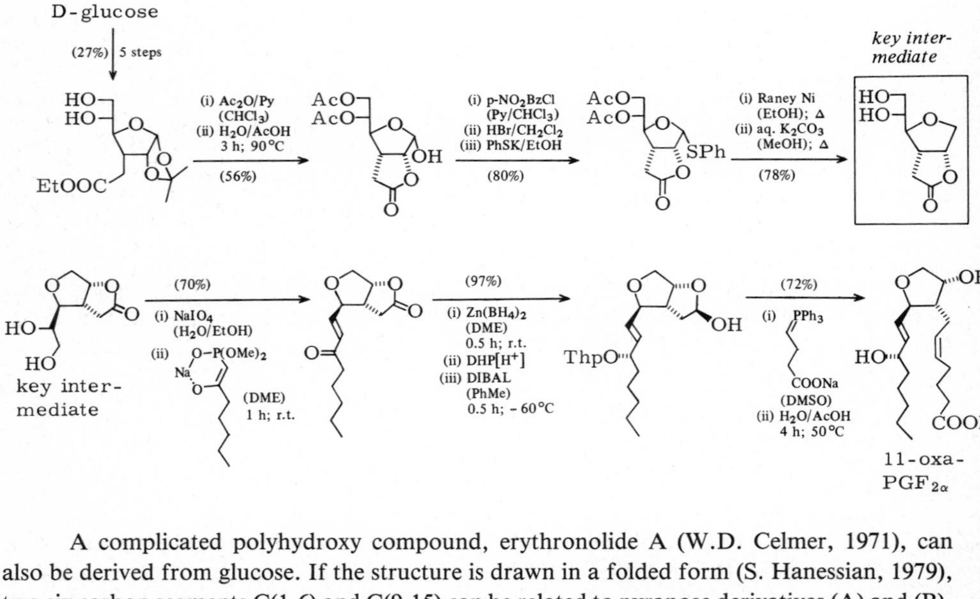

A complicated polyhydroxy compound, erythronolide A (W.D. Celmer, 1971), can also be derived from glucose. If the structure is drawn in a folded form (S. Hanessian, 1979), two six carbon segments C(1-6) and C(9-15) can be related to pyranose derivatives (A) and (B).

erythronolide A

R = protecting groups

The two precursors comprise eight of the ten chiral centres present in the erythronolide A *seco* acid. We shall describe the published syntheses of (A) and (B) again with the main objective of demonstrating some glucose chemistry. The general plan was to synthesize the ketone (C) as a key intermediate first, in which C-2 and C-3 have the functionalities required in both (A) and (B).

Synthesis of ketone (C) starts with the 2,3-epoxide of the benzylidene protected α-methyl glucoside (see p. 244). Treatment with lithium dimethylcuprate gave the 2-methyl derivative. Oxidation with DMSO-acetic anhydride yielded the 3-ketone, and by simple treatment with sodium methoxide epimerization at the neighboring C-2 was achieved in essentially quantitative yield. The methyl group changed from the axial into the less energy-rich equatorial position. The first chiral centre common to target molecules (A) and (B) was thus obtained. Reduction of the 3-keto group with sodium borohydride gave the axial alcohol, since the hydride is added to the sterically less hindered equatorial position. Methylation of this alcohol, hydrogenolysis of the benzylidene derivative, and treatment with trityl chloride yielded the 6-O-trityl derivative. Now the chirality at C-3 must be inverted, and the chirality centre at C-4 has to be converted. This was accomplished by oxidation of the only free hydroxyl group at C-4 with pyridinium chlorochromate to give the ketone and treatment with strong base to rearrange the neighboring axial methoxyl group into the less hindered equatorial position. Both common chiral centres at C-2 and C-3 of compounds (A) and (B) have now been introduced. The ketone is identical to the desired key intermediate (C).

Alkylation of (C) with methyl lithium puts the "small" methyl group into the axial, the "large" hydroxy group into the equatorial positions, whereas divalent magnesium tends to form complexes with the oxygen atom on C-3 and therefore directs the C-4 oxygen into the axial position. Both intermediates are then converted by standard reactions into precursors (A) and (B).

The last steps of the synthesis of (A), however, required extensive experimentation to raise the yields. Mild β-elimination of methanol with dilute aqueous calcium hydroxide and oxidation with manganese dioxide in the presence of NaCN and methanol were found as optimal procedures for the preparation of the α,β-unsaturated methyl ester. Catalytic hydrogenation led to *cis*-addition from less hindered side (S. Hanessian, 1977, 1978, 1979).

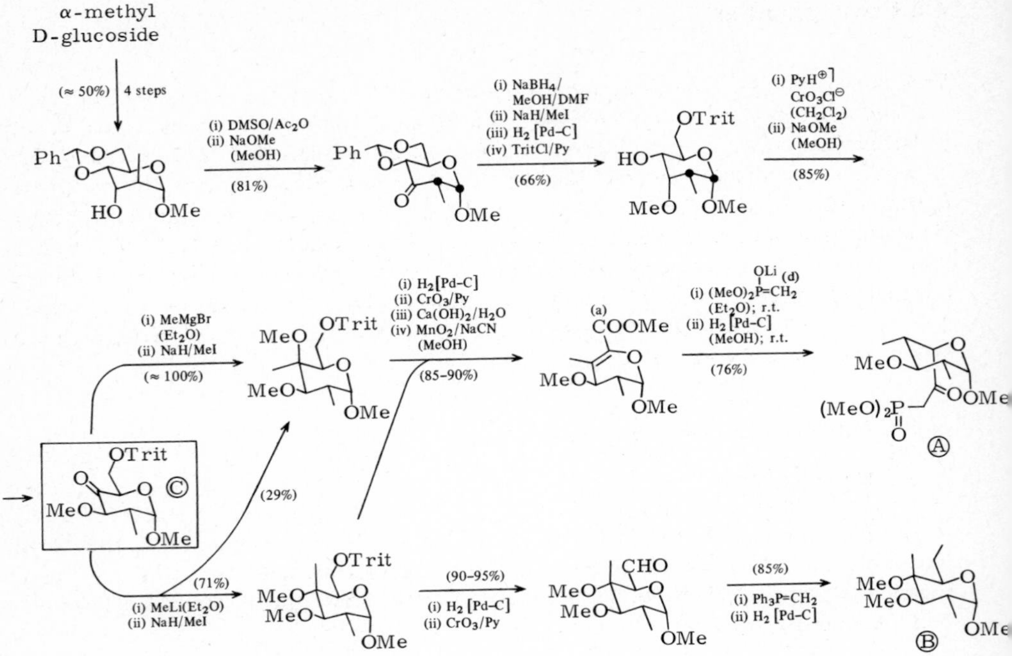

α-methyl
D-glucoside

(≈ 50%) | 4 steps

(i) DMSO/Ac₂O
(ii) NaOMe
(MeOH)

(81%)

(i) NaBH₄/
MeOH/DMF
(ii) NaH/MeI
(iii) H₂ [Pd-C]
(iv) TritCl/Py

(66%)

(i) PyH⊕]
CrO₃Cl⊖
(CH₂Cl₂)
(ii) NaOMe
(MeOH)

(85%)

(i) MeMgBr
(Et₂O)
(ii) NaH/MeI

(≈ 100%)

(i) H₂[Pd-C]
(ii) CrO₃/Py
(iii) Ca(OH)₂/H₂O
(iv) MnO₂/NaCN
(MeOH)

(85–90%)

(i) (MeO)₂P=CH₂ (d)
(Et₂O); r.t.
(ii) H₂ [Pd-C]
(MeOH); r.t.

(76%)

(MeO)₂P—COOMe
‖
O
Ⓐ

Ⓒ

(29%)

(71%)

(i) MeLi(Et₂O)
(ii) NaH/MeI

(90–95%)

(i) H₂ [Pd-C]
(ii) CrO₃/Py

(85%)

(i) Ph₃P=CH₂
(ii) H₂ [Pd-C]

Ⓑ

The cyclic acetal of (B) was then hydrolyzed with aqueous acetic acid. The liberated aldehyde was transformed to an allylic alcohol with vinylmagnesium bromide and, after protection by benzylation, ozonolyzed to give the α-benzyloxyaldehyde (D). The connection of both precursors (A) and (D) was performed using a Horner reaction. Michael type alkylation with lithium dimethylcuprate and alkylation of the ketone with methyllithium introduced the last methyl groups required in the target molecule.

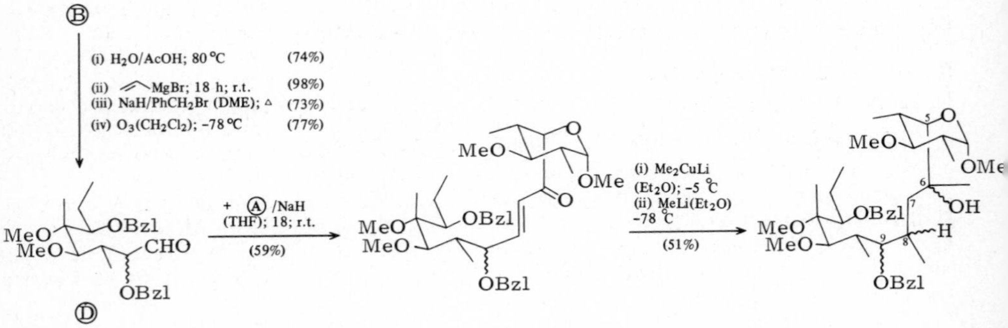

Ⓑ

(i) H₂O/AcOH; 80 °C   (74%)

(ii) ⌇MgBr; 18 h; r.t.   (98%)
(iii) NaH/PhCH₂Br (DME); △ (73%)
(iv) O₃(CH₂Cl₂); −78 °C   (77%)

+ Ⓐ /NaH
(THF); 18; r.t.

(59%)

(i) Me₂CuLi
(Et₂O); −5 °C
(ii) MeLi(Et₂O)
−78 °C

(51%)

Ⓓ

4.4 Prostaglandins

Prostaglandins (= PG) constitute a class of hormones that are present in almost all human tissues and fluids in minute concentrations and are thought to play a dominant role in the control of pregnancy, hypertension, ulcers, asthma, and pain. Since they may be useful as pharmaceuticals, several total syntheses been developed, and the term to ,,prostaglandize'' a class of compounds, which means to exhaust its synthetic potentials, has become known. The principal structures of prostaglandins are shown below. The capital letters A,B,C,E,F denote the state of oxidation and the position of double bonds in the cyclopentane or cyclopentene ring; the numeral subscript refers to the number of double bonds in the side-chains: 1 = (trans)-13; 2 = (cis,trans)-6,13; 3 = (cis,trans,cis)-6,13,17; α(= below) and β(= above) indicate the position of the hydroxyl group on C-9.

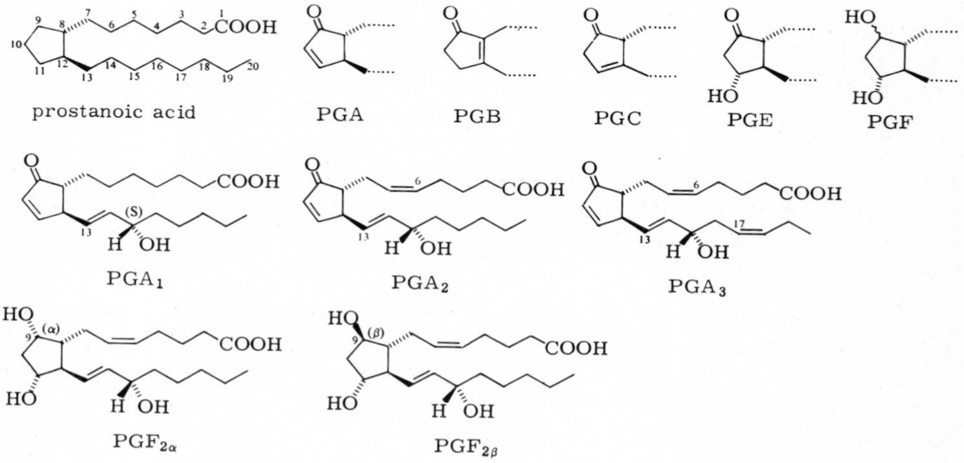
prostanoic acid
PGA
PGB
PGC
PGE
PGF
PGA1
PGA2
PGA3
PGF2α
PGF2β

The partial or full structures given above indicate the following difficulties which one has to face in synthesis of pharmacologically interesting PGE and PGF derivatives:

(i) The 11-alcohol group of PGE_{1-3} (β-ketol grouping) is sensitive to acids and bases which will cause dehydration (= PGA_{1-3} formation) and isomerization (= PGB_{1-3}) formation. The β-ketol should therefore be introduced at the end of the synthesis.

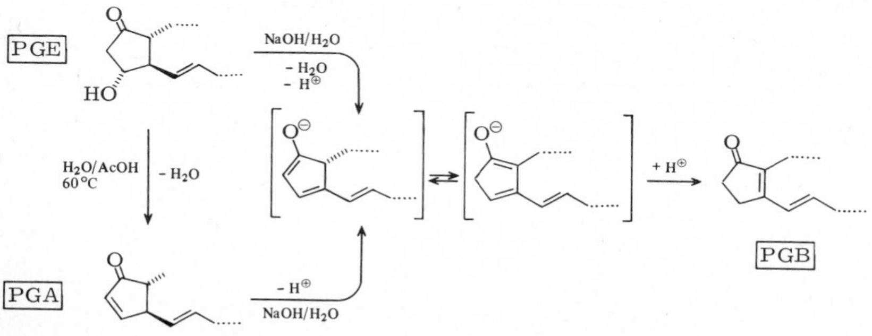
PGE
NaOH/H2O
- H2O
- H⊕
H2O/AcOH
60°C
- H2O
+ H⊕
PGB
PGA
- H⊕
NaOH/H2O

(ii) PGFs will be relatively stable towards acids and bases, but their natural 9α-configuration produces an extra problem not present in the other prostaglandins.

We shall concentrate in our short account of PG syntheses on solutions to both problems as provided in the literature. Some fundamentals of cyclopentane synthesis have been discussed in section 1.13.3 (see also 4.5.1).

4.4.1 Partial Synthesis of PGE$_2$ and PGF$_2$ from PGA$_2$

PGA$_2$ derivatives can be obtained in relatively large amounts from gorgonian corals. Since PGAs are of little pharmaceutical interest, they are converted commercially to the highly active PGF$_{2\alpha}$ and PGE$_2$ (G.L. Bundy, 1971, 1972).

The regioselective hydration of the endocyclic double bond between C-10 and C-11 in the presence of two exocyclic double bonds can be achieved, if one utilizes the activating effect of the keto group. Hydrogen peroxide in the presence of small amounts of potassium hydroxide attacks only the electrophilic β-carbon atom of the conjugated enone system. The ratio of the desired α-epoxide to the β-isomer is approximately 7:3. Reductive opening of the epoxide mixture by chromium(II) acetate in buffered solution gave PGE$_2$ derivatives regioselectively presumably because the Cr(II) ion was first bound by the epoxyketone and reduction led to the enol. The alcohol was protected as trimethylsilyl ether and the carbonyl group on C-9 reduced with NaBH$_4$. The desired 9α-compound, PGF$_{2\alpha}$, was formed as the major product since the large α-trimethylsilyl ether group directed the hydride reagent to the β-position (G.L. Bundy, 1972).

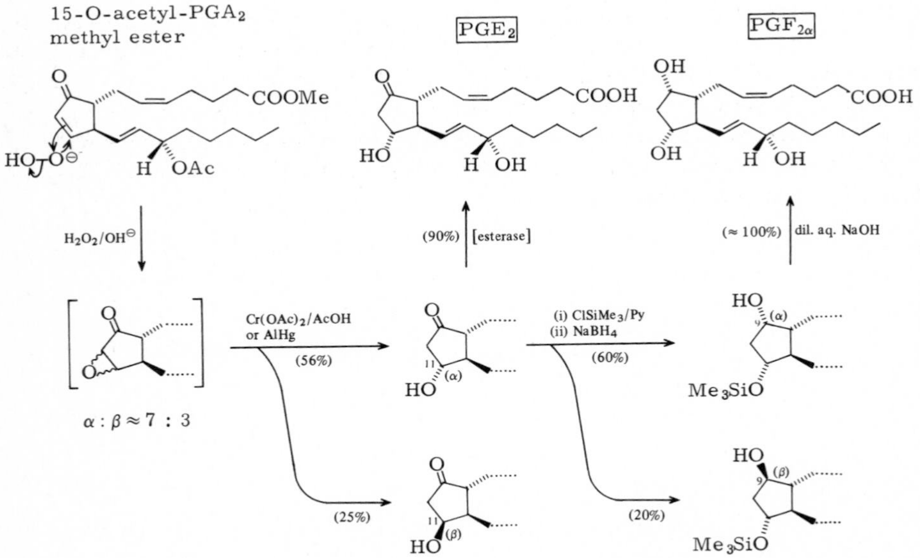

4.4.2 Total Syntheses of PGF Intermediates

Cyclopentene derivatives with carboxylic acid side-chains can be stereoselectively hydroxylated by the "iodolactonization" procedure (E.J. Corey, 1969, 1970). To the trisubstituted cyclopentene described on p. 191 a large iodine cation is added stereoselectively to the less hindered β-side of the 9,10 double bond*. Lactone formation occurs on the intermediate iodonium ion specifically at C-9α*. Later the iodine is reductively removed with tri-*n*-butyltin hydride. The cyclopentane ring now bears all oxygen and carbon substituents in the right stereochemistry, and the carbon chains can be built starting from the C-8 and C-12 substituents.

(opt. pure)

PGF$_{2\alpha}$ precursor

A more recent synthesis (M.J. Dimsdale, 1977) leads in three highly stereoselective steps from a cyclopentadiene-ketene adduct to a bridged cyclohexanone which can be transformed to (±)-(15 R/S)-PGF$_{2\alpha}$ by Baeyer-Villiger oxidation (see p. 122) and chain elongation. The first step involves a highly regio- and stereoselective formation of a bromohydrin with N-bromoacetamide. The bromonium cation is only formed at the less hindered β-side of the C=C double bond and opened selectively at the less hindered 9α-position. After protection of the alcohol group with *t*-butyldimethylsilyl chloride only one enolate is formed with the non-nucleophilic base potassium *tert*-butoxide. The enolate to the bridgehead carbon does not form (Bredt's rule). The carbanion then substitutes the bromine atom and a tricycloheptanone is formed in the second step. This compound is susceptible to homo-Michael type attack at the most labile bond of the highly strained cyclopropane. The cuprate reagent with the strongly nucleophilic alkenyl substituent and the less nucleophilic acetylide reacts at -78 °C to give the desired norbornanone in the third step. The following conventional steps of the synthesis involve Baeyer-Villiger oxidation of the ketone, DIBAL reduction of the lactone, and Wittig reaction with the resulting aldehyde.

---

* Prostaglandin numbering is used for synthetic intermediates.

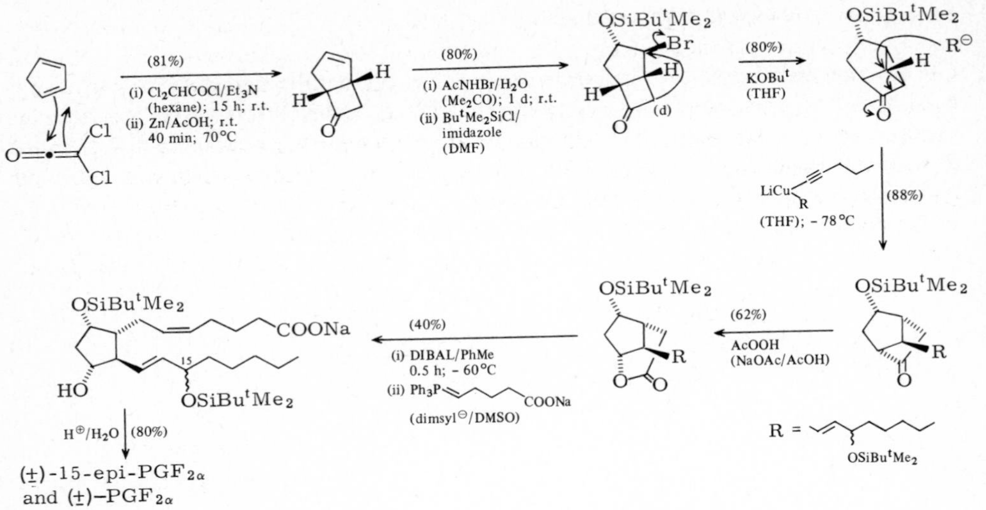

(±)-15-epi-PGF$_{2\alpha}$
and (±)-PGF$_{2\alpha}$

4.5 Steroids

Pharmaceutically useful steroids may be either obtained by total synthesis or by degradation and functional group conversions from inexpensive natural steroids. Both approaches will be discussed in this section (H. Langecker, 1977; R.T. Blickenstaff, 1974).

4.5.1 Total Syntheses

Classical syntheses of steroids consist of the stepwise formation of the four rings with or without angular alkyl groups and the final construction of the C-17 side-chain. The most common reactions have been described in chapter 1, e.g. Diels-Alder (p. 78) and Michael additions (p. 64ff.), Robinson anellation (p. 64ff.), Torgov reaction (p. 65), Knoevenagel condensation (p. 52f.), and regioselective alkylation (p. 24f.). Although many of the classical syntheses are highly stereoselective, they yield only racemic products, and separation of enantiomers at an early stage is mandatory, if an acceptable industrial process for pharmaceutical steroids is desired.

The commercially most successful steroid total synthesis is that of "norgestrel", the gestagen of most contraceptives. It starts with the stereospecific microbiological reduction of the Torgov adduct of tetralone, vinylmagnesium chloride, and 2-ethylcyclopentane-1,3-dione (p. 65). Both "prochiral" carbon atoms C-13 and C-17* become chiral in this enzyme-catalyzed reaction, and since all following reactions are highly stereoselective only

---

* Steroid numbering is used for synthetic intermediates.

(+)-norgestrel is formed. The acid-catalyzed cyclization of the enone does not produce new asymmetric centres but hydrogenation of the 14,15-double bond occurs on the less hindered α-side. Birch reduction of the 8,9-double bond and of the aromatic ring produces the thermodynamically most stable *trans*-configuration of rings B and C. The 17-alcohol is oxidized to the ketone, and the chiral centre is introduced by ethynylation of the prochiral keto group. Acid-catalyzed cleavage and rearrangement of the enol ether produces norgestrel in high yield (H. Smith, 1964; C. Rufer, 1967; H. Langecker, 1977).

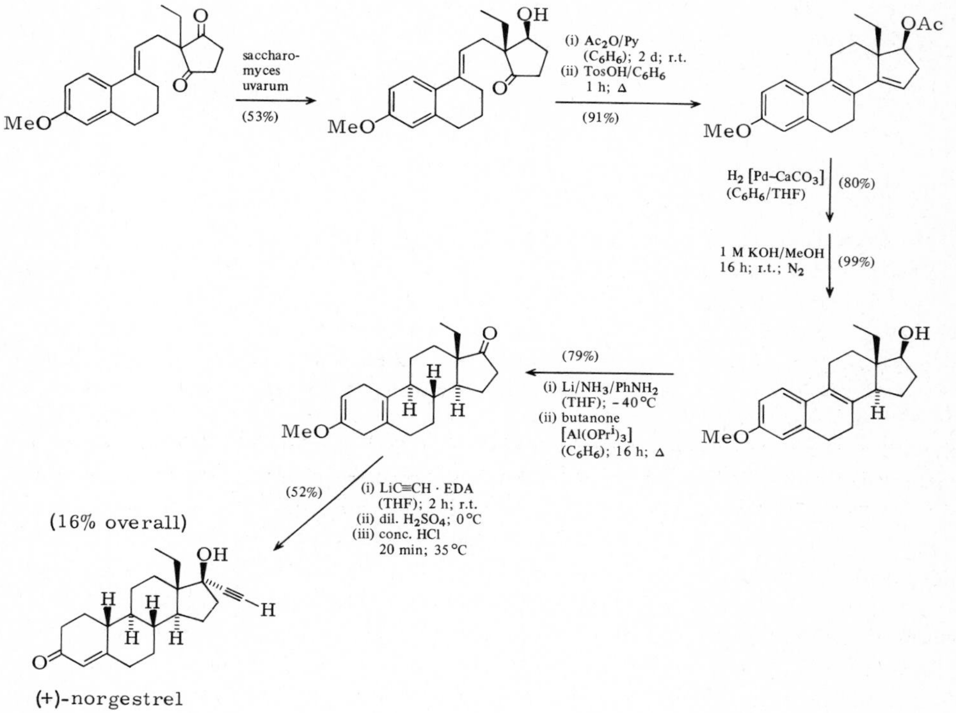

Recent syntheses of steroids apply efficient strategies in which open-chain or monocyclic educts with appropiate side-chains are stereoselectively cyclized in one step to a tri- or tetracyclic steroid precursor. These procedures mimic the biochemical synthesis scheme where acyclic, achiral squalene is first oxidized to a 2,3-epoxide containing one chiral carbon atom and then enzymatically cyclized to lanosterol with no less than seven asymmetric centres (W.S. Johnson, 1968, 1976; E.E. van Tamelen, 1968).

Non-enzymatic cyclizations of educts containing chiral centres can lead to products with additional "asymmetric" centres. The underlying effect is called "asymmetric induction". Its systematic exploration in steroid syntheses started when G. Saucy discovered in 1971 that a chiral carbon atom in a cyclic educt induces a stereoselective Torgov condensation several carbon atoms away (M. Rosenberger, 1971, 1972).

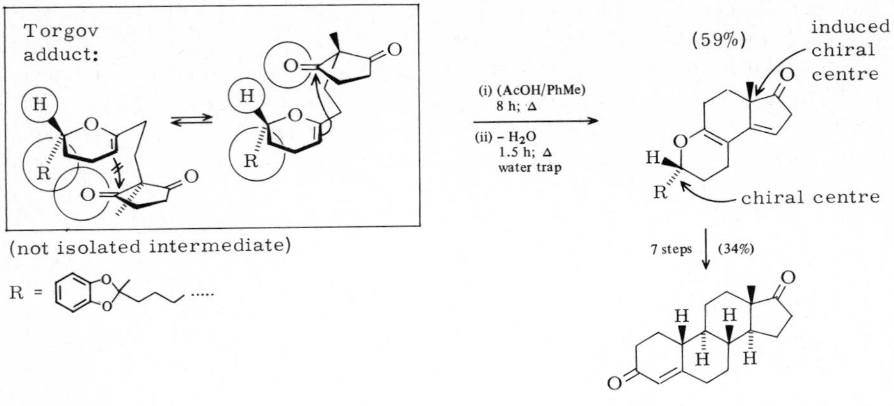

Torgov adduct:

(not isolated intermediate)

R =

(+)-Estr-4-ene-3,17-dione

Proton-catalyzed olefin cyclizations of open-chain educts may give tri- or tetracyclic products but low yields are typical (E.E. van Tamelen, 1968, 1977; see p. 83). More useful are cyclizations of monocyclic educts with appropriate side-chains. The chiral centre to which the chain is attached may direct the steric course of the cyclization, and several asymmetric centres may be formed stereoselectively since the cyclizations usually lead to *trans*-fused rings.

The following acid-catalyzed cyclizations leading to steroid hormone precursors exemplify some important facts: an acetylenic bond is less nuleophilic than an olefinic bond; acetylenic bonds tend to form cyclopentane rather than cyclohexane derivatives, if there is a choice; in proton-catalyzed olefin cyclizations the thermodynamically most stable *trans* connection of cyclohexane rings is obtained selectively; electroneutral nucleophilic agents such as ethylene carbonate can be used to terminate the cationic cyclization process forming stable enol derivatives which can be hydrolyzed to carbonyl compounds; without this nucleophile and with trifluoroacetic acid the corresponding enol ester may be obtained (M.B. Gravestock, 1978, A,B; P.E. Peterson, 1969).

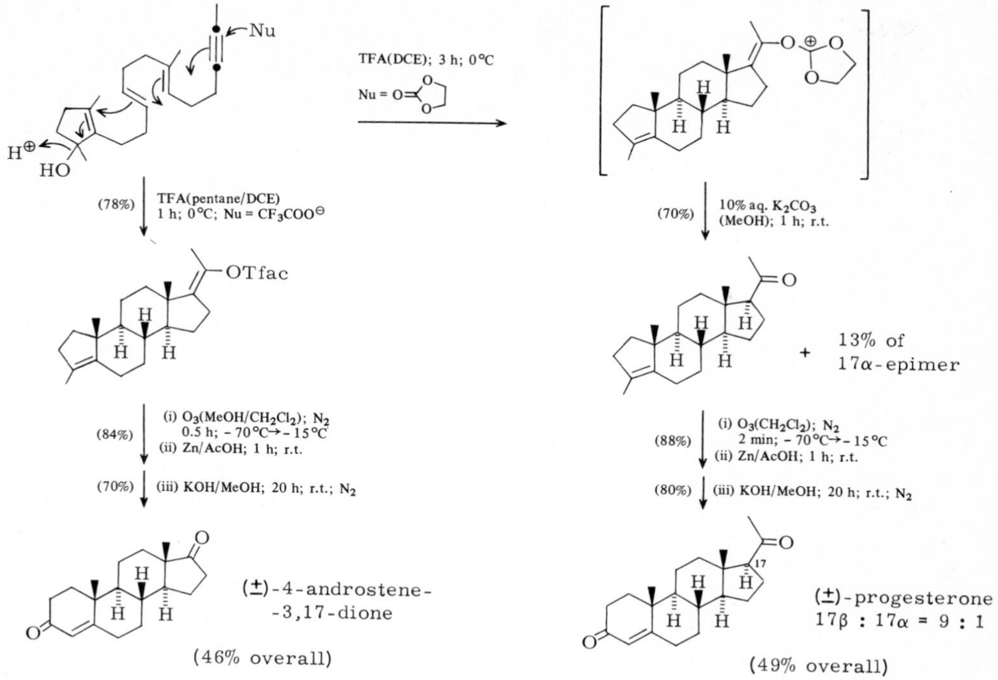

Two approaches to *convergent* steroid syntheses are based on the thermal opening of benzocyclobutenes to the *o*-quinodimethane derivatives (see p. 74; W. Oppolzer, 1978 A) and their stereoselective intramolecular Diels-Alder cyclizations. T. Kametani (1977 B, 1978) obtained (+)-estradiol in a six-step synthesis. The final Diels-Alder reaction occurred regio- and stereoselectively in almost quantitative yield, presumably because the *exo* transition state given below is highly favored over the *endo* state in which rings A and D would sterically interact.

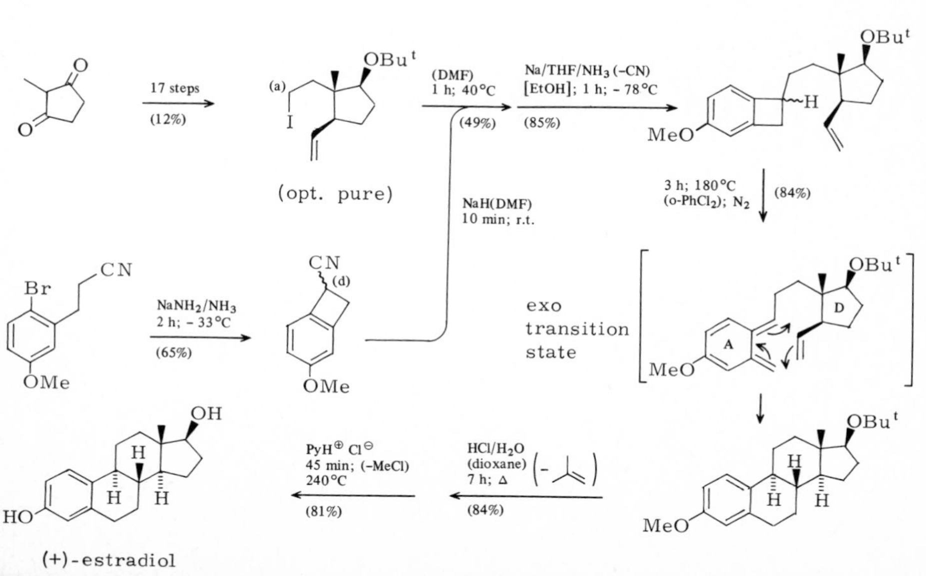

A similar synthesis starts from commercially available 1,5-hexadiyne and 2-methyl-cyclopent-2-enone. The benzocyclobutene is obtained from a bis-acetylene in a cobalt-catalyzed reaction. It rearranges regio- and stereoselectively to a 3-deoxy steroid derivative. The overall yield from the cyclopentenone was 40% (R.L. Funk, 1977).

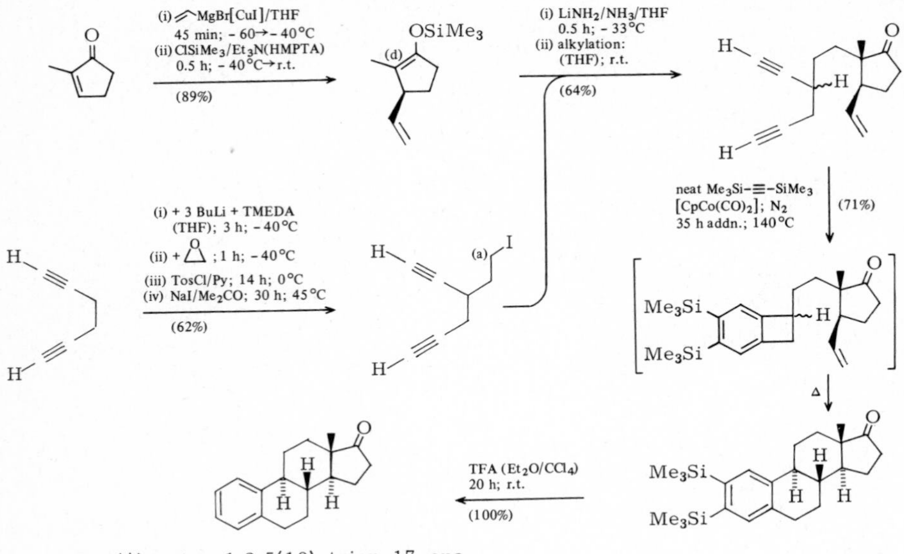

(±)-estra-1,3,5(10)-trien-17-one

4.5.2 Oxidation, Dehydrogenation, and Fluorination of Steroids

Some plant and animal steroids occur in large quantities and can be used as inexpensive starting materials for pharmaceutically useful steroid hormones (see table 20).

The major problem in such conversions is the degradation of the branched carbon side-chain on C-17 which is present in all abundant steroids and lacking in all steroid hormones. The most important starting material used in industry today is diosgenin from the

Table 20. Inexpensive steroids.

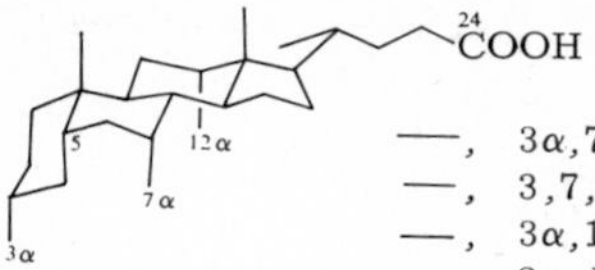

5β-Cholan-24-oic acid	

—, $3\alpha,7\alpha,12\alpha$,-Trihydroxy-(*Cholic acid*)   [4]
—, $3,7,12$-Trioxo-(*Dehydrocholic acid*)   [4]
—, $3\alpha,12\alpha$-Dihydroxy-(*Deoxycholic acid*)   [4]
—, 3α-Hydroxy-(*Lithocholic acid*)   [4]

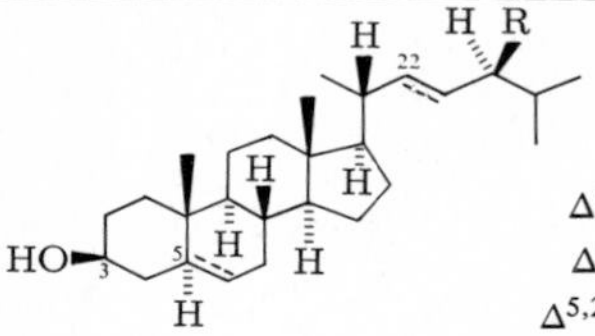

R = H 3β-Cholestanol   [10]
R = Et 3β-Stigmastanol   [3]

Δ^5; R = H   5-Cholesten-3β-ol (*Cholesterol*) [2]
Δ^5; R = Et   5-Stigmasten-3β-ol (*Sitosterol*) [2]
$\Delta^{5,22}$; R = Et   5,22-Stigmastadien-3β-ol (*Stigmasterol*) [29]

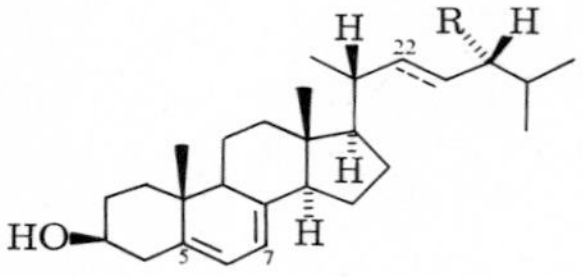

$\Delta^{5,7}$; R = H   5,7-Cholestadien-3β-ol   [20]
(*Provitamin D$_3$*)

$\Delta^{5,7,22}$; R = Me   5,7,22-Ergostatrien-3β-ol [17]
(*Provitamin D$_2$, Ergosterol*)

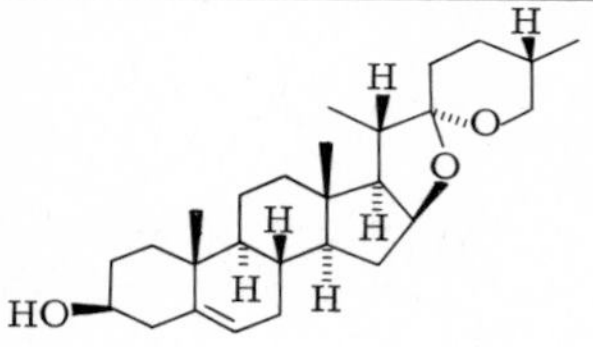

 Diosgenin [24]	 4,22-Stigmastadien-3-one [12]
 3β-Hydroxy- -5-pregnen-20-one [24] (*Pregnenolone*) acetate: [24]	 X = H$_2$ 4-Pregnene-3,20-dione [22] (*Progesterone*) X = O 4-Pregnene-3,11,20-trione [22]
 8,24-Lanostadien-3β-ol [2] (*Lanosterol*)	 Glycyrrhizic acid, ammonium salt [19]

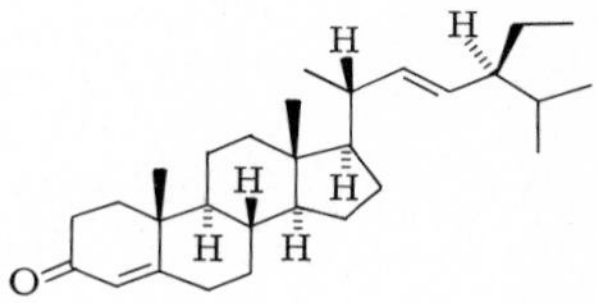

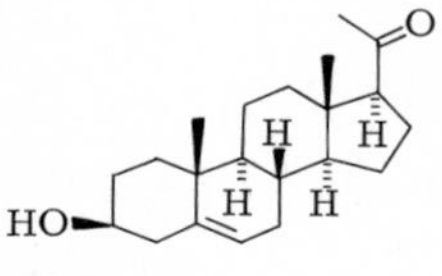

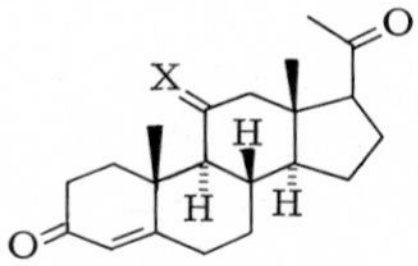

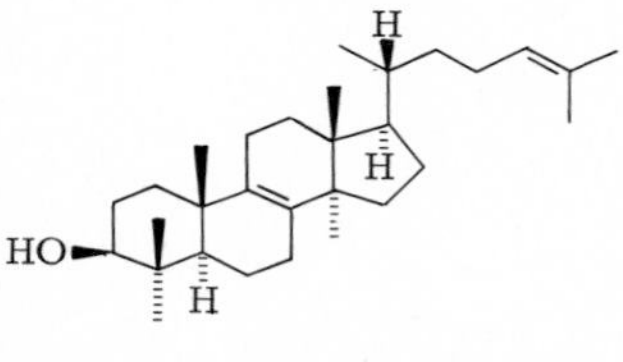

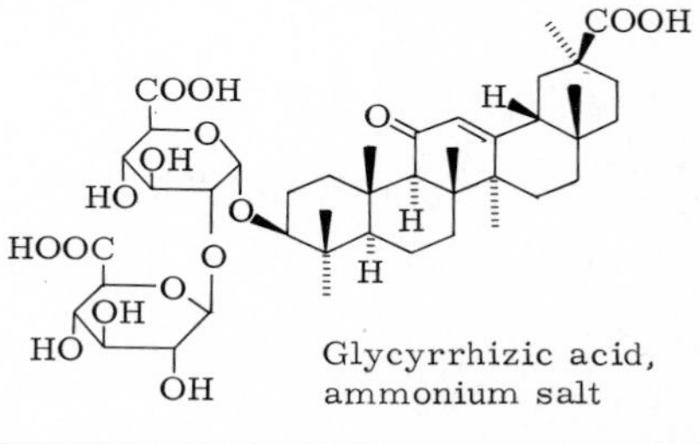

* The numbers in brackets indicate the approximate prices in German marks per 10 g.

Mexican dioscorea plant. It is degraded by the method of Marker to 16-dehydropregnenolone in 45% total yield. This compound is a key substance in the production of several hormones with anabolic*, catabolic**, and sexual*** effects.

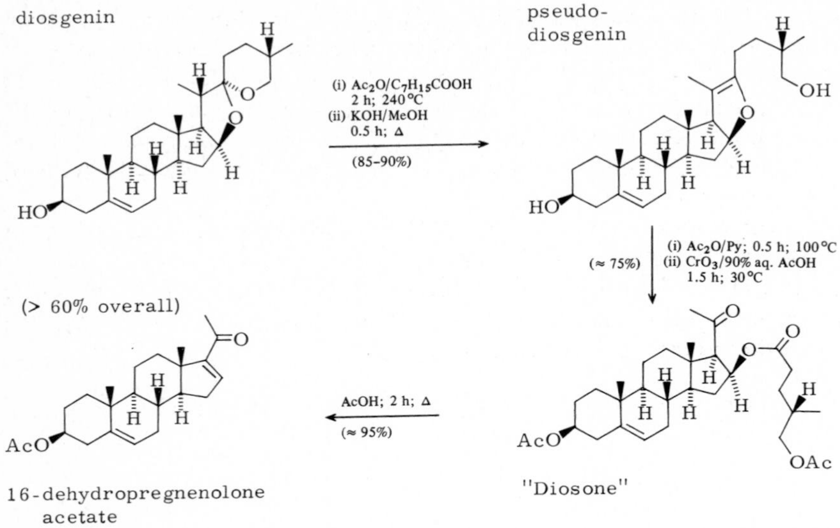

diosgenin

(i) Ac$_2$O/C$_7$H$_{15}$COOH
   2 h; 240 °C
(ii) KOH/MeOH
   0.5 h; Δ

(85–90%)

pseudo-
diosgenin

(≈ 75%)

(i) Ac$_2$O/Py; 0.5 h; 100 °C
(ii) CrO$_3$/90% aq. AcOH
   1.5 h; 30 °C

(> 60% overall)

AcOH; 2 h; Δ

(≈ 95%)

"Diosone"

16-dehydropregnenolone
acetate

Stigmasterol from soy bean extracts can be selectively ozonolyzed on the side-chain double bond. The 20-formyl group formed is converted to the enamine with piperidine. This can be oxidized to progesterone.

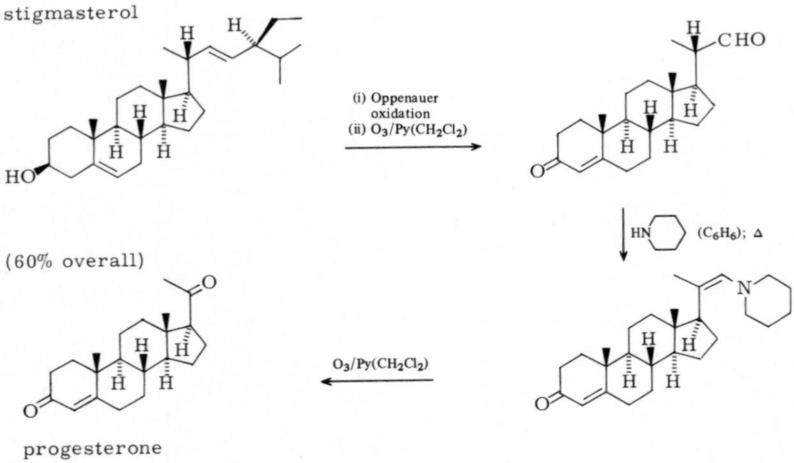

stigmasterol

(i) Oppenauer
   oxidation
(ii) O$_3$/Py(CH$_2$Cl$_2$)

HN⟨⟩ (C$_6$H$_6$); Δ

(60% overall)

O$_3$/Py(CH$_2$Cl$_2$)

progesterone

---

* Promotion of protein synthesis, e.g. in muscle development.
** Retardation of protein synthesis, e.g. in treatment of inflammations.
*** Mainly as contraceptives.

The most abundant natural steroid is cholesterol. It can be obtained in large quantities from wool fat (15%) or from brain or spinal chord tissues of fat stock (2-4%) by extraction with chlorinated hydrocarbons. Its saturated side-chain can be removed by chromium trioxide oxidation, but the yield of such reactions could never be raised above 8% (see page 108).

Yields up to 50% are obtained in a "remote oxidation" procedure developed by Breslow. If, for example, the hydroxyl group of 3α-stigmastanol is esterified with 4'-iodobiphenyl-3-carboxylic acid, the biphenyl substituent arranges itself parallel to the steroid plane. The iodine is situated close to C-17 of the steroid. It is chlorinated; upon irradiation the chlorine is cleaved off and attacks selectively at C-17. HCl elimination and rearrangement reactions lead to a double bond between C-17 and C-20 which can be ozonolyzed (R. Breslow, 1977).

A similar intramolecular oxidation, but for the methyl groups C-18 and C-19 was introduced by D.H.R. Barton (1979). Axial hydroxyl groups are converted to esters of nitrous or hypochlorous acid and irradiated. Oxyl radicals are liberated and selectively attack the neighboring axial methyl groups. Reactions of the methylene radicals formed with nitrosyl or chlorine radicals yield oximes or chlorides.

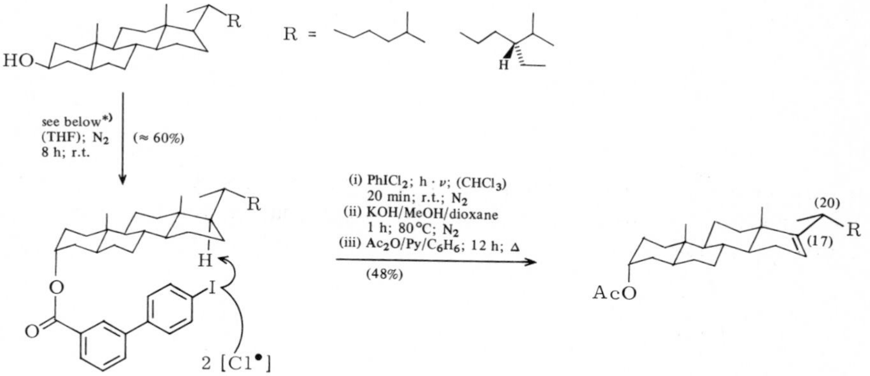

* inversion-esterification:

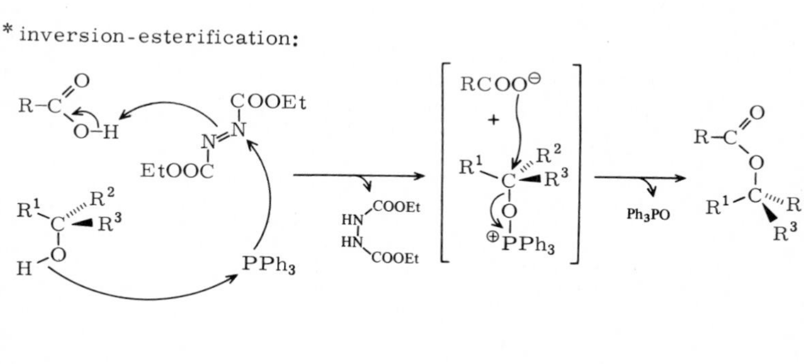

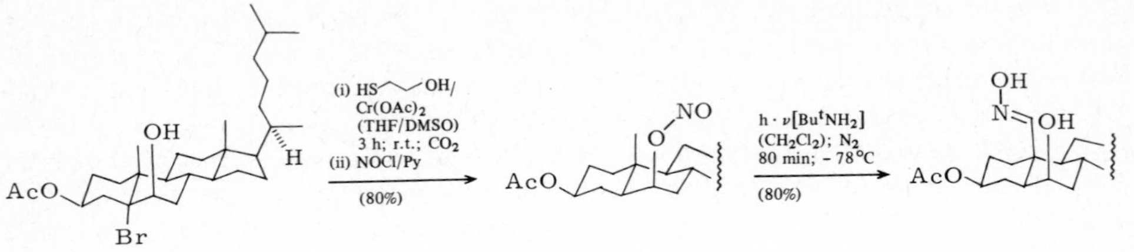

(i) HS⌒OH/
Cr(OAc)$_2$
(THF/DMSO)
3 h; r.t.; CO$_2$
(ii) NOCl/Py

(80%)

h·ν[Bu^{t}NH$_2$]
(CH$_2$Cl$_2$); N$_2$
80 min; − 78°C

(80%)

bromohydrin of
cholesteryl acetate

The most difficult pharmaceutically relevant oxidation of steroids is the introduction of a 14β-hydroxyl group. This functional group is found in heart-active steroids (cardenolides) such as digitoxigenin, which also contain a 17β-butenolide substituent. The 14β-hydroxyl group is easily cleaved off by dehydration and must therefore not be treated with Lewis or protic acids. One approach is highly stereo- and regioselective bromohydrination (see also p. 114f.) and subsequent reductive debromination (W. Fritsch, 1969, 1974). Nowadays microbiological hydroxylation is preferred.

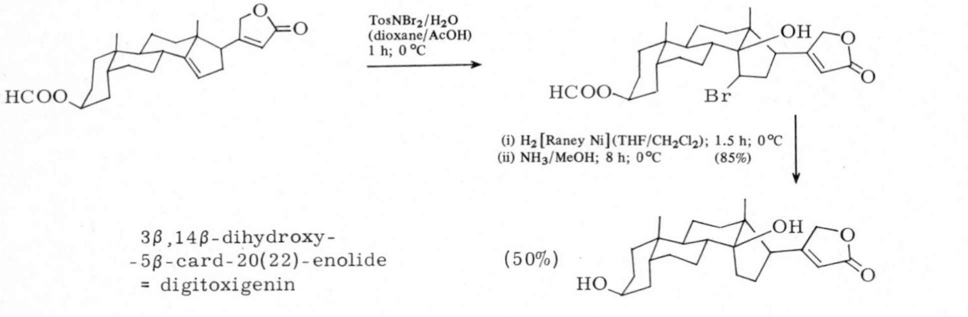

TosNBr$_2$/H$_2$O
(dioxane/AcOH)
1 h; 0°C

(i) H$_2$[Raney Ni](THF/CH$_2$Cl$_2$); 1.5 h; 0°C
(ii) NH$_3$/MeOH; 8 h; 0°C   (85%)

3β,14β-dihydroxy-
-5β-card-20(22)-enolide
= digitoxigenin

(50%)

Several cortisone derivatives with glucocorticoid effects* are most active, if they contain fluorine in the 9α-position together with an 11β-OH group. Both substituents are introduced by the cleavage of a 9,11-epoxide with hydrogen fluoride. The regio- and stereoselective formation of the β-epoxide is achieved by bromohydrination of a 9,11-double bond and subsequent alkali treatment (J. Fried, 1954).

---

* It affects the metabolisms of glucose and proteins; used in rheumatoid arthritis, inflammatory conditions, etc.

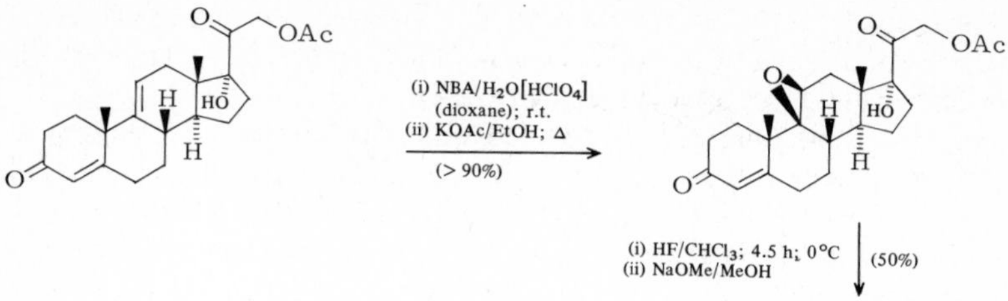

(i) NBA/H$_2$O[HClO$_4$]
   (dioxane); r.t.
(ii) KOAc/EtOH; Δ

(> 90%)

(i) HF/CHCl$_3$; 4.5 h; 0°C   (50%)
(ii) NaOMe/MeOH

9α-fluorohydrocortisone
= 9α-fluorocortisol

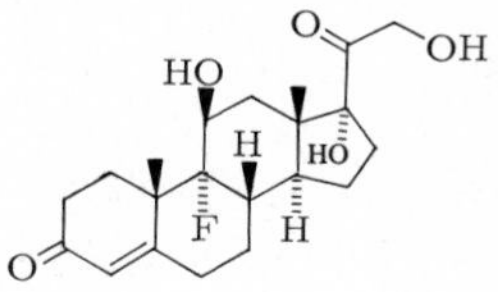

Finally the chemical aromatization of Ring A which occurs in nature in the biosynthesis of estrogens must be mentioned. It can be done by thermal cleavage of the C-19 methyl group in 1,4-dien-3-ones (H.H. Inhoffen, 1940; C. Djerassi, 1950) and was later achieved at lower temperatures with lithium — biphenyl in THF (H.L. Dryden, Jr., 1964).

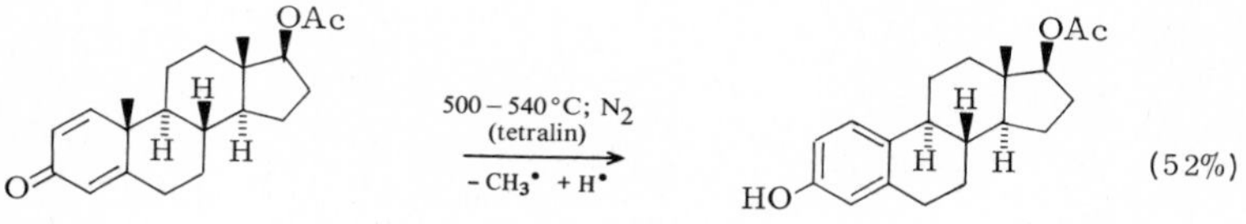

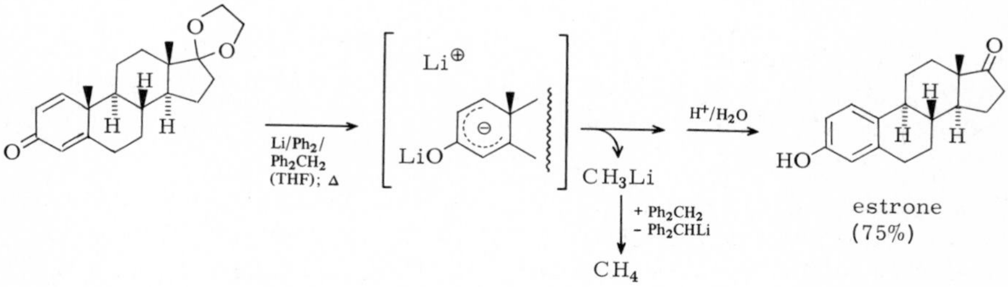

500 – 540°C; N$_2$
(tetralin)

– CH$_3$$^\bullet$ + H$^\bullet$

(52%)

Li/Ph$_2$/
Ph$_2$CH$_2$
(THF); Δ

H$^+$/H$_2$O

estrone
(75%)

CH$_3$Li

+ Ph$_2$CH$_2$
– Ph$_2$CHLi

CH$_4$

4.5.3 Chemical Reactivity

Many stereoselective reactions have been most thoroughly studied with steroid examples because the rigidity of the steroid nucleus prevents conformational changes and because enormous experience with analytical procedures has been gathered with this particular class of natural products (J. Fried, 1972). The name "steroids" (*stereos* (gr.) = solid, rigid) has indeed been selected very well, if one considers stereochemical problems. We shall now briefly point to some other interesting, more steroid-specific reactions.

3-Hydroxy steroids can be isomerized by heating with alkali alcoholates to give C-3 epimers. A/B *trans*-fused steroids, e.g. cholestanol, yield selectively β-isomers, *cis*-fused steroids give α-isomers. In both cases the equatorial OH group is favored. β-Epimers form insoluble complexes with the natural steroid glycoside digitonin. This complex formation is highly stereoselective for steroids with a 3β-OH, provided the 10-CH₃ has the natural β-configuration.

5α-cholestan-3β-ol   epicholestanol

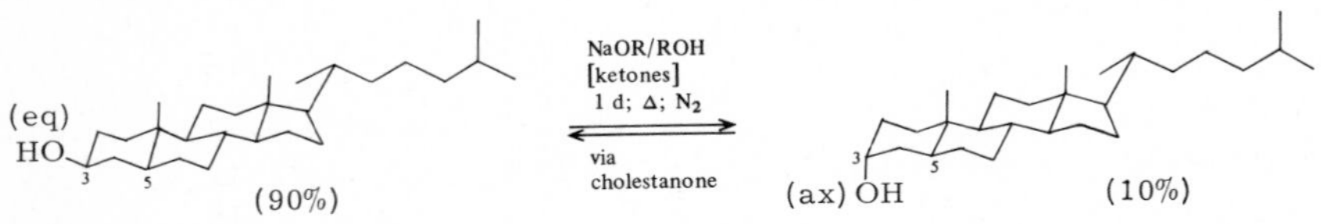

5β-cholestan-3β-ol
= coprostanol   epicoprostanol

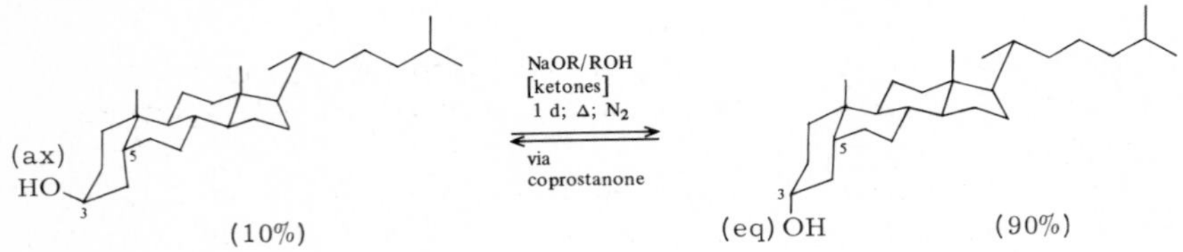

3β-Tosyloxy Δ^5-steroids, e.g. O-tosylcholesterol, give 3,5-cyclosteroids (= *i*-steroids) on treatment nucleophiles. Internal hydroxyl displacement, e.g. with PCl₅, leads to 3β-substituted products or overall retention of configuration at C-3 by rearrangement of the 6β substituent (E.M. Kosower, 1956).

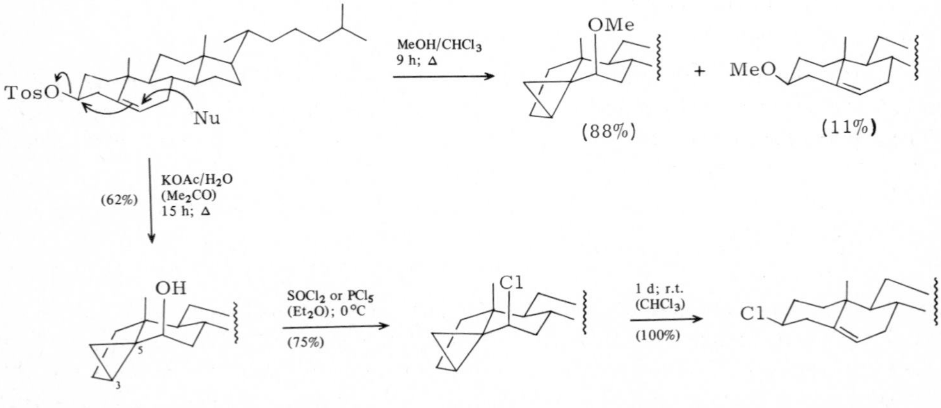

Irradiation of steroidal 5,7-dienes with ultraviolet light may result in ring opening and formation of various trienes. The most important reaction of this type is the conversion of ergosterol to previtamin D_2.

ergosterol
(provitamin D_2)

tachysterol
(previtamin D_2)

ergocalciferol
(vitamin D_2)

D_2: R =

4.6 Alkaloids

Alkaloids are basic nitrogen compounds of great structural complexity which occur in plants. About two dozen of different classes of alkaloids are covered in the "Specialist Periodical Reports" of the English Chemical Society.

In spite of the diverse nature of alkaloid structures, two structural units, i.e. fused pyrrolidine and piperidine rings in different oxidation states, appear as rather common denominators. We therefore chose to give several examples for four types of synthetic reactions which have frequently been used in alkaloid total synthesis and which provide generally useful routes to polycyclic compounds with five- or six-membered rings containing one nitrogen atom. These are:

(i) Nucleophilic substitution with amines.
(ii) Mannich reactions.
(iii) Oxidative and photochemical coupling of phenol derivatives.
(iv) Electrocyclic reactions.

Partial synthesis is relatively unimportant in the field of alkaloid synthesis, since only a few compounds are available at low price (see table 21). An exception is the derivatization of the morphine base, which leads to codeine, heroin, and other important compounds. These trivial reactions, however, are covered in elementary text books.

4.6.1 Nucleophilic Substitutions with Amines

Amines are powerful nucleophiles which react under neutral or slightly basic conditions with several electron-accepting carbon reagents. The reaction of alkyl halides with amines is useful for the preparation of tertiary amines or quaternary ammonium salts. The conversion of primary amines into secondary amines is usually not feasible since the secondary amine tends towards further alkylation.

Table 21. Inexpensive alkaloids.*

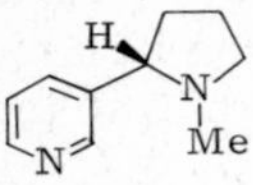

(−)-(S)-Nicotine [3] Piperine [20]	(−)-Sparteine [13] •H_2SO_4 [7]
Tropine [15] (−)-(S)-Hyoscyamine [24] •H_2SO_4 [20] (+)-Atropine•H_2SO_4•H_2O [14]	Tryptamine [15] Gramine [4]
Ephedrine, rac- [3]; (+)-(1S, 2R)- [5]; •HCl [3] (−)-(1R, 2S)- [4]; •HCl [8] Dopamine •HCl [14]	(+)-Yohimbine•HCl [33]
Papaverine Narcotine •HCl [17] •HCl • H_2O: [7] Berberine X = OH [16] X = Cl [9]	R = MeO: (−)-Brucine •2 H_2O: [8] R = H: (−)-Strychnine [14] •HCl: [12]

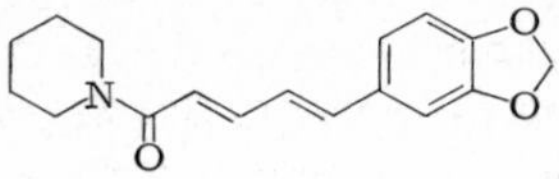

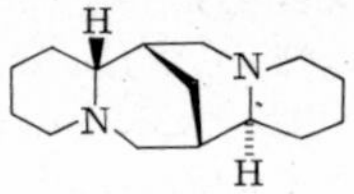

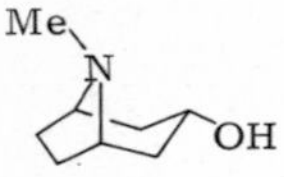

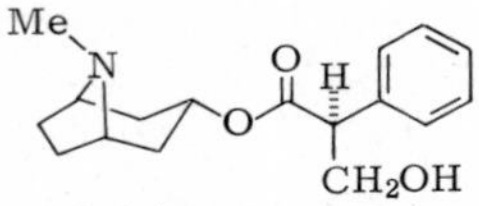

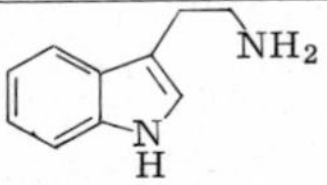

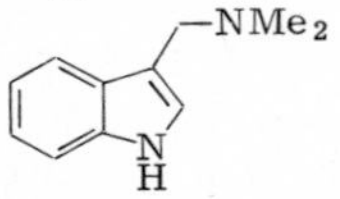

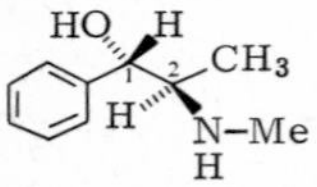

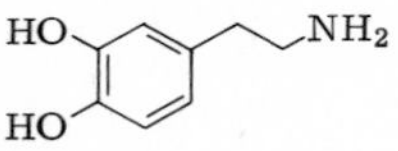

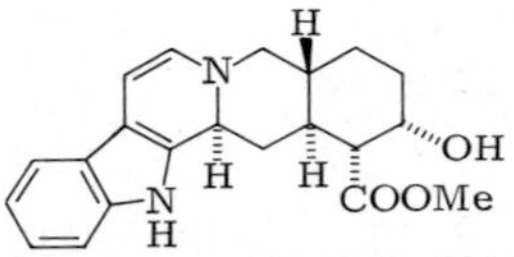

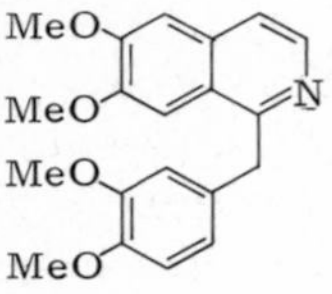

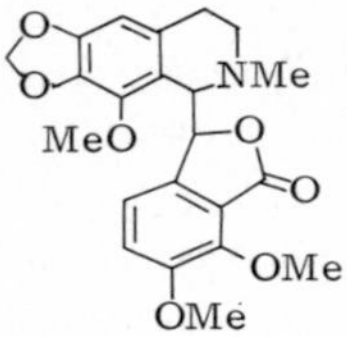

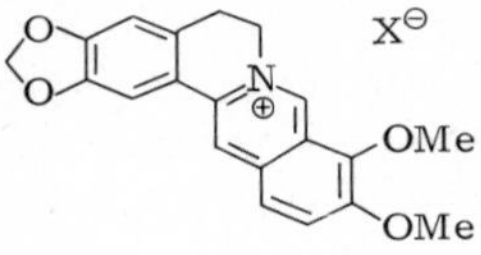

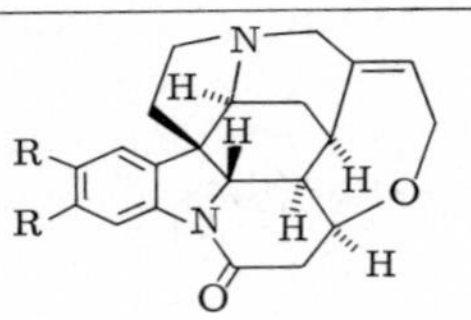

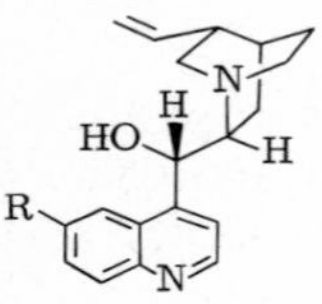

	R = MeO: (+)-Quinidine [19] •H_2SO_4 [13] R = H: (+)-Cinchonine [6] •HCl [5]	(−)-Quinine [15] •HBr [12] (−)-Cinchonidine [5] •HCl [3]			

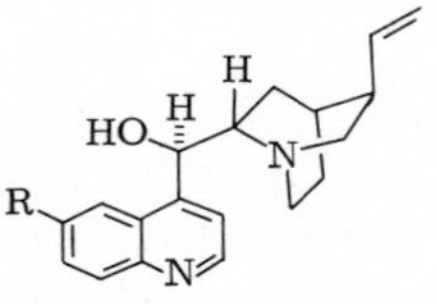

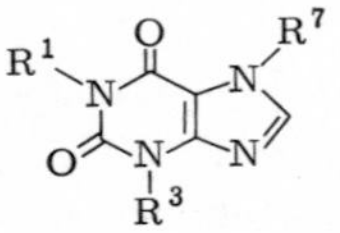

R^1	R^3	R^7	
Me	Me	H	Theophylline [2]
H	Me	Me	Theobromine [1]
Me	Me	Me	Caffeine [1]

* The numbers in brackets indicate the approximate prices in German marks per 10 g.

Primary and secondary amines also react with epoxides (or *in situ* produced episulfides or aziridines) to β-hydroxyamines (or β-mercaptoamines or 1,2-diamines). The Michael type addition of amines to activated C=C double bonds is also a useful synthetic reaction. Finally amines react readily with carbonyl compounds to form imines and enamines and with carboxylic acid chlorides or esters to give amides which can be reduced to amines with LiAlH$_4$ (p. 102). All these reactions are often applied in synthesis to produce polycyclic alkaloids with nitrogen bridgeheads (J.W. Huffman, 1967; G. Stork, 1963; S.S. Klioze, 1975).

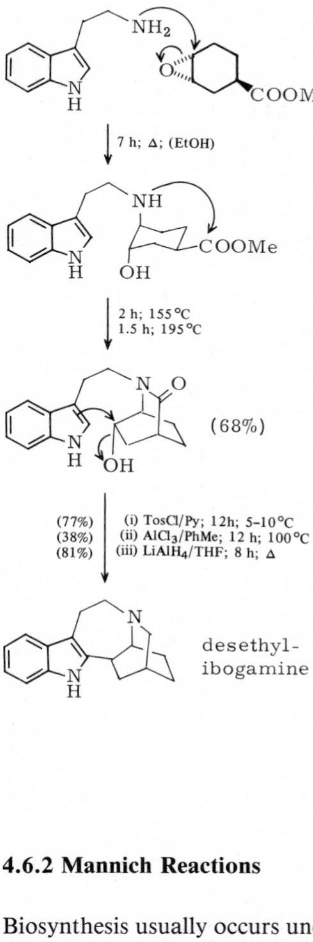

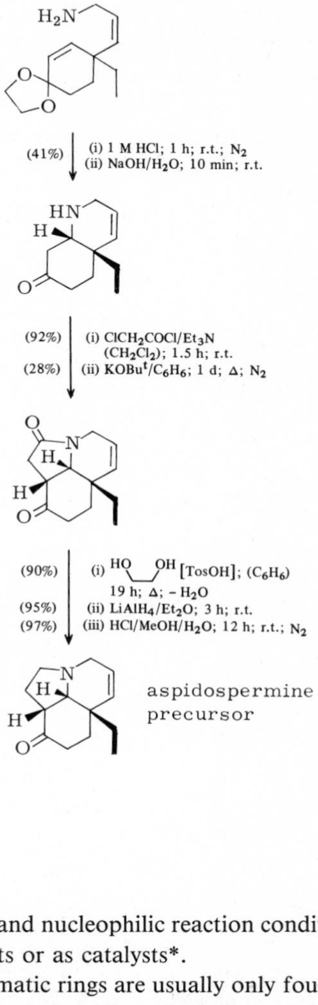

4.6.2 Mannich Reactions

Biosynthesis usually occurs under reductive and nucleophilic reaction conditions. Weak bases and acids are employed in nature as reagents or as catalysts*.

Reactions of aromatic and heteroaromatic rings are usually only found with highly reactive compounds containing strongly electron donating substituents or hetero atoms (e.g. phenols, anilines, pyrroles, indoles). Such molecules can be substituted by weak electrophiles, and the reagent of choice in nature as well as in the laboratory is usually a Mannich reagent or

---

* An exception is, of course, metal-catalyzed oxidation with molecular oxygen.

a related compound. Many diverse skeletons of the alkaloids have thus been (and still are) constructed by the linking of phenols or indoles with an aldehyde and an amine. One may perhaps say that, of all classic synthetic reactions, the Mannich reaction is the one which survived all modern developments as the least "harmed" (or least improved) one. The most common variant of the Mannich reaction in alkaloid synthesis is the Pictet-Spengler condensation of a β-aminoethyl phenol or indole derivative with an aldehyde to yield tetrahydroisoquinolines or tetrahydro-β-carbolines (T. Kametani, 1977 A; J.P. Kutney, 1977).

The reaction conditions applied are usually heating the amine with a slight excess of aldehyde and a considerable excess of 20-30% hydrochloric acid at 100 °C for a few hours, but much milder ("physiological") conditions can be used with good success. Diols, olefinic double bonds, enol ethers, and glycosidic bonds survive a Pictet-Spengler reaction very well, since phenol and indole systems are much more reactive than any of these acid sensitive functional groups (W.M. Whaley, 1951; J.E.D. Barton, 1965; A.R. Battersby, 1969).

Pictet-Spengler cyclization

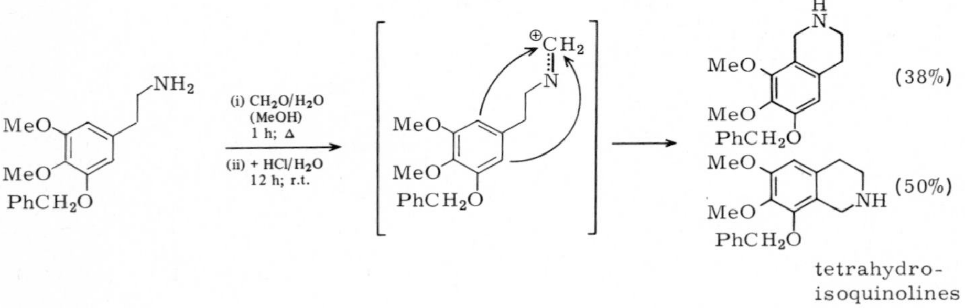

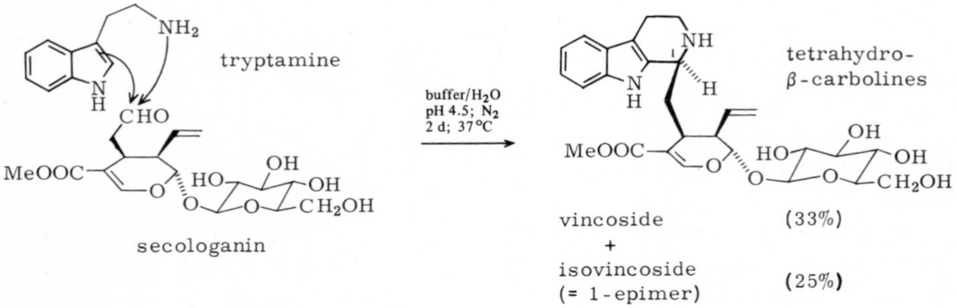

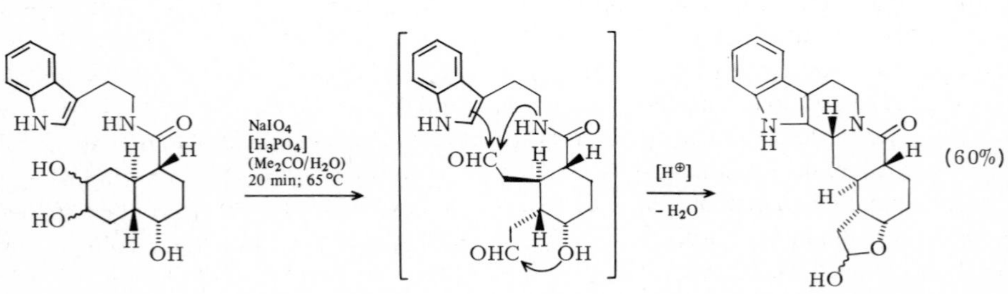

Even amide carbonyl groups can be activated to undergo Mannich type reactions, when phosphorus oxychloride, pentoxide, or pentachloride is used as condensing reagent. This is the basis of the popular Bischler-Napieralski reaction which leads to 3,4-dihydro-isoquinolines. An N-(β-arylethyl)carboxamide is phosphorylated on the amide oxygen, and the resulting iminium cation system is as reactive as a Mannich reagent (Vilsmeier type reagent, see p. 231).

The reaction is again most successful with phenol and indole derivatives. Reaction conditions are also quite mild. Phosphorus oxychloride may be used in refluxing chloroform or higher boiling hydrocarbons. Duration of the reaction is usually less than 6 hours (T. Kametani, 1977 A; E.E. van Tamelen, 1969).

Bischler-Napieralski cyclization

4.6.3 Coupling of Phenol Derivatives

Chemical or electrochemical oxidation of phenols affords phenol radicals. These are relatively stable when compared with nonaromatic radicals since the odd electron is spread over the electronegative oxygen atom and the *ortho-* and *para-*positions of the aromatic ring. Suitable oxidizing agents are ferric chloride, potassium ferricyanide, manganese dioxide, thallium(III) trifluoroacetate, vanadyl trichloride, and several others. Unsubstituted phenol radicals dimerize favorably by forming new carbon-carbon bonds. The reaction often proceeds regioselectively in the *para-*positions and cleanly in one direction when the coupling is a cyclization of two already linked phenols. In alkaloid syntheses, however, the situation is complicated by the fact that amines are nucleophilic. C—N coupling as well as several side reactions are observed. The most common high-yield intramolecular coupling reaction is the attack of an amine to a quinone or semiquinone radical (B. Franck, 1971). Phenol-phenol coupling occurs, if the amine is quaternized (B. Franck, 1962, 1966, 1967) or protected by complexation with palladium salts (L.L. Miller, 1971, 1973). It has quite often been found that electrochemical oxidations on platinum electrodes give better results in intramolecular reactions than chemical oxidations in solution.

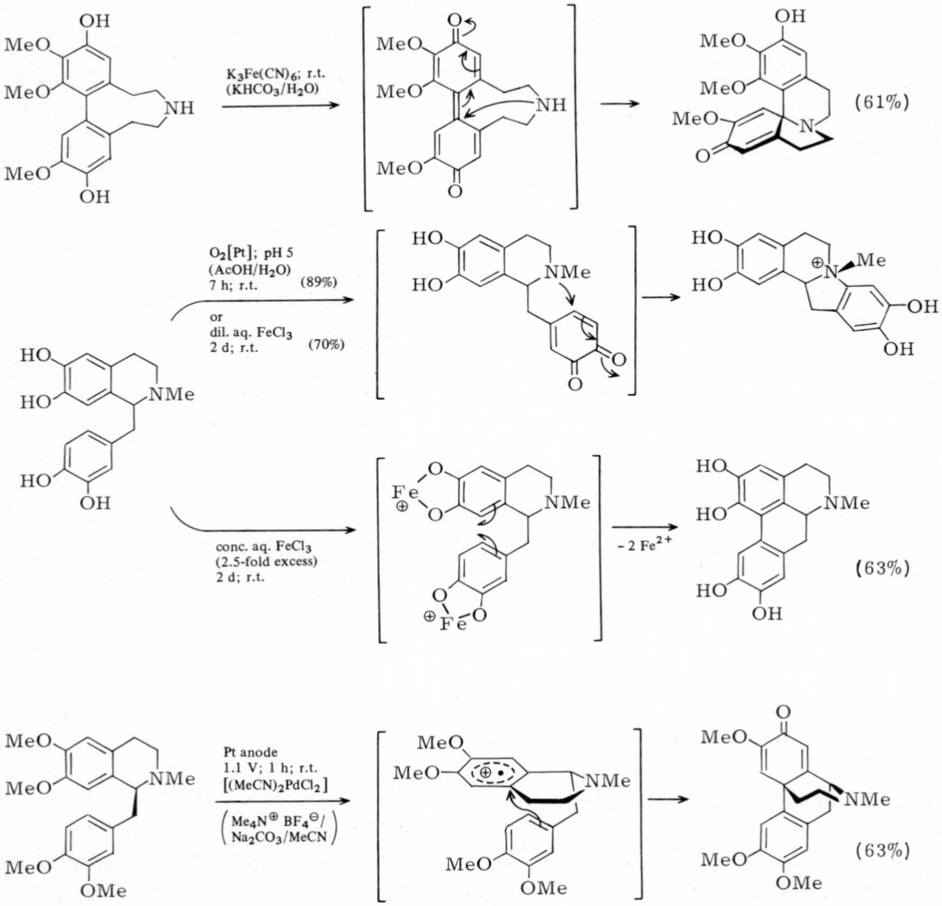

Carbon-oxygen bonds are formed by the Ullmann reaction (= coupling of aryl halides with copper) which has been varied in alkaloid chemistry to produce diaryl ethers instead of biaryls. This is achieved by the use of CuO in basic media (T. Kametani, 1969; R.W. Doskotch, 1971).

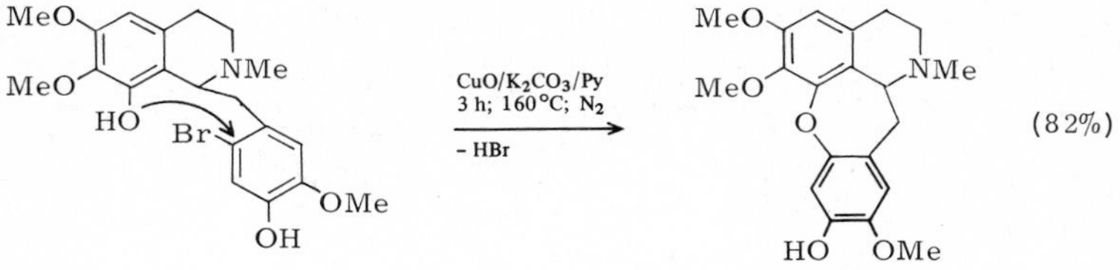

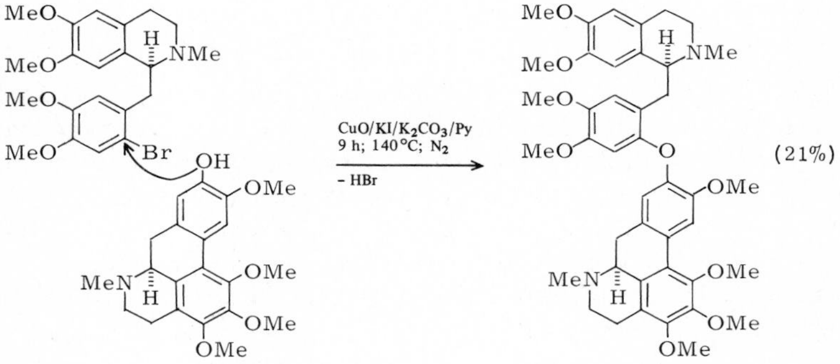

(82%)

(21%)

(S,S)-adiantifoline

4.6.4 Electrocyclic Reactions

Reaction that can be carried out by the oxidative coupling of radicals may also be initiated by irradiation with UV light. This procedure is especially useful if the educt contains olefinic double bonds since they are vulnerable to the oxidants used in the usual phenol coupling reactions. Photochemically excited benzene derivatives may even attack ester carbon atoms which is g enerally not observed with phenol radicals (I. Ninomiya, 1973; N.C. Yang, 1966).

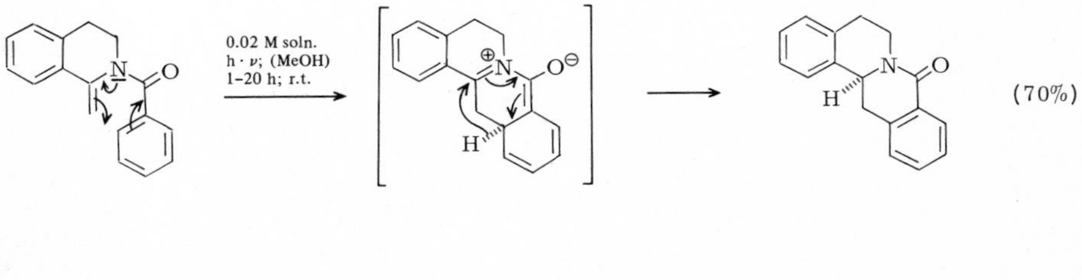

(70%)

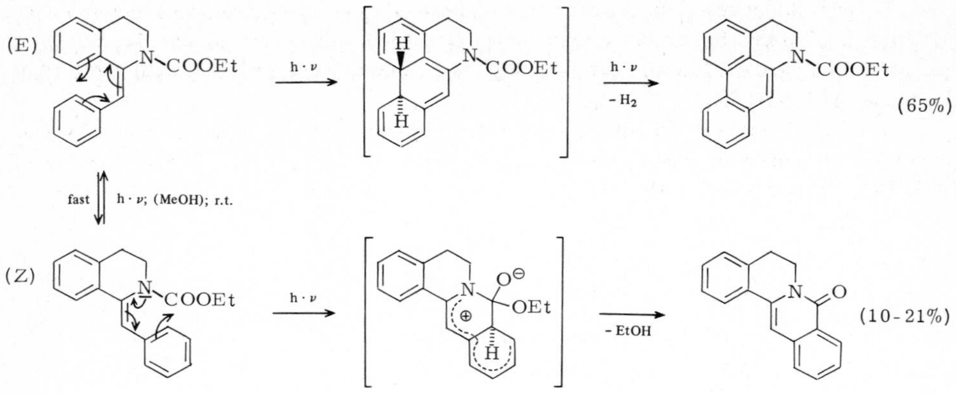

(E)

h·ν →

[ ]

h·ν
−H₂
→

(65%)

fast ⇅ h·ν; (MeOH); r.t.

(Z)

h·ν →

[ ]

−EtOH
→

(10−21%)

Indole derivatives are often synthesized by the method of E. Fischer (see p. 138). The key step of this synthetic reaction is an (intramolecular) [3,3]-sigmatropic rearrangement and may therefore be used to produce sterically hindered products otherwise difficult to obtain. In the first example given below the (planar) lactam group causes a sterical strain which prevents enolization of the bridgehead carbon neighboring the phenylhydrazone carbon. In the second example a non-planar amine nitrogen does not have such effect and a highly hindered spiro-fused indolenine is formed besides the thermodynamically favored indole (G. Stork, 1963; Y. Ban, 1965;, S. Klioze, 1975).

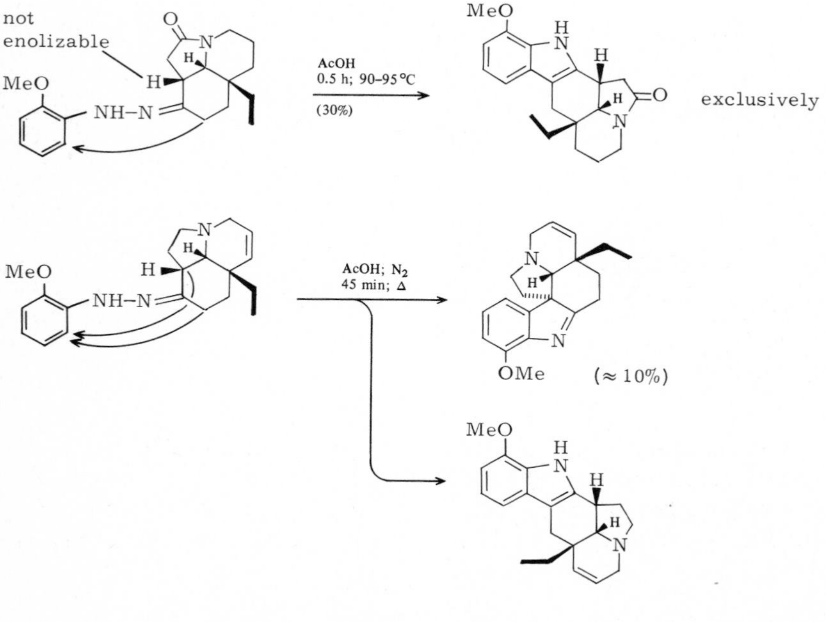

not
enolizable

AcOH
0.5 h; 90–95 °C
(30%)
→

exclusively

AcOH; N₂
45 min; Δ
→

(≈10%)

In Diels-Alder reactions a nitroolefin may function as an electron-deficient ene component or a 1,2-dihydropyridine derivative may be used as a diene component. Both types of reactants often yield cyclic amine precursors in highly stereoselective manner (R.K. Hill, 1962; G. Büchi, 1965, 1966A).

The Diels-Alder reaction of *o*-quinodimethanes from benzocyclobutenes with nitrogen-substituted enes has also been applied to alkaloid synthesis (see p. 253f.; T. Kametani, 1972, 1973, 1974; W. Oppolzer, 1978 A).

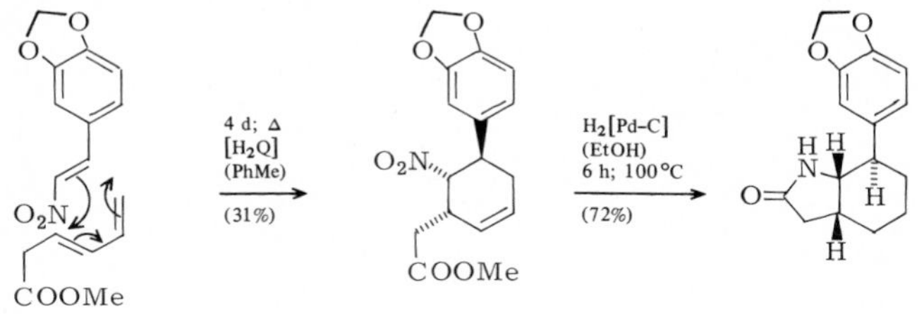

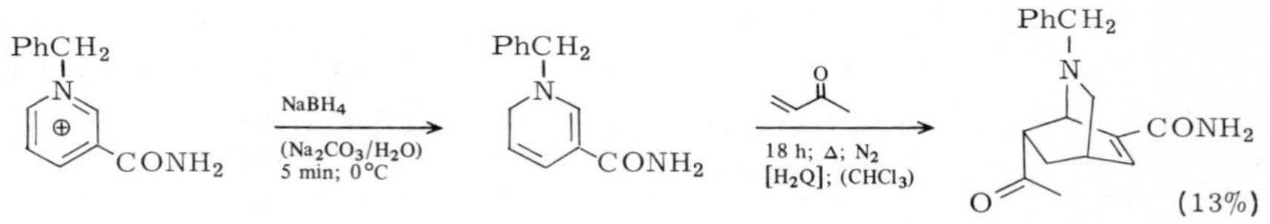

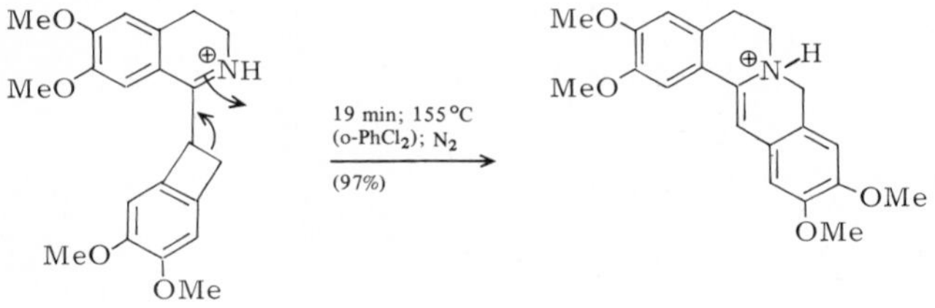

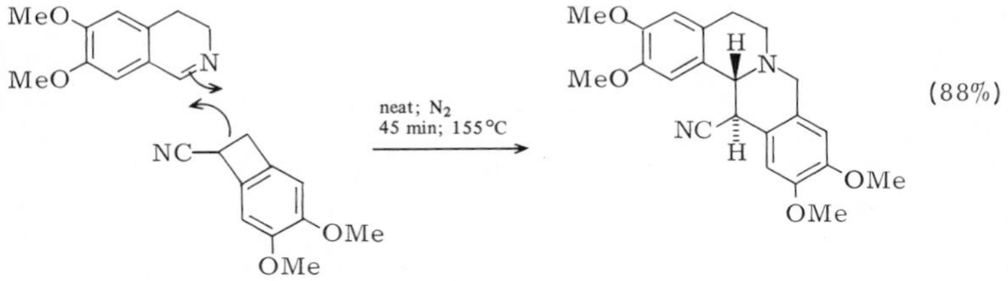

Thermal and photochemical electrocyclic reactions are particularly useful in the synthesis of alkaloids (W. Oppolzer, 1973, 1978 B; K. Wiesner, 1968). A high degree of regio-and stereoselectivity can be reached, if cyclic enamine components are used. It is usually found, that the nitrogen of the cyclic enamine and the electronegative substituent of the second ene component tend to avoid each other in photochemical [2 + 2]-cycloadditions, if the latter is an α,β-unsaturated carbonyl compound.

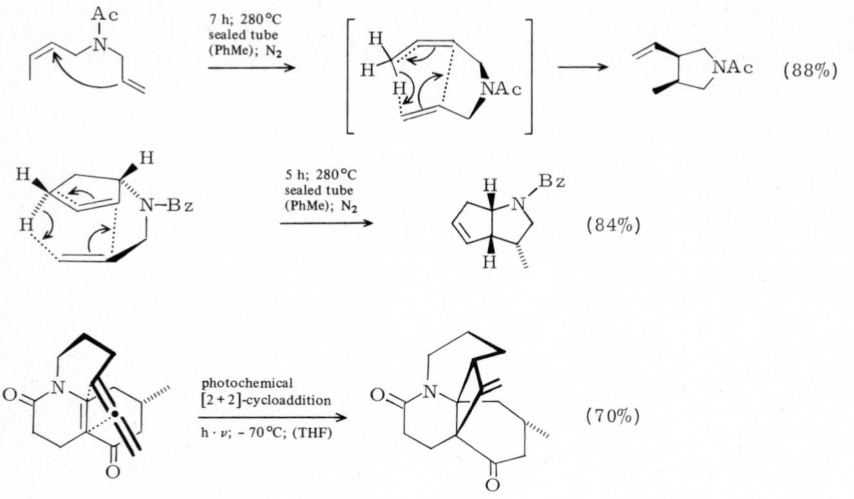

Finally a general approach to synthesize Δ²-pyrrolines must be mentioned. This is the acid-catalyzed (NH₄Cl or catalytic amounts of HBr) and thermally (150 °C) induced rearrangement of cyclopropyl imines. These educts may be obtained from commercial malonitriles, cyclopropyl cyanide, or benzyl cyanide derivatives by the routes outlined below. The rearrangement is reminiscent to the rearrangement of 1-silyloxy-1-vinyl cyclopropanes (p. 70) but since it is acid-catalyzed it occurs at much lower temperatures. Δ²-Pyrrolines constitute reactive enamines and may be used in further addition reactions such as the Robinson anellation with methyl vinyl ketone (R.V. Stevens, 1967, 1968, 1971).

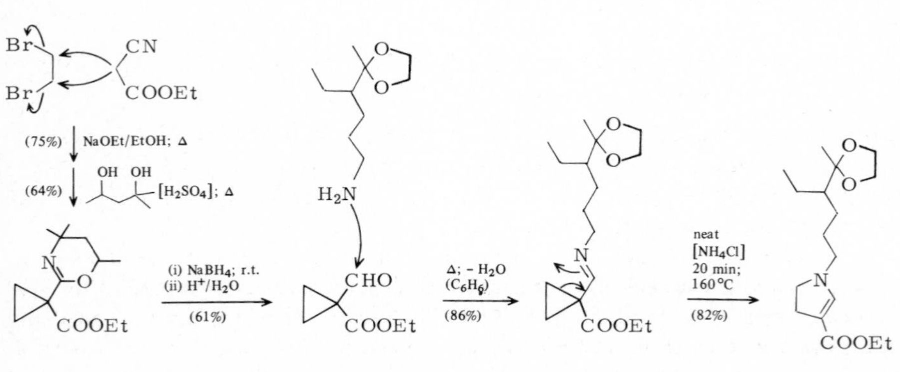

(i)–(v) (8%)   neat [HBr] 20 min; 148 °C (76%)   2.5 h; 57 °C + 1 h; 119 °C; N₂ (56%)

(±)-mesembrine

(i) BuLi/THF; 2 h; r.t.; N₂

(ii) Br Br (THF); 2 h; –78 °C → r.t. + 5 h; r.t. } (24%)

(iii) LiAlH₄/THF; 4 h; –78 °C → r.t. + 4 h; r.t.
(iv) + dil. aq. HCl } (38%)

(v) MeNH₂/MgSO₄(C₆H₆); 20; r.t.; –H₂O (91%)

(35%)
(i) Li
(Et₂O); –70 °C
(ii) + Na₂SO₄ · 10 H₂O

(75%)
neat [HCl] 20 min; 100 °C   apoferro-rosamine

4.7 Synthetic Drugs

The one hundred ethical drugs most frequently prescribed in the USA in 1974 (J.C. Stokes, 1975) may be divided into three chemical classes:

(i) Benzene derivatives with nitrogen containing side-chains.
(ii) Nitrogen heterocycles, sometimes also containing oxygen and/or sulfur.
(iii) Natural products and their synthetic analogues:
   a) Steroids (see section 4.5)
   b) Alkaloids (see section 4.6)
   c) Antibiotics (penicillins, cephalexin, tetracyclines, erythromycins, nystatin, gramicidins; see section 4.8).

Compounds of class (iii) are either totally synthetic (a few steroids) or derivatives of natural products isolated from plants or microorganisms. They are discussed within the indicated sections. We will briefly describe published syntheses of the most common compounds of classes (i) and (ii). Since most synthetic schemes contain only simple reactions, commentaries will be short. The difficulties in the development of low-cost, low-risk, large-scale, large-profit commercial syntheses will not and cannot be evaluated. Such discussions, within another context, can be found in K. Weissermel's (1978) most informative book. Most syntheses mentioned in this section together with several others are described in more detail in three recent books (G. Ehrhart, 1972; E. Schröder, 1975; D. Lednicer, 1977, 1980). We are trying to give a general impression of "medicinal synthetic chemistry".

4.7.1 Benzene Derivatives with a Nitrogen Containing Side-Chain

The major task in the synthesis of benzene derived drugs is the attachment of nitrogen containing side-chains to functional benzene derivatives. Two special properties of this class of compounds are generally exploited: (i) both, benzylic protons and nucleofuges (e.g. Br) can be substituted under relatively mild conditions; (ii) nucleophilic reactions at functional groups attached to a benzene nucleus are often highly regioselective, because the benzene nucleus is usually inert to nucleophiles and to bases.

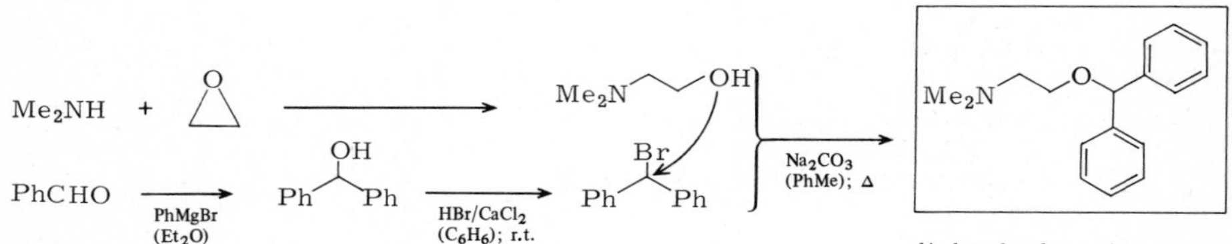

diphenhydramine
(antihistaminic)

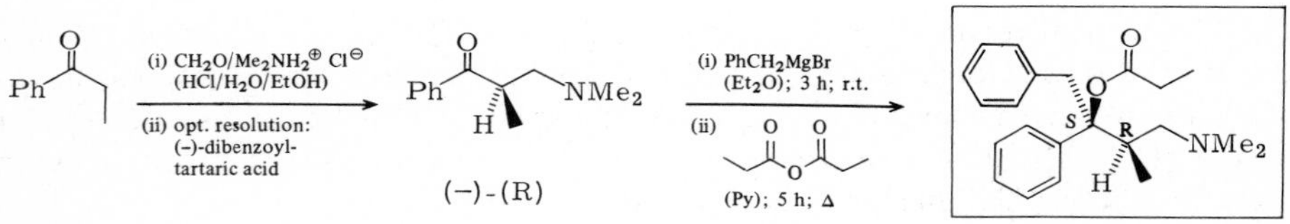

(+)-α-propoxyphene
(analgesic)

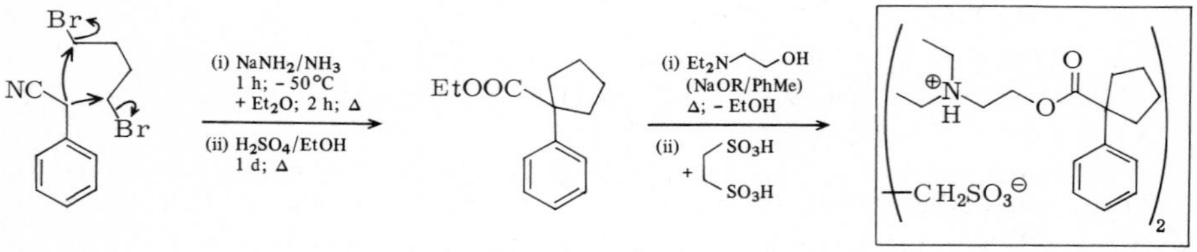

caramiphene ethanedisulfonate
(anticholinergic, antitussive)

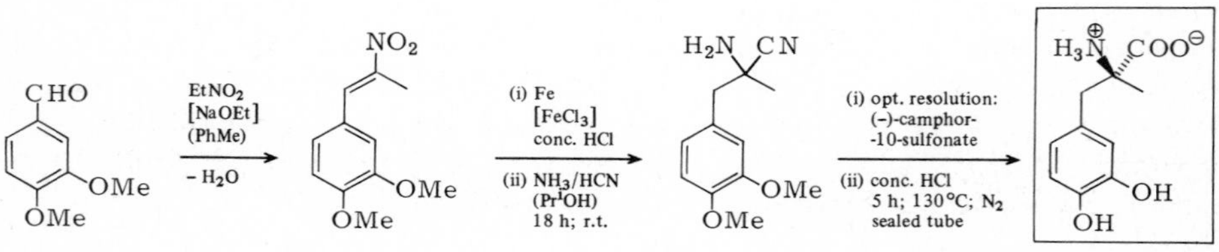

(−)-L-methyldopa
(antihypertensive)

A special problem arises in the preparation of secondary amines. These compounds are highly nucleophilic, and alkylation of an amine with alkyl halides cannot be expected to stop at any specific stage. Secondary amides, however, can be monoalkylated and hydrolyzed or be reduced to secondary amines (p. 102). In the elegant synthesis of phenylephrine an intermediate β-hydroxy isocyanate (from a hydrazide and nitrous acid) cyclizes to give an oxazolidinone which is monomethylated. Treatment with strong acid cleaves the cyclic urethan.

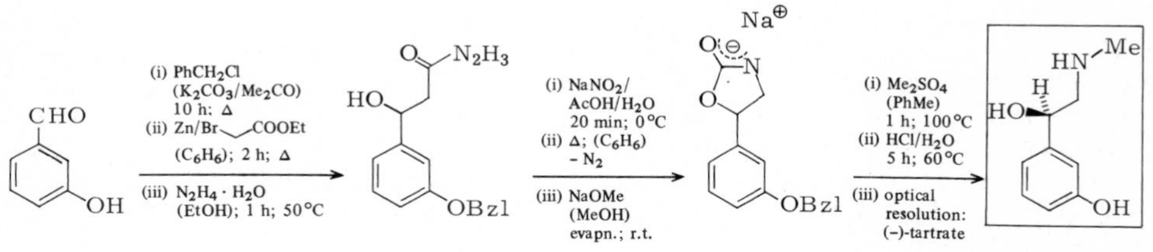

(−)-(R)-phenylephrine
(adrenergic)

Urea derivatives are of general interest in medicinal chemistry. They may be obtained either from urea itself (barbiturates, see p. 277) or from amines and isocyanates. The latter are usually prepared from amines and phosgene under evolution of hydrogen chloride. Alkyl isocyanates are highly reactive in nucleophilic addition reactions. Even amides, e.g. sulfonamides, are nucleophilic enough to produce urea derivatives.

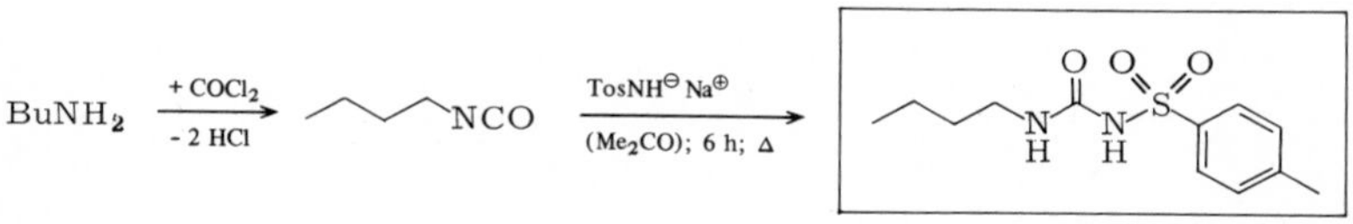

tolbutamide (hypoglycemic)

Other aromatic best-sellers in the pharmacy are given below. Syntheses of these compounds are simple and may be outlined by the interested reader himself. The only common "open-chain" synthetic drug is meprobamate. Its conventional synthesis is given.

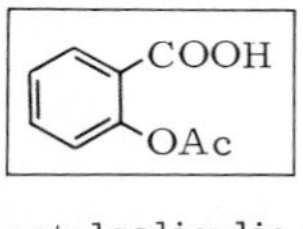

O-acetylsalicylic acid
= aspirin
(analgesic, antipyretic,
anti-inflammatory)

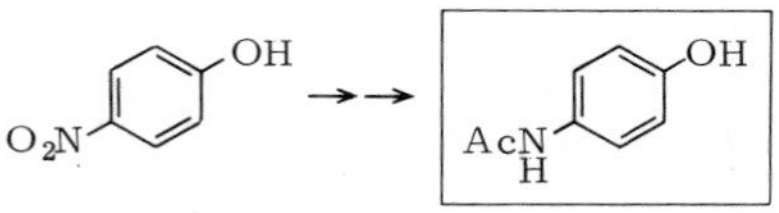

acetaminophen

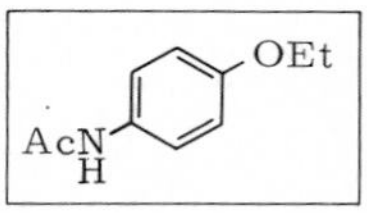

acetophenetidin
= phenacetin

(analgesic, antipyretic)

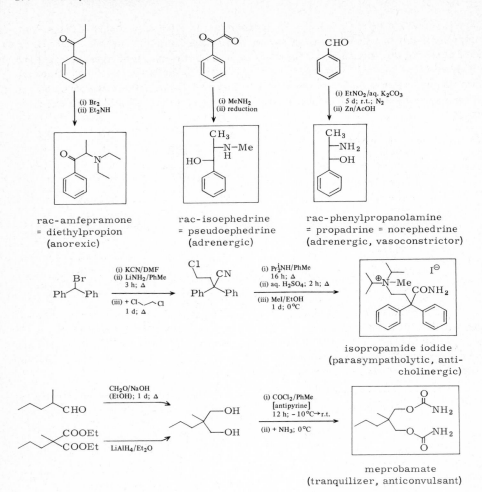

rac-amfepramone
= diethylpropion
(anorexic)

rac-isoephedrine
= pseudoephedrine
(adrenergic)

rac-phenylpropanolamine
= propadrine = norephedrine
(adrenergic, vasoconstrictor)

isopropamide iodide
(parasympatholytic, anti-
cholinergic)

meprobamate
(tranquilizer, anticonvulsant)

4.7.2 Nitrogen Heterocycles

Although the structures of heterocyclic drugs often look complicated, their synthesis is usually of the outmost simplicity. We start with syntheses in which benzylic anions or halides are again the key synthons.

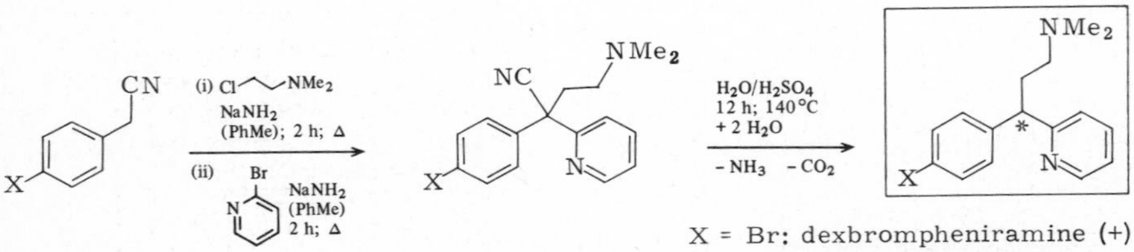

X = Br: dexbrompheniramine (+)
        rac-brompheniramine
X = Cl: rac-chlorpheniramine
        (antihistaminics)

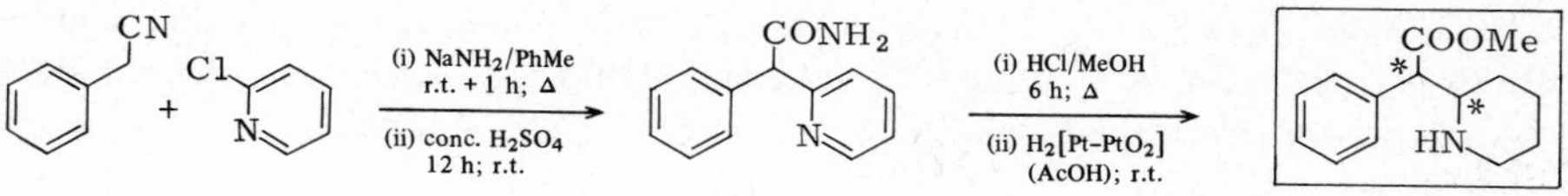

rac-methylphenidate
(central stimulant)

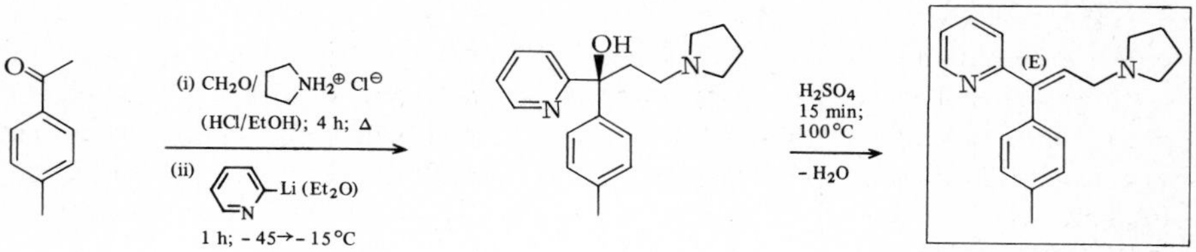

triprolidine
(antihistaminic)

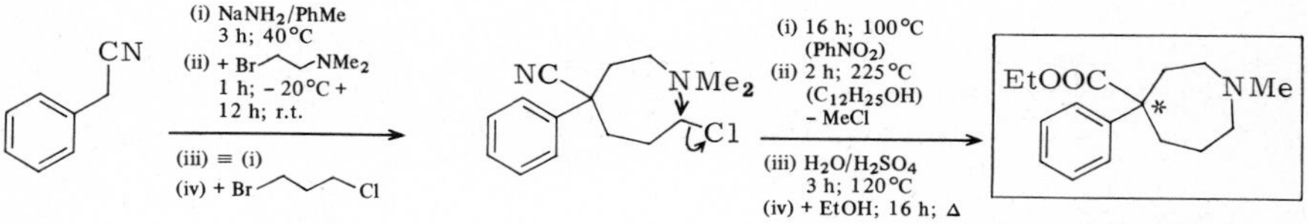

rac-ethoheptazine
(analgesic)

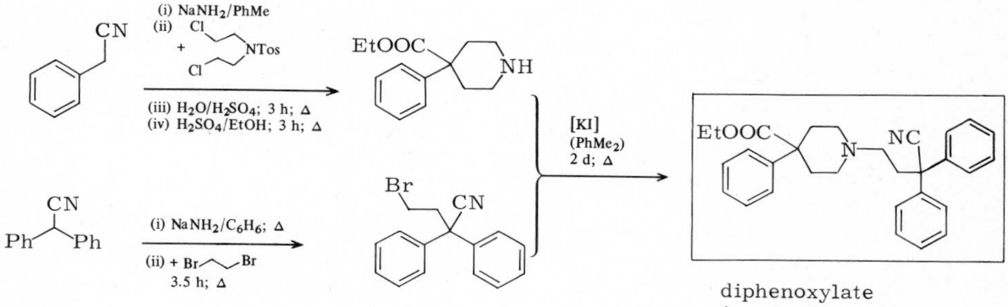

diphenoxylate
(antiperistaltic)

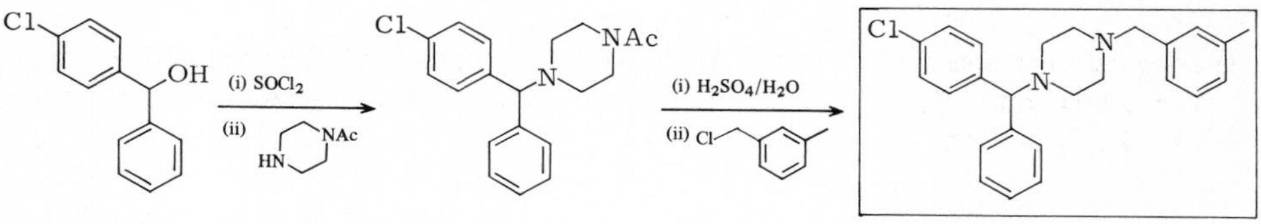

rac-meclizine = meclozine
(antinauseant)

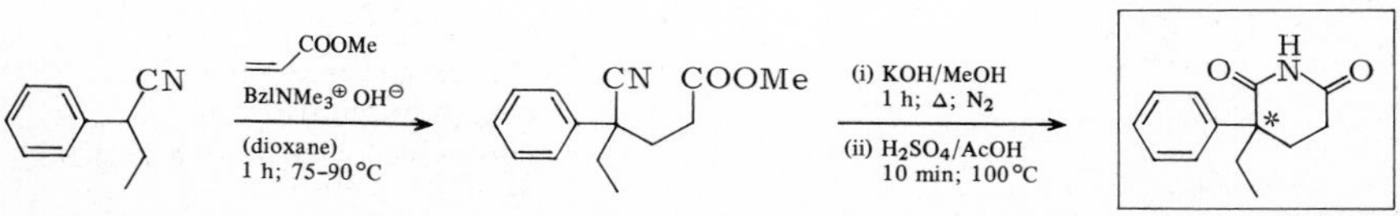

rac-glutethimide
(sedative, hypnotic)

The following examples give some insight into typical syntheses of heterocyclic benzene derivatives. Nitrobenzene forms radicals on reduction with metals, e.g. iron powder, and the dimerization product hydrazobenzene can be isolated in high yield under appropriate conditions. Cyclization of the hydrazine with alkylated malonic esters yields useful drugs, e.g. phenylbutazone. Electron-withdrawing substituents such as carboxyl and sulfamoyl activate the benzene ring towards nucleophilic substitution. Chlorine may be replaced by primary amines. The regioselectivity observed in the synthesis of furosemide may be caused by steric effects of the large sulfamoyl group. The exclusive chloromethylation of carbon atom 5 in the synthesis of oxymetazoline is caused by the strong sterical hindrance exerted by the bulky *tert*-butyl group. This is especially severe in planar cyclic compounds. Finally, the benzene nucleus stabilizes the diazonium group. Such salts are useful reactants in electrophilic substitutions. Azo compounds are generally used as components of dye synthesis, but a drug example also exists.

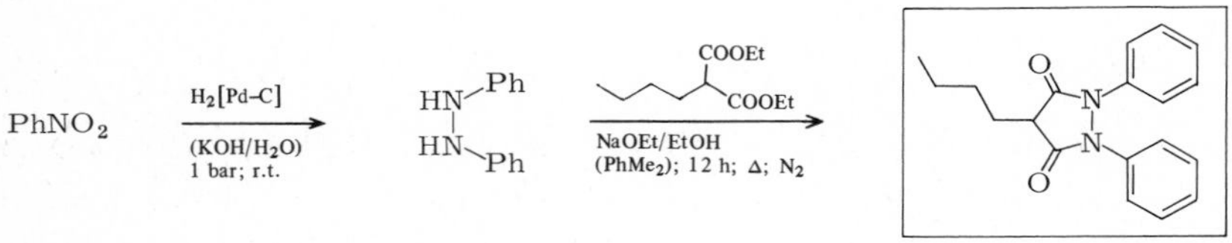

phenylbutazone
(anti-inflammatory)

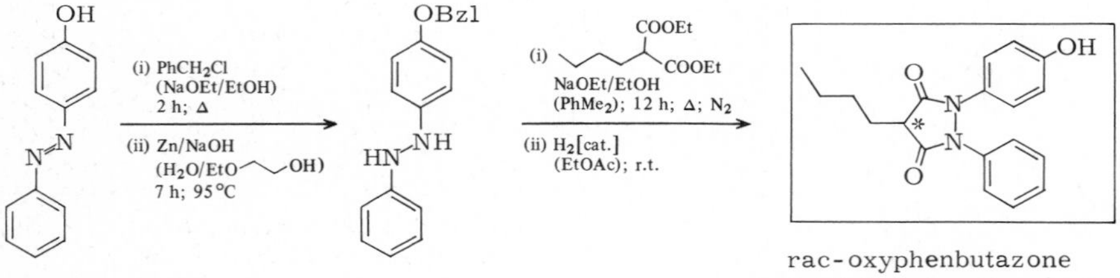

rac-oxyphenbutazone
(anti-inflammatory)

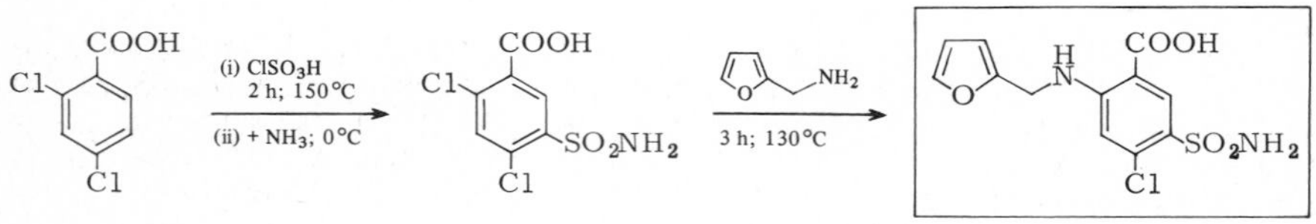

furosemide
(diuretic, antihypertensive)

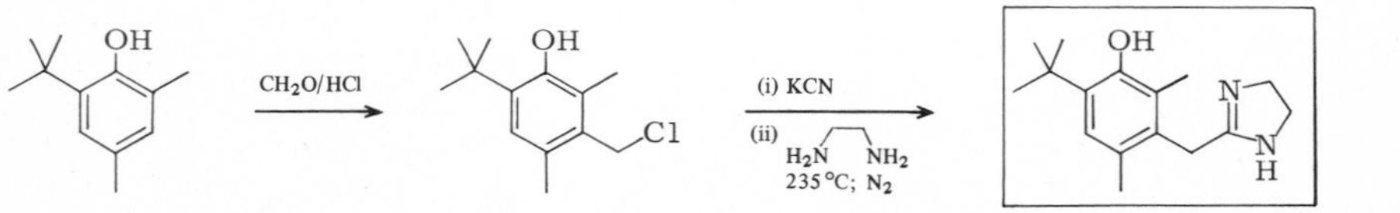

oxymetazoline
(adrenergic, vasoconstrictor)

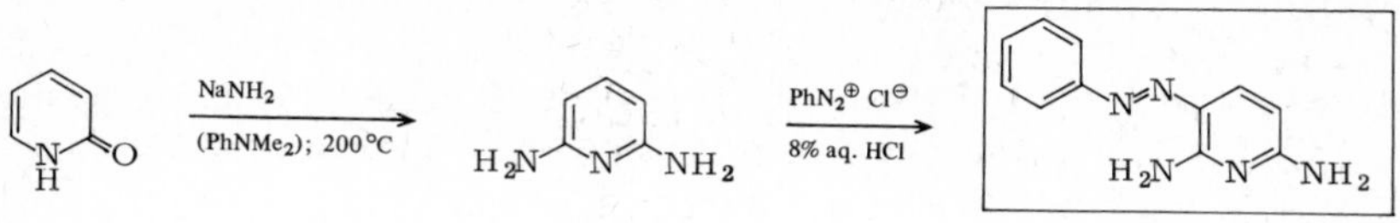

phenazopyridine
(urinary analgesic)

The nucleophilicity of the nitrogen atom survives in many different functional groups, although its basicity may be lost. Reactions of non-basic, but nucleophilic urea nitrogens provide, for example, an easy entry to sleeping-pills (barbiturates) as well as to stimulants (caffeine). The nitrogen atoms of imidazoles and indole anions are also nucleophilic and the NH protons can be easily substituted.

barbiturates:

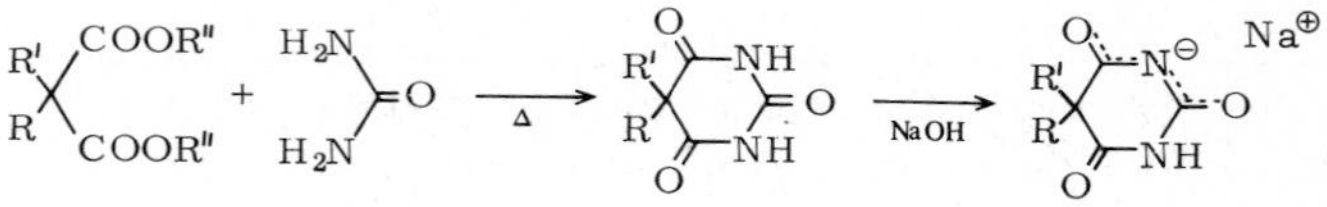

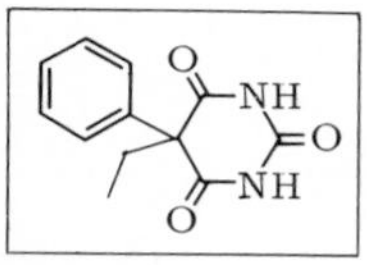

phenobarbital
(anticonvulsant,
hypnotic, sedative)

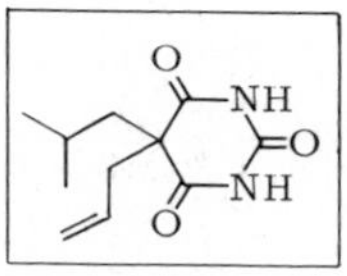

butalbital
(sedative)

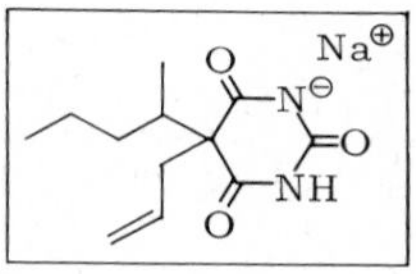

secobarbital
sodium
(hypnotic)

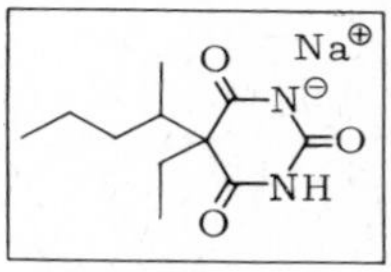

pentobarbital
sodium
(hypnotic, sedative)

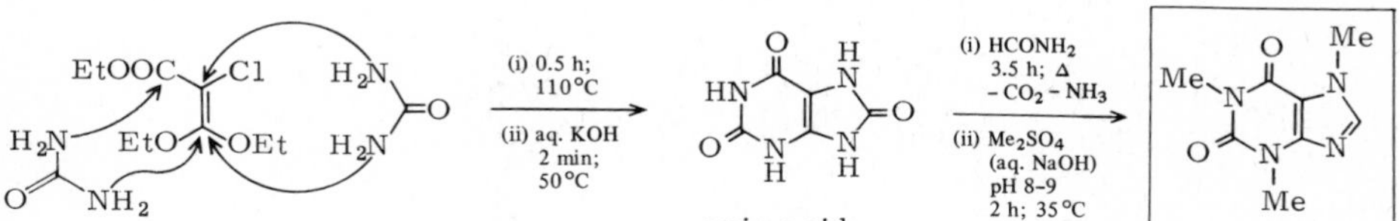

uric açid

caffeine
(central stimulant)

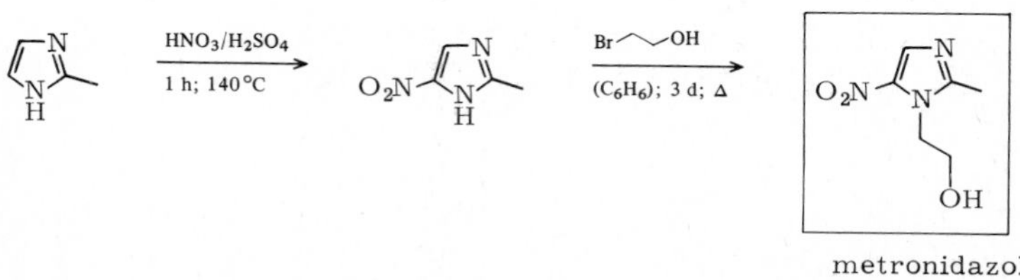

metronidazole
(antiprotozoal, trichomonacide)

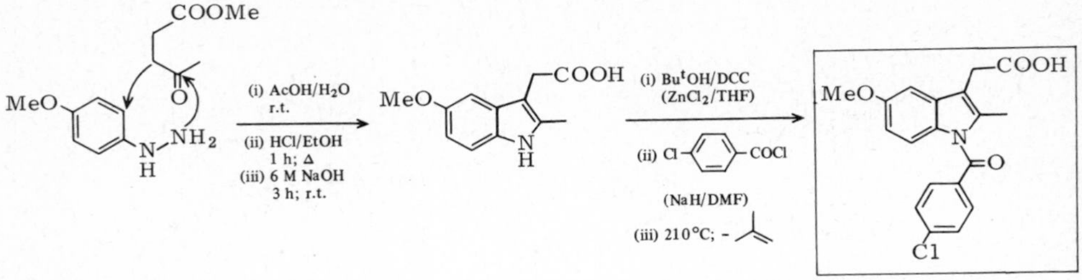

indomethacin
(anti-inflammatory,
antipyretic, analgesic)

The use of oximes as nucleophiles can be quite perplexing in view of the fact that nitrogen or oxygen may react. Alkylation of hydroxylamines can therefore be a very complex process which is largely dependent on the steric factors associated with the educts. Reproducible and predictable results are obtained in intramolecular reactions between oximes and electrophilic carbon atoms. Amides, halides, nitriles, and ketones have been used as electrophiles, and various heterocycles such as quinazoline N-oxide, benzodiazepines, and isoxazoles have been obtained in excellent yields under appropriate reaction conditions.

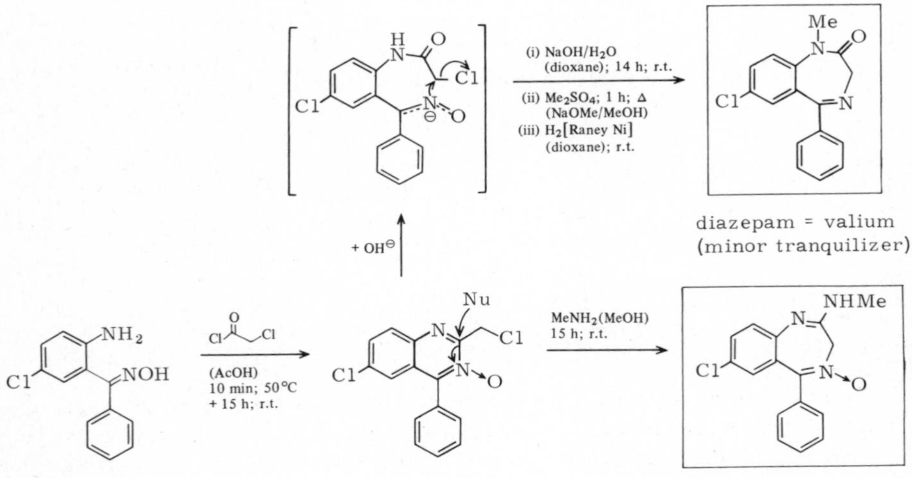

diazepam = valium
(minor tranquilizer)

chlordiazepoxide = librium
(minor tranquilizer)

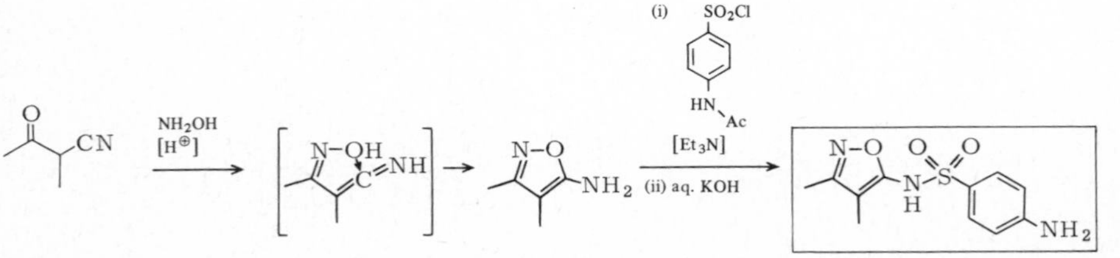

sulfisoxazole = sulfafurazole
(antibacterial)

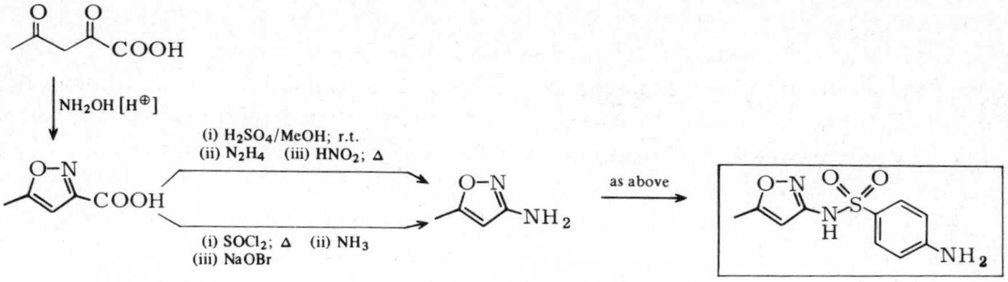

sulfamethoxazole
(antibacterial)

The nucleophilicity of amine nitrogens is also differentiated by their environments. In 2,4,5,6-tetraaminopyrimidine the most basic 5-amino group can be selectively converted to a Schiff base. It is *meta* to both pyrimidine nitrogens and does not form a tautomeric imine as do the *ortho-* and *para*-amino groups. This factor is the basis of the commercial synthesis of triamterene.

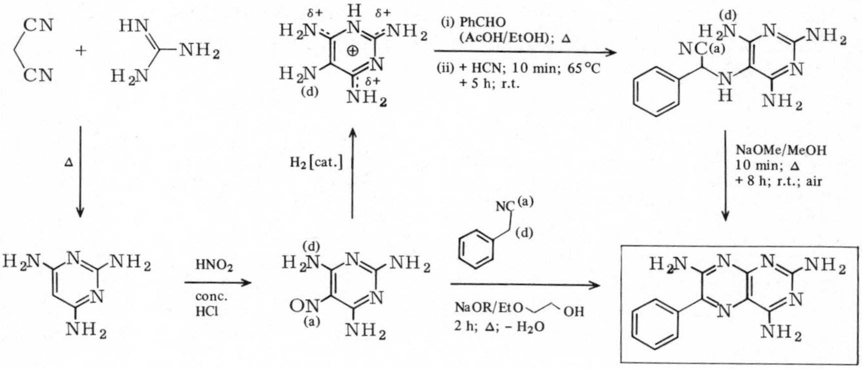

triamterene
(diuretic)

Other interesting regioselective reactions are carried out within the synthesis of nitrofurantoin. Benzaldehyde semicarbazone substitutes chlorine in chloroacetic ester with the most nucleophilic hydrazone nitrogen atom. Transamidation of the ester occurs with the diprotic outer nitrogen atom. Only one nucleophilic nitrogen atom remains in the cyclization product and reacts exclusively with carbonyl compounds.

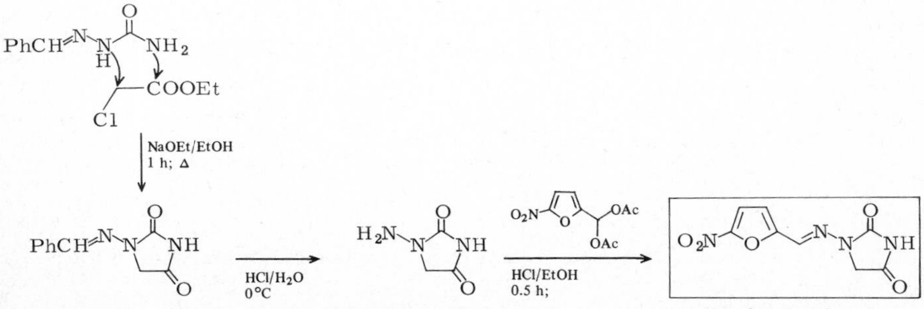

1-aminohydantoin

nitrofurantoin
(urinary antibacterial)

Two synthetic bridged nitrogen heterocycles are also prepared on a commercial scale. The pentazocine synthesis consists of a reductive alkylation of a pyridinium ring, a remarkable and puzzling addition to the most hindered position, hydrogenation of an enamine, and acid-catalyzed substitution of a phenol derivative. The synthesis is an application of the reactivity rules discussed in the alkaloid section. The same applies for clidinium bromide.

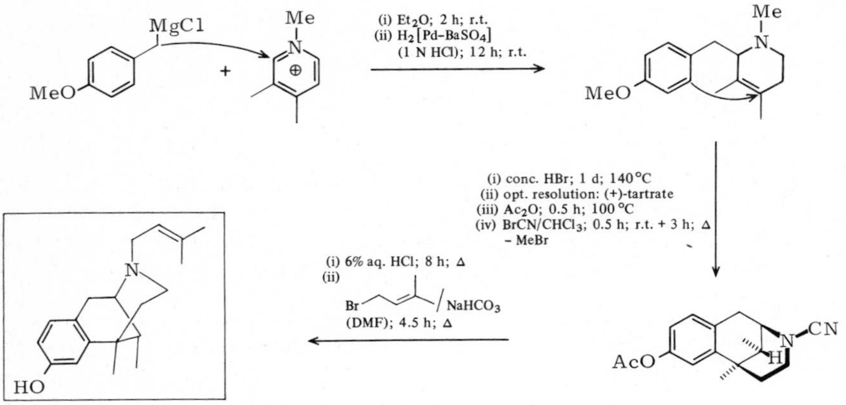

(−)-pentazocine (analgesic)

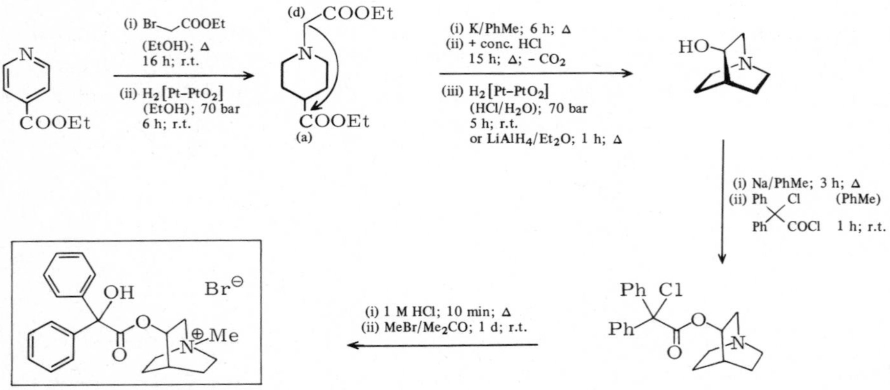

clidinium bromide
(anticholinergic)

In the synthesis of commercial sulfur-heterocycles two interesting reactions are used: (i) diphenylamines may be connected by a sulfur bridge in the *ortho*-positions; (ii) the amino grouping of sulfonamides undergoes condensation reactions with neighboring imino- and amide groups.

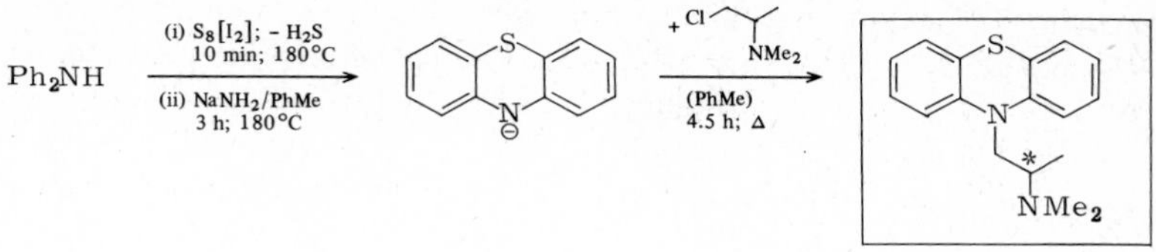

rac-promethazine
(antihistaminic)

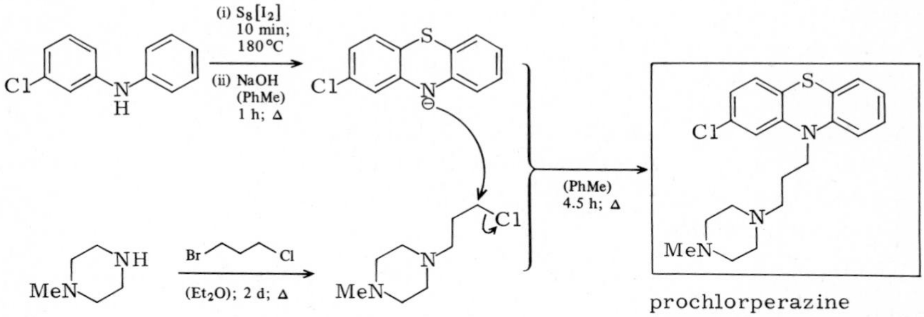

prochlorperazine
(tranquilizer, antiemetic)

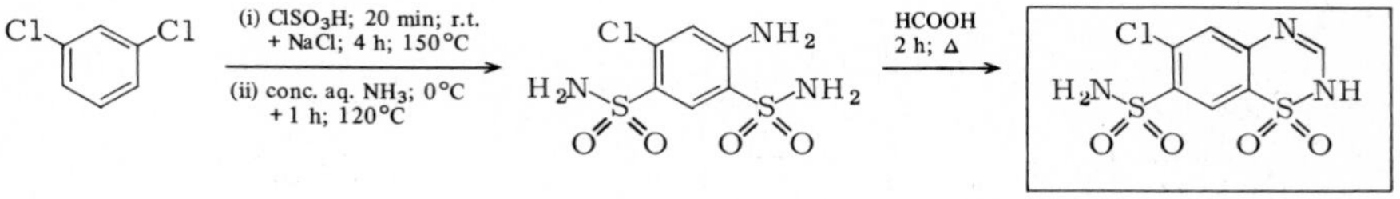

chlorothiazide
(diuretic, antihypertensive)

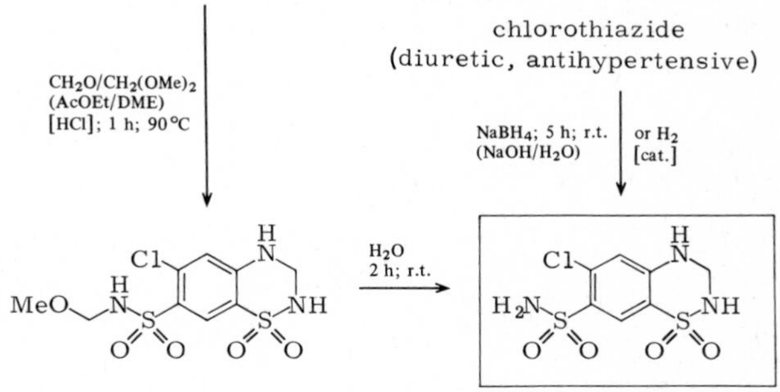

hydrochlorothiazide
(diuretic)

4.8 Antibiotics

Antibiotics are toxic water-soluble compounds produced by molds or bacteria which inhibit the growth of other microorganisms. For an antibiotic to be useful in medicine it should have a high order of selective toxicity to microorganisms which are pathogenic to man. Although several hundred antibiotics are known, only four types are used on a large scale in therapy:

(i) β-Lactams (penicillins, cephalexin).
(ii) Tetracyclines (tetracyclines, doxycycline, oxytetracycline).
(iii) Macrolides (erythromycin, nystatin).
(iv) Peptides (gramicidin, see p. 220f.).

All these highly complex compounds are produced on the industrial scale from microorganisms. Only the β-lactams are modified chemically after the isolation from the fermentation broth. We shall describe these "partial syntheses" as well as some typical synthetic sequences of academic total syntheses.

4.8.1 β-Lactams

The most commonly prescribed penicillins are phenoxymethylpenicillin, ampicillin, and penicillin G.

"penam"
= 4-Thia-1-azabicyclo[3.2.0]-heptan-7-one*

(3S,6R,7R)--6-amino--penicillanic acid (6-APA)

penicillins

R =
PhCH$_2$– benzylpenicillin
= penicillin G
(usual fermentation product)

PhOCH$_2$– phenoxymethylpenicillin
= penicillin V

Ph—|—NH$_2$ (with H above) ampicillin

Under natural conditions various strains of *Penicillium* fungi produce either penicillin G or free 6-aminopenicillanic acid (= 6-APA). The techniques used to prepare analogues such as the ones given above have been: (i) fermentation in the presence of an excess of appropriate acids which may be incorporated as side-chain; (ii) chemical acylation of 6-APA with activated acid derivatives.

6-APA may be either obtained directly from special *Penicillium* strains or by hydrolysis of penicillin G with the aid of amidase enzymes. A major problem in the synthesis of different amides from 6-APA is the acid-sensitivity of its β-lactam group. One synthesis of ampicillin applies the condensation of 6-APA with a mixed anhydride of N-protected phenylglycine.

---

* The formula shows the traditional (nonsystematic) numbering used in this section.

Catalytic hydrogenation removes the N-protecting group. Yields are low ($\leqq$ 30%) (*without scheme*). Another synthesis avoids the isolation of 6-APA and starts directly with penicillin G. Reaction with chloromethyl pivalate gives its pivaloyloxymethyl ester. This reacts with PCl_5 to an imidoyl chloride which may be solvolyzed with propanol. The acid chloride of R-phenylglycine is added to yield ampicillin pivaloyloxymethyl ester, or "pivampicillin", an ampicillin for oral application. After careful optimization of reaction conditions the cleavage of the β-lactam unit could be totally avoided (W. von Daehne, 1970).

pivampicillin

A similar cleavage of the amide linkage in natural cephalosporin C has been developed to synthesize 7-amino-cephalosporanic acid (= 7-ACA). In this case no biochemical hydrolysis procedure has been described so far. Treatment with nitrosyl chloride directly yields a cyclic imidic ester (presumably via an unstable diazonium salt) which can be hydrolyzed in the presence of weak acids. In this case the carboxyl group needs not to be protected since it reacts only slowly with nitrosyl chloride (R.B. Morin, 1962). A more recent method applies phosphorus pentachloride at low temperature. The carboxyl groups are protected by O-silylation (B. Fechtig, 1968) and deblocked upon acidic hydrolysis. Condensation with N-protected phenylglycine, e.g. by the mixed anhydride method, and removal of the protecting group by mild acid treatment after hydrogenolysis of the allylic 3'-acetoxy group yields the commercial antibiotic cephalexin (J.L. Spencer, 1966; C.W. Ryan, 1969).

"cepham"

= 5-Thia-1-aza-
-bicyclo[4.2.0]-
-octan-7-one

"Δ^3-cephem"

7-aminocephalosporanic
acid (7-ACA): R = H

cephalosporin C:

R = $^\ominus OOC$ ... (aminoadipyl-7-ACA)

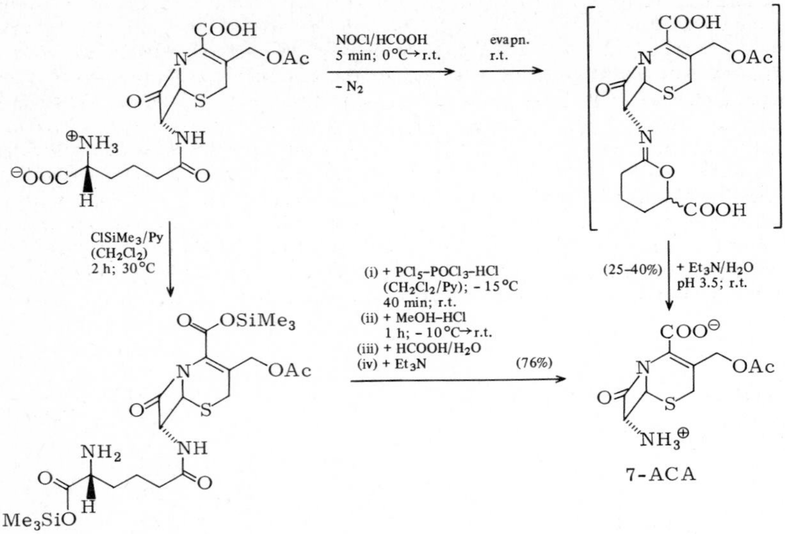

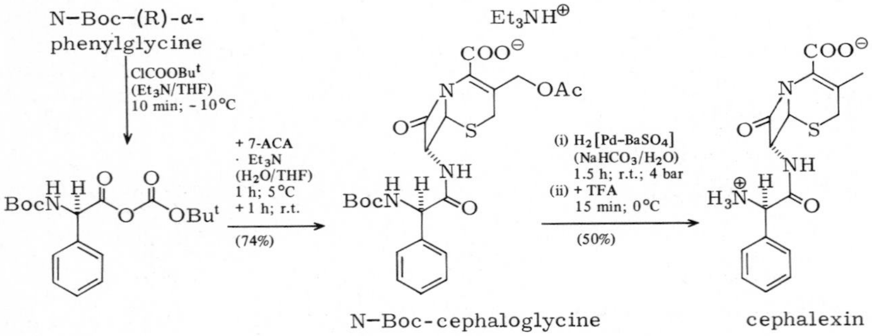

N-Boc-cephaloglycine   cephalexin

Although in β-lactam antibiotics the four-membered ring is highly acid-labile, this is not generally the case with substituted β-lactams. Only the attachment of double bonds or small rings render the β-lactams unstable (R.B. Woodward, 1977). The final ring closure in total synthesis may therefore be either on the sulfur-containing heterocycle or on the β-lactam unit. Both reactions should strictly avoid acidic catalysts, and examples of such ring closures have been given earlier (p. 31f. and 131). The classical total syntheses of penicillins and cephalosporins have been described in I. Fleming's (1973) book and need not to be repeated here.

J.E. Baldwin (1975, 1976A) has developed a "biomimetic" synthesis which is loosely analogous to the biosynthetic pathway which starts with the so-called Arnstein tripeptide. Baldwin used bicyclic dipeptides more suitable for stereoselective *in vitro* syntheses.

In his cephalosporin synthesis methyl levulinate was condensed with cysteine in acidic medium to give a bicyclic thiazolidine. One may rationalize the regioselective formation of this bicycle with the assumption that in the acidic reaction mixture the thiol group is the only nucleophile present, which can add to the ketone. Intramolecular amide formation from the methyl ester and acid-catalyzed dehydration would then lead to the thiazolidine and γ-lactam rings. The stereochemistry at the carboxylic acid α-carbon is defined by the chiral educt. The assignment of the position of the angular methyl group has been based on the fact that only a single isomer was formed and on the assumption that a *trans*-position to the carboxyl group should be favored. This bicyclic acid was connected with (R)-isodehydrovaline benzyl ester to give a dipeptide. Stereoselective (because of the large neighboring carboxamide group) chlorination of the thioether with dibenzoyl peroxide and HCl provided the necessary *trans*-chloro leaving group for β-lactam formation. Treatment of the chloride with sodium hydride gave one tricyclic β-lactam (A) in quantitative yield. In (A) the thioether bears two lactam N-atoms in the α-positions and has therefore a rather symmetric environment. Reaction with *tert*-butyl hypochlorite, however, opened the central five-membered ring only between S and C(1) of the γ-lactam and not between S and C(2) of the β-lactam. The intermediate sulfenyl chloride attacked the carbon-carbon double bond to yield the 3β-chloromethyl-penam (B) in 55% yield. In DMF solution the penicillin-like (B) isomerized to the cephalosporin analogue chlorocepham (C).

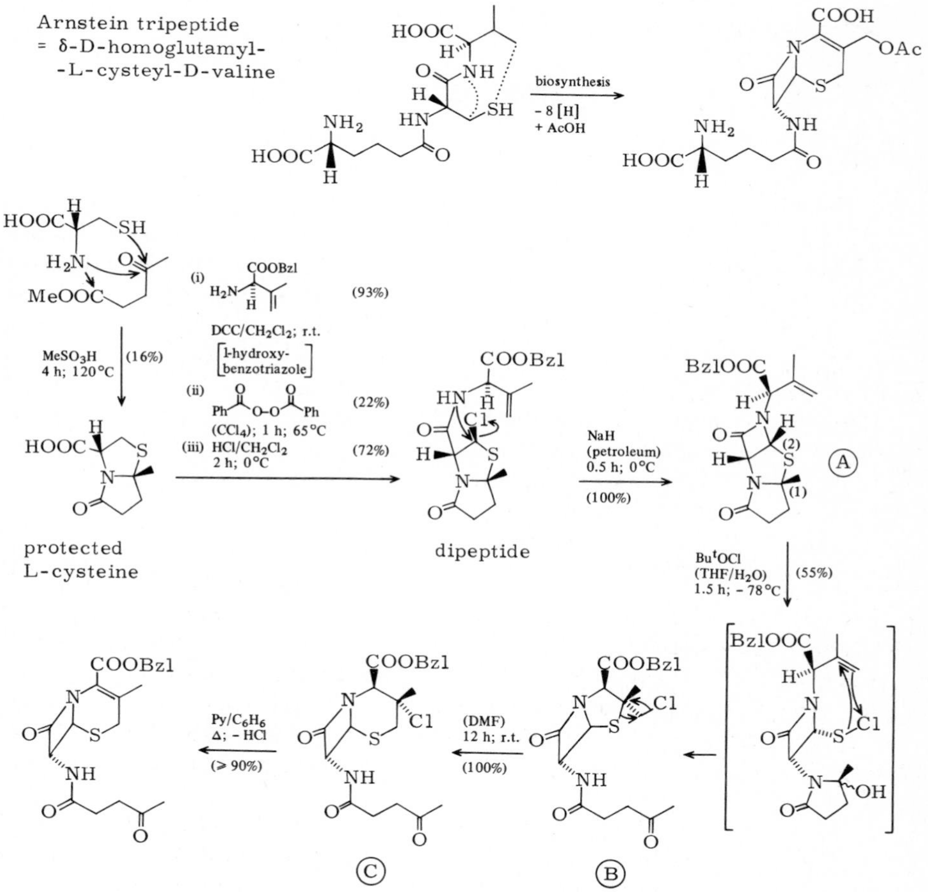

The high regioselectivity (''stereoelectronic control'') in the ring cleavage by chlorination of sulfur was anticipated. It had been found before that in corresponding bicyclic systems such as in the scheme below oxidation of the sulfur atom always led to the undesired cleavage of the $S-C_B$ bond. This was rationalized through the observation on molecular models that $S-C_B$ is perpendicular to the amide plane of the β-lactam and therefore weakened. The $S-C_A$ bond, on the other hand, is not affected by electronic interactions with the benzamide plane. It was now thought, that a bridging of the thiazolidine moiety would bring the C_A-S bond into a more orthogonal position with respect to the amide plane of the new lactam and make this bond more fragile. The tricyclic thiazolidine was synthesized as described above and fulfilled the predictions (J.E. Baldwin, 1978).

The high reactivity of β-lactam antibiotics in penicillins and cephalosporins has been explained on a similar basis by R.B. Woodward (1977; I. Ernest, 1978). Fusion of the β-lactam with a five-membered ring (penicillins) results in a pyramidal geometry of the β-lactam nitrogen. The presence of conjugated double bonds (cephalosporins) tends to withdraw electron density from the nitrogen atom. Both effects destabilize the amide bond, since the usual electron distribution of the nitrogen electrons between the nitrogen, carbon, and oxygen atoms of the amide group is disturbed. In spite of the instability of their four-membered lactam rings, it is possible to convert penicillins to cephalexins by skeletal rearrangements. One example, namely thioether cleavage with t-butyl hypochlorite, rearrangement, and dehydrohalogenation, has been described above. This reaction is analogous to the thermal cleavage of sulfoxides (p. 60), which has been used to convert ''penams'' to ''cephams''. Trapping of the intermediate sulfenic acids with trimethylsilyl chloride improves the yields (T.S. Chou, 1974).

7β-phthalimido-
penicillanic acid
methyl ester

J.E. Baldwin (1976 B) also introduced a set of rules for closure reactions of 3- to 7-membered rings which are derived from similar arguments about "stereoelectronic control". The specific rules are clearly explained in the original paper and need not to be repeated here since their utility seems to be restricted to some specialized cases. The assumed physical basis of the rules, however, is of general interest and should be outlined. Cyclization reaction usually involve the intramolecular attack of a nucleophile at an electron acceptor. This may be tetrahedral (e.g. a bromide), trigonal (e.g. a carbonyl group or activated C=C double bond) or digonal (an alkyne). A transition state is first formed in which the attacking nucleophile and the electron accepting group approach bond angles of 180°, 109°, and 60° resp. The chain that links the nucleophile and the electron accepting group may, however, restrict the relative motion of both reactants. The attainment of the transition state may require severe distortion of bond angles and distances within the linking chain. In such cases, the ring closure should be disfavored. Favored ring closures, on the other hand, are those in which the length and nature of the linking chain enables the terminal atoms to achieve the required trajectories to form the final ring bond. This generalization about selectivity in intramolecular reactions can be of great help in synthesis design.

An example (J.E. Baldwin, 1976 C) is again given without referring to formal rules. The γ-hydroxy ester with a carbon-carbon double bond in conjugation with the ester group can undergo either lactonization or intramolecular Michael type addition. The addition of the hydroxy group to the carbon-carbon double bond, however, is unlikely because the oxygen atom must approach the terminal carbon atom on a trajectory perpendicular to the plane of the C=C double bond (O—C—C angle $\approx$ 109°, see p. 29). The connecting chain of two methylene groups allows the correct distance of the reacting groups, but by no means the correct bond angles of the transition state. Lactonization, on the other hand, is without problems since the carbomethoxy group can be turned into a position which allows a transition-state-like trajectory of the hydroxyl group and the correct distance for bond formation. Experiments show that upon treatment with base only the expected lactone was formed, although the double bond is very susceptible to Michael type additions with nucleophilic reagents, e.g. methoxide.

········> possible trajectory of the reacting groups

4.8.2 Tetracyclines

Tetracyclines are produced by various *Streptomyces* strains and are extensively applied in human and veterinary medicine. They display a broad spectrum of antimicrobial activity in combination with low toxicity and can be applied orally. The most commonly prescribed tetracycline drugs are tetracycline itself and oxytetracycline, an oxygenated derivative, which are directly isolated from fermentation liquors, and doxycycline whose partial synthesis from oxytetracycline (C.R. Stephens, 1963) indicates some typical neighboring group effects found in tetracyclines. If C-11a in ring C is chlorinated, hemiketal formation between C=O(12) and OH(6) is observed. Acid-catalyzed dehydration and sulfite-dechlorination produce the 6-methylene compound. The 5- and 12a-alcohols do not react in this dehydration procedure. The methylene group is finally reduced by catalytic hydrogenation.

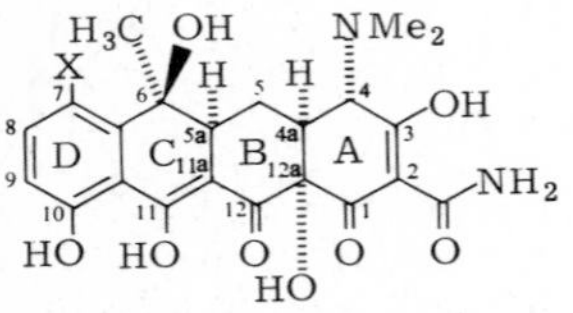

X = H: tetracycline = achromycin
      (streptomyces viridifaciens,
       streptomyces aureofaciens)
X = Cl: chlortetracycline = aureomycin
      (streptomyces aureofaciens)

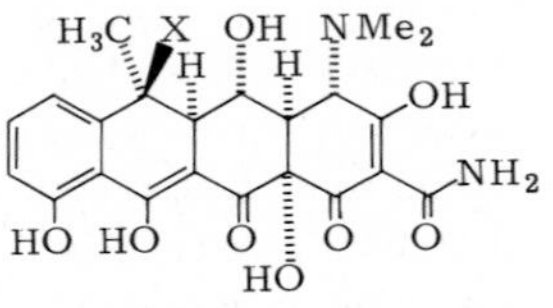

X = OH: oxytetracycline = terramycin
       (streptomyces rimosus)
X = H: doxycycline = vibramycin
       (more stable artificial
        derivative)

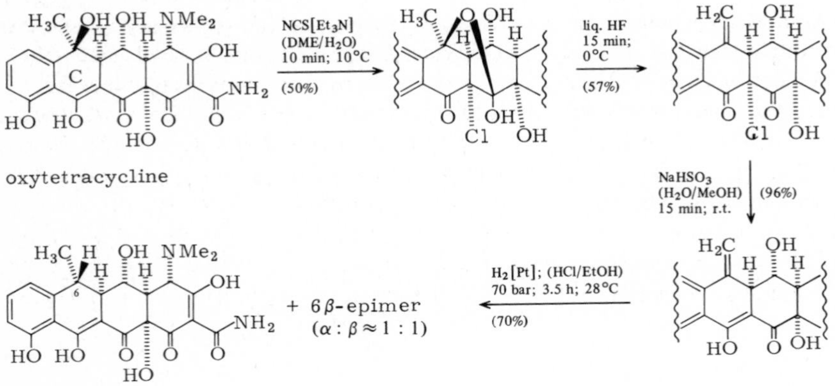

Tetracyclines except doxycycline easily dehydrate to give a phenolic ring C. Another mode of degradation is epimerization at C-4. The 4α-dimethylamino function interacts strongly with the 5α- and 12aα-OH groups . Enolization of C-4 at the β-tricarbonyl system in ring A is easily possible, and in aqueous solution at pH 3-5 an equilibrium mixture of about equal amounts of both 4-epimers is rapidly formed. Both reactions, dehydration and epimerization, lead to compounds which are inactive as antibiotics and may be more toxic. Isolation and storage of unaltered tetracyclines is therefore a major problem in their medicinal chemistry. It should also be mentioned that ring A contains a conjugated keto-enol which is reminiscent of the structure found in ascorbic acid. The observed pK_a of 3 of this system indicates the validity of this comparison. The 2-carboxamide group inbetween is extraordinarily electronrich and reactive. Its nitrogen atom is nucleophilic enough to react with Mannich reagents (W.J. Gottstein, 1959).

rolitetracycline

The total synthesis of several tetracyclines has been worked out by H. Muxfeldt (1979). One approach, the thiazolone method, is briefly outlined. Acetylation of commercial juglone and subsequent Diels-Alder reaction with 1-acetoxybutadiene led stereoselectively to an adduct with the 5-acetate* group in β-position. The adduct was methylated at C-6 from the less hindered α-side. Several standard reactions were used to transform the cyclohexenol ring to aldehyde (A). This was condensed with a thiazolone carbanion (see related anion on p. 45) to (B) in high yield. Michael addition of the anion of 3-oxoglutaric acid monoamide monoester and subsequent treatment with lithium *t*-butoxide yielded the tetracyline derivative (C) in 27% yield in a "one-pot reaction". The three condensation steps occurred in a regioselective way because the acceptor and donor properties of all six reaction centres were differentiated in the right order, although the discriminations between the three donor and five acceptor centres are not very large. If, however, aI and dI have been combined, only the given order I,II,III would produce the six-membered rings favored by thermodynamics. Hydroxylation of the 12a-carbanion with O_2 in the presence of triethyl phosphite (see p. 111) and standard deprotection and N-alkylation reactions gave racemic oxytetracycline in 10% yield from (C). The most stable, natural keto-enol tautomer of the product was formed during the final steps in protic solvents.

---

* The numbering system of tetracyclines is used for synthetic intermediates.

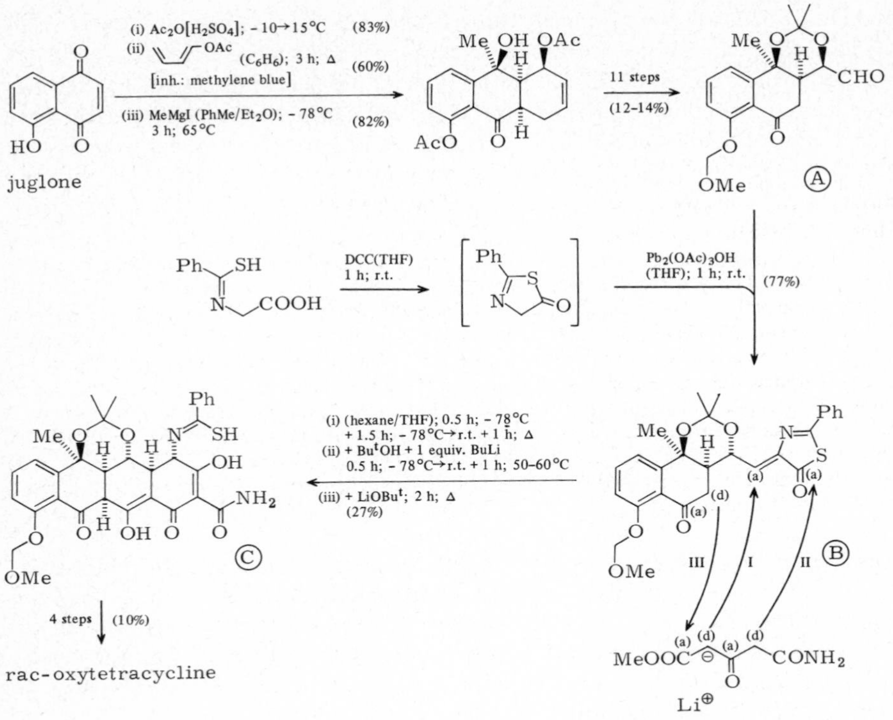

(i) $Ac_2O[H_2SO_4]$; $-10\rightarrow15\,°C$ (83%)

(ii) $\diagup\!\!\diagdown\!\!\diagup\!\!-OAc$ (C_6H_6); 3 h; Δ (60%)

[inh.: methylene blue]

(iii) MeMgI (PhMe/Et_2O); $-78\,°C$ (82%)
3 h; $65\,°C$

juglone

Me OH OAc

AcO

11 steps
(12–14%)

Me

CHO

OMe   Ⓐ

Ph—SH
|
N—COOH

DCC(THF)
1 h; r.t.

[Ph—S / N—O (thiazolone)]

$Pb_2(OAc)_3OH$
(THF); 1 h; r.t.   (77%)

Me O O

Ph

N—SH
OH

NH₂

Ⓒ

OH OH O O

OMe

(i) (hexane/THF); 0.5 h; $-78\,°C$
+ 1.5 h; $-78\,°C\rightarrow$r.t. + 1 h; Δ

(ii) + Bu^tOH + 1 equiv. BuLi
0.5 h; $-78\,°C\rightarrow$r.t. + 1 h; $50–60\,°C$

(iii) + LiOBu^t; 2 h; Δ
(27%)

Me O O

Ph

N S
O

OMe   III   I   II   Ⓑ

MeOOC⊖   CONH₂
O
Li⊕

4 steps | (10%)

rac-oxytetracycline

4.8.3 Macrolides

Erythromycins from *Streptomyces erythreus* strains are probably the structurally most complex of the best selling drugs. Erythromycin A costs only about 5 DM per gram. The two principal erythromycins A and B differ only with respect to hydroxylation at C-12.

streptomyces erythreus antibiotics: erythromycin A (R = OH) /B (R = H)

aglycones:
erythronolide A (R = OH)
           B (R = H)

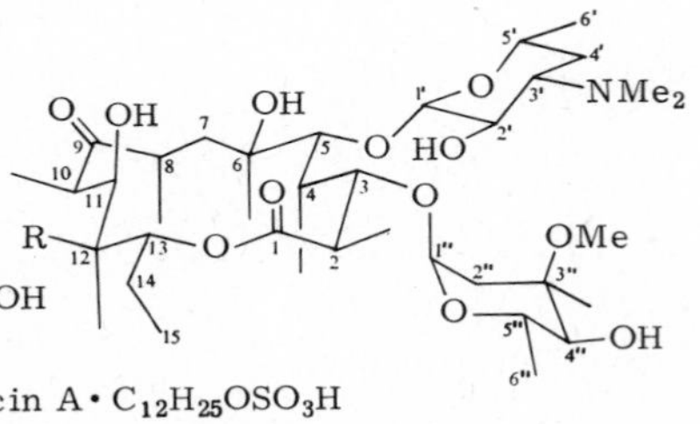

desosamine

cladinose

erythrocin stearate:
erythromycin A · $C_{17}H_{35}COOH$

erythromycin estolate:
2'-O-propionyl erythromycin A · $C_{12}H_{25}OSO_3H$

One total synthesis of erythronolide A starting from glucose was described in section 4.3. Other syntheses have been reported by E.J. Corey (1978B, 1979). We outline only the stereoselective synthesis of a protected fragment (A) which contains carbon atoms 1-9. This fragment was combined with fragment (B) by a Grignard reaction and cyclized by one of the methods typical for macrolide formation (see p. 132).

Synthesis of (A) started with the combination of 2,4,6-trimethylphenol and allyl bromide to give the *ortho*-allyl dienone. Acid-catalyzed rearrangement and oxidative hydroboration yielded the dienone with a propanol group in *para*-position. Oxidation of the alcohol and bromolactonization (see p. 249) followed. Bromine and the lactone ring were *trans* in the product as expected (see p. 249). Treatment with aqueous potassium hydroxide gave the epoxy acid, which formed a crystalline salt with (R)-1-(α-naphthyl)ethylamine. This was recrystallized to constant rotation.

The now optically active epoxy acid was again stereoselectively bromolactonized, and the bromine was removed with 87% inversion and 13% retention of configuration by tributyltin hydride. Reduction of the epoxide by catalytic hydrogenation and of the ketone using zinc borohydride led to a diol which was benzoylated and purified by crystallization. Methylation of the lactone introduced the last carbon atom at C-8 again stereoselectively from the α-side of the bicyclic system. Base-catalyzed opening of the lactone and Jones oxidation of the alcohol yielded a protected cyclohexanone derivative with all substituents in the correct positions. Baeyer-Villiger oxidation and treatment with 2,2′-dipyridyl disulfide concluded the synthesis of fragment (A). It contains a good leaving group at C-9 which could be substituted by a carbanionic fragment (B).

4.9 Esoteric Polycyclic Hydrocarbons

"Esoteric" is meant to indicate that we are considering polycyclic hydrocarbons which do not relate to natural compounds or fulfill society needs, but which are of interest only to chemists, to the "insiders", themselves. This interest is generally derived from anticipated special properties of specific arrangements of chemical bonds. At the end of this treatise on modern organic synthesis these compounds provide an opportunity to demonstrate some fascinating rearrangements of carbon skeletons, which have so far only been mentioned occasionally. We chose conceptually simple examples, which only use metal compounds, heat, or light as initiators (J.F. Liebman, 1976; L.T. Scott, 1972).

4.9.1 (CH)$_4$

Cyclobutadiene itself is not stable at room temperature. Several derivatives with stabilizing groups have been prepared by the acid-catalyzed dimerization of alkynes (R. Gompper, 1975). Less substituted cyclobutadienes could be obtained by photolytic reactions in solid matrix at low temperatures (G. Maier, 1973, 1974).

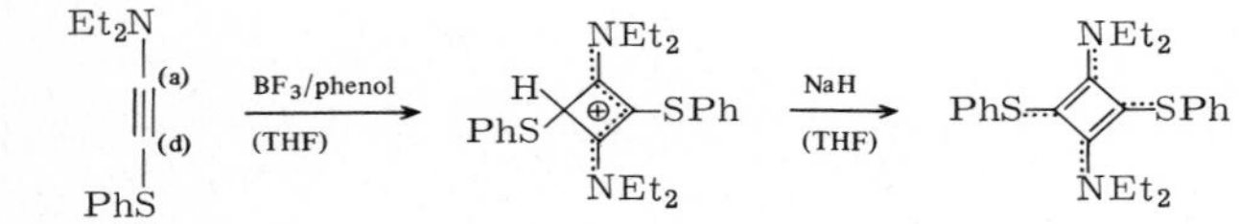

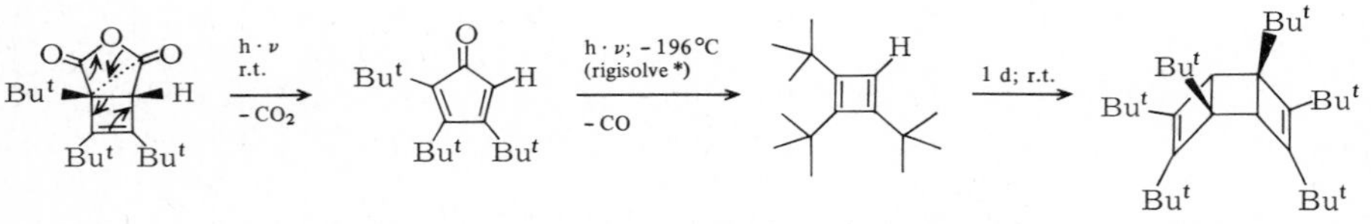

*rigisolve = pentane + neohexane 3 : 8   (82%)

The irradiation of tetra-t-butylcyclopentadienone with 254 nm light at 77 K produced a tricyclopentanone which, upon extended irradiation, lost carbon monoxide. Tetra-t-butyltetrahedrane was formed. This derivative of the second fundamental hydrocarbon of molecular formula (CH)$_4$, namely tetrahedrane, is stable at room temperature and could be isolated after chromatography on silica gel in crystalline form (G. Maier, 1978). The exceptional stability of tetralithiotetrahedrane (G. Rauscher, 1978) was predicted from MO calculations.

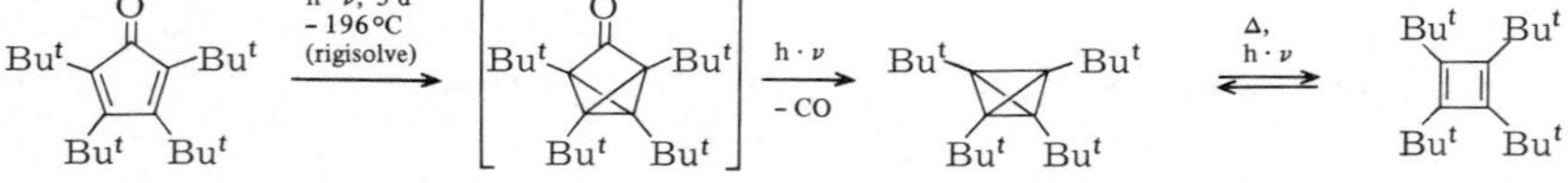

(35%)

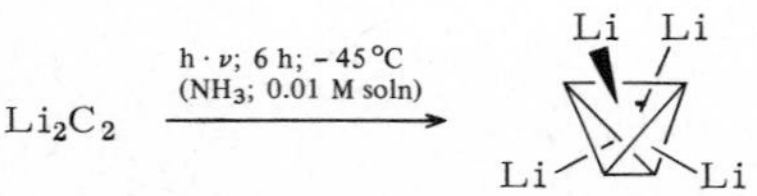

4.9.2 (CH)$_6$

2-Butyne trimerizes in the presence of aluminum chloride to give hexamethyl "Dewarbenzene" (W. Schäfer, 1967). Its irradiation leads not only to aromatization but also to hexamethylprismane (D.M. Lemal, 1966). Highly substituted prismanes may also be obtained from the corresponding benzene derivatives by irradiation with 254 nm light. The rather stable prismane itself was synthesized via another C_6H_6 hydrocarbon, namely benzvalene, a labile molecule (T. J. Katz, 1971, 1972).

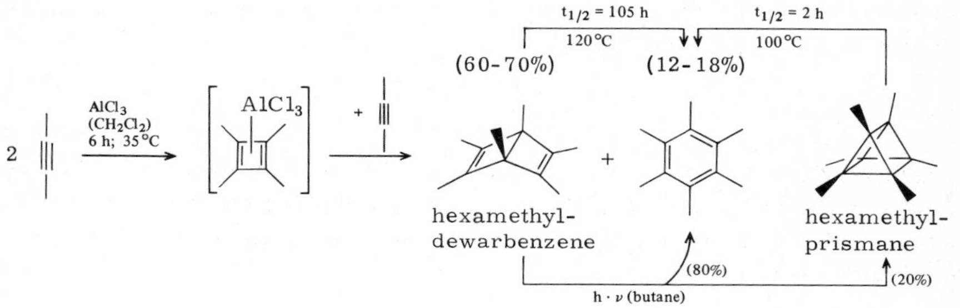

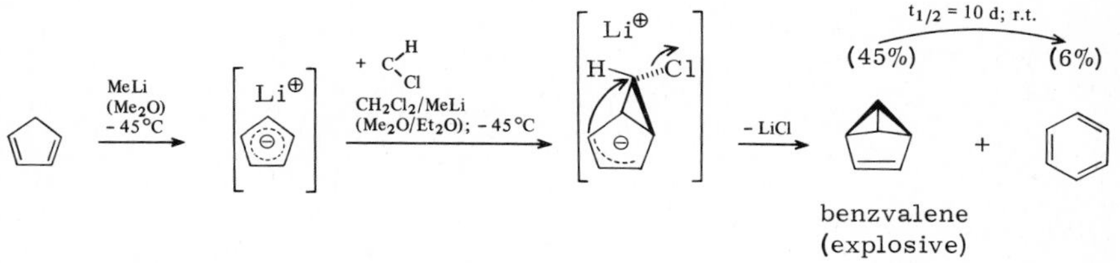

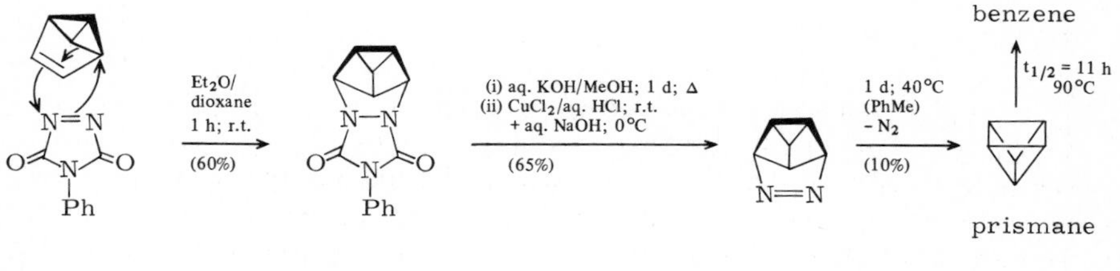

hexamethyl-
dewarbenzene

hexamethyl-
prismane

benzvalene
(explosive)

benzene

prismane

4.9.3 (CH)$_8$

Cyclooctatetraene can be obtained on an industrial scale by metal carbonyl catalyzed thermal tetramerization of acetylene. If cyclooctatetraene is UV-irradiated at low temperature in the presence of acetone, it is reversibly rearranged to form semibullvalene (H.E. Zimmerman, 1968, 1970).

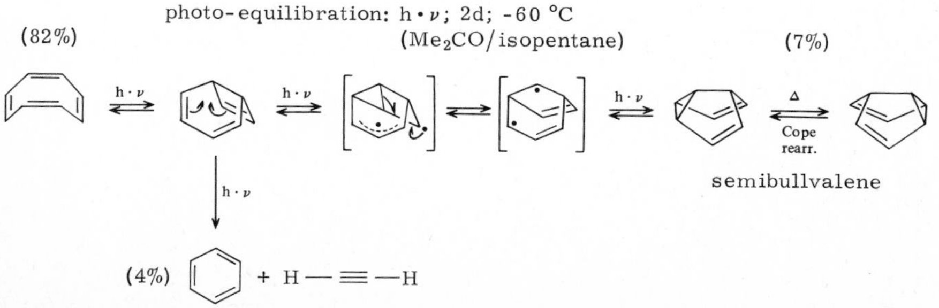

photo-equilibration: h·ν; 2d; −60 °C
(Me$_2$CO/isopentane)

semibullvalene

Barrelene was obtained via a double Diels-Alder reaction from α-pyrone with methyl acrylate (H.E. Zimmerman, 1969A). The primarily forming bicyclic lactone decarboxylates in the heat, and the resulting cyclohexadiene rapidly undergoes another Diels-Alder cyclization. Standard reactions have then been used to eliminate the methoxycarbonyl groups and to introduce C=C double bonds. Irradiation of barrelene produces semibullvalene and cyclooctatetraene (H.E. Zimmerman, 1969B).

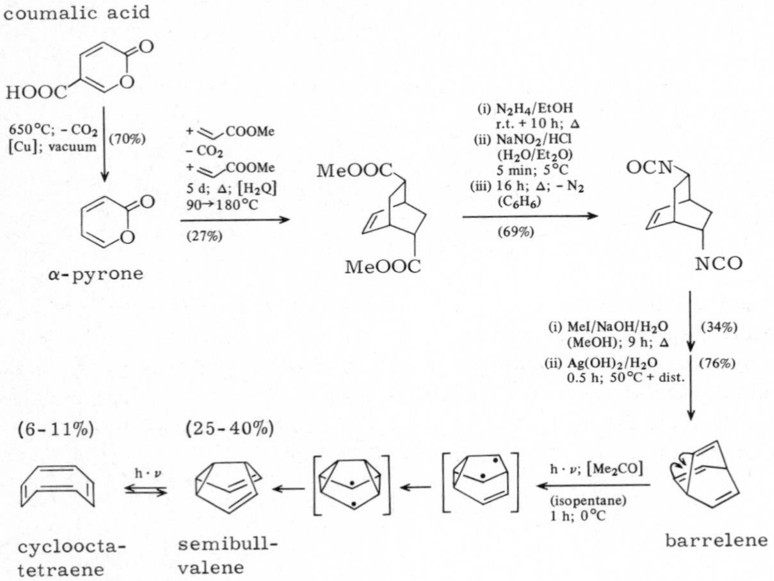

The synthesis of cubane has been outlined on p. 72. Cuneiform cuneanes are formed by silver(I)-catalyzed isomerization of cubanes in almost quantitative yield. Rhodium(I), on the other hand, effects isomerization of cubane to a *syn*-tricyclooctadiene (L. Cassar, 1970).

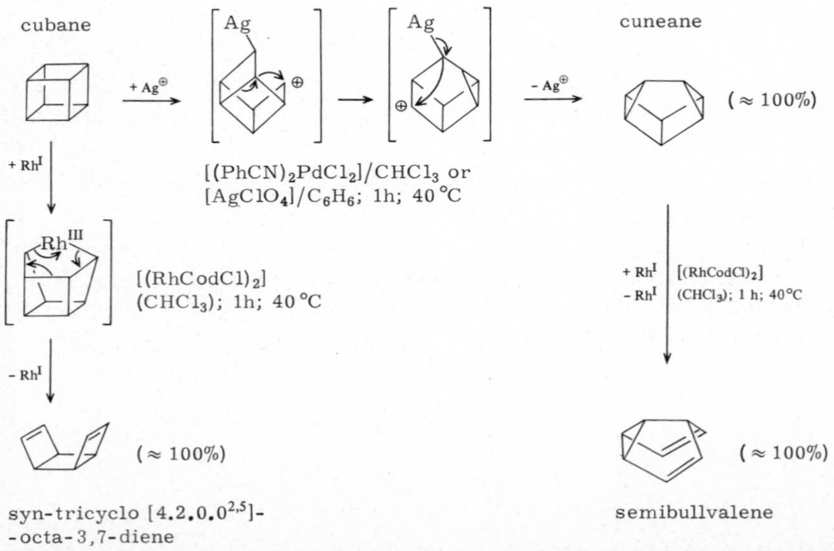

4.9.4 (CH)$_{10}$

(All-cis)-[10]annulene has been obtained in low yield from *cis*-9,10-dihydronaphthalene by UV-irradiation at -60 °C. This reaction was reversed at room temperature (S. Masamune, 1969).

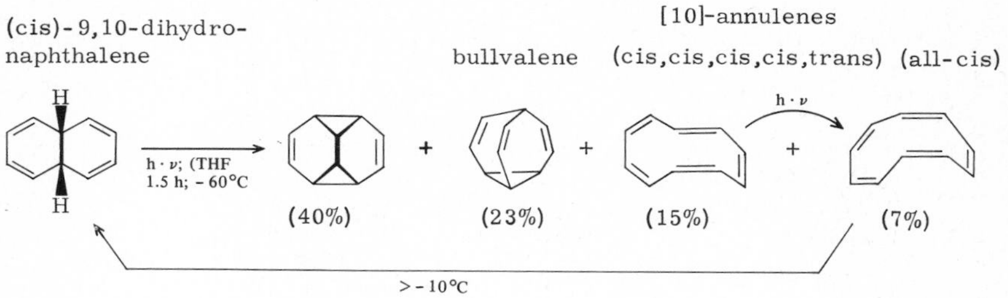

(cis)-9,10-dihydro-   [10]-annulenes
naphthalene   bullvalene (cis,cis,cis,cis,trans) (all-cis)

The most intriguing hydrocarbon of this molecular formula is named bullvalene, which is found in the mixture of products of the reaction given above. G. Schröder (1963, 1964, 1967) synthesized it by a thermal dimerization presumably via diradicals of cyclooctatetraene and the photolytical cleavage of a benzene molecule from this dimer. The carbon-carbon bonds of bullvalene fluctuate extremely fast by thermal Cope rearrangements. 10!/3 = 1,209,600 different combinations of the carbon atoms are possible.

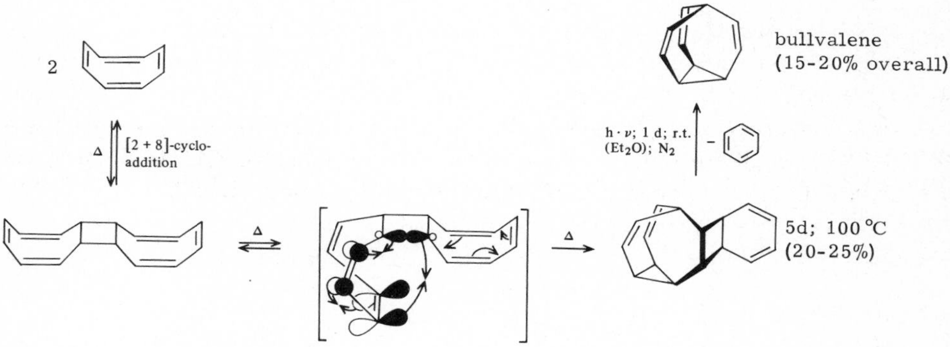

Several interesting polycyclic C$_{10}$H$_{10}$-hydrocarbons have been obtained from cyclooctatetraene and maleic anhydride. Thermal cycloaddition, photochemical [2 + 2]-cycloaddition, and oxidative decarboxylation yield basketene, which may be rearranged in almost quantitative yield to snoutene. Irradiation with UV light and silver(I)-catalyzed rearrangement give triquinacene, an interesting tricyclic triene which was made with the anticipation that it could be dimerized to dodecahedrane (E.P. Eaton, 1979; L.A. Paquette, 1977, 1979; see p. 297f.).

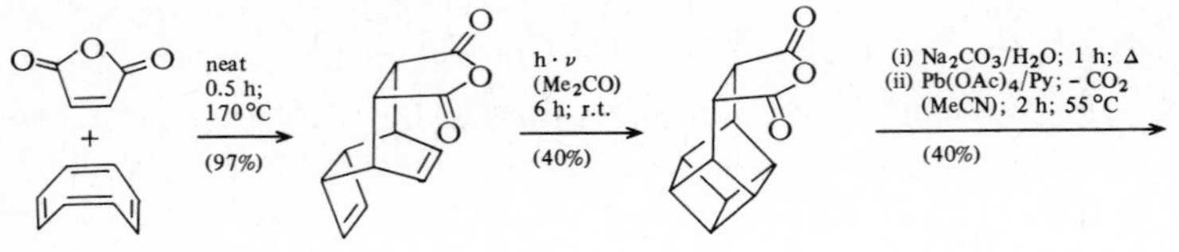

basketene

basketene

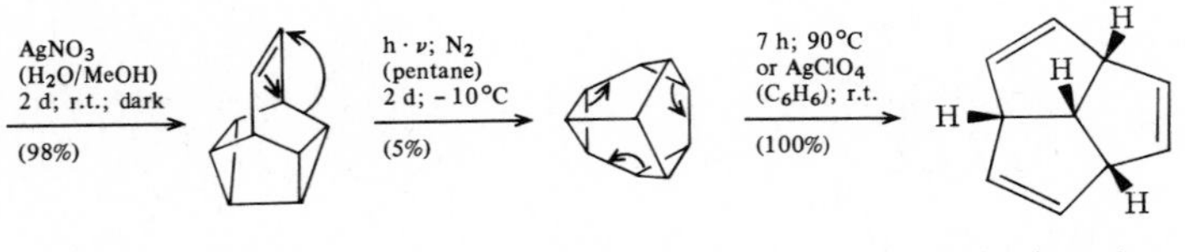

snoutene　　　　　diademane　　　　triquinacene

4.9.5 Methylene-Bridged $[4n+2]$-Annulenes

Large annulenes tend to undergo conformational distortion, *cis-trans* isomerizations, and sigmatropic rearrangements (p. 39 and p. 92). Methylene-bridged conjugated $(4n+2)$-π cyclopolyenes were synthesized with the expectation that these almost planar annulenes should represent stable Hückel arenes (E. Vogel, 1970, 1975).

An interesting general synthesis using repeating anellation procedures has been developed by E. Vogel (W. Wagemann, 1978). First cyclohepta-3,5,7-triene-1,3-dicarboxaldehyde was synthesized from the Diels-Alder adduct of butadiene and butynedioic ester by conventional methods. The anellation of this dialdehyde with a pentanedioic acid bis-Horner reagent led to the bicyclic and tricyclic dialdehydes. Each of these dialdehydes was cyclized, e.g. with sulfide bis-Wittig reagents to give the aromatic bridged annulene systems after the extrusion of sulfur (p. 37) or by reaction with one mole both of methylene and of chloromethylene Wittig reagents and subsequent thermal 18π-electrocyclization and dehydrohalogenation.

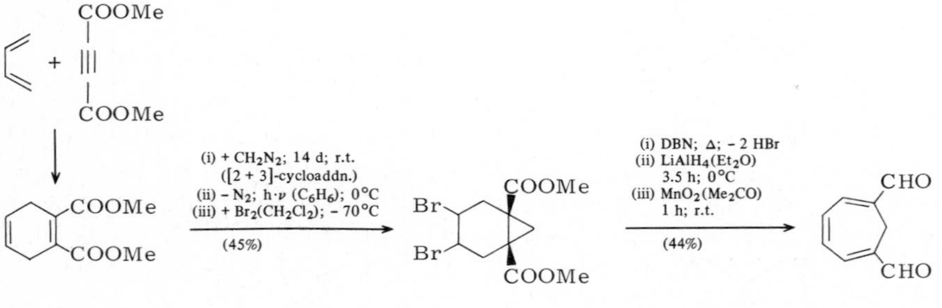

CHO → (i)(ii)(iii)*) (≈ 30%) → CHO CHO → (i)(ii)(iii)*) (≈ 30%) → CHO CHO

CHO

(34%) │ (i) ⟨PPh₃⊕ Br⊖⟩ ⟨PPh₃⊕ Br⊖⟩
LiOEt/DMF; 60 °C
(ii) Ph₃C⊕BF₄⊖
(CH₂Cl₂); r.t.

(6%) │ (i) ⟨S⟩ PPh₃⊕ Br⊖ / PPh₃⊕ Br⊖
LiOMe/DMF; 4 h; r.t.
(ii) Ph₃P(C₆H₆)
2 h; Δ; – Ph₃PS

(32%) │ (i) 1 mol Ph₃P=CH₂
(THF); 3 h; r.t.
(ii) Ph₃P=CHCl
(THF); 3 h; r.t.
(iii) 1.5 h; Δ; (DMF)
– HCl

[⊕] BF₄⊖

*) (i)

$$EtOOC \quad COOEt$$
$$(EtO)_2PO \quad PO(OEt)_2$$

NaH; 3h; Δ
(CH₂Cl₂/C₆H₆)

(ii) DIBAL(C₆H₆); r. t.
(iii) DDQ (dioxane); 6h; r. t.

4.9.6 Dodecahedrane (CH)₂₀

Spherical, pentagonal dodecahedrane is the thermodynamically most stable $C_{20}H_{20}$-polycycloalkane. It is the so-called "$C_{20}H_{20}$ stabilomer". It should therefore be available by thermodynamically controlled, e.g. acid-catalyzed, isomerization of less stable $C_{20}H_{20}$-isomers. Experiments along this line, e.g. treatment of the basketene photo-dimer with Lewis acids, have, however, been unsuccessful (E.P. Eaton, 1979). Attempts to dimerize triquinacene, $C_{10}H_{10}$, or to combine a C_{15}-unit ("peristylane") with a C_5-unit (L.A. Paquette, 1979; W. Grahn, 1981) also failed so far, because the carbon-carbon bonds which have to be formed are all in the unfavorable *endo* position. Strong preference for *exo,exo* carbon-carbon bond formation was evident in reductive coupling reactions (L.A. Paquette, 1979).

basketene → photo-dimer → (Lewis acids ╫)

undecacyclo [9.9.0.0²,⁹.
0³,⁷.0⁴,²⁰.0⁵,¹⁸.0⁶,¹⁶.
0⁸,¹⁵.0¹⁰,¹⁴.0¹²,¹⁹.0¹³,¹⁷]
eicosane = dodecahedrane

triquinacene   + → (coupling ╫)

peristylane
(•Corresponds to functionalized carbon atoms)   + ╫

correct name (IUPAC rules):
hexadecahydro-5,2,1,6,3,4-
[2,3]butanediyl [1,4]diylidene-
dipentaleno [2,1,6-cde: 2',1',6'-
gha]pentalene

L.A. Paquette recently accomplished an extremely difficult and elegant stepwise synthesis of 1,16-dimethyldodecahedrane. It starts with a "domino Diels-Alder reaction" between 9,10-dihydrofulvalene and dimethyl acetylenedicarboxylate. The reaction involves initial intermolecular cycloaddition of the dienophile to the first 1,3-diene moiety and subsequent reaction of the residual double bond of the acetylenic dienophile in intramolecular bond formation with the second diene (L.A. Paquette, 1974). The resulting diester (A) hydrolyzed to the diacid. Its iodolactonization (see p. 249) proceeded with high efficiency to give (B). Cleavage with methanolic sodium methoxide at room temperature, Jones oxidation, and reductive deiodination with zinc-copper couple furnished isomerically pure diketoester (C) (L.A. Paquette, 1976, 1979). To the dimethyl ester of this "cross-corner oxidized" C_{14}-frame six more carbon atoms were added symmetrically by spiroanellation with cyclopropyldiphenylsulfonium ylide (see p. 72f.). The dispiro-diketodiester (D) therefore contains all twenty carbon atoms of the dodecahedrane molecule. Sixteen more steps were performed to build up all the five-membered rings necessary for the final closure of the polycyclic sphere. In this sequence of delicate reactions it was necessary to replace two acidic hydrogen atoms neighboring carboxylic ester groups by methyl substituents. The final acid-catalyzed rearrangement of olefin (E) gave 1,16-dimethyldodecahedrane which was characterized by an X-ray structure determination (L.A. Paquette, 1982; G.G. Christoph, 1982).

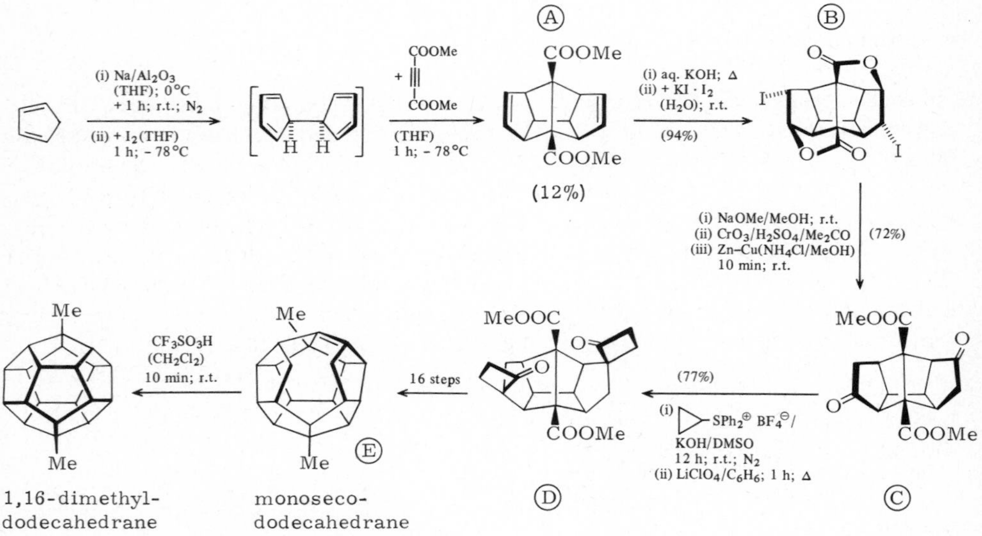

4.9.7 Kekulene, $C_{48}H_{24}$

The kekulene macrocycle consists of twelve anellated benzene rings and may be considered as an [18]annulene (inside) or a [30]annulene (outside). H. Staab (F. Diederich, 1978) called it a "superbenzene", since it has the same D_{6h} symmetry as benzene.

2,19-Dimethyl-6,7,9,10,23,24,26,27-octahydro-2,19-dithionia-[3,3](3,11)dibenz[a,j]-anthracenophane bis(fluorosulfonate) (A) was synthesized by procedures similar to those described earlier for other cyclophanes (p. 37). (A) was treated with potassium *t*-butoxide (Stevens rearrangement), methyl fluorosulfonate (methylation), and again potassium *t*-butoxide (elimination of dimethyl sulfide), and gave the vinylene-bridged octahydrodibenz-anthracenophane (B). Short irradiation with iodine caused cyclodehydrogenation of both *cis*-stilbene units. The resulting octahydrokekulene was dehydrogenated with DDQ to kekulene.

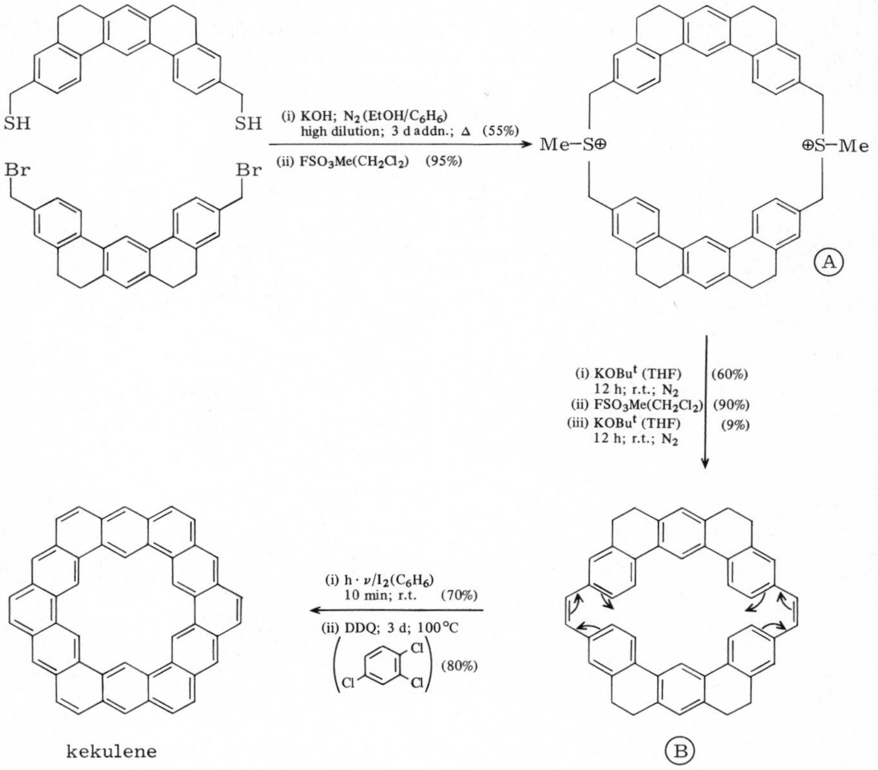

References

Aalbersberg, W. G., Barkovich, A. J., Funk, R. L., Hillard III, R. L., Vollhardt, K. P. C. **1975,** J. Am. Chem. Soc. *97,* 5601

Abe, Y., Harukawa, T., Ishikawa, H., Miki, T.; Sumi, M., Toga, T. **1956,** J. Am. Chem. Soc. *78,* 1422

Acheson, R. M., Paglietti, G. **1976 A,** J. Chem. Soc. Perkin I *1976,* 45

Acheson, R. M. **1976 B,** *An Introduction to the Chemistry of Heterocyclic Compounds,* Wiley, New York/London/Sydney

Achmad, S. A., Cavill, G. W. K. **1963, 1965,** Austral. J. Chem. *16,* 858, *18,* 1989

Adam, W., Erden, I. **1978,** Angew. Chem. *90,* 223 (Int. Ed. *17,* 210)

Adams, M. A., Duggan, A. J., Smolanoff, J., Meinwald, J. **1979,** J. Am. Chem. Soc. *101,* 5364

Adickes, H. W., Politzer, I. R., Meyers, A. I. **1969,** J. Am. Chem. Soc. *91,* 2155

Agarwal, K. L., Berlin, Y. A., Fritz, H. J., Gait, M. J., Kleid, D. G., Lees, R. G., Norris, K. E., Ramamoorthy, B., Khorana, H. G. **1976,** J. Am. Chem. Soc. *98,* 1065

Akhrem, A. A., Reshetova, I. G., Titov, Yu. A. **1972,** *Birch Reduction of Aromatic Compounds* Plenum, New York

Allen, W. S., Bernstein, S., Littell, R. **1954,** J. Am. Chem. Soc. *76,* 6116

Almog, J., Baldwin, J. E., Dyer, R. L., Peters, M. **1975,** J. Am. Chem. Soc. *97,* 226, 227

Ames, D. E., Goodburn, T. G.; Jevans, A. W.; McGhie, J. F. **1968,** J. Chem. Soc. (C) *1968,* 268

Ananchenko, S. N., Limanov, V. Ye.; Leonov, V. N.; Rzheznikov, V. N.; Torgov, I. V. **1962,** Tetr. *18,* 1355

Ananchenko, S. N., Torgov, I. V. **1963,** Tetr. Lett. *1963,* 1553

Angst, C., Kajiwara, M., Zass, E., Eschenmoser, A. **1980,** Angew. Chem. *92,* 139 (Int. Ed. *19,* 140)

Arigoni, D., Vasella, A., Sharpless, K. B., Jensen, H. P. **1973,** J. Am. Chem. Soc. *95,* 7917

Arndt, D. **1975** in: Houben-Weyl, *Methoden der Organischen Chemie,* Vol. IV/1 b, p. 505 ff. Thieme, Stuttgart

Arzoumanian, H., Metzger, J. **1971,** Synthesis *1971,* 527

Asinger, F., Vogel, H. H., **1970** in: Houben-Weyl, *Methoden der Organischen Chemie,* Vol V/1a, p. 327, Thieme, Stuttgart

Atherton, E., Clive, D. L. J., Sheppard, R. C. **1975,** J. Am. Chem. Soc. *97,* 6584

Augustine, R. L. **1965,** *Catalytic Hydrogenation: Techniques and Applications in Organic Synthesis,* M. Dekker, New York

Augustine, R. L. **1968** *Reduction: Techniques and Applications in Organic Synthesis,* M. Dekker, New York

Augustine, R. L. **1969,** *Oxidation: Techniques and Applications in Organic Synthesis,* Vol. 1, M. Dekker, New York

Augustine, R. L., Trecker, D. J. **1971,** *Oxidation: Techniques and Aplications in Organic Synthesis,* Vol. 2, M. Dekker, New York

Augustine, R. L. **1976,** Catal. Rev. *13,* 285

Avram, M., Nenitzescu, C. D. **1964,** Chem. Ber. *97,* 372

Ayer, W. A., Bowman, W. R., Joseph, T. C., Smith, P. **1968,** J. Am. Chem. Soc. *90,* 1648

Ayres, D. C., Raphael, R. A. **1958,** J. Chem. Soc. *1958,* 1779

Bailey, E. J., Barton, D. H. R., Ellis, J., Templeton, J. F. **1962,** J. Chem. Soc. *1962,* 1578

Bailey, P. S. **1978,** *Ozonation in Organic Chemistry,* Part I: *Olefinic Compounds,* Academic Press, New York/London

Baker, B. W., Linstead, R. P., Weedon, B. C. L. **1955,** J. Chem. Soc. *1955,* 2218

Baker, D. C., Horton, D., Tindall, C. G., Jr. **1976** in: Whistler, R. L., BeMiller, J. N. (eds), *Methods in Carbohydrate Chemistry,* Vol. VII, p. 3, Academic Press, New York, London

Baker, R., Blackett, B. N. Cookson, R. C. **1972** Chem. Comm. *1972,* 802

Baker, R., Cookson, R. C., Vinson, J. R. **1974,** Chem. Comm. *1974,* 515
Baldwin, J. E., Au, A., Christie, M. A., Haber, S. B., Hesson, D. **1975,** J. Am. Chem. Soc. *97,* 5957
Baldwin, J. E., Christie, M. A., Haber, S. B., Kruse, L. I. **1976,** J. Am. Chem. Soc. *98,* 3045
Baldwin, J. E. **1976 B,** Chem. Comm. *1976,* 734, 738
Baldwin, J. E., Cutting, J., Dupont, W., Kruse, L., Silberman, L., Thomas, R. C. **1976 C,** Chem. Comm. *1976,* 736
Baldwin, J. E., Christie, M. A. **1978,** J. Am. Chem. Soc. *100,* 4597
Ban, Y., Sato, Y., Inoue, I., Nagai, M., Oishi, T., Terashima, M., Yonemitsu, O., Kanaoka, Y. **1965,** Tetr. Lett. *1965,* 2261
Barborak, J. C., Watts, L., Pettit, R. **1966,** J. Am. Chem. Soc. *88,* 1328
Barth, W. E., Lawton, R. G. **1971,** J. Am. Chem. Soc. *93,* 1730
Bartlett, P. A., Green, F. R. **1978,** J. Am. Chem. Soc. *100,* 4858
Bartlett, P. A. **1980,** Tetr. *36,* 1
Bartlett, P. D.; Staufer, C. H. **1935,** J. Am. Chem. Soc. *57,* 2580
Barton, D. H. R., Seoane, E. **1956,** J. Chem. Soc. *1956,* 4150
Barton, D. H. R., Lier, E. F., McGhie, J. F. **1968,** J. Chem. Soc. (C) *1968,* 1031
Barton, D. H. R., Willis, B. J. **1972,** J. Chem. Soc. Perkin I *1972,* 305
Barton, D. H. R., Guziec, F. S., Shahak, I. **1794,** J. Chem. Soc. Perkin I *1974,* 1794
Barton, D. H. R. **1975** in: Stirling, C. J. M. (ed), *Organic Sulphur Chemistry,* p. 181, Butterworths, London
Barton, D. H. R., Hesse, R. H., Markwell, R. E., Pechet, M. M., Rozen, S. **1976,** J. Am. Chem. Soc. *98,* 3034/6
Barton, D. H. R., Hesse, R. H., Pechet, M. M., Smith, L. C. **1979,** J. Chem. Soc. Perkin I *1979,* 1159
Barton, J. E. D., Harley-Mason, J. **1965** Chem. Comm. *1965,* 298
Bastús, J. B. **1963,** Tetr. Lett. *1963,* 955
Battersby, A. R., Turner, J. C. **1960,** J. Chem. Soc. 1960, 717
Battersby, A. R., Burnett, A. R., Parsons, P. G. **1969,** J. Chem. Soc. (C) *1969,* 1193
Battersby, A. R., McDonald, E. **1979,** Accts. Chem. Res. *12,* 14
Bélanger, A., Ponpart, J., Deslongchamps, P. **1968,** Tetr. Lett. *1968,* 2127
Bellet, P., Nominé, G., Mathieu, J., Velluz, L. **1966,** Comptes rendus Sér. C *263,* 88
Bergbreiter, D. E., Whiteside, G. M. **1975,** J. Org. Chem. *40,* 779
Bergelson, L. D., Shemyakin, M. M. **1964,** Angew. Chem. *76,* 113 (Int. Ed. *3,* 250)
Bernstein, S., Lenhard, R. H., Allen, W. S., Heller, M., Littell, R., Stolar, S. M., Feldmann, L. I., Blank, R. H. **1956,** J. Am. Chem. Soc. *78,* 5693
Bestmann, H. J., Stransky, W., Vostrowsky, O. **1976,** Chem. Ber. *109,* 1694
Bestmann, H. J. **1979,** Pure Appl. Chem. *51,* 515
Birch, A. J., Subba Rao, G. **1972,** Adv. Org. Chem. *8,* 1
Birch, A. J., Williamson, D. H. **1976,** Org. React. *24,* 1
Blackburn, G. M., Ollis, W. D., Smith, C., Sutherland, I. O. **1969,** Chem. Comm. *1969,* 99
Blickenstaff, R. T., Ghosh, A. C., Wolf, G. C. **1974,** *Total Synthesis of Steroids,* Academic Press, New York
Bloomfield, J. J., Owsley, D. C., Nelke, J. M. **1976,** Org. React. *23,* 259
Blumbergs, P., LaMontagne, M. P., Stevens, J. I. **1972,** J. Org.Chem. *37,* 1248
Bly, R. S., DuBose, C. M., Konizer, G. B. **1968,** J. Org. Chem. *33,* 2188
Bodanszky, M., Klausner, Y. S., Ondetti, M. A. **1976,** *Peptide Synthesis,* Wiley, New York/London
Boeckman, R. K., Jr., Silver, S. M. **1973,** Tetr. Lett. *1973,* 3497
Boeckman, R. K., Jr. **1974,** J. Am. Chem. Soc. *96,* 6179
Bohlmann, F., Eickeler, E. **1979,** Chem. Ber. *112,* 2811
Bonse, G., Metzler, M. **1978,** *Biotransformationen organischer Fremdsubstanzen,* Thieme, Stuttgart
Borch, R. F. **1968,** Tetr. Lett. *1968,* 61
Borch, R. F. **1969,** J. Org. Chem. *34,* 627
Bosche, H. G. **1975** in: Houben-Weyl, *Methoden der Organischen Chemie,* Vol. IV/1b: *Oxidation II,* p. 429, Thieme, Stuttgart
Boutigue, M. H., Jacquesy, R. **1973,** Bull. Soc. Chim. France *1973* (II), 750, 3062

Brady, S. F., Ilton, M. A., Johnson, W. S. **1968,** J. Am. Chem. Soc. *90,* 2882

Brady, W. T., Hoff, E. F. **1970,** J. Org. Chem. *35,* 3733

Bredereck, G., Gompper, R., Schuh, H. G., Theilig, G. **1959,** Angew. Chem. *71,* 753

Bredereck, H., Effenberger, F., Simchen, G. **1963, 1965,** Chem. Ber. *96,* 1350, *98,* 1078

Bredereck, H., Simchen, G., Rebsdat, S., Kantlehner, W., Horn, P., Wahl, R., Hoffmann, H., Grieshaber, P. **1968,** Chem. Ber. *101,* 41

Breslow, R., Corcoran, R. J., Snider, B. B., Doll, R. J., Khanna, P. L., Kaleya, R. **1977,** J. Am. Chem. Soc. *99,* 905

Brockmann, H., Kluge, F., Muxfeldt, H. **1957** Chem. Ber. *90,* 2302

Brown, H. C., Heim, P. **1964,** J. Am. Chem. Soc. *86,* 3566

Brown, H. C., Kawakami, J. H., Ikegami, S. **1970,** J. Am. Chem. Soc. *92,* 6914

Brown, H. C., Negishi, E. **1972 A,** J. Am. Chem. Soc. *94,* 3567

Brown, H. C., Krishnamurthy, S. **1972 B,** J. Am. Chem. Soc., *94,* 7159

Brown, H. C. **1972 C,** *Boranes in Organic Chemistry,* Cornell Univ. Press, Ithaca N. Y.

Brown, H. C. **1975,** *Organic Synthesis via Boranes,* J. Wiley, New York/London

Brown, H. C. **1980,** Angew. Chem. *92,* 675

Brown, W. G. **1951,** Org. React. *6,* 469

Buchanan, J. G., Clode, D. M., Vethaviyasar, N. **1976,** J. Chem. Soc. Perkin I *1976,* 1449

Büchel, K. H., Röchling, H., Korte F. **1965,** Liebigs Ann. *685,* 10

Büchi, G., Coffen, D. L., Kocsis, K., Sonnet, P. E., Ziegler, F. E. **1965/1966,** J. Am. Chem. Soc. *87,* 2073/*88,* 3099

Büchi, G., Hofheinz, W., Paukstelis, J. V. **1966,** J. Am. Chem. Soc. *88,* 4113

Büchi, G., Carlson, J. A. **1968,** J. Am. Chem. Soc. *90,* 5336

Büchi, G., Carlson, J. A., Powell, J. E., Tietze, L. F. **1973,** J. Am. Chem. Soc. *95,* 540

Bucourt, R., Pierdet, A., Costerousse, G., Toromanoff, E. **1965,** Bull. Soc. Chim. France *1965,* 645

Bull, J. R., Tuinman, A. **1975,** Tetr. *31,* 2151

Bundy, G. L., Lincoln, F. H., Nelson, N. A., Pike, J. E., Schneider, W. P. **1971,** Ann. N. Y. Acad. Sci. *180,* 76

Bundy, G. L., Schneider, W. P., Lincoln, F. H., Pike, J. E. **1972,** J. Am. Chem. Soc. *94,* 2123

Bürgi, H. B., Dunitz, J. D., Shefter, E. **1973,** J. Am. Chem. Soc. *95,* 5065

Burke, S. D., Grieco, P. A. **1979,** Org. React. *26,* 361

Cadogan, J. I. G. (ed.) **1979;** *Organophosphorus Reagents in Organic Synthesis,* Academic Press, London

Cain, E. N., Vukov, R., Masamune, S. **1969,** Chem. Comm. *1969,* 98

Caine, D. **1976,** Org. React. *23,* 1

Cambie, R. C., Potter, G. J., Rutledge, P. S., Woodgate, P. D. **1977,** J. Chem. Soc. Perkin I *1977,* 530

Carlson, R. M., Oyler, A. R. **1974,** Tetr. Lett. *1974,* 2615

Carlson, R. M., Oyler, A. R., Peterson, J. R. **1975,** J. Org. Chem. *40,* 1610

Carpino, L. A. **1957,** J. Am. Chem. Soc. *79,* 98

Carpino, L. A. **1973,** Accts. Chem. Res. *6,* 191

Carruthers, W. **1973,** Chem. & Ind. *1973,* 931

Carruthers, W. **1978,** *Some Modern Methods of Organic Syntheses,* 2nd edn., Cambridge Univ. Press

Cason, J. A., Allen, C. F. **1949,** J. Org. Chem. *14,* 1036

Cassar, L., Eaton, P. E., Halpern, J. **1970,** J. Am. Chem. Soc. *92,* 6366

Cava, M. P., Pohl, R. J. **1960,** J. Am. Chem. Soc. *82,* 5242

Cava, M. P., Mitchell, M. J. **1967,** *Cyclobutadiene and Related Compounds,* Academic Press, New York/London

Cavill, G. W. K., Hall, C. D. **1967,** Tetr. *23,* 1119

Celmer, W. D. **1971,** Pure Appl. Chem. *28,* 413

Červinka, O., Fusek, J. **1973,** Coll. Czech. Chem. Comm. *38,* 441

Chan, T. H., Chang, E. **1974,** J. Org. Chem. *39,* 3264

Chan, T. H., Ong, B. S. **1976,** Tetr. Lett. *1976,* 319

Chapman, O. L., Mattes, K., McIntosh, C. L., Pacansky, J. **1973,** J. Am. Chem. Soc. *95,* 6134

Chapman, O. L., Chang, C.-C., Kolc, J., Rosenquist, N. R., Tomioka, H. **1975,** J. Am. Chem. Soc. *97,* 6586

Chérest, M., Felkin, H., Prudent, N. **1968,** Tetr. Lett. *1968,* 2199, 2205

Chou, T. S., Burgtorf, J. R., Ellis, A. L., Lammert, S. R., Kukolja, S. P. **1974,** J. Am. Chem. Soc. *96,* 1609; see also:

Chou, T. S. **1974,** Tetr. Lett. *1974,,* 725

Christoph, G. G., Engel, P., Usha, R., Balogh, D. W., Paquette, L. A., **1982,** J. Am. Chem. Soc., *104,* 784

Clark, R D.; Kozar, L. G.; Heathcock, C. H. **1975,** Synth. Comm. *5,* 1

Clark, R. D., Heathcock, C. H. **1976,** J. Org. Chem. *41,* 1396

Coates, G. E., Green, M. L. H., Powell, P., Wade, K. **1977,** *Principles of Organometallic Chemistry,* Chapman&Hall, London

Coates, R. M., Shaw, J. E. **1968,** Chem. Comm. *1968,* 47

Coates, R. M., Shaw, J. E. **1970,** J. Org. Chem. *35,* 2597, 2601

Cole, J. E., Johnson, W. S., Robins, P. A., Walker, J. **1962,** J. Chem. Soc. *1962,* 244

Collman, J. P., Winter S. R., Clark, D. R. **1972,** J. Am. Chem. Soc. *94,* 1788

Collman, J. P. **1975,** Accts. Chem. Res. *8,* 342 [$Na_2Fe(CO)_4$]

Collman, J. P., Gagne, R. R., Reed, C. A., Halbert, T. R., Lang, G., Robinson, W. T. **1975,** J. Am. Chem. Soc. *97,* 1427 [picket-fence porphyrin]

Collman, J. P. **1977,** Accts. Chem. Res. *10,* 265

Collman, J. P., Finke, R. G., Cawse, J. N., Brauman, J. I. **1978,** J. Am. Chem. Soc. *100,* 4766

Conia, J. M. **1975,** Pure Appl. Chem. *43,* 317

Cook, A. F. **1968,** J. Org. Chem. *33,* 3589

Cook, C. E., Corley, R. C., Wall, M. E. **1968,** J. Org. Chem. *33,* 2789

Cook, G. A. **1969,** *Enamines,* M. Dekker, New York/London

Cooke, M. P., Parlman, R. M. **1975,** J. Am. Chem. Soc. *97,* 6863

Cookson, R. C., Wariyar, N. S. **1956,** J. Chem. Soc. *1956,* 2302

Cope, A. C., Moore, P. T., Moore, W. R. **1960 A,** J. Am. Chem. Soc. *82,* 1744

Cope, A. C., Turnbull, E. R. **1960 B,** Org. React. *11,* 317

Coppola, G. M. **1978** J. Heterocyclic Chem. *15,* 645

Corey, E. J., Burke, H. J. **1956A,** J. Am. Chem. Soc. *78,* 174

Corey, E. J., Burke, H. J., Remers, W. **1956B,** J. Am. Chem. Soc. *78,* 180

Corey, E. J., Mitra, R. B., Uda, H. **1963A,** J. Am. Chem. Soc. *85,* 362

Corey, E. J., Nozoe, S. **1963B,** J. Am. Chem. Soc. *85,* 3527

Corey, E. J., Mitra, R. B., Uda, H. **1964A,** J. Am. Chem. Soc. *86,* 485 [cyclobutane]

Corey, E. J., Ohno, M., Vatakencherry, P. A., Mitra, R. B. **1964B,** J. Am. Chem. Soc. *86,* 478 [OsO_4]

Corey, E. J., Hortmann, A. G. **1965,** J. Am. Chem. Soc. *87,* 5736

Corey, E. J., Chaykovsky, M. **1965A,** J. Am. Chem. Soc. *87,* 1353

Corey, E. J., Nozoe, S. **1965B,** J. Am. Chem. Soc. *87,* 5728

Corey, E. J. **1967A,** Pure Appl. Chem. *14,* 19

Corey, E. J., Semmelhack, M. F. **1967B,** J. Am. Chem. Soc. *89,* 2756

Corey, E. J., Hamanaka, E. **1967C,** J. Am. Chem. Soc. *89,* 2757

Corey, E. J., Gilman, N. W., Ganem, B. E. **1968 A,** J. Am. Chem. Soc. *90,* 5616

Corey, E. J., Posner, G. H. **1968B,** J. Am. Chem. Soc. *90,* 5615

Corey, E. J., Shulman, J. I. **1968 C,** Tetr. Lett. *1968,* 3655

Corey, E. J., Katzenellenbogen, J. A., Gilman, N. W., Roman, S. A., Erickson, B. W. **1968D,** J. Am. Chem. Soc. *90,* 5618

Corey, E. J., Weinshenker, N. M., Schaaf, T. K., Huber, W. **1969,** J. Am. Chem. Soc. *91,* 5675

Corey, E. J., Noyori, R., Schaaf, T. K. **1970,** J. Am. Chem. Soc. *92,* 2586

Corey, E. J., Schaaf, T. K., Huber, W., Koelliker, U., Weinshenker, N. M. **1970,** J. Am. Chem. Soc. *92,* 397

Corey, E. J. **1971,** Quart. Rev. *25,* 455

Corey, E. J., Kirst, H. A. **1972A,** J. Am. Chem. Soc. *94,* 667

Corey, E. J., Venkateswarlu, A. **1972B,** J. Am. Chem. Soc. *94,* 6190

Corey, E. J., Nicolaou, K. C., Toru, T. **1975A,** J. Am. Chem. Soc. *97,* 2287 [DIBAL]

Corey, E. J., Ulrich, P. **1975B,** Tetr. Lett. *1975,* 3685 [cyclopropanes]
Corey, E. J., Danheiser, R. L., Chandrasekaran, S. **1976,** J. Org. Chem. *41,* 260
Corey, E. J., Enders, D. **1978,** Chem. Ber. *111,* 1337 [dimethylhydrazones]
Corey, E. J., Trybulski, E. J., Melvin, L. S., Jr., Nicolaou, K. C., Secrist, J. A., Lett, R., Sheldrake, P. W., Falck, J. R., Brunelle, D. J., Haslanger, M. F., Kim, S., Yoo, S. **1978,** J. Am. Chem. Soc. *100,* 4618, 4622
Corey, E. J., Hopkins, P. B., Kim, S., Yoo, S., Nambiar, K. P., Falck, J. R. **1979,** J. Am. Chem. Soc., *101,* 7131
Cram, D. J., Abd El-Hafez, F. A. **1952,** J. Am. Chem. Soc. *74,* 5828
Cram, D. J., Sahyun, M. R. V., Knox, G. R. **1962,** J. Am. Chem. Soc. *84,* 1734
Creaser, I. I., Harrowfield, J. MacB., Herlt, A. J., Sargeson, A. M., Springborg, J., Gene, R. J., Snow, M. R. **1977,** J. Am. Chem. Soc., *99,* 3181
Creger, P. L. **1972,** J. Org. Chem. *37,* 1907
Cregge, R. J., Herrmann, J. L., Lee, C. S., Richman, J. E., Schlessinger, R. H. **1973,** Tetr. Lett. *1973,* 2425
Criegee, R., Kropf, H. **1979** in: Houben-Weyl, *Methoden der organischen Chemie,* Vol. VI/1a–1: *Alkohole I,* p. 592, Thieme, Stuttgart
Crosby, D. J., Berthold, R. V. **1962,** J. Org. Chem. *27,* 3083
Cuilleron, Cl. Y., Fétizon, M., Golfier, M. **1970,** Bull. Soc. Chim. France *1970,* 1193

Danieli, N., Mazur, Y., Sondheimer, F. **1966,** Tetr. *22,* 3189
Danishefsky, S. **1974,** J. Am. Chem. Soc. *96,* 1256
Danishefsky, S., Hirama, M., Gombatz, K., Harayama, T., Berman, E., Schuda, P. F. **1979,** J. Am. Chem. Soc. *101,* 7020
Danishefsky, S., Walker, F. J. **1979A,** J. Am. Chem. Soc. *101,* 7018
Dauben, H. J., Jr., Löken, B., Ringold, H. J. **1954,** J. Am. Chem. Soc. *76,* 1359
Dauben, W. G., Lorber, M., Fullerton, D. S. **1969,** J. Org. Chem. *34,* 3587
Dauben, W. G., Williams, R. G., McKelvey, R. D. **1973,** J. Am. Chem. Soc. *95,* 3932
Dear, R. E. A., Pattison, F. L. M **1963,** J. Am. Chem. Soc. *85,* 622
Deem, M. L., **1972,** Synthesis *1972,* 675
deMeijere, A. **1979,** Angew. Chem. *91,* 867 (Int. Ed. *18,* 809)
Denis, J. M., Conia, J. M. **1972,** Tetr. Lett. *1972,* 4593
Després, J.-P., Greene, A. E. **1980,** J. Org. Chem. *45,* 2036
Dicker, I. D., Grigg, R., Johnson, A. W., Pinnock, H., Richardson, K., van den Broek, P. **1971,** J. Chem. Soc. (C) *1971,* 536
Diederich, F., Staab, H. A. **1978,** Angew. Chem. *90,* 383 (Int. Ed. *17,* 372)
Dietrich, B., Lehn, J.-M., Sauvage, J. P. **1969,** Tetr. Lett. *1969,* 2885, 2889, see also **1973,** Tetr. *29,* 1629, 1647
Dimsdale, M. J., Newton, R. F., Rainey, D. K., Webb, C. F., Lee, T. V., Roberts, S. M. **1977,** Chem. Comm. *1977,* 716
Djerassi, C., Rosenkranz, G., Romo, J., Kaufmann, S., Pataki, J. **1950,** J. Am. Chem. Soc. *72,* 4534 (see also ibid. 4531, 4540 and **1950,** J. Org. Chem. *15,* 1289)
Djerassi, C. **1951,** Org. React. *6,* 207
Djerassi, C., Batres, E., Romo, J., Rosenkranz, G. **1952,** J. Am. Chem. Soc. *74,* 3634
Djerassi, C., Engle, R. R., Bowers, A. **1956,** J. Org. Chem. *21,* 1547
Djerassi, C., Shamma, M., Khan, T. Y. **1958,** J. Am. Chem. Soc. *80,* 4723
Dobson, N. A., Raphael, R. A. **1955,** J. Chem. Soc. *1955,* 3358,
Dobson, N. A., Eglinton, G., Krishnamurti, M., Raphael, R. A., Willis, R. G. **1961,** Tetr. *16,* 16
Doskotch, R. W., Phillipson, J. D., Ray, A. B., Beal, J. L. **1971,** J. Org. Chem. *36,* 2409
Doyle, P., McLean, I. R., Murray, R. D. H., Parker, W., Raphael, R. A. **1965,** J. Chem. Soc. *1965,* 1344
Dryden, H. L., Jr., Webber, G. M., Wieczorek, J. J. **1964,** J. Am. Chem. Soc. *86,* 742
Durst, T. **1979,** *Sulphoxides,* in: Barton, D., Ollis, W. D. (eds.), *Comprehensive Organic Chemistry,* Vol. 3, p. 121 ff., Pergamon Press, Oxford
Dye, J. L., Lok, M. T., Tehan, F. J., Ceraso, J. M., Vorhees, K. J. **1973,** J. Org. Chem. *38,* 1773
Dyke, S. F. **1972,** Adv. Heterocyclic Chem. *14,* 279

Dyke, S. F. **1973,** *The Chemistry of Enamines,* Cambridge Univ. Press

Eaton, E. P. **1979,** Tetr. *35,* 2189
Ebel, H. F., Lüttringhaus, A. **1970,** in: Houben-Weyl, *Methoden der Organischen Chemie,* Vol. XIII/1, p. 621 ff, Thieme, Stuttgart
Eckstein, F. **1967,** Chem. Ber. *100,* 2228, 2236
Edwards, J. A., Sundeen, J., Salmond, W., Iwadare, T., Fried, J. H. **1972,** Tetr. Lett. *1972,* 791
Eglington, G., McCrae, W. **1963,** Adv. Org. Chem. *4,* 225
Ehrhart, G., Ruschig, H. **1972,** *Arzneimittel: Entwicklung, Wirkung, Darstellung,* 2nd ed., Verlag Chemie, Weinheim, Ger.
Eisner, U., Kuthan, J. **1972,** Chem. Rev. *72,* 1
Ellison, R. A. **1973,** Synthesis *1973,* 397
Emerson, G. F., Watts, L., Pettit, R. **1965,** J. Am. Chem. Soc. *87,* 131
Emmons, W. D., Lucas, G. B. **1955,** J. Am. Chem. Soc., *77,* 2287
Enders, D., Eichenauer, H. **1979,** Chem. Ber. *112,* 2933
Erickson, B. W., Merrifield, R. B. **1976,** in: Neurath, H., Hill, R. L. (eds), *The Proteins,* 3rd ed., Vol. II, p. 257–493, Academic Press, New York/London
Ernest, I., Gosteli, J., Woodward, R. B. **1979,** J. Am. Chem. Soc., *101,* 6301
Eschenmoser, A., Felix, D., Gut, M., Meier, J., Stadler, P. **1959,** in G. E. W. Wolstenholme, O'Connor, M. (eds), *Biosynthesis of Terpenes and Sterols,* Churchill, London
Eschenmoser, A. **1970,** Quart. Rev. *24,* 366; see also Dubs, P., Götschi, E., Roth, M., Eschenmoser, A. **1970,** Chimia, *24,* 34
Eschenmoser, A. **1974,** Naturwiss. *61,* 513
Evans, D. A., Scott, W. L., Truesdale, L. K. **1972,** Tetr. Lett. *1972,* 121
Evans, D. A., Vogel, E., Nelson, J. V. **1979,** J. Am. Chem. Soc. *101,* 6120
Evans, M. E. **1980,** in: Whistler, R. L., BeMiller, J. N. (eds), *Methods in Carbohydrate Chemistry,* Vol. VIII, p. 313, Academic Press, New York, London

Fechtig, B., Peter, H., Bickel, H., Fischler, E. **1968,** Helv. Chim. Acta *51,* 1108
Felix, D., Schreiber, J., Piers, K., Horn, U., Eschenmoser, A. **1968,** Helv. Chim. Acta **51,** 1461
Felix, D., Schreiber, J., Ohloff, G., Eschenmoser, A. **1971,** Helv. Chim. Acta, *54,* 2896
Felner, I., Fischli, A., Wick, A., Pesaro, M., Bormann, D., Winnacker, E. L., Eschenmoser, A. **1967,** Angew. Chem. *79,* 863 (Int. Ed. *6,* 864)
Fenselau, A. H., Moffatt, J. G. **1966,** J. Am. Chem. Soc., *88,* 1762
Ferles, M., Pliml. J. **1970,** Adv. Heterocyclic Chem., *12,* 43
Ferrier, R. J. **1965,** Adv. Carbohydr. Chem. Biochem. *20,* 67
Ferrier, R. J., Sankey, G. H. **1966,** J. Chem. Soc. (C) *1966,* 2339
Fieser, L. F., Smuszkovicz, J. **1948,** J. Am. Chem. Soc. *70,* 3352
Fieser, L. F., Stevenson, R. **1954,** J. Am. Chem. Soc. *76,* 1728
Fieser, L. F., Fieser, M. **1959,** *Steroids,* p. 507; Reinhold, New York/Chapman & Hall, London
Fieser, L. F., Fieser, M. **1959,** *Steroide,* Verlag Chemie, Weinheim, Ger.
Finn, F. M., Hofmann, K. **1976,** in: Neurath, H., Hill, R. L. (eds), *The Proteins,* 3rd ed., Vol. II, Chapter 2 (p. 106–237), Academic Press, New York/London
Fischer, E. **1914,** Chem. Ber. *47,* 196
Fischer, H., Stangler, G. **1927,** Chem. Ber. *459,* 53
Fischer, H., Neber, M. **1932,** Liebigs Ann. *496,* 1
Fischer, H., Stern, A. **1940,** *Die Chemie des Pyrrols,* Band II, Akademische Verlagsgesellschaft, Leipzig
Fleming, I. **1973,** *Selected Organic Syntheses,* Wiley, New York/London
Fleming, I. **1979,** *Organic Silicon Chemistry, in:* Barton, D., Ollis, W. D. (eds) *Comprehensive Organic Chemistry,* Vol. 3, p. 539, Pergamon Press, Oxford
Fletcher, H. G., Jr. **1963,** in: Whistler, R. L., Wolfrom, M. L., BeMiller, J. N., (eds), *Methods in Carbohydrate Chemistry,* Vol. II, p. 307, Academic Press, New York/London
Franck, B., Schlingloff, G. **1962,** Liebigs Ann. *659,* 123
Franck, B., Blaschke, G. **1966,** Liebigs Ann. *695,* 144
Franck, B., Dunkelmann, G., Lubs, H. J. **1967,** Angew. Chem. *79,* 1066, (Int. Ed. *6,* 1075)

Franck, B., Teetz, V. **1971,** Angew. Chem. *83,* 509 (Int. Ed. *10,* 411)

Fraser, R. R., Schuber, F. J., Wigfield, Y. Y. **1972,** J. Am. Chem. Soc. *94,* 8795

Freifelder, M., Robinson, R. M., Stone, G. R. **1962** J. Org. Chem. *27,* 284

Freifelder, M. **1963,** Adv. Catalysis *14,* 203

Fridkin, M., Patchornik, A., Katchalski, E. **1965,** J. Am. Chem. Soc. *87,* 4646

Fried, J., Szabo, E. F. **1954,** J. Am. Chem. Soc. *76,* 1455

Fried, J., Heim, S., Etheredge, S. J., Sunder-Plassmann, P., Santhanakrishnan, T. S., Himizu, J.-I., Lin., C. H. **1968,** Chem. Comm. *1968,* 634

Fried, J., Edwards, J. A. **1972,** *Organic Reactions in Steroid Chemistry,* Vol. 1 and 2, Van Nostrand, New York

Friedrich, A. **1975** in: Houben-Weyl, *Methoden der Organischen Chemie,* Vol. IV/1b: *Oxidation II,* p. 82, Thieme, Stuttgart

Fritsch, W., Stache, U., Haede, W., Radscheit, K., Ruschig, H. **1969,** Liebigs Ann. *721,* 168

Fritsch, W., Haede, W., Radscheit, K., Stache, U., Ruschig, H. **1974,** Liebigs Ann. *1974,* 621

Fuhrer, H., Ganguly, A. K., Gopinath, K. W., Govindachari, T. R., Nagarajan, K., Pai, B. R., Parthasarathy, P. C. **1969, 1970,** Tetr. Lett. *1969,* 133, Tetr. *26,* 2371

Fuhrhop, J.-H. **1974,** Angew. Chem. *86,* 363 (Int. Ed. *13,* 321)

Fuhrhop, J.-H., Baccouche, M. **1976,** Liebigs Ann. *1976,* 2058

Fuhrhop, J.-H., Witte, L., Sheldrick, W. S. **1976,** Liebigs Ann. *1976,* 1537

Fujita, K. **1961,** Bull. Chem. Soc. Japan *34,* 968

Funk, R. L., Vollhardt, K. P. C. **1977,** J. Am. Chem. Soc. *99,* 5483

Fusco, R. **1967** in: R. H. Wiley (ed.), *Pyrazoles, Pyrazolines, Pyrazolidines, Indazoles, and Condensed Rings,* p. 3, Interscience, New York

Gait, M. J., Sheppard, R. C. **1976,** J. Am. Chem. Soc. *98,* 8514

Gardner, J. N., Carlon, F. E., Gnoj, O. **1968 A,** J. Org. Chem. *33,* 3294

Gardner, J. N., Popper, T. L., Carlon, F. E., Gnoj, O., Herzog, H. L. **1968 B,** J. Org. Chem. *33,* 3695

Gauthier, J., Deslongchamps, P. **1967,** Can. J. Chem. *45,* 297

Gaylord, N. G. **1956,** *Reduction with Complex Metal Hydrides,* Interscience, New York

Gensler, W. J., Bruno, J. J. **1963,** J. Org. Chem. *28,* 1254

Gerlach, H. **1966,** Helv. Chim. Acta *49,* 1291

Gibson, T. W., Erman, W. F. **1969,** J. Am. Chem. Soc. *91,* 4771

Gilbert, E. E. **1965,** *Sulphonation and Related Reactions,* Wiley, New York

Gilchrist, T. L., Gymer, G. E. **1974,** Adv. Heterocyclic Chem. *16,* 33

Girard, C., Amice, P., Barnier, J. P., Conia, J. M. **1974,** Tetr. Lett. *1974,* 3329

Gisin, B. F., Merrifield, R. B. **1972,** J. Am. Chem. Soc. *94,* 3102, 6165

Glotter, E., Schwartz, E. **1976,** J. Chem. Soc. Perkin I *1976,* 1660

Gompper, R., Mensch, S., Seybold, G. **1975,** Angew. Chem. *87,* 711 (Int. Ed. *14,* 704)

Gopal, H., Gordon, A. J. **1971,** Tetr. Lett. *1971,* 2941

Gossauer, A. **1974,** *Die Chemie der Pyrrole,* Springer, Berlin/New York

Gottstein, W. J., Minor, W. F., Cheney, L. C. **1959,** J. Am. Chem. Soc. *81,* 1198

Graf, E., Lehn, J.-M. **1975,** J. Am. Chem. Soc. *97,* 5022

Grahn, W. **1981,** Chemie in unserer Zeit *15,* 52

Gravestock, M. B., Johnson, W. S., Myers, R. F., Bryson, T. A., Miles, D. H., Ratcliffe, B. E. **1978** J. Am. Chem. Soc. *100,* 4268

Gravestock, M. B., Johnson, W. S., McCarry, B. E., Parry, R. J., Ratcliffe, B. E. **1978,** J. Am. Chem. Soc. *100,* 4274

Greene, A. E., Deprés, J.-P. **1979,** J. Am. Chem. Soc. *101,* 4003

Grenda, V. J., Lindberg, G. W., Wendler, N. L., Pines, S. H. **1967,** J. Org. Chem. *32,* 1236

Griffin, B. E., Reese, C. B., Stephenson, G. F., Trentham, D. R. **1966,** Tetr. Lett. *1966,* 4349

Grimme, W., Reissdorff, J., Jünemann, W., Vogel, E. **1970** J. Am. Chem. Soc. *92,* 6335

Grimmett, M. R. **1970,** Adv. Heterocyclic Chem. *12,* 104

Grob, C. A., Ohta, M., Renk, E., Weiss, A. **1958,** Helv. Chim. Acta *41,* 1191

Gröbel, B.-T., Seebach, D. **1974,** Angew. Chem. *86,* 102 (Int. Ed. *13,* 83)

Gröbel, B.-T., Seebach, D. **1977 A,** Chem. Ber. *110,* 852, 867

Gröbel, B.-T., Seebach, D. **1977B,** Synthesis *1977,* 357
Grundon, M. F., Henbest, H. B., Scott, M. D. **1963,** J. Chem. Soc. *1963,* 1855
Guha, M., Nasipuri, D. **1973,** Org. Synth. Coll. Vol. *V,* 384
Gutte, B., Merrifield, R. B. **1969,** J. Am. Chem. Soc. *91,* 502

Haines, A. H. **1976,** Adv. Carbohydr. Chem. Biochem. *33,* 11
Hajós, A. **1966,** *Komplexe Hydride und ihre Anwendung in der organischen Chemie,* VEB Deutscher Verlag der Wissenschaften, Berlin
Hajós, A. **1979,** *Complex Hydrides and Related Reducing Agents in Organic Synthesis,* Elsevier, Amsterdam/New York
Hampton, K. G., Hauser, C. R., **1965,** J Org. Chem. *30,* 2934
Hampton, K. G., Harris, T. M., Hauser, C. R. **1973,** Org. Synth. Coll. Vol. *V,* 848
Hanessian, S., Rancourt, G. **1977,** Can. J. Chem. *55,* 1111, Pure Appl. Chem. *49,* 1201
Hanessian, S., Rancourt, G., Guindon, Y. **1978,** Can. J. Chem. *56,* 1843
Hanessian, S. **1979,** Accts. Chem. Res. *12,* 159
Hankinson, B., Heaney, H., Sharma, R. P. **1972,** J. Chem. Soc. Perkin I, *1972,* 2372
Hardcastle, G. A., Jr., Johnson, D. A., Panetta, G. A., Scott, A. I., Sutherland, S. A. **1966,** J. Org. Chem. *31,* 897
Harrison, W. A., Curcumelli-Rodostamo, M., Carson, D. F., Barclay, L. R. C., MacLean, D. B. **1961,** Can. J. Chem. *39,* 2086
Hartmann, W., Fischer, H.-M., Heine, H.-G. **1972,** Tetr. Lett. *1972,* 853
Hassall, C. H. **1957,** Org. React. *9,* 73
Hattersley, P. J., Lockhart, I. M., Wright, M. **1969,** J. Chem. Soc. (C) *1969,* 217
Hauser, C. R., Harris, T. M. **1958,** J. Am. Chem. Soc. *80,* 6360
Hauthal, H. G., Lorenz, D. **1971,** in: Martin, D., Hauthal, H. G., *Dimethylsulfoxid,* p. 344 ff., Akademie Verlag, Berlin
Hazen, G. G., Rosenburg, D. W. **1964,** J. Org. Chem. *29,* 1930
Heathcock, C. H. **1966,** Tetr. Lett. *1966,* 2043
Heathcock, C. H., Badger, R. A., Patterson, J. W., Jr. **1967,** J. Am. Chem. Soc. *89,* 4133
Heimbach, P., Traunmüller, R. **1970,** *Die Chemie der Metall-Olefin-Komplexe,* Verlag Chemie, Weinheim, Ger.
Heldt, W. Z., **1960,** Org. React. *11,* 1
Helferich, B. **1952,** Adv. Carbohyd. Chem. *7,* 209
Hellmann, H., Opitz, G. **1960,** α-Aminoalkylierung, Verlag Chemie, Weinheim, Ger.
Henbest, H. B., Wilson., R. A. L. **1956,** J. Chem. Soc. *1956,* 3289
Hill, R. K., Joule, J. A., Loeffler, L. L. **1962,** J. Am. Chem. Soc. *84,* 4951
Hill, R. K., Ladner, O. W. **1975,** Tetr. Lett. *1975,* 989
Hiroi, K., Yamada, S.-I. **1975,** Chem. Pharm. Bull *23,* 1103
Hirschmann, R., Nutt, R. F., et al. **1969,** J. Am. Chem. Soc. *91,* 508
Hoffmann, R. W. **1967,** *Dehydrobenzene and Cycloalkynes,* Verlag Chemie, Weinheim – Academic Press, New York/London
Höfle, G., Steglich, W., Vorbrüggen, H. **1978,** Angew. Chem. *90,* 602 (Int. Ed. *17,* 569)
Hofmann, A., Frey, A. J., Ott, H. **1961,** Experienta *17,* 206
Hofmann, A., Ott, H., Griot, R., Stadler, P. A., Frey, A. J. **1963,** Helv. Chim. Acta *46,* 2306
Hortmann, A. G., Martinelli, J. E., Wang, Y.-S. **1969** J. Org.Chem. *34,* 732
Hortmann, A. G., Robertson, D. A. **1972,** J. Am. Chem. Soc. *94,* 2758
Hough, L., Richardson, A. C. **1979,** in: Barton, D., Ollis, W. D., *Comprehensive Organic Chemistry,* Vol. 5, Pergamon Press, Oxford
Houghton, R. P. **1979,** *Metal Complexes in Organic Chemistry,* Cambridge Univ. Press
House, H. O., Schellenbaum, M. **1963,** J. Org. Chem. *28,* 34
House, H. O., Trost, B. M. **1965,** J. Org. Chem. *30,* 2502
House, H. O. **1967,** Rec. Chem. Progr. *28,* 99
House, H. O., Czuba, L. J., Gall, M., Olmstead, H. D. **1969,** J. Org. Chem. *34,* 2324
House, H. O. **1972,** *Modern Synthetic Reactions,* W. A. Benjamin, Menlo Park (Calif.)
House, H. O., **1976,** Accts. Chem. Res. *9,* 59
Hückel, W., Schlee, H. **1955,** Chem. Ber. *88,* 346

Hückel, W., Rücker, D. **1963,** Liebigs Ann. *666,* 30

Hünig, S., Lendle, W. **1960,** Chem. Ber. *93,* 913

Hünig, S., Müller, H. R., Thier, W. **1965,** Angew. Chem. *77,* 368 (Int. Ed. *4,* 271)

Huffman, J. W., Rao, C. B. S., Kamiya, T. **1967,** J. Org. Chem. *32,* 697

Hutchins, R. O., Kandasamy, D. **1973 A,** J. Am. Chem. Soc. *95,* 6131

Hutchins, R. O., Milewski, C. A., Maryanoff, B. E. **1973 B,** J. Am. Chem. Soc. *95,* 3662

Hutchins, R. O., Kacher, M., Rua, L. **1975,** J. Org. Chem. *40,* 923

Igarishi, K. **1977,** Adv. Carbohydr. Chem. Biochem. *34,* 243

Inhoffen, H. H., Köster, H. **1939,** Chem. Ber. *72 (B),* 595

Inhoffen, H. H. **1940,** Angew. Chem. *53,* 471

Inhoffen, H. H., Irmscher, K., Hirschfeld, H., Stache, U., Kreutzer, A. **1958A,** Chem. Ber. *91,* 2309

Inhoffen, H. H., Schütz, S., Rossberg, P., Berger, O., Nordsiek, K.-H., Plenio, H., Höroldt, E. **1958B,** Chem. Ber. *91,* 2626

Inhoffen, H. H., Burkhardt, H., Quinkert, G. **1959,** Chem. Ber. *92,* 1564

Ireland, R. E., Marshall, J. A. **1962,** J. Org. Chem. *27,* 1615, 1620

Irwin, W. J., Mc Quillin, F. J. **1968,** Tetr. Lett. *1968,* 2195

Itakura, K. **1977,** Science *198,* 1056

Ito, Y., Konoike, T., Saegusa, T. **1975,** J. Am. Chem. Soc. *97,* 649

Jackson, A. H., Smith, K. M. **1973,** in: ApSimon, J. (ed.) *The Total Synthesis of Natural Products,* Vol. I, p. 167, Wiley, New York/London

Jacquet, I., Vigneron, J. P. **1974,** Tetr. Lett. *1974,* 2065

Jäger, V., Viehe H. G. **1977,** in: Houben-Weyl, *Methoden der Organischen Chemie,* Vol. V/2a, p. 509 ff., Thieme, Stuttgart

Johansen, J. E., Angst, C., Kratky, C., Eschenmoser, A. **1980,** Angew. Chem. *92,* 141 (Int. Ed. *19,* 141)

Johnson, C. R., Herr, R. W., Wieland, P. M. **1973 A,** J. Org. Chem. *38,* 4263

Johnson, C. R., Schroeck, C. W. **1973 B,** J. Am. Chem. Soc. *95,* 7418

Johnson, C. R. **1979,** *Sulphur Ylides,* in: Barton, D.; Ollis, W. D. (eds.), *Comprehensive Organic Chemistry,* Vol. 3: D. N. Jones (ed.), p. 247 ff., Pergamon Press, Oxford

Johnson, R. A. **1978,** in: W. S. Trahanovsky (ed.), *Oxidation in Organic Chemistry,* Part C, (Organic Chemistry, Vol. 5—C) Chapter II, p. 131, Academic Press, New York/London

Johnson, W. S., Petersen, J. W., Gutsche, C. D. **1945,** J. Am. Chem. Soc. *67,* 2274

Johnson, W. S., Petersen, J. W., Gutsche, C. D. **1947,** J. Am. Chem. Soc. *69,* 2942

Johnson, W. S., Szmuszkovicz, S., Rogier, E. R., Hadler, H. I., Wynberg, H. **1956,** J. Am. Chem. Soc. *78,* 6285

Johnson, W. S., Allen, D. S., Jr. **1957,** J. Am. Chem. Soc. *79,* 1261, 1995

Johnson, W. S., Allen, D. S., Jr., Hindersinn, R. R., Sausen, G. N., Pappo, R. **1962,** J. Am. Chem. Soc. *84,* 2181

Johnson, W. S., Collins, J. C., Jr., Pappo, R., Rubin, M. B., Kropp, P. J., Johns, W. F., Pike, J. E., Bartmann, W. **1963,** J. Am. Chem. Soc. *85,* 1409

Johnson, W. S. **1968,** Accts. Chem. Res. *1,* 1

Johnson, W. S. **1976,** Angew. Chem. *88,* 33 (Int. Ed. *15,* 9)

Jones, G. **1970,** J. Chem. Soc. (C), *1970,* 1230; see also: Collington, E. W., Jones, G., Chem. Comm. *1968,* 958, J. Chem. Soc. (C) *1969,* 2656

Jorgenson, M. J. **1970,** Org. React. *18,* 1

Julia, M., Julia, S., Tchen, S. Y. **1961,** Bull. Soc. Chim. France *1961,* 1849

Julia, M., Paris, J. M. **1974,** Tetr. Lett. *1974,* 3445

Kagan, H. B., Fiaud, J. C. **1978,** Tropics Stereochem. *10,* 175

Kametani, T., Iida, H., Kikuchi, T. **1969,** Chem. Pharm. Bull. *17,* 709

Kametani, T., Ogasawara, K., Takahashi, T. **1972/1973,** Chem. Comm. *1972,* 675/Tetr. *29,* 73

Kametani, T., Takahashi, T., Ogasawara, K., Fukumoto, K. **1974,** Tetr. *30,* 1047

Kametani, T. **1977 A,** in Ap Simon, J. (ed.), *The Total Synthesis of Natural Products,* Vol. 3, p. 1, Wiley, New York

Kametani, T., Nemoto, H., Ishikawa, H., Shiroyama, K., Matsumoto, H., Fukumoto, K. **1977B,** J. Am. Chem. Soc. 99, 3461

Kametani, T., Matsumoto, H., Nemoto, H., Fukumoto, K. **1978,** J. Am. Chem. Soc. *100,* 6218

Katagiri, N., Itakura, K., Narang, S. A. **1975,** J. Am. Chem. Soc. *97,* 7332

Katz, T. J., Wang, E. J., Acton, N. **1971,** J. Am. Chem. Soc. *93,* 3782

Katz, T. J., Roth, R. J. **1972,** J. Am. Chem. Soc. *94,* 4770

Katzenellenbogen, J. A., Bowlus, S. B. **1973,** J. Org. Chem. *38,* 627

Kaufmann, D., de Meijere, A. **1979,** Tetr. Lett. *1979,* 779, 783, 787

Kawana, M., Emoto, S. **1966,** Bull. Chem. Soc. Japan *39,* 910

Kenner, G. W., Mc Combie, S. W., Smith, K. M. **1973 A,** Lieb. Ann. *1973,* 1329

Kenner, G. W., Smith, K. M., Unsworth, J. F. **1973 B,** Chem. Comm. *1973,* 43

Khan, M. M. T. **1974,** *Homogeneous Catalysis by Metal Complexes, Vol. II, Activation of Alkenes and Alkynes,* Academic Press, New York/London

Khorana, H. G., Jacob, T. M., Moon, M. W., Narang, S. A., Ohtsuka, E. **1965,** J. Am. Chem. Soc. *87,* 2954–95

Kieslich, K. **1976,** *Microbial Transformations of Non-Steroid Cyclic Compounds,* Thieme, Stuttgart

Kimel, W., Surmatis, J. D., Weber, J., Chase, G. O., Sax, N. W., Ofner, A. **1957,** J. Org. Chem. *22,* 1611; see also: Royals, E. E. **1946,** Ind. Eng. Chem. *38,* 546

Kimel, W., Sax, N. W., Kaiser, S., Eichmann, G. G., Chase, G. O., Ofner, A. **1958,** J. Org. Chem. *23,* 153

Kitahara, Y., Yoshikoshi, A., Oida, S. **1964,** Tetr. Lett. *1964,* 1763

Kitatani, K., Hiyama, T., Nozaki, H. **1976,** J. Am. Chem. Soc. *98,* 2362

Kleschick, W. A., Buse, C. T., Heathcock, C. H., **1977,** J. Am. Chem Soc. *99,* 247

Klioze, S. S., Darmory, F. P. **1975,** J. Org. Chem. *40,* 1588

Knöll, W., Tamm, C. **1975,** Helv. Chim. Acta *58,* 1162

Knowles, W. S., Sabacky, M. J., Vineyard, B. D., Weinkauff, D. J. **1975,** J. Am. Chem. Soc. *97,* 2567

Köbrich, K., Werner, W. **1969,** Tetr. Lett. *1969,* 2181

Kosower, E. M., Winstein, S. **1956,** J. Am. Chem. Soc. *78,* 4347, 4354

Köster, H. **1979,** Nachr. Chem. Tech. Lab. *27,* 694

Köster, R., Arora, S., Binger, P. **1971,** Synthesis *1971,* 322

Kozikowski, A. P., Wetter, H. F. **1976,** Synthesis, *1976,* 561

Kramer, U., Guggisberg, A., Hesse, M., Schmid, H. **1978,** Angew. Chem. *90,* 210 (Int. Ed. *17,* 200)

Kramer, U., Schmid, H., Guggisberg, A., Hesse, M. **1979,** Helv. Chim. Acta *62,* 811

Krishnamurthy, S., Brown, H. C. **1976,** J. Am. Chem. Soc. *98,* 3383

Kropf, H. **1980** in: Houben-Weyl, *Methoden der Organischen Chemie,* Vol. IV/1c: *Reduktion I,* Thieme, Stuttgart/New York

Kuivila, H. G. **1968,** Accts. Chem. Res. *1,* 299

Kuo, C. H., Taub, D., Wendler, N. L. **1968,** J. Org. Chem. *33,* 3126

Kutney, J. P. **1977,** in Ap Simon, J. (ed.), *The Total Synthesis of Natural Products,* Vol. 3, p. 273, Wiley, New York

Kuwajima, I., Nakamura, E. **1975,** J. Am. Chem. Soc. *97,* 3257

Kuwajima, I., Sato, T., Arai, M., Minami, N. **1976,** Tetr. Lett. *1976,* 1817

Landor, S. R., Punja, N. **1967,** J. Chem. Soc. (C) *1967,* 2495

Langecker, H., Scheiffele, E., Geiger, R., Prezewowsky, K., Stache, U., Schmitt, K. **1977,** in Bartholomé, E. (ed.) *Ullmanns Encyklopädie der technischen Chemie,* Vol. 13, 4th ed., p. 1, Verlag Chemie, Weinheim

Latimer, W. M. **1952,** *Oxidation Potentials,* Prentice-Hall, Englewood Cliffs (N. J.)

Lednicer, D., Mitscher, L. A. **1977/1980,** *The Organic Chemistry of Drug Synthesis,* Vol. I/II, Wiley, New York/London

Lee, D. G. **1969,** in: Augustine, R. L. (ed.), *Oxidation,* Vol. 1, chapters 1 and 2, M. Dekker, New York

Lee, D. G., van den Engh, M. **1973,** in: Trahanovsky, W. S. (ed.), *Oxidation in Organic Chemistry,* Part B, Academic Press, New York, London

Lehmann, H. **1975** in: Houben-Weyl, *Methoden der Organischen Chemie,* Vol. IV/1b: *Oxidation II,* p. 905, Thieme, Stuttgart

Lehmann, J. **1976,** *Chemie der Kohlenhydrate,* Thieme, Stuttgart

Lemal, D. M., Lokensgard, J. P. **1966,** J. Am. Chem. Soc. *88,* 5934

Lemieux, R. U., Gunner, S. W., Nagabhushan, T. L. **1965,** Tetr. Lett. *1965,* 2143, 2149

Lemieux, R. U., Nagabhushan, T. L., O'Neill, I. K. **1968,** Canad. J. Chem. *46,* 413

Letsinger, R. L., Lunsford, W. B. **1976,** J. Am. Chem. Soc. *98,* 3655

Levin, R. H. **1978,** *Arynes,* in: Jones, M., Moss, R. A. (eds.), *Reactive Intermediates,* Vol. 1, Part 1, Wiley, New York

Lewis, S. N. **1969** in: Augustine, R. L. (ed.), *Oxidation,* Vol. 1, chapter 5 (p. 213), M. Dekker, New York

Liebman, J. F., Greenberg, A. **1976,** Chem. Rev. *76,* 311

Lindlar, H., Dubuis, R. **1973,** Org. Synth. Coll. Vol. V, 880

Loewenthal, H. J. E., **1959,** Tetr. *6,* 269

Lohmar, R., Steglich, W. **1980,** Chem. Ber. 113. 3706

Losse, G., Zeidler, D., Grieshaber, T. **1968,** Liebigs Ann. *715,* 196

Lourens, G. J., Koekemoer, J. M. **1975,** Tetr. Lett. *1975,* 3719

Lübke, K., Schröder, E., Kloss, G. **1975,** *Chemie und Biochemie der Aminosäuren, Peptide und Proteine,* Vol. I, Thieme, Stuttgart

Luche, J.-L., Rodriguez-Hahn, L., Crabbé, P. **1978,** Chem. Comm. *1978,* 601

McCloskey, C. M. **1957,** Adv. Carbohydrate Chem. *12,* 137

McCormick, J. P., Barton, D. L. **1975,** Chem. Comm. *1975,* 303

McDonald, P. D., Hamilton, G. A. **1973,** in: Trahanovsky, W. S. (ed.), *Oxidation in Organic Chemistry,* Part B, p. 97, Academic Press, New York/London

McElvain, S. M. **1948,** Org. React. *4,* 256

McKillop, A., Oldenziel, O. H., Swann, B. P., Taylor, E. C., Robey, R. L. **1973,** J. Am. Chem. Soc. *95,* 1296

McMurry, J. E. **1974,** Accts. Chem. Res. *7,* 281

McMurry, J. E., Fleming, M. P. **1974,** J. Am. Chem. Soc. *96,* 4708

McMurry, J. E., Fleming, M. P. **1976A,** J. Org. Chem. *41,* 896

McMurry, J. E., Krepski, L. R. **1976B,** J. Org. Chem. *41,* 3929

McOmie, J. F. W. **1973,** *Protective Groups in Organic Chemistry,* Plenum, New York

Maercker, A. **1965,** Org. React. *14,* 270

Magnusson, G. **1977,** Tetr. Lett. *1977,* 2713

Maier, G., Alzérreca, A. **1973,** Angew. Chem. *85,* 1056, (Int. Ed. *12,* 1015)

Maier, G., Pfriem, S., Schäfer, U., Matusch, R. **1978,** Angew. Chem. *90,* 552 (Int. Ed. *17,* 520)

Majerski, Z., von R. Schleyer, P. **1968,** Tetr. Lett. *1968,* 6195

Maki, Y., Kikuchi, K., Sugiyama, H., Seto, S. **1977,** Tetr. Lett. *1977,* 263

Málek, J., Černý, M. **1972,** Synthesis *1972,* 217

Malpass, J. R., Tweddle, N. J. **1977,** J. Chem. Soc. Perkin I *1977,* 874

Manegold, D. **1975** in: Houben-Weyl, *Methoden der Organischen Chemie,* Vol. IV/1b: *Oxidation II,* p. 76, Thieme, Stuttgart

Mangoni, L., Adolfini, M., Barone, G., Parilli, M. **1973,** Tetr. Lett. *1973,* 4485

Mareš, F., Roček, J. **1961,** Coll. Czech. Chem. Comm. *26,* 2370

Marino, J. P., Landick, R. C. **1975,** Tetr. Lett. *1975,* 4531

Marshall, J. A., Fanta, W. I., **1964,** J. Org. Chem. *29,* 2501

Marshall, J. A., Bundy, G. L. **1966,** J. Am. Chem. Soc. *88,* 4291

Marshall, J. A., Bundy, G. L., Fanta, W. I. **1968,** J. Org. Chem. *33,* 3913

Marshall, J. A., Roebke, H. **1969,** J. Org. Chem. *34,* 4188

Martin, D., **1971** in: Martin, D., Hauthal, H. G. (eds.), *Dimethylsulfoxid,* chapter 8, p. 306, Akademie Verlag, Berlin

Martin, S. F., Pesai, S. R., Phillips, G. W., Miller, A. C., **1980,** J. Am. Chem. Soc. *102,* 3294

Marx, J. N., Argyle, J. C., Norman, L. R. **1974,** J. Am. Chem. Soc. *96,* 2121

Masamune, S. **1961,** J. Am. Chem. Soc. *83,* 1009

Masamune, S. **1964,** J. Am. Chem. Soc. *86,* 288, 290

Masamune, S., Cuts, H., Hogben, M. G. **1966** Tetr. Lett. *1966,* 1017

Masamune, S., Seidner, R. T. **1969,** Chem. Comm. *1969,* 542
Mathias, L. J. **1979,** Synthesis *1979,* 561
Matthews, R. S., Meteyer, T. E. **1971,** Chem. Comm. *1971,* 1576
Mauzerall, D. **1960,** J. Am. Chem. Soc. *82,* 2605
Mazur, Y., Danieli, N., Sondheimer, F. **1960,** J. Am. Chem. Soc. *82,* 5889
Meerwein, H., Bodenbrenner, K., Borner, P., Kunert, F. **1960,** Liebigs Ann. *632,* 38
Meinwald, J., Frauenglass, E. **1960,** J. Am. Chem. Soc. *82,* 5235
Melson, G. A. (ed.) **1979,** *Coordination Chemistry of Macrocyclic Compounds,* Plenum, New York
Menapace, L. W. **1964,** J. Am. Chem. Soc. *86,* 3047
Mercier, C., Soucy, P., Rosen, W., Deslongchamps, P. **1973,** Synth. Comm. *3,* 161
Merrifield, R. B. **1969,** in: Nord, F. F. (ed.), *Advances in Enzymology,* Interscience, New York
Meyer, W. L., Biesalski, B. S. **1963,** J. Org. Chem. *28,* 2896
Meyers, A. I., Knauss, G., Kamata, K., Ford, M. E. **1976,** J. Am. Chem. Soc. *98,* 567
Meyer zu Reckendorf, W., Wassiliadou-Micheli, N., Bischof, E. **1971,** Chem. Ber. *104,* 1
Michelson, A. M. **1963,** *The Chemistry of Nucleosides and Nucleotides,* Academic Press, New York/London
Miller, C. E. **1965,** J. Chem. Educ. *42,* 254
Miller, L. L., Stermitz, F. R., Falck, J. R. **1971/1973,** J. Am. Chem. Soc. *93,* 5941/*95,* 2651
Minami, T., Niki, I., Agawa, T. **1974,** J. Org. Chem. *39,* 3236
Mitchell, R. H., Boeckelheide, V. **1974,** J. Am. Chem. Soc. *96,* 1547, 1558
Miyaura, N., Itoh, M., Sasaki, M. **1975,** Synthesis *1975,* 317
Miyaura, N., Itoh, M., Suzuki, A. **1976,** Tetr. Lett. *1976,* 255
Moffatt, J. G. **1971** in: Augustine, R. L., Trecker, D. J. (eds.), *Oxidation,* Vol. 2, chapter 1 (p. 1), M. Dekker, New York
Montanari, F. **1975,** in: Stirling, C. J. M., *Organic Sulphur Chemistry,* Butterworths, London
Mori, K., **1975,** Tetr. *31,* 3011
Mori, K., Oda, M., Matsui, M. **1976,** Tetr. Lett. *1976,* 3173
Mori, T., Nakahara, T., Nozaki, H. **1969,** Can. J. Chem. *47,* 3266
Morin, R. B., Jackson, B. G., Flynn, E. H., Roeske, R. W., **1962,** J. Am. Chem. Soc. *84,* 3400
Morrison, J. D., Mosher, H. S. **1971,** *Asymmetric Organic Reactions,* Prentice-Hall, Englewood Cliffs (N. J.)
Mousseron, M., Jacquier, R., Christol. H. **1957,** Bull. Soc. Chim. France *1957,* 346
Mukaiyama, T., Sato, T., Hanna, J. **1973,** Chem. Lett. *1973,* 1041
Mukaiyama, T., Banno, K., Narasaka, K. **1974,** J. Am. Chem. Soc. *96,* 7503
Mulzer, J., Brüntrup, G., Finke, J., Zippel, M. **1979,** J. Am. Chem. Soc. *101,* 7723
Mulzer, J., Zippel, M., Brüntrup, G., Segner, J., Finke, J. **1980,** Liebigs Ann. Chem. *1980,* 1108
Mundy, B. F. **1972,** J. Chem. Educ. *49,* 91
Muxfeldt, H., Haas, G., Hardtmann, G., Kathawala, F., Mooberry, J. B., Vedejs, E. **1979,** J. Am. Chem. Soc. *101,* 689

Nace, H. R. **1962,** Org. React. *12,* 57
Nagarkatti, J. P., Ashley, K. R. **1973,** Tetr. Lett. *1973,* 4599
Nagata, W., Sugasawa, T., Narisada, M., Wakabayashi, T., Hagase, Y., **1967,** J. Am. Chem. Soc. *89,* 1483
Nagata, W., Hirai, S., Okumura, T., Kawata, K. **1968** J. Am. Chem. Soc. *90,* 1650
Nakamura, E., Kuwajima, I. **1977,** J. Am. Chem. Soc. *99,* 7360
Nakatsubo, F., Kishi, Y., Goto, T. **1970,** Tetr. Lett. *1970,* 381
Nakazaki, M., Naomura, K. **1966,** Tetr. Lett. *1966,* 2615
Narang, S. A., Wightman, R. H. **1973,** in: Ap Simon, J. (ed.), *The Total Synthesis of Natural Products,* Vol. I, p. 279, Wiley, New York/London
Nazarov, I. N., Torgov, I. V., Verkholetova, G. **1957,** Dokl. Akad. Nauk SSSR *112,* 1067
Nedelec, L., Gasc, T., Bucourt, R. **1974,** Tetr. *30,* 3263
Nefkens, G. H. L. **1960,** Nature *185,* 309
Nefkens, G. H. L., Tesser, G. I., Nivard, K. J. F. **1960,** Rec. Trav. Chim. Pays-Bas *79,* 688
Negishi, E., Yoshida, T. **1973,** Chem. Comm. *1973,* 606
Nerdel, F., Frank, D., Lengert, H.-J., Weyerstahl, P. **1968,** Chem. Ber. *101,* 1850

Newhall, W. F. **1958,** J. Org. Chem. *23,* 1274
Newman, M. S., Mekler, A. B. **1960,** J. Am. Chem. Soc. *82,* 4039
Nicolaou, K. C. **1977,** Tetr. *33,* 683
Nielsen, A. T., Houlihan, W. J. **1968,** Org. React. *16,* 1
Nigh, W. G. **1973** in: Trahanovsky, W. S. (ed.), *Oxidation in Organic Chemistry,* Part B, p. 35, Academic Press, New York/London
Ninomiya, I., Naito, T. **1973,** Chem. Comm. *1973,* 137
Nishimura, J., Kawabata, N., Furukawa, J. **1969,** Tetr. *25,* 2647
Noland, W. E. **1955,** Chem. Rev. *55,* 137
Nutt, R. F., Veber, D. F., Saperstein, R. **1980,** J. Am. Chem. Soc. *102,* 6539

Olah, G. A., Brydon, D. L. **1970,** J. Org. Chem. *35,* 313
Oldenziel, O. H., van Leusen, A. M. **1973, 1974,** Tetr. Lett. *1973,* 1357, *1974,* 163, 167
Oppolzer, W., Pfenninger, E., Keller, K. **1973,** Helv. Chim. Acta *56,* 1807
Oppolzer, W., Snowden, R. L. **1976,** Tetr. Lett. *1976,* 4187
Oppolzer, W. **1978 A,** Synthesis *1978,* 793
Oppolzer, W., Snieckus, V. **1978 B,** Angew. Chem. *90,* 506, (Int. Ed. *17,* 476)
Ouellette, R. J. **1973** in: Trahanovsky, W. S. (ed.), *Oxidation in Organic Chemistry,* Part B, p. 135, Academic Press, New York/London
Overberger, C. G., Kaye, H., **1967,** J. Am. Chem. Soc. *89,* 5640
Overberger, G. G., Weise, J. K. **1968,** J. Am. Chem. Soc. *90,* 3525

Pappo, R., Allen, D. S., Jr., Lemieux, R. U., W. S. Johnson **1956,** J. Org. Chem. *21,* 478
Paquette, L. A., Wyvratt, M. J. **1974,** J. Am. Chem. Soc. *96,* 4671
Paquette, L. A., Wyvratt, M. J., Schallner, O., Schneider, D. F., Begley, W. J., Blankenship, R. M. **1976,** J. Am. Chem. Soc. *98,* 6744
Paquette, L. A., Lavrik, P. B., Summerville, R. H. **1977,** J. Org. Chem. *42,* 2659
Paquette, L. A. **1979,** Topics Curr. Chem. *79,* 41
Paquette, L. A., Balogh, D. W., Usha, R., Koumtz, D., Christoph, G. G. **1981,** Science *211,* 575
Paquette, L. A., Balogh, D. W., **1982,** J. Am. Chem. Soc. *104,* 774
Parham, W. E., Anderson, E. L. **1948,** J. Am. Chem. Soc. *70,* 4187
Paulsen, H., Sinnwell, V., Stadler, P. **1972,** Chem. Ber. *105,* 1978
Paulsen, H. **1977,** Pure Appl. Chem. *49,* 1169
Pelter, A., Smith, K. **1979,** *Organic Boron Compounds,* in: Barton, D., Ollis, W. D. (eds.), *Comprehensive Organic Chemistry* Vol. 3: D. N. Jones (ed.), p. 687 ff., Pergamon Press, Oxford
Peterson, D. J. **1968,** J. Org. Chem. *33,* 781
Peterson, P. E., Kamat, R. J. **1969,** J. Am. Chem. Soc. *91,* 4521
Pettit, G. R., van Tamelen, E. E. **1962 A,** Org. React. *12,* 356
Pettit, G. R., Piatak, D. M. **1962 B** J. Org. Chem. *27,* 2127
Pfaltz, A., Bühler, N., Neier, R., Hirai, K., Eschenmoser, A. **1977,** Helv. Chim. Acta *60,* 2653
Pfau, M, Ughetto-Monfrin, J., Joulain, D. **1979** Bull. Soc. Chim. France II *1979,* 627
Pfeiffer, F. R., Cohen, S. R., Williams, K. R., Weisbach, J. A. **1968,** Tetr. Lett. *1968,* 3549
Pfeiffer, F. R., Miao, C. K., Weisbach, J. A. **1970** J. Org. Chem. *35,* 221
Pfitzner, K. E., Moffatt, J. G. **1965,** J. Am. Chem. Soc. *87,* 5661, 5670
Plesničar, B. **1978** in: Trahanovsky, W. S. (ed.), *Oxidation in Organic Chemistry,* Part C, chapter III, p. 211, Academic Press, New York/London
Pohmakotr, M., Seebach, D. **1977,** Angew. Chem. *89,* 333 (Int. Ed. *16,* 320)
Pommer, H. **1960,** Angew. Chem. *72,* 811
Pommer, H., Nürrenbach, A. **1975,** Pure Appl. Chem. *43,* 527
Pommer, H. **1977,** Angew. Chem. *89,* 437 (Int. Ed. *16,* 423).
Posner, G. H. **1972,** Org. React. *19,* 1; **1975,** Org. React. *22,* 253

Rabjohn, N. **1949/1976,** Org. React. *5,* 331/*24,* 261
Ramachandran, S., Newman, M. S. **1961,** Org. Synth. *41,* 39
Ramirez, F., Marecek, J. F., Ugi, I. **1975,** Synthesis *1975,* 99
Ramirez, F., Evangelidou-Tsolis, E., Jankowski, A., Marecek, J. F. **1977,** J. Org. Chem. *42,* 3144

Rasmussen, J. K. **1977,** Synthesis *1977,* 91

Rauscher, G., Clark, T., Poppinger, D., von R. Schleyer, P. **1978,** Angew. Chem. *90,* 306 (Int. Ed. *17,* 276)

Ravid, U., Silverstein. R. M. **1977,** Tetr. Lett. *1977,* 423

Reese, C. B., Trentham, D. R. **1965,** Tetr. Lett. *1965,* 2459

Reich, H. J., Renga, J. M., Reich, I. L. **1975,** J. Am. Chem. Soc. *97,* 5434

Reich, H. J. **1978** in: Trahanovsky, W. S., (ed.), *Oxidation in Organic Chemistry,* Part C, p. 1, Academic Press, New York/London

Reich, H. J. **1979,** Accts. Chem. Res. *12,* 23

Reichert, B. **1959,** *Die Mannich-Reaktion,* Springer Verlag, Berlin/New York

Reif, W., Grassner, H. **1973,** Chem. Ing. Tech. *45,* 646

Reinefeld, E., Heincke, K. D. **1971,** Chem. Ber. *104,* 265

Reist, E. J., Bartuska, V. J., Goodman, L., **1964,** J. Org. Chem. *29,* 3725

Rerick, M. N., Eliel, E. L. **1962,** J. Am. Chem. Soc. *84,* 2356

Riley, R. G., Silverstein, R. M. **1974,** Tetr. *30,* 1171

Robinson, B., **1963/1969,** Chem. Rev. *63,* 373, *69,* 227

Roček, J. **1957,** Coll. Czech. Chem. Comm. *22,* 1509

Roček, J., Mareš, F. **1959,** ibid. *24,* 2741

Romo, J., Rosenkranz, G., Djerassi, C. **1951** J. Am. Chem. Soc. *73,* 4961

Rosenberger, M., Fraher, T. P., Saucy, G. **1971,** Helv. Chim. Acta *54,* 2857

Rosenberger, M., Duggan, A. J., Borer, R., Müller, R., Saucy, G. **1972,** Helv. Chim. Acta *55,* 2663

Rosenthal, A., Benzing-Nguyen, L. **1969,** J. Org. Chem. *34,* 1029

Rotermund, G. W. **1975** in: Houben-Weyl, *Methoden der Organischen Chemie,* Vol. IV/1b: *Oxidation II,* Thieme, Stuttgart

Ruden, R. A., Bonjouklian, R. **1974,** Tetr. Lett. *1974,* 2095

Rufer, C., Kosmol, H., Schröder, E., Kieslich, K., Gibian, H. **1967,** Liebigs Ann. *702,* 141

Rühlmann, K. **1971,** Synthesis *1971,* 236

Ruppert, J. F., White, J. D. **1976,** J. Org. Chem. *41,* 550

Russo, R., Lambert, Y., Deslongchamps, P. **1971,** Can. J. Chem. *49,* 531

Ryan, C. W., Simon, R. L., van Heyningen, E. M. **1969,** J. Med. Chem. *12,* 310

Rylander, P. N. **1979,** *Catalytic Hydrogenation in Organic Syntheses,* Academic Press, New York/London

Sakan, T., Fujino, A., Murai, F., Suzui, A., Butsugan, Y. **1960,** Bull. Chem. Soc. Japan *33,* 1737

Salaun, J., Garnier, B., Conia, J. M. **1974,** Tetr. *30,* 1413

Sammes, P. G. (ed.) **1979,** *Heterocyclic Compounds,* (Vol. 4 of Barton, D., Ollis, W. D., eds., Comprehensive Organic Chemistry) Pergamon Press, Oxford

Satoh, T., Suzuki, S., Suzuki, Y., Miyaji, Y., Imai, Z. **1969,** Tetr. Lett. *1969,* 4555

Saucy, G., Borer, R. **1971,** Helv. Chim. Acta *54,* 2121, 2517

Sauers, R. R., Ahearn, G. P. **1961,** J. Am. Chem. Soc. *83,* 2759

Schaefer, J. P., Bloomfield, J. J. **1967,** Org. React. *15,* 1

Schäfer, W., Hellmann, H. **1967,** Angew. Chem. *79,* 566 (Int. Ed. *5,* 669)

Schering process **1957, 1977** in: *Ullmanns Encyklopädie der technischen Chemie,* 3rd edn: Vol. 8, p. 648, Urban & Schwarzenberg, Munich, 4th edn.: Vol. 13, p. 27, Verlag Chemie, Weinheim

Schick, H., Lehmann, G., Hilgetag, G. **1969,** Chem. Ber. *102,* 3238

Schlosser, M. **1970,** Topics Stereochem. *5,* 1

Schlosser, M. **1973,** *Struktur und Reaktivität polarer Organometalle,* Springer, Berlin

Schlosser, M., Schaub, B. **1982,** J. Am. Chem. Soc. *104,* 5821

Schmidlin, J., Anner, G., Billeter, J.-R., Heusler, K., Ueberwasser, H., Wieland, P., Wettstein, A. **1957,** Helv. Chim. Acta *50,* 2101

Schreiber, J., Leimgruber, W., Pesaro, M., Schudel, P., Eschenmoser, A. **1961,** Helv. Chim. Acta *44,* 540

Schreiber, J., Felix, D., Eschenmoser, A., et al. **1967,** Helv. Chim. Acta *50,* 2101

Schröder, E., Lübke, K. **1963,** Experientia *19,* 57

Schröder, E., Rufer, C., Schmiechen, R. **1975,** *Arzneimittelchemie,* Vol. 1–3, Thieme, Stuttgart

Schröder, G. **1963/1964,** Angew Chem. *75,* 722 / Chem. Ber. *97,* 3131, 3140

Schröder, G., Oth, J. F. M. **1967,** Angew. Chem. *79,* 458, (Int. Ed. *6,* 414)

Scott, L. T., Jones, M., Jr. **1972,** Chem. Rev. *72,* 181
Seebach, D. **1969,** Synthesis *1969,* 17
Seebach, D., Kolb, M., Gröbel, B.-T. **1973,** Chem. Ber. *106,* 2277
Seebach, D., Daum, H. **1974,** Chem. Ber. *107,* 1748
Seebach, D., Hoekstra, M. S., Protschuk, G. **1977,** Angew. Chem. *89,* 334 (Int. Ed. *16,* 321)
Seebach, D. **1979,** Angew. Chem. *91,* 259 (Int. Ed. *18,* 239)
Semmelhack, M. F., Heinsohn, G. E. **1972,** J. Am. Chem. Soc, *94,* 5139
Shamma, M., Jones C. D. **1970,** J. Am. Chem. Soc. *92,* 4943
Shapiro, R. H. **1976,** Org. React. *23,* 405
Sharpless, K. B. **1970,** J. Am. Chem. Soc. *92,* 6999
Sharpless, K. B., Lauer, R. F. **1972,** J. Am. Chem. Soc. *94,* 7154
Sharpless, K. B., Teranishi, A. Y. **1973,** J. Org. Chem. *38,* 185
Sharpless, K. B., Teranishi, A. Y., Bäckvall, J.-E. **1977,** J. Am. Chem. Soc. *99,* 3120
Sheehan, J. C., Honery-Logan, K. R. **1959,** J. Am. Chem. Soc. *81,* 3089
Sheldon, R. A., Kochi, J. K. **1972,** Org. React. *19,* 279
Shemyakin, M. M., Vinogradova, E. I., Feigina, M. Yu., Aldanova, N. A., Oladkina, V. A., Shchukina, L. A. **1961,** Dokl. Akad. Nauk S. S. S. R. *140,* 387 (C. A. *56,* 536 b)
Siddall, J. B., Marshall, J. P., Bowers, A., Cross, A. D., Edwards, J. A., Fried, J. H. **1966,** J. Am. Chem. Soc. *88,* 379
Simmons, H. E., Cairns, T. L., Vladuchik, S. A., Hoiness, C. M. **1973,** Org. React. *20,* 1
Smith, H., Watson, D. H. P., et al. **1964,** J. Chem. Soc. *1964,* 4472
Smith, K. M. (ed.) **1975,** *Porphyrins and Metalloporphyrins,* Elsevier, Amsterdam/New York
Smith, M., Khorana, H. G. **1959,** J. Am. Chem. Soc. *81,* 2911
Sondheimer, F. **1950,** J. Chem. Soc. *1950,* 877
Sondheimer, F., Wolovsky, R., Amiel, Y., **1962,** J. Am. Chem. Soc. *84,* 274
Sondheimer, F. **1963,** Pure Appl. Chem. *7,* 363
Sondheimer, F., McQuilkin, Garratt, P. J. **1970,** J. Am. Chem. Soc. *92,* 6682
Sowada, R. **1971,** *Dimethylsulfoniummethylid und Dimethyloxosulfoniummethylid,* in: Martin, D., Hauthal, H. G., *Dimethylsulfoxid,* p. 367 ff., Akademie Verlag, Berlin
Spangler, R. J., Kim, J. H. **1973,** Synthesis *1973,* 107
Spangler, R. J., Beckmann, B. G. **1976,** Tetr. Lett. *1976,* 2517
Spangler, R. J., Beckmann, B. G., Kim, J. H. **1977,** J. Org. Chem. *42,* 2989
Spencer, J. L., Flynn, E. H., Roeske, R. W., Siu, F. Y., Chauvette, R. R. **1966,** J. Med. Chem. *9,* 746
Staab, H. A., Schwendemann, M. **1979,** Liebigs Ann. *1979,* 1258
Stadler, P. A., Nechvatal, A., Frey, A. J., Eschenmoser, A. **1957,** Helv. Chim. Acta *40,* 1373
Stallberg, G., Stallberg, S., Stenhagen, E. **1956,** Acta Chem. Scand. *6,* 313
Stechl, H. H. **1975** in: Houben-Weyl, *Methoden der Organischen Chemie,* Vol. IV/1b: *Oxidation II,* p. 873, Thieme, Stuttgart
Stephens, C. R., Beereboom, J. J., Rennhard, H. R., Gordon, P. N., Murai, K., Blackwood, R. K., Schach von Wittenau, M. **1963,** J. Am. Chem. Soc. *85,* 2643; see also: **1962,** ibid. *84,* 2645 and **1961,** ibid. *83,* 2773
Stevens, J. D. **1972,** in: Whistler, R. L., BeMiller, J. N. (eds.), *Methods in Carbohydrate Chemistry,* Vol. VI, p. 124, Academic Press, New York/London
Stevens, R. V., Ellis, M. C. **1967,** Tetr. Lett. *1967,* 5185
Stevens, R. V., Wentland, M. P. **1968,** J. Am. Chem. Soc. *90,* 5580
Stevens, R. V., Fitzpatrick, J. M., Kaplan, M., Zimmerman, R. L. **1971,** Chem. Comm. *1971,* 857
Stewart, R. **1965** in: Wiberg, K. B. (ed.), *Oxidation in Organic Chemistry,* Part A, chapter I (p. 2), Academic Press, New York/London
Still, W. C., Mac Donald, T. L. **1976,** Tetr. Lett. *1976,* 2659
Stokes, J. C., Esslinger, W. G. **1975,** J. Chem. Educ. *52,* 784
Storey, H. T., Beacham, J., Cernosek, S. F., Finn, F. M., Yanaihara, C., Hofmann, K. **1972,** J. Am. Chem. Soc. *94,* 6170
Stork, G., van Tamelen, E. E., Friedmann, L. J., Burgstahler, A. W. **1953,** J. Am. Chem. Soc. *75,* 384
Stork, G., Terrell, R., Szmuszkovicz, J. **1954,** J. Am. Chem. Soc. *76,* 2029

Stork, G., Burgstahler, A. W. **1955,** J. Am. Chem. Soc. *77,* 5068
Stork, G., Schulenberg, J. W. **1962,** J. Am. Chem. Soc. *84,* 284
Stork, G., Darling, S. D., Harrison, I. T., Wharton, P. S. **1962A,** J. Am. Chem. Soc. *84,* 2018
Stork, G., Tomasz, M. **1962B,** J. Am. Chem. Soc. *84,* 310
Stork, G., Dolfini, J. E., **1963,** J. Am. Chem. Soc., *85,* 2872
Stork, G., Tomasz. M. **1964,** J. Am. Chem. Soc. *86,* 471 [spiroanellation]
Stork, G., Darling, S. D. **1964,** J. Am. Chem. Soc. *86,* 1761 [Li-NH$_3$ reduction]
Stork, G. **1968A,** Pure Appl. Chem. *17,* 383
Stork, G., Kretchmer, R. A., Schlessinger, R. H. **1968B,** J. Am. Chem. Soc. *90,* 1647
Stork, G., Hudrlik, P. F. **1968 C,** J. Am. Chem. Soc. *90,* 4462, 4464
Stork, G., Ganem, B. **1973,** J. Am. Chem. Soc. *95,* 6152
Stork, G., Kraus, G. A., Garcia, G. A. **1974A,** J. Org. Chem. *39,* 3459
Stork, G., Singh, J. **1974B,** J. Am. Chem. Soc. *96,* 6181
Sum, F. W., Weiler, L. **1979,** J. Am. Chem. Soc. *101,* 4401
Sundberg, R. J. **1970,** *The Chemistry of Indoles,* Academic Press, New York/London
Svoboda, M., Závada, J., Sicher, J. **1965,** Coll. Czech. Chem. Comm. *30,* 413
Swern, D. **1953,** Org. React. *7,* 378
Swern, D. **1970,** *Organic Peroxides,* Vol. I, chapter 6, p. 313, Wiley, New York
Szantay, C., Töke, L., Honti, K. **1965,** Tetr. Lett. *1965,* 1665

Taguchi, H., Tanaka, S., Yamamoto, H., Nozaki, H. **1973,** Tetr. Lett. *1973,* 2465
Tanabe, M., Crowe, D. F., Dehn, R. L. **1967,** Tetr. Lett. *1967,* 3739, 3943
Tanis, S. P., Nakanishi, K. **1979,** J. Am. Chem. Soc. *101,* 4398
Tarbell, D. S., Williams, K. I. H., Sehm, E. J. **1959,** J. Am. Chem. Soc. *81,* 3443
Taylor, E. C., McKillop, A. **1970,** Accts. Chem. Res. *3,* 338
Taylor, E. C., Robey, R. L., Liu, K.-T., Favre, B., Bozimo, H. T., Conley, R. A., Chiang, C.-S, **1976,** J. Am. Chem. Soc. *98,* 3037
Taylor, E. J., Djerassi, C. **1976,** J. Am. Chem. Soc. *98,* 2275
Tejima, S., Fletcher, H. G., Jr. **1963,** J. Org. Chem. *28,* 2999
Thummel, R. P. **1974,** Chem. Comm. *1974,* 899
Thummel, R. P. **1980,** Accts. Chem. Res. *13,* 70
Todd, D. **1948,** Org. React. *4,* 378
Trachtenberg, E. N. **1969** in: Augustine, R. L. (ed.), *Oxidation,* Vol. 1, chapter 3 (p. 125), M. Dekker, New York
Trahanovsky, W. S. (ed.), **1973/1978,** *Oxidation in Organic Chemistry, Parts B/C* Academic Press, New York/London
Traynelis, V. J., Hergenrother, W. L. et al **1962/1964,** J. Org. Chem. *27,* 2377/*29,* 123
Treibs, A. **1971,** *Das Leben und Wirken von Hans Fischer,* Hans-Fischer-Gesellschaft, München
Trost, B. M., Bogdanowicz, M. J. **1973,** J. Am. Chem. Soc. *95,* 289, 5311
Trost, B. M., Melvin, L. S., Jr. **1975A,** *Sulfur Ylides,* Academic Press, New York
Trost, B. M., Preckel, M., Leichter, L. M. **1975B,** J. Am. Chem. Soc. *97,* 2224
Truex, T. J., Holm, R. H. **1972,** J. Am. Chem. Soc. *94,* 4529
Turro, N. J., Hammond, W. B. **1966,** J. Am. Chem. Soc. *88,* 3672
Tyrlik, S., Wolochowicz, I. **1973,** Bull. Soc. Chim. France *1973,* 2147

Ulrich, H. (ed.) **1967,** *Cycloaddition Reaction of Heterocumulenes,* Academic Press, New York/London
Umino, N., Iwakuma, T., Itoh, N. **1976,** Tetr. Lett. *1976,* 763
Utimoto, K., Tanaka, T., Furubayashi, T., Nozaki, H. **1973,** Tetr. Lett. *1973,* 787

Valentine, D., Jr., Scott, J. W. **1978,** Synthesis *1978,* 329
van Tamelen, E. E., Spencer, T. A., Allen, D. S., Orvis, R. L. **1961,** Tetr. *14,* 8
van Tamelen, E. E. **1968,** Accts. Chem. Res. *1,* 111
van Tamelen, E. E., Placeway, C., Schiemenz, G. P., Wright, I. G. **1969,** J. Am. Chem. Soc. *91,* 7359
van Tamelen, E. E., James, D. R. **1977,** J. Am. Chem. Soc. *99,* 950

Vedejs, E. **1975,** Org. React. *22,* 401
Vedejs, E., Engler, D. A., Telschow, J. E., **1978** J. Org. Chem. *43,* 188
Vernon, L. P., Seely, G. R. (eds.) **1966,** *The Chlorophylls,* Academic Press, New York/London
Vineyard, B. D., Knowles, W. S., Sabacky, M. J.; Bachman, G. L., Weinkauff, D. J. **1977,** J. Am. Chem. Soc. *99,* 5946
Vogel, E., Biskup, M., Pretzer, W., Böll, W. A. **1964,** Angew. Chem. *76,* 785 (Int. Ed. *3,* 642)
Vogel, E., Feldmann, R., Düwel, H. **1970,** Tetr. Lett. *1970,* 1941
Vogel, E., Sombroek, J., Wagemann, W. **1975,** Angew. Chem. *87,* 591 (Int. Ed. *14,* 564)
von Daehne, W., Frederiksen, E. Gundersen, E. Lund, F. Mørch, P., Petersen, H. J., Roholt, K., Tybring, L., Godtfredsen, W. O. **1970,** J. Med. Chem. *13,* 607

Wadsworth, W. S. **1977,** Org. React. *25,* 73
Wagemann, W., Iyoda, M., Deger, H. M., Sombroek, J., Vogel, E. **1978,** Angew. Chem. *90,* 988 (Int. Ed. *17,* 956)
Walker, B. J. **1972,** *Organophosphorus Chemistry,* Penguin, London
Walker, D., Hiebert, J. D. **1967,** Chem. Rev. *67,* 153
Warren, S. **1978,** *Designing Organic Synthesis – A Programmed Introduction to the Synthon Approach,* Wiley, New York/London
Wasserman, H. H., Glazer, E. **1975,** J. Org. Chem. *40,* 1505
Weissermel, K., Arpe, H.-J. **1978,** *Industrielle organische Chemie,* 2nd ed., Verlag Chemie, Weinheim
Welch, S. C., Chayabunjonglerd, S. **1979,** J. Am. Chem. Soc. *101,* 6768
Wendisch, D. **1971,** in: Houben-Weyl, *Methoden der Organischen Chemie,* Vol. IV/3, p. 126 ff. Thieme, Stuttgart
Wenkert, E., Mueller, R. A., Reardon, E. J., Sathe, S. S., Scharf, D. J., Tosi, G. **1970A,** J. Am. Chem. Soc. *92,* 7428
Wenkert, E., Yoder, J. E. **1970B,** J. Org. Chem. *35,* 2985
West, R., Glaze, W. H. **1961,** J. Org. Chem. *26,* 2096
Whaley, W. M. **1951 A/B,** Org. React. *6,* 74/151
Wharton, P. S., Hiegel, G. A. **1965,** J. Org. Chem. *30,* 3254
Whitesides, G. M., Fischer, W. F., SanFilippo. J., Bashe, R. W., House, H. O. **1969,** J. Am. Chem. Soc. *91,* 4871
Wiberg, K. B. (ed.) **1965,** *Oxidation in Organic Chemistry,* Part A, Academic Press, New York/London
Wiesner, K., Valenta, Z., Ayer, W. A., Fowler, L. R., Francis, J. E. **1958,** Tetr. *4,* 87
Wiesner, K., Musil, V., Wiesner, K. J. **1968,** Tetr. Lett. *1968,* 5643
Wigfield, D. C., Taymaz, K. **1973,** Tetr. Lett. *1973,* 4841
Williams, J. M., Richardson, A. C., **1967,** Tetr. *23,* 1369
Williams, N. R. **1970,** Adv. Carbohydr. Chem. Biochem. *25,* 109
Wilson, C. V., **1957,** Org. React. *9,* 332, 350, 380
Wilson, G. E., Jr., Huang, M. G., Schlomann, W. W., Jr. **1968,** J. Org. Chem. *33,* 2133
Winterfeldt, E. **1975,** Synthesis *1975,* 617
Wittig, G., Davis, P., Koenig, G. **1951,** Chem. Ber. *84,* 627
Wittig, G., Pohmer, L. **1956,** Chem. Ber. *89,* 1334
Wittig, G., Reiff, H. **1968,** Angew. Chem. *80,* 8 (Int. Ed. *7,* 7)
Wittig, G. **1980,** Angew. Chem. *92,* 671
Wolf, F. J., Weijlard, J. **1963,** Org. Synth. Coll. Vol. *IV,* 124
Wolff, M. E., Kerwin, J. F., Owings, F. F., Lewis, B. B., Blank, B. **1963,** J. Org. Chem. *28,* 2729
Wolfrom, M. L., Thompson, A. **1963** in: Whistler, R. L., Wolfrom, M. L. (eds.), *Methods in Carbohydrate Chemistry,* Vol. 2, p. 211, Academic Press, New York/London
Wolfrom, M. L., Bhat, H. B. **1967,** J. Org. Chem. *32,* 1821
Wolinsky, J., Chan, D. **1965,** J. Org. Chem. *30,* 41
Woodward, R. B., Sondheimer, F., Taub, D., Heusler, Mc Lamore, W. M. **1952,** J. Am. Chem. Soc. *74,* 4223
Woodward, R. B., Patchett, A. A., Barton, D. H. R., Ives, D. A. J., Kelly, R. B. **1957,** J. Chem. Soc. *1957,* 1131

Woodward, R. B., Bader, F. E., Bickel, H., Frey, A. J. Kierstead, R. W. **1958,** Tetr. *2,* 1

Woodward, R. B. **1960,** Angew. Chem. *72,* 651

Woodward, R. B. **1961,** Pure Appl. Chem. *2,* 383; see also: R. B. Woodward et al. **1960,** J. Am. Chem. Soc. *82,* 3800

Woodward, R. B., Cava, M. P., Ollis, W. D., Hunger, A., Daeniker, H. U., Schenker, K. **1963,** Tetr. *19,* 247

Woodward, R. B. **1967,** Spec. Publ. Chem. Soc. *21,* 217

Woodward, R. B., Pachter, I. J., Scheinbaum, M. L. **1971,** J. Org. Chem. *36,* 1137

Woodward, R. B. **1977,** Spec. Publ. Chem. Soc. *28,* 167

Woodworth, C. W., Buss, V., von R. Schleyer, P. **1968,** Chem. Comm. *1968,* 569

Yang, N. C., Shani, A., Lenz, G. R. **1966,** J. Am. Chem. Soc. *88,* 5369

Zahn, H., Schmidt, G. **1970,** Liebigs Ann. *731,* 91, 101

Ziegler, K., Krupp, F., Zosel, K. **1960,** Liebigs Ann. *629,* 241

Zimmerman, H. E., Iwamura, H. **1968,** J. Am. Chem. Soc. *90,* 4763

Zimmerman, H. E., Grunewald, G. L., Paufler, R. M., Sherwin, M. A. **1969 A,** J. Am. Chem. Soc. *91,* 2330

Zimmerman, H. E., Binkley, R. W., Givens, R. S., Grunewald, G. L., Sherwin, M. A. **1969 B,** J. Am. Chem. Soc. *91,* 3316

Zimmerman, H. E., Iwamura, H. **1970,** J. Am. Chem. Soc. *92,* 2015

Zorbach, W. W., Tio, C. O. **1961,** J. Org. Chem. *26,* 3543

Zurflüh, R., Wall, E. N., Siddall, J. B., Edwards, J. A. **1968,** J. Am. Chem. Soc. *90,* 6224

Zweifel, G., Brown, H. C. **1963,** Org. React. *13,* 1

Zweifel, G., Arzoumanian, H., Whitney, C. C. **1967,** J. Am. Chem. Soc. *89,* 3652, 5086

Zweifel, G., Polston, N. L., Whitney, C. C. **1968,** J. Am. Chem. Soc. *90,* 6243

Index

pr. = *price of fine chemicals*

Abbreviations

a	electron-pair **a**cceptor site
Ac	**ac**etyl
Ac$_2$O	**ac**etic anhydride
AcOEt	ethyl **a**cetate
AcOH	**ac**etic acid
AcOOH	per**a**cetic acid
Adams' catalyst	pre-hydrogenated platinum dioxide (= Pt on PtO$_2$), hydrogenation catalyst
addn.	**add**ition
AIBN	α,α'-**a**zob**i**siso**b**utyro**n**itrile
Am	**am**yl = pentyl
Am^t	tert-**am**yl = 1,1-dimethylpropyl
Am^{t}OH	t-**am**yl alcohol = 2-methyl-2-butanol
anh.	**anh**ydrous
aq.	**aq**ueous
Ar	**ar**yl (mostly also heteroaryl)
az.dist.	**a**zeotropic **dist**illation
9-BBN	9-**b**ora**b**icyclo[3.3.1]**n**onane
Boc	t-**b**ut**o**xy**c**arbonyl (= „t-box")
Btea$^+$	**b**enzyl**t**ri**e**thyl**a**mmonium cation
Btma$^+$	**b**enzyl**t**ri**m**ethyl**a**mmonium cation
BTMSA	N,O-**b**is(**t**ri**m**ethyl**s**ilyl)**a**cetamide
Bu	**bu**tyl
Bu^i	**i**so**bu**tyl = 2-methylpropyl
Bu^s	**s**ec-**bu**tyl = 1-methylpropyl
Bu^t	**t**ert-**bu**tyl = 1,1-dimethylethyl
Bz	**b**en**z**oyl
Bzl	**b**en**z**y**l** (= CH$_2$Ph)
BzOH	**b**en**z**oic acid
BzOOH	per**b**en**z**oic acid
cat.	**cat**alyst
Cb, Cbz	benzoxy**c**ar**b**onyl (= „**c**ar**b**o**b**en**z**oxy")
CC	**c**olumn **c**hromatography
Ce	2-**c**yano**e**thyl
Cet	**cet**yl = hexadecyl
Ch	**c**yclo**h**exyl
C$_6$H$_6$	benzene
C$_4$H$_8$NH	pyrrolidine
C$_5$H$_{10}$NH	piperidine
C$_4$H$_8$SO$_2$	sulfolane = thiolane 1,1-dioxide
chloranil	tetrachlorobenzoquinone (ortho or para)
3-ClBzOOH	3-**c**hloroper**b**en**z**oic acid (= „MCPBA")
C$_2$O$_4^{2-}$	oxalate anion
Cod	1,5-**c**yclo**o**cta**d**iene (ligand)
Collins' reagent	chromium trioxide in pyridine (CrO$_3$Py$_2$) for oxidation of secondary alcohols
conc.	**conc**entrated
Cot	1,3,5,7-**c**yclo**o**cta**t**etraene (ligand)
Cp	**c**yclo**p**entyl
Cpd	**c**yclo**p**enta**d**ienyl ligand
CSI	**c**hloro**s**ulfonyl **i**socyanate (ClSO$_2$NCO)
CTAB	**c**etyl**t**rimethyl**a**mmonium **b**romide
d	electron-pair **d**onor site

d	**d**extrorotatory = (+)
Δ	reflux, heat
DABCO	1,4-**d**i**a**za**b**icyclo[2.2.2]**o**ctane
DBN	1,5-**d**ia**z**a**b**icyclo[4.3.0]**n**on-5-ene
DBPO	**d**i**b**enzoyl **p**eroxide
DCC	**d**i**c**yclohexyl **c**arbodiimide (Ch – N = C = N – Ch)
DCE	1,2-**d**i**c**hloro**e**thane
DCU	1,3-**d**i**c**yclohexyl**u**rea (Ch – NH – CO – NH – Ch)
DDQ	2,3-**d**ichloro-5,6-**d**icyano-1,4-benzo**q**uinone
Deae	2-(**d**i**e**thyl**a**mino)**e**thyl
Dec	**dec**yl
DEG	**d**i**e**thylene **g**lycol = 3-oxapentane-1,5-diol
deriv.	**deriv**ative
DHP	3,4-**d**i**h**ydro-2H-**p**yran
DIBAH, DIBAL	**d**i**i**so**b**utyl**a**luminum **h**ydride = hydrobis(2-methylpropyl)aluminum
diglyme	2,5,8-trioxanonane
dil.	**dil**ute
diln.	**dil**utio**n**
DISIAB	**di**s**i**amyl**b**orane = di-sec-isoamylborane = bis(1,2-dimethylpropyl)borane
dist.	**dist**illation
dl	racemic (= rac) mixture of **d**extro- and **l**evorotatory form
DMA	**d**i**m**ethyl**a**cetamide
DMAP	4-(**d**i**m**ethyl**a**mino)**p**yridine
DME	1,2-**d**i**m**eth**o**xy**e**thane (= „glyme")
DMF	**d**i**m**ethyl**f**ormamide
DMP	N,N-**d**i**m**ethyl-**p**ropionamide
DMSO	**d**i**m**ethyl **s**ulf**o**xide (Me$_2$SO)
Dmso$^-$	anion of DMSO, „dimsyl" anion
Dnp	**d**i**n**itro**p**henyl
DNPH	(2,4-**d**i**n**itro**p**henyl)**h**ydrazine
Dod	**dod**ecyl
DODAC	**d**i**o**cta**d**ecyl**d**imethyl**a**mmonium **c**hloride
EDA	**e**thylene **d**i**a**mine = 1,2-diaminoethane
EDTA	**e**thylene**d**iamine-N,N,N',N'-**t**etra**a**cetate
ee	**e**nantiomeric **e**xcess: 0%ee = racemization 100%ee = stereospecific reaction
EG	**e**thylene **g**lycol = 1,2-ethanediol
en	**e**thyle**n**ediamine as ligand
Et	**et**hyl (e. g. EtOH, Et$_2$O, EtOAc)
gas.	**gas**eous
GC	**g**as **c**hromatography (= v. p. c.)
GLC	**g**as-**l**iquid **c**hromatography
glyme	1,2-dimethoxyethane (= DME)
Hal	**hal**o, **hal**ides
H$_2$C$_2$O$_4$	oxalic acid
Hep	**hep**tyl